Bell Beaker Copper Use in Central Europe: A Distinctive Tradition?

Matthias B. Merkl

BAR International Series 2267
2011

Published in 2016 by
BAR Publishing, Oxford

BAR International Series 2267

Bell Beaker Copper Use in Central Europe: A Distinctive Tradition?

ISBN 978 1 4073 0838 8

BAR Publishing is the trading name of British Archaeological Reports (Oxford) Ltd. British Archaeological Reports was first incorporated in 1974 to publish the BAR Series, International and British. In 1992 Hadrian Books Ltd became part of the BAR group. This volume was originally published by Archaeopress in conjunction with British Archaeological Reports (Oxford) Ltd / Hadrian Books Ltd, the Series principal publisher, in 2011. This present volume is published by BAR Publishing, 2016.

Printed in England

BAR
PUBLISHING

BAR titles are available from:

BAR Publishing
122 Banbury Rd, Oxford, OX2 7BP, UK
EMAIL info@barpublishing.com
PHONE +44 (0)1865 310431
FAX +44 (0)1865 316916
www.barpublishing.com

FOR CHRISTIANE

ACKNOWLEDGEMENT

This book is based on my doctoral research which took place between 2006 and 2009 at the Department of Archaeology at the University of Edinburgh. It is an edited version of my PhD-thesis, which was examined by Prof. Dr. Barbara Ottaway and Prof. Dr. Roger Mercer in 2010.

I am indebted to numerous people who provided help and support during the writing of this thesis. First of all, I would like to express my appreciation for my two supervisors Dr. Magdalena Midgley (University of Edinburgh) and Dr. Alison Sheridan (National Museum of Scotland) for their guidance and encouragement.

I would also like to thank Prof. Dr. Christian Strahm (University of Freiburg) for his inspiration and guidance. He contributed the first inspiration to my archaeo-metallurgical studies and counselled me throughout the course of this research.

Furthermore, I would like to express my appreciation to Frances Provan from the Statistical Systems Support team at the University of Edinburgh. Her expertise and her ideas enormously encouraged my statistical evaluation. I feel very grateful to Frances Provan who spent so many hours discussing with me my statistics.

The thesis was completed with the financial support of the 'Dr. Margaret Stewart Bequest'. Without this my research would not have been successful. Therefore, I would like to thank the trustees – in particular Prof. Dr. Dennis Harding (University of Edinburgh).

I am grateful for the grants given by the 'Abercromby Fund' and the School of History, Classics and Archaeology which enabled me to attend several national and international conferences, and for supporting a lengthy research trip in 2007 during which I visited museums and sites throughout the area of my research. I would also like to thank all colleagues who supported this research trip and my collection of data – in particular Prof. Dr. Josef Batorá (Institute of Archaeology of the Academy of the Sciences of the Slovak Republic, Nitra), Dr. Katalin Bíro (National Museum, Budapest), Dr. Miroslav Dobeš (Institute of Archaeology of the Academy of Sciences of the Czech Republic, Prague), Dr. Pétr Dvořak (Břeclav), Dr. Ernst Lauermann (Urgeschichtemuseum Nieder-österreich, Asparn/Z.), Dr. Jaroslav Peška (Archaeological Centre, Olomouc), Dr. Knut Rassmann (Roman-Germanic Commission, Frankfurt/M.) and Dr. Samuel van Willingen (Schweizerisches Nationalmuseum, Zürich). I am indebted to all scholars who carried out trace element analyses and published their results. Their research provided the data base for the present study.

Thanks must also go to all friends and colleagues who provided their support during the last years. Just to name a few: Markus Rembold, Åsa Holmgren, Veronika Kallus, Dr. Vasiliki Koutrafouri and Dr. Peter Bray. Especially, I am indebted to Dr. Megan Jones and Ben Mewburn, Graham Richie, Dr. Ben Roberts, Dr. Jeff Sanders, Dr. Shelly Werner and Neil Wilkins who spent their invaluable time to proof read my thesis.

Last but not least, I am very much indebted to my friend and colleague Marc Heise who told me about the 'Dr. Margaret Stewart Bequest' and encouraged me to apply for the scholarship in Edinburgh. Thank you for your invaluable support and for a great time with you in Edinburgh.

Finally, I would like to thank my family and, in particular, Christiane Schmid for their many years of support and understanding. It is to them that I dedicate this book.

TABLE OF CONTENTS

TEXT VOLUME

CD-ROM

1. APPENDICES

APPENDIX A: Principle Component Analyses of the Double-Analyses

APPENDIX B: Output of the Two-Step Cluster Analyses

APPENDIX C: Output of the Principal Component Analysis

APPENDIX D: Output of the Principal Component Analysis (only using Bell Beaker objects)

APPENDIX E: Graphs and Plots of the Statistical Evaluation

2. CATALOGUE

Catalogue of the entire Database.

Catalogue of the dataset of trace element analyses of Bell Beaker metal finds

Please note that the CD-ROM has now been replaced with a download available at www.barpublishing.com/additional-downloads.html

1. Introduction, Research Questions and Methodology

1.1 Research Questions and Methodological Approach

It has often been argued that, during the 3rd millennium BC[1], various changes, especially social and economic changes, were taking place all over Europe (e.g. Shennan 1986; Thomas 1991, 138 ff.; Strahm 2002; Harrison and Heyd 2007, 193 – 207). The so-called Bell Beaker phenomenon has been central to the debate surrounding the nature of these changes[2]. In central Europe, there has been particular emphasis on chronological and chorological studies as regards the overlap between the Bell Beaker and Corded Ware phenomena, which show similarities concerning their archaeological remains. In this context, questions of 'culture-defining' social, religious and economic aspects have often been discussed, and metallurgy has been considered to be one reason for that diversity, as it seems to gain greater social importance over that period (cf. Heyd 2007).

Metal objects are frequently found in association with Bell Beakers and thus the Bell Beaker phenomenon has often been linked to metallurgy[3] (cf. Vander Linden 2004, 43 with further references). The majority of studies carried out to date on the metallurgy of the Eastern Bell Beaker group have placed the main emphasis entirely on copper objects directly associated with Bell Beakers, and have examined their compositions either in connection with artefact typology or in regard to the location of the raw material (Kuna and Matoušek 1978; Matuschik 2004; Metzinger-Schmitz 2004). In contrast, the present study will consider a wide range of Chalcolithic trace element analyses – not just those directly associated with Bell Beakers – in order to clarify whether there was a specific Bell Beaker metal. It also aims to provide new insights regarding the interpretation of trace element analyses and thus to re-interpret Chalcolithic metallurgy. In short, this research will investigate the technical level – metallurgical skills and expertise – of Bell Beaker metallurgists in central Europe in comparison with that of previous and contemporaneous metal using communities. In particular, the technological purpose underpinning a broader use of fahlore-coppers – copper types that contains significant levels of several impurities such as arsenic, antimony, silver and nickel – during the 3rd millennium BC has not hitherto been clarified.

Based on a dataset of trace element compositions of copper finds from central Europe dating between around 4500 and 2000 BC, this research aims to explain whether the metallurgy of the Eastern Bell Beaker group indicates a specific knowledge that distinguishes the metal work of this community from that of other archaeological groups. The present study also aims to explain the motivation underlying the use of chronologically and chorologically varying types of copper by communities during the 4th and 3rd millennia BC. Since previous studies had been unable to clarify why these types had been used, it will be questioned whether the selection of copper types was based on the knowledge of their specific material. Those properties could certainly have been enhanced by working for instance, but the choice of a particular copper by the metal worker is the basis for further processes. Consequently, this book will focus on a possible selection

[1] If not stated otherwise, all absolute chronological dates that are used in this study are calibrated dates and refer to the mentioned literature. These dates are coded with *BC*.

[2] The term *Bell Beaker phenomenon*, introduced by Burgess and Shennan (1976), does not reflect the concept of an archaeological 'culture' as Childe had defined it (Childe 1929, vi; see also Chapter 3 of this study). It does not describe a fully coherent assemblage of archaeological remains – settlement pattern, burial rites and artefact types. Although the typical Bell Beaker and some related objects are widely distributed, they appear in different archaeological contexts associated with a variety of other artefacts. Thus, the Bell Beaker phenomenon cannot be regarded as an archaeological 'culture'. Instead, the concept of the Bell Beaker phenomenon appears to express that there is, behind the variation in the archaeological record across Europe, a certain 'idea' or 'ideology'. This is distributed throughout a large area, but differs at the regional level, because of local influences (cf. Burgess and Shennan 1976).

[3] In this study, the term *metallurgy* encompasses activities involved in the so-called *chaîne opératoire* of metal production and circulation. Here, the *chaîne opératoire* consists of all relevant technological procedures and social aspects that are involved in the production and distribution of raw copper and copper artefacts, from prospection and locating the copper deposits, through to their extraction by mining, their benefication, smelting, casting and working processes. The term also includes the exchange and transportation of both raw material and finished products. Detailed descriptions of these procedures will be presented in Chapter 6.2.
The terms *metallurgists* and *metal workers* are used to refer to all the people who were involved in the metallurgical chaîne opératoire. For the present, these terms should not be taken to imply a particular group of people, or a particular division of labour.
As will be shown later, throughout the study area there is a dearth of archaeological evidence for metallurgy in the form of copper mines, slag, crucibles, moulds and other metalworking tools. Therefore, finished metal objects are seen as indirect evidence of metallurgy, because these may indicate that metallurgy had been known and accepted by the particular community. However, these finds do not necessarily mean that metal was mined, produced and traded by the group who possessed the object.

of copper types in order to improve both the manufacture processes and the object's properties.

The research will concentrate on a review of published trace element analyses of copper objects. Therefore, the results of other archaeometallurgical studies will be considered to support the interpretations of trace element analyses; it was not possible to undertake fresh analysis for the present study. Previous metallurgical projects have shown that trace element impurities affect the physical and chemical properties of copper (e.g. Archbutt and Prytherch 1937; Dies 1967).

This study attempts to answer two major research questions:

1. Did Chalcolithic metal workers use specific types of copper (e.g. fahlore-copper) for specific purposes relating to the properties of the metal? If so, how do various trace element compositions affect the properties of copper, such as hardness or ease of casting? The results of previous trace element analyses will be used to investigate this question.

2. The second question concerns impurities in Bell Beaker metal. Do these differ from those in the copper found in metals used by other archaeological groups (e.g. Corded Ware communities) in central Europe? If so, how do they differ? Did the use of different copper types provide a technological advantage over other communities? Could this be one of the reasons behind the extensive spread of the Bell Beaker phenomenon, and for its relative diversity compared with other contemporary archaeological groups?

The chronological development of the use of various types of copper has already been established (Ottaway 1978a; Krause 2003). The present research aims to study whether these types differ in their metallurgical properties. Especially, it will investigate whether types of fahlore-copper that were widely used during the 3rd millennium BC have beneficial properties in comparison with other types (e.g. [almost] pure or arsenical copper), and whether the metallurgists of the Bell Beaker phenomenon played a key role in the distribution and establishment of fahlore metallurgy.

1.2 STRUCTURE OF THE STUDY

The following section offers a brief outline of the chapters and structure of this study. After defining the framework of this study, describing the archaeological evidence for metallurgy dating from the emergence of the first copper finds (c. 4500 BC) to the establishment of tin-bronze technology in central Europe, this book will provide an overview of the projects that provided the trace element analyses for this research (SAM[4] I and SAM II; Ottaway 1982; Krause 2003) (cf. also Chapters 2 – 4). The database of this study will be assembled, according to several criteria, from those projects which have compiled c. 35 000 trace element analyses of copper and bronze objects from all over the 'Old World' (Chapter 5). Various criteria, such as the location and date of each analysed object, must be confirmed by literature, so that a trace element analysis can be added to the database of this study. The database is attached to the book in the catalogue (cf. CD-Rom).

Chapter 6 aims to ascertain the impact of both ore formation and the technical processes of early copper metallurgy, such as smelting and casting, on the trace element composition of copper objects. In spite of the fact that metallurgical techniques which tended to eliminate impurities from copper ores, it has been demonstrated, in the aforementioned projects, that prehistoric copper objects usually contain significant concentrations of impurities (e.g. arsenic, antimony, iron, nickel and silver).

Earlier studies have tried to locate copper deposits that were exploited in prehistoric times by studying the trace element analyses of finished objects (cf. Otto and Witter 1952; Pittioni 1957) or by developing chronological and chorological sequences for the use of particular types of copper (e.g. SAM I and II; Ottaway 1978a; Krause 2003). This study attempts to find out whether or not different copper types were used due to their material properties (Chapter 7). Apart from the metallurgical literature (e.g. Archbutt and Prytherch 1937), this study will also refer to archaeometallurgical studies that have investigated the properties of prehistoric copper artefacts (Northover 1989; Budd 1991a; Junk 2003; Kienlin et al. 2006). Whereas these key studies focused mainly on the trace elements arsenic and antimony, a broader selection of impurities will be included here.

Following the discussion of the possible effects (e.g. on strength or colour) of impurities on copper and their importance for the prehistoric metallurgist have been discussed, the database of trace element analyses will be investigated in respect of the research questions. As it is necessary to ease the manageability and interpretability of the extensive dataset, the trace element analyses will be compared and clustered according to their similarities through combining of two statistical methods. In Chapter 8 these methods and the statistical evaluation and results are presented. Some of the output, graphs and diagrams resulting from the statistics are also displayed in the appendices (cf. CD-Rom).

[4] 'SAM' is the abbreviation of the archaeometallurgical research project 'Studien zu den Anfängen der Metallurgie'. The publications of its results is referred as SAM I and SAM II. SAM II consists of four volumes (SAM II/1 – SAM II/4) (cf. Bibliography).

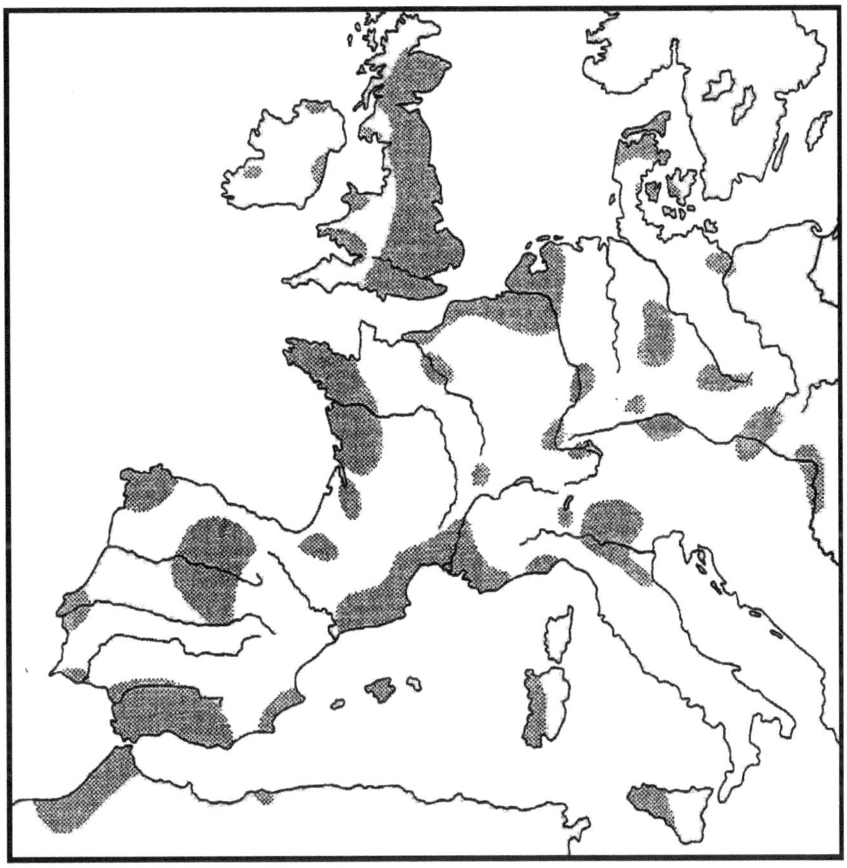

Fig. 1.1 Spread of the pan-European Bell Beaker phenomenon (grey shaded area) (Benz et al. 1997, Fig. 1).

At first, the data are grouped by so-called two-step cluster analysis (TSC). This is a relatively new technique of cluster analysis, used to facilitate the classification of multivariate groups. This method has not yet been used in archaeometallurgy. In contrast to commonly applied methods of hierarchical cluster analysis, the TSC method automatically indicates how many clusters a dataset contains by using a specific algorithm (Zhang et al. 1996; Zhang et al. 1997). Hence, the disadvantage of the hierarchical cluster analysis – the subjective selection of the number of clusters – is avoided. Because statistical grouping does not necessarily reflect the actual or the only situation during prehistory, a second method, principle component analysis (PCA), was also used. This method does not divide data into fixed clusters, but compares them and arranges the information according to their degree of similarity. The results can be displayed as scatter plots which effectively visualise the correspondence of the objects. PCA also demonstrates the influence of certain variables on the similarities between cases (cf. Shennan 1997, 265 – 307; Baxter 2003, 73 – 83). The results of both methods will be compared in order to minimise possible errors in the statistical calculations and to group the types of copper according to which their compositional similarity. However, statistical approaches can only simplify and structure the dataset and analyses based on archaeological perspectives must be made.

Following the statistical findings, the evaluated copper types with similar trace element composition will be brought into an appropriate metallurgical and archaeological context. Those copper types will also be reviewed to ascertain whether they reflect certain metallurgical properties and, furthermore, they will be contrasted with the archaeological background of the respective finds.

Finally, whether the metal workers of the Eastern Bell Beaker group did use specific types of copper will be explored, and the critical issues of whether they had an advantage of metallurgical expertise, and whether the Bell Beaker network itself is based on metallurgical networking, will be answered (Chapter 9).

1.3 THE BELL BEAKER PHENOMENON

Ever since the uniform, often comb-impressed and zone decorated, S-profiled Bell Beakers were defined for the first time by Cartailhac (1886), the so-called Bell Beaker phenomenon has received much attention. According to the current state of research, Bell Beakers were used between c. 2700 and 2000 BC throughout Western Europe – from Portugal to Poland and from Sicily to Scotland (Fig. 1.1) (Müller and van Willigen 2001; Czebreszuk 2004b). The wide distribution of these uniform Bell Beakers, however, is only one facet of several aspects (such as typical objects or features) that appeared in connection with Bell Beakers. Furthermore, it is assumed that associated cultural changes were taking place throughout Europe at that time. The Europe-wide emergence and establishment of single burial customs and gender- differentiated burial rites have been interpreted as expressions of individualism. The combination of burial customs with both qualitative and quantitative individual variations in grave goods and grave constructions are also seen as indicators of the formation of a hierarchical, structured society (Kim 2005; Harrison and Heyd 2007, 132 f.).

The origin and diffusion of Bell Beakers have often been discussed. Over more than 100 years of Bell Beaker research, different theories concerning the origin of the Bell Beaker phenomenon have been debated. Whereas early investigations had already postulated an area of origin for Bell Beakers in the Iberian Peninsula and tried to explain their diffusion on ceramic sequences (e.g. Schmidt 1909; Reinecke 1930, 23 & 28), after World War II interpretations based on migration or diffusion were avoided and several scholars focused on a local typological sequence of Beaker pottery, in order to clarify regional chronology (e.g. Sangmeister 1966a; Hájek 1966; Bill 1973). These analyses described the origin and development of Beakers from a local pre-Beaker substratum. As these studies were not based on independent dating, Lanting and van der Waals (1976) brought a new approach to research. They suggested a local, Dutch development of Bell Beakers (All Over Ornamented beakers) from the Protruding Foot Beakers of Single Grave/Corded Ware contexts, by confirming the Beaker typology with radiocarbon dates. This model, the so-called *Dutch Model*, has often been adopted for the whole Bell Beaker area, although both authors pointed out that their method described a local, Dutch development which would need to be verified in relation other regions[5] (Lanting and van der Waals 1976, 72). Other analyses followed which examined the local, typological, succession of Bell Beakers and compared the results with radiocarbon dates (e.g. Heyd 2000). Thus today, according to absolute chronology and typological analogies, it is suggested that the origin of the Maritime Bell Beakers – the assumed earliest type of Bell Beakers – is located in the Portuguese Levant c. 2700 BC and many archaeologists tend to support this model of diffusion (cf. Müller and van Willigen 2001; Czebreszuk 2004b; Harrison and Heyd 2007, 203). However, the origin of the entire Bell Beaker phenomenon, its ideological background and the rôle that local pre-Bell Beaker groups played in the genesis of the Bell Beaker phenomenon is not yet fully understood.

Bell Beaker single graves are associated with a rather standardised package of artefacts, consisting of some or all of the following: Beakers, bowls, flint arrowheads, wristguards, bow-shaped pendants, V-perforated conical buttons, tanged daggers and Palmela points (Fig. 1.2); there may well have been organic items as well, such as wooden bows, that have not survived. These are often interpreted as an expression of a certain ideology that links all areas of the Bell Beaker phenomenon (Shennan 1975b). For example, the Bell Beaker package, and particularly the Beaker vessels themselves, have been interpreted as components of a drinking ritual involving alcohol (Childe 1925, 223; Sherratt 1987, 90 ff.). The

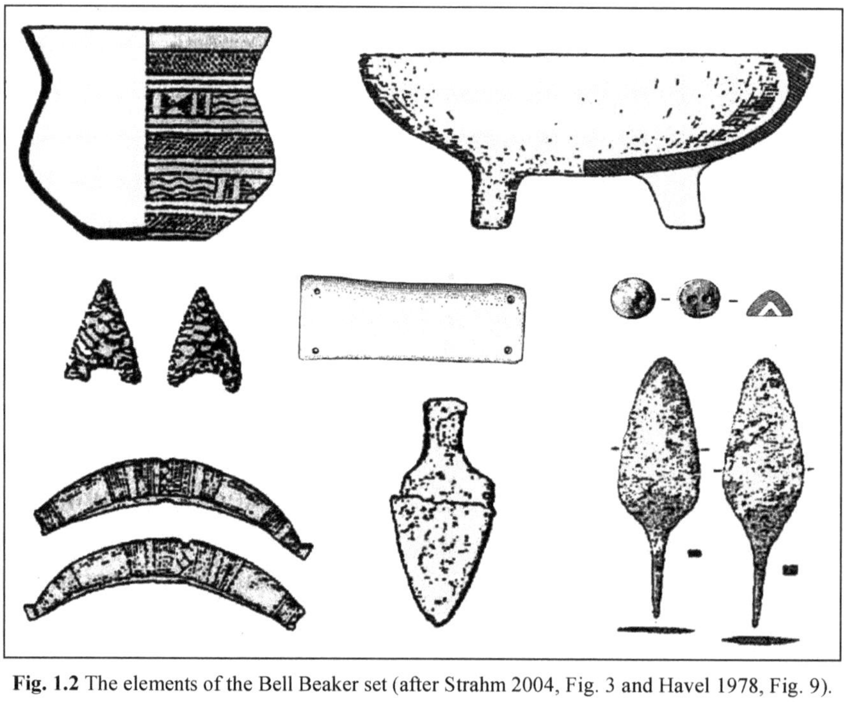

Fig. 1.2 The elements of the Bell Beaker set (after Strahm 2004, Fig. 3 and Havel 1978, Fig. 9).

[5] Strahm criticised this model, because Lanting and van der Waals did not use calibrated radiocarbon dates. Regarding the latter, however, a plateau at the calibration curve between 2950 and 2600 cal. BC means that many dates will appear to fall within that (pre-Beaker) period (Strahm 1979).

Fig. 1.3 Provinces of the Bell Beaker phenomenon according to Vander Linden (circles) (map: Benz et al. 1997, Fig. 1).

prominence of arrowheads, wrist-guards and daggers have been interpreted as symbolisation of a male hunter or warrior caste, whereby the Bell Beakers may have been used for catching and drinking the blood of slain animal or enemy (e.g. Case 2004, 29; Fokkens et al. 2008, 123 f.).

In general, the Bell Beaker using communities are considered to be an elite group in a hierarchically-structured society. Therefore, the Bell Beaker set, in its various combinations in single graves, is regarded as a status symbol of the buried individual and an accentuation of individuality (cf. Clarke 1976; Kim 2005; Heyd 2007). There is some evidence to support these various interpretations: for example, several herbal residues found inside some Bell Beakers from Spain confirm that these were filled with alcoholic drinks (Guerra-Doce 2006, 248 ff.). Furthermore, weapons placed in the grave are seemingly linked with the maiming of someone, something, or possibly even spirits in the 'Otherworld' (Case 2004, 29). Harrison and Heyd (2007, 206), therefore, suggested that the Bell Beaker phenomenon may integrate all of these interpretations. It is hard to say with certainty which elements or interpretations are correct and while weapons suggest hunting or warfare and Beakers imply drinking, whether this was intended to present the dead as a warrior or hunter, and whether or the funeral involved drinking ritual, is unclear. Whereas the first interpretation points to the individual identity of a single person, the second one rather indicates a wider social interaction. Does a high quality and quantity of grave goods automatically demonstrate an exceptional status in life? Or do people pretend to be 'better', 'richer' or 'more successful' in the 'Otherworld' than during life?

Considering all these hypotheses it is essential to keep in mind that burial rites do not necessarily reflect an individual's everyday life and rituals, but rather represent an idealised identity (Thomas 1991, 129). The individual arranged with grave goods in a grave neither indicates the social status of that person nor the role of his or her personal possessions in life (cf. Härke 1993; Härke 1997). To ask what the so-called entire Bell Beaker phenomenon in central Europe means is probably the wrong question, because most of the Bell Beaker evidence has been discovered in graves and we do not yet know much about the everyday life of the people behind the Beakers. From a modern perspective it is hard to clarify prehistoric culture never mind spiritual habits. Taking that into account, an ensemble of objects which appear rather standardised in archaeological contexts can be recognise over a large part of Europe.

Although it is worth trying to explain why the distribution and character of aspects of the Bell Beaker phenomenon are uniform across Europe, due to the regional variation inherent within the Beaker phenomenon, no single explanatory interpretation can be offered. Perhaps in central Europe the idea of differentiation, especially from Corded Ware people, was much more important than a drinking ritual. It seems that drinking was also important for the 'Corded Ware rites', because a cord-decorated Beaker played a fundamental part in funerary rites. In central Europe, this has often been seen in dialectic with the roughly contemporary Corded Ware complex, which is characterised by similar aspects, albeit differently-expressed aspects. For example, Corded Ware burials, like those of the Bell Beaker tradition, feature a gender dichotomy in the posture of the dead body, but the precise details of posture and composition of the grave good packages varies between the two traditions (Fischer 1976; Strahm 2002).

Even though the Bell Beaker package spread throughout the entire area of the Bell Beaker phenomenon, regional influences and variations can be seen in its contexts. Different regions, however, show similar traits in terms of burial rites, settlement pattern and artefact types, which can be identified as provinces[6] (Strahm (ed.) 1995; Vander Linden 2004). The eastern province equates to the Eastern Bell Beaker group, the area on which this study is particularly focused (Fig. 1.3).

The regional component of the Eastern group is revealed, for instance, by domestic pottery which especially in that area indicates a continuous local tradition (Besse 2004, 142 f.). Thus, the Bell Beaker phenomenon can be said to consist of several elements. Apart from the typical Bell Beaker and directly related finds (i.e. the Bell Beaker set) other associated items, such as 'domestic' pottery, suggest regional characteristics and influences that have to be included in the Bell Beaker phenomenon (Besse and Strahm 2001). In these regions, the Bell Beaker phenomenon interacted with the local substratum (Burgess and Shennan 1976, 324 ff.), so that either Bell Beaker people would have adopted certain elements (i.e. the rite of single burials in crouched position) or the local population accepted ideas that were related to Bell Beakers[7].

The actual Bell Beaker is considered to be the central symbol of an ideology, whilst other elements reflect certain regional peculiarities that may be affected by various influences from the social or ritual sphere or through domestic or economic activities. The distribution of similar archaeological remains indicates that the people who used Bell Beakers communicated over large areas, so that we can speak of a Bell Beaker 'network'. The so-called '*Cremade*-Model' tries to conceptualise this network which is based on a Bell Beaker 'ideology', but which is reflected differently in the provinces (Strahm (ed.) 1995, 389 – 396; Benz et al. 1998). It has been suggested that various social, economic and/or religious influences were conducive to the dissemination of the Bell Beaker ideology or ideals (e.g. Burgess and Shennan 1976). However, instead of long-distance relationships, vander Linden (2007b) recently argued for small-scale, regional networks. His reasoning is mainly based on a summary of several regional studies on various aspects (e.g. domestic pottery) of the Bell Beaker phenomenon, but he also remarked that it is the coherent element of the Bell Beaker phenomenon, "which asks for an explanation" (Vander Linden 2007b, 347). In conclusion, it seems that the formation of the Bell Beaker phenomenon was characterised by the impact of an idea which supported long-distance spread, but at the particular provinces it also interacted with local traditions.

What were the motives for either adopting the 'Bell Beaker idea' or for spreading it? This question is omnipresent in Bell Beaker research. At a very early stage it was suggested that Bell Beaker users were nomads or travelling hawkers (Schliz 1906, 334; Childe 1925, 222). This was based on the fact that most of the Bell Beaker record was found in graves and there was no evidence of settlements, especially in the central European Eastern Bell Beaker group which is the focus of this study. In contrast, Burgess for instance, argued for a diffusion of ideology, as had been adopted by indigenous communities (Burgess and Shennan 1976, 313). The question of migration or diffusion of the Bell Beaker 'culture' has been discussed for many decades (see, for example, Brodie on Britain: 1994, 3 ff.). That at least some of the Bell Beaker people were mobile has recently been proved by several strontium isotope analyses. It is confirmed that some individuals were moving and did not grow up in the same region as that in which they were buried (e.g. Grupe et al. 1998; Price et al. 2004). In contrast, however, by analysing the cross-section geometry of the femur, which may be affected by long-distance walking, a difference between Late Neolithic and Early Bronze Age populations could not be proved for samples from the Czech Republic and Austria (Sládek et al. 2006). In reality, the Bell Beaker package probably reflects a mixture of migrating people *and* the diffusion of an ideology, but it definitely shows long-distance interaction.

1.4 BELL BEAKERS AND METALLURGY

From the beginning of Bell Beaker research, metallurgy has often been seen as one of the principal factors affecting the distribution of the Bell Beaker phenomenon (e.g. Schmidt 1909, 131 f.; Childe 1929, 196). The interpretation of Bell Beaker people as metallurgists – copper prospectors, smiths or traders – is founded on the fact that, after a period of relatively poor evidence of metalwork in central Europe, copper and especially gold artefacts increase with the appearance of the Bell Beaker phenomenon (Sangmeister 1972, 196). Copper objects

[6] Although the reginal distribution and the criteria for defining the particular Bell Beaker provinces vary between Strahm and Vander Linden, both chiefly agree in the definition of the Eastern Bell Beaker group (cf. Chapter 4.1).

[7] The term *Bell Beaker people* considers people who have left the archaeological remains that are, form modern prehistorians' perspective, regarded as the Bell Beaker phenomenon. The terms *Bell Beaker metallurgist/metal worker* are used in a similar mannter. That name covers metal workers who have produced artefacts which are closely related to the Bell Beaker phenomenon, although it is not yet clear whether they were also part of the Bell Beaker community and their expertise was specifically related to that group.

discovered within Bell Beaker contexts are represented by tanged daggers, Palmela points, oar-headed pins with S-profiled heads, awls and halberds, and gold objects, such as spiral finger rings (*Noppen-* or *Lockenrings*), sheet metals and 'basket-shaped' hair ornaments, although not all of these artefacts are present in every Bell Beaker province, with the exception of tanged daggers. Palmela points, for instance, are mainly distributed in the Iberian Peninsula and in south-western France, whereas halberds and 'basket-shaped' hair ornaments are exclusively found in north-western Europe. The deposition of copper and gold artefacts in single graves indicates that these objects were to symbolise a higher status of the buried person and the exceptional value of the goods (e.g. Strahm 2002; Heyd 2007).

Apart from the great number of copper and gold objects within Bell Beaker contexts, objects used as metalworking tools have been excavated from graves. This supports the fact that the Bell Beaker phenomenon is connected with metal work. Butler and van der Waals described these finds for the first time and, through ethnographic comparisons and metal analyses, they were able to identify the so-called 'cushion stones' as anvils (Butler and van der Waals 1966, 63 – 75; Bátora 2002; Armbruster 2006; Freudenberg 2009). Despite of an ongoing discussion no analysis of cushion stones had been published for a long time, despite intentions to do so (Bertemes and Heyd 2002, 216 f.), and even in their key article Butler and van der Waals (1966, 72) could not confirm their explanation by analyses. However, Freudenberg (2009) studied recently several cushion stones from northern Germany. She was able to reveal traces of copper on some of her samples. One small axe with a blunted blade, found in Bell Beaker Grave 9 in Künzing 'Bruck' (D)[8], is the only stone tool from contexts of the Eastern Bell Beaker group on which traces of metal have been detected (75% gold and 25% copper). This artefact is meant to be a tool for hammering metal (Bertemes et al. 2000).

Additionally, these authors' interpretations are based on experimental archaeology. It was possible to show that those tools can be successfully used for metal working, and, therefore, these artefacts are, in fact, metal workers' tools (Hundt 1975; Freudenberg 2009). Cushion stones can be used as anvils and axes as hammers.

These graves which contain metal working tools are often labelled as metallurgists' or craftsmen's graves (cf. Bátora 2002), although it is not clear whether the buried individuals really dealt with metal during their lifetime. As already addressed above, grave goods and the arrangement in the grave do not necessarily reflect the social position or roles that the buried person occupied in everyday life. Whether buried people were actually metal workers, or whether they have been portrayed as 'metallurgists' in order to underline the importance of metallurgy in Bell Beaker ideology, can still not be proved by analysis of bones (Pike and Richards 2002). Nevertheless, these graves clearly document the importance and status of metal work within Bell Beaker society, or at least show that metal working was a part of the Bell Beaker ideology.

Despite certain regional variations in artefact types, indicative for the entire Bell Beaker phenomenon is the fact that metal- or metallurgy-related finds seem to express a value or importance for people who were 'supporters' of the Bell Beaker ideology.

This indicates that copper metallurgy may have been one of the factors that allowed the spread of the Bell Beaker ideology, and it is suggested that Bell Beaker metallurgists may have been involved in its dispersal. In their first comprehensive publication of research on trace element analyses, the SAM-group devoted a chapter to Bell Beaker metal (SAM I, 187 – 197). They chiefly focused on the question of whether the spread of the Bell Beaker phenomenon and the *Rückstromtheorie* (*Reflux-theory*) could be confirmed by the distribution of the kinds of copper which were used for Bell Beaker objects[9]. If Bell Beaker metal was uniform throughout Europe, this would suggest that Bell Beaker people were metal traders or prospectors. Therefore, not only central Europe, but the entire Bell Beaker area was studied. Finally, the SAM-group found evidence of both an artefact and metal typological Europe-wide contact – in particular with the Early Bronze Age groups.

In the Netherlands and on the British Isles the introduction of metal work and the first metal objects were probably linked to the Bell Beaker phenomenon, as both appear for the first time in archaeological contexts that are dated as being more or less contemporary. This assumption is also based on the usage of a specific Bell Beaker metal in this area (Butler and van der Waals 1966; Needham 2002). In the Iberian Peninsula knowledge of copper metallurgy was present in pre-Beaker periods, and the abundant evidence for copper metallurgy in Bell Beaker contexts at many sites indicates the importance of metal work for Bell Beaker people (Montero-Ruiz and

[8] In Grave 9 at Künzing 'Bruck' (D) has been found – beside the small stone axe with blunted blade – one comb decorated Bell Beaker, one wrist-guard and one copper awl, two flint points and two flint arrowheads and various other stone artefacts, e.g. one arrowshaft smoother. The body was buried with the head pointing towards north-northwest in the foetal position, lying on the left side (Schmotz 1992).

[9] Sangmeister's so-called *Rückstromtheorie* sees an expansion of various elements related to the earliest phase of Bell Beaker phenomenon (e.g. Maritime Bell Beakers or copper metallurgy) from the Iberian Peninsula westwards. In central Europe these elements interacted and then became associated with the local substratum and, subsequently, central European components: for example, cord-decorated Bell Beakers (All-Over-Cord Beakers) and wrist-guard 'drifted' back to Spain. This theory is based on the assumption that material culture should reflect the existence of copper traders (Sangmeister 1957, 263 – 266). Later, Sangmeister contributed to his approach a mass of trace element analyses and tried to put it into a relative chronological framework (SAM I, 187 ff.; Sangmeister 1966b). Soon after its publication, Sangmeister's model was criticized, especially because Sangmeister did not synchronise his ideas with radiocarbon dates. Taking radiocarbon dates into account it has been argued that, first of all, the phases of *Rückstrom* do not correspond with the radiocarbon dates and there is no chronological difference between Maritime, All-Over-Cord and other Beakers (Butler and van der Waals 1966, 56 f.; Clarke 1970, 45 ff.).

Rodríguez de la Esperanza 2004; Rovira Llorens and Delibes de Castro 2005, 513 f.). Recent residue analyses have revealed traces of copper slag inside Bell Beaker sherds, which supports that connection. But these very few samples have only been discovered in Spain and not in central Europe (Guerra-Doce 2006, 252 f.).

Due to the connections seen between the general contemporary emergence of Bell Beakers and metallurgy in the British Isles and the direct relation of metal work and Bell Beaker sites in the Iberian Peninsula, it is clear that, at least in these regions, metallurgy and the Bell Beaker phenomenon were closely linked. Whether this linkage also existed in the Eastern Bell Beaker group is yet to be verified. Due to the lack of direct evidence of copper mining, smelting and casting from contexts of the Eastern European Bell Beaker, it is necessary to employ other methods. Here the archaeological significance of trace element analyses could be consulted.

In the case of the Eastern Bell Beaker group, which is mainly known from grave contexts, the question of what the Bell Beaker ideology meant to prehistoric people is unclear and several scholars have interpreted this differently: whether the grave assemblages symbolise a drinking symposium (Sherratt 1987), an elite warrior caste (Heyd 2007) or a group of metallurgists (Sangmeister 1972, 196). We see similarities and differences in archaeological remains spread over a large area, giving the impression that some areas are more closely connected than others. Even if it is not possible to understand completely the ideological sense of the Bell Beaker phenomenon (which may vary in different regions) we may nevertheless be able to explain the distribution of this idea throughout Europe.

While metallurgy may not explain the entire Bell Beaker phenomenon, it can provide one reason for the distribution and acceptance of the ideology in various regions of Bell Beaker distribution and, in particular, may support the assumption that the Bell Beaker phenomenon is based on social, economic or religious networks (cf. Vander Linden 2007b). Nevertheless, the archaeological record of the Eastern Bell Beaker group indicates that there was a link between Bell Beakers and metal – even if metallurgy was not brought to central Europe by Bell Beaker users. This study attempts shed light on the relationship between users of Bell Beakers and metallurgy in central Europe: to investigate whether specific types of copper were used in relation to Bell Beakers; and to clarify whether metallurgy or metallurgical expertise can be seen as a 'culture-defining' element of the Eastern Bell Beaker group.

2. Definition of the Geographical and Chronological Framework of the Area of Research

The following section describes the geographical and chronological framework of this study. After the geographical boundaries have been mapped out, the chronological structure will be defined. Finally it will be explained why this area is ideal for discussing the issue of the intentional use of copper types.

2.1 Geographical Framework of the Area of Research

The geographical limits of the research area are primarily set by natural boundaries. We can assume that, in the past, natural boundaries such as rivers or mountain ranges had divided up the landscape. It is, however, necessary to keep in mind that rivers and even high mountains were never impregnable obstacles, and exchange along and across rivers and mountains has always happened. Nevertheless it is necessary to define the geographical framework quite strictly in order to present an appropriate outline for the compilation of the data set.

In general, the area of research includes the northalpine region and the Carpathian basin (Fig. 2.1). The natural boundaries of the study region will be itemised starting at the Serbian-Romanian frontier – the so-called Iron Gates. The southern border is chiefly bounded by the Danube and Drava Rivers, the main chain of the Alps, the upper Rhône Valley and Lake Geneva. The western border is defined by the Swiss Jura Mountains, the Vosges and the Moselle River. The northern border, however, cannot be defined so clearly. For archaeological reasons the study has to include not only the central European low mountain ranges, such as the Harz and the Ore Mountains, but also areas further to the north. As will be demonstrated below, archaeological remains from the Middle-Elbe-Saale region and from southern Poland are closely connected with Bohemian or Moravian sites which might be important for a subsequent comparison and interpretation of copper compositions. Therefore, the periphery of the research area encompasses the Rhenish Slate Mountains and the Harz. From there the borderline follows the Elbe River and then, eastwards, the area includes Silesia as far as the Oder River, the triangle between the Vistula and the San River and the Slovak Carpathians. The eastern part of the research area is easier to define. It includes the Carpathian Basin, but not the Carpathian Mountains. The outline continues along the foot of Carpathian Mountains.

Overall, the area of research covers roughly the region from the Carpathian Basin to the Rhine valley and from the northern Alpine region to the low mountain ranges of the Rhenish Slate Mountains, the Harz, the Ore Mountains and the Western Carpathians. This area will be called 'Central Europe' in this study.

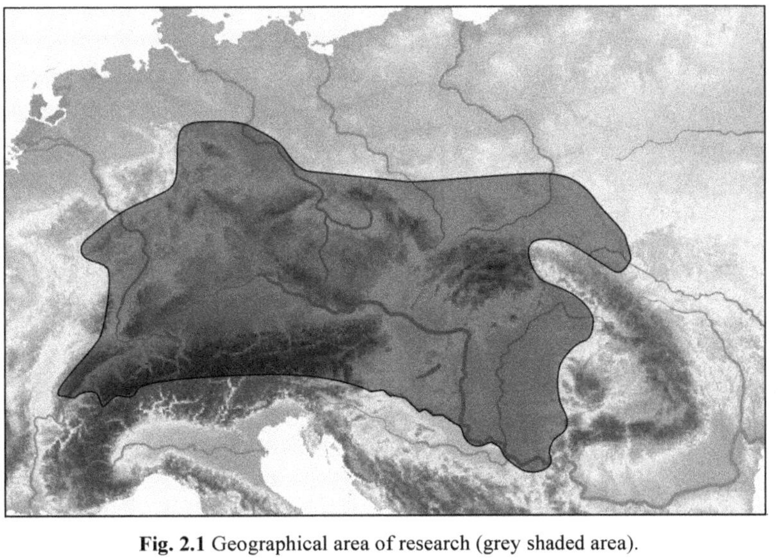

Fig. 2.1 Geographical area of research (grey shaded area).

2.2 CHRONOLOGICAL FRAMEWORK OF THE AREA OF RESEARCH

Although the main focus of the study is the 3rd millennium BC – and in particular the Bell Beaker period – this research is not chronologically restricted to that period. Antecedent archaeological groups are also taken into account, as this provides an opportunity to consider pre-Bell Beaker metallurgy and its possible influence on metal workers' knowledge during the second half of the 3rd millennium BC. Thus, the earliest copper objects must be included in this analysis in order to consider any potential technological upheavals, transformations and tendencies which might have affected the copper technology of the 3rd millennium BC. Since the chronological development of central European metallurgy has already been examined in several previous studies (e.g. SAM I & II; Ottaway 1982; Krause 2003), this study will rely on these investigations[10]. It will analyse technological changes in copper metallurgy and their influence on society. The chronological framework is therefore characterised not only technologically but also archaeologically. In other words it is defined by the period when copper was the main metal in use, in contrast to later periods when tin-bronze or iron were in use.

The earliest, and sporadic, finds of copper objects in Europe are from the Balkans and date to the 6th millennium BC. The earliest known copper mines date to the 5th millennium BC (e.g. Rudna Glava [SRB]) (Bourgarit 2007; Ottaway and Roberts 2008).

Within the area of research, the archaeological contexts from which the earliest copper objects came can be dated to the second half of the 5th millennium BC (Cevey et al. 2006, 25). However, there is no fixed point marking the beginning of the timeframe for the whole research area; the start point is simply taken to be whenever – according to the current state of research – copper objects appear in each respective region. While around the mid-5th millennium BC copper artefacts emerged throughout the Carpathian basin, in the western part of the research area metal objects appeared only rarely at that time (Cevey et al. 2006, 25). The mechanisms of transmission and adoption of metallurgy are complex and rely on various technological, social and economic factors (cf. Roberts 2008a). Copper finds emerged in the area of research often only sporadically and at different times in different regions between around 4300 and 3800 BC.

By contrast, the end of the chronological framework circumscribing the present study is not well defined. In general, this study ought to end with the implementation of a fully developed tin-bronze technology. However, the new technology did not become established in every region at the same time; thus tin-bronze technology appears in one area earlier than in another, and it can be very difficult to pinpoint its first use in a specific area.

[10] A detailed description of the development of central European copper metallurgy and the archaeological background will be given in Chapter 4.

Two seminal works focused on the appearance of tin-bronze objects in central Europe (Spindler 1971; Liversage 1994). Spindler (1971, 252) found that between the relative-chronological phases Reinecke Bronze Age A1 and A2[11] the number of bronze objects containing an appreciable percentage of tin rose rapidly in central Europe. This result has been confirmed by Liversage (1994, 59 ff.), who used a slightly different computer-based methodology. He looked at a range of radiocarbon dated sites in the Carpathian Basin and also detected that the number of artefacts containing over 4% tin increases in the later part of period A1; from the Hajdúsámson horizon, a yet higher concentration tin-copper alloy predominates[12]. It seems that a break at a concentration of 4% tin marks a barrier between, native copper and "low-level alloyed" bronze on the one hand, and a deliberately alloyed tin-bronze on the other[13]. Whereas some authors argue that only objects containing more than 4% tin have been intentionally alloyed (e.g. Pare 2000, 2), others have claimed that even a concentration of more than just 0.1% tin indicates that the copper has already been alloyed (e.g. Friedman et al. 1966, 1506). As it has not yet been possible to clarify beyond doubt how much tin occurs naturally in copper ores (cf. Rovira and Montero 2003, 15), the combination of metallurgical knowledge with the chronological difference between low- and high-alloyed tin-copper is decisive (cf. Liversage 1994; Müller, J. 2002). Subsequently, during the earliest Bronze Age phase (e.g. Bronze Age A1 in southern Germany and FB I-II in the Carpathian Basin) proper tin-bronze was not been commonly used[14] (Fig. 2.2; Liversage 1994, 77 ff.).

As the Reinecke periodisation is relative, being based on typological artefact analyses, it is essential to compare those typo-chronological phases with independent dating in order to ascertain the transition from the Copper Age to the Bronze Age. Recent radiocarbon and dendro-chronological dates indicate that the transition from Reinicke A1 to A2 occurred regularly throughout the area of research between 2200 BC and 2000 BC (Becker et al. 1989, 441; Müller, J. 1999, 77; Hafner and Suter 2004b, 246). J. Müller (2002, 276) listed several larger absolutely-dated archaeological complexes of the Early Bronze Age, such as cemeteries, and found out that the

[11] In his article Spindler used a deviant chronological system. His phases 2 and 3, however, correlate with the commonly used periodisation, Bronze Age A1 and A2 respectively, of Reinecke's chronological system (Spindler 1971, 210, note 25).

[12] The Hajdúsámson horizon was roughly contemporary to Reinecke's Bronze Age period A2 or the *Frühdanubische Bronzezeit III* in Hänsel's Hungarian chronology (Liversage 1994, 62). In comparison with current radiocarbon dating the Hungarian Early Bronze Age phase III (FB III) is even contemporary with Reinecke's Bronze Age period A1 (Maran 1998, 313 ff.). As a result, tin-bronze was already established in the Carpathian Basin around 2200 BC.

[13] Because the present research only looks at copper metallurgy (including arsenical copper) and not tin-bronze, objects exceeding a concentration of more than 4.5% of Sn are excluded from the database (cf. Chapter 5.2.1).

[14] There are, however, a few earlier objects that contain over 4.5% tin; these appeared sporadically (cf. Schickler 1981). For example a tanged dagger from the Bell Beaker grave 1 at Bylany (CZ) contains 6.1% tin (SAM II/1, 78 f., An.Nr. 3238). Within the area covered by the current study, those samples are very rare – less than 1% of the entire available data – and can be excluded as outliers.

tin concentration in copper increased in most parts of the research area from c. 2000 BC onwards. Nevertheless there have been a few, sporadic finds of tin-bronze objects dating to earlier periods (Schickler 1981; Krause 2003, 210 ff.), but they are too randomly distributed to be regarded as evidence for a widespread and fully established tin-bronze technology. Consequently, we can assume that, in central Europe, the transition from the usage of copper to tin-bronze dates to around 2000 BC (+/- 200 years) (Spindler 1971; Liversage 1994; Pare 2000, 16 ff.; Müller, J. 2002, 276).

Consequently, this research covers a period from around 4500 to 2000 BC in central Europe. This area offers the best opportunity to address all the issues at the core of this study, for the following reasons:

Firstly, a large number of central European copper artefacts have been subject to trace element analysis and the results have been published (cf. Krause 2003). Not only does this dataset cover the area of distribution of the Eastern Bell Beaker group with its homogeneous features; it also fulfils a number of other conditions which are necessary for investigating all issues of this study.

Secondly, within this geographical region several, well-defined archaeological 'cultures' overlap with each other, both spatially and chronologically. In central Europe, the distribution of Bell Beakers overlaps with that of the Corded Ware complex as well as a range of Earliest Bronze Age groups, such as the Hungarian Makó-Kosiny-Čaka group. Thus, interactions between these groups are to be expected (Bertemes 2000, 26; Heyd 2007, 362 ff.).

These two factors offer us a firm foundation for a further investigation of copper metallurgy and its influence on societies in this part of Europe.

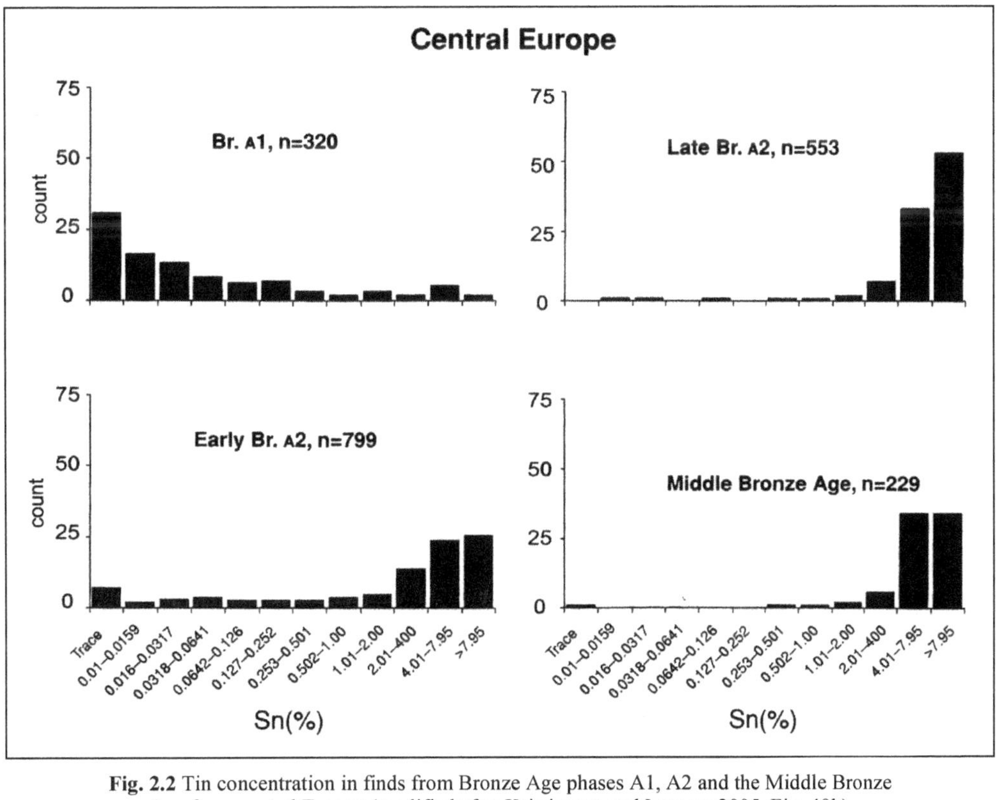

Fig. 2.2 Tin concentration in finds from Bronze Age phases A1, A2 and the Middle Bronze Age from central Europe (modified after Kristiansen and Larsson 2005, Fig. 49b).

3. THE CONCEPT OF AN 'ARCHAEOLOGICAL CULTURE'

Throughout the history of prehistoric research terms such as *archaeological culture*, *archaeological group* or *archaeological complex* have been used. This vocabulary emerged out of the need to classify and compare archaeological material. What, however, do archaeologists comprehend as *culture*? An early and, perhaps, the most famous definition was given by Childe:

> "We find certain types of remains – pots, implements, ornaments, burial rites and house forms – constantly recurring together. Such a complex of associated traits we shall term a 'cultural group' or just a 'culture'. We assume that such a complex is the material expression of what today would be called a 'people' [...]"
> (Childe 1929, v f.)

This quote therefore posits that similar archaeological remains may directly reflect a prehistoric people and its spatio-temporal distribution. Earlier, Kossinna had carried this idea to the extreme and said:

> "Streng umrissene, scharf sich heraushebende, geschlossene archäologische Kulturprovinzen fallen unbedingt mit bestimmten Völker- und Stammesgebieten zusammen"
> (Kossinna 1926, 21).

Translated, this means that similarities in material remains strictly classified as archaeological cultures represent a prehistoric folk or tribe. Consequently this idea was adopted and archaeologists concluded that finds – chiefly ceramics – defined archaeological cultures, which themselves define people: Pots equal People and, thus, the distribution of artefact styles directly reflects to prehistoric people.

In our area of research, this issue is especially reflected in the criteria used to define archaeological 'cultures', which may vary from period to period and region to region. Whereas Neolithic cultures have generally been typified by a specific pottery spectrum, larger archaeological phenomena have been characterised by more complex patterns. Archaeological complexes are typified by a certain broader uniformity (e.g. similar artefact types), but with a slight chronological and/or local diversity (e.g. TRB or Corded Ware complex).

Over the decades of the 20th century, and under the influence of ethnological and anthropological models of culture, researchers became more and more dissatisfied with the aforementioned definition, because from an anthropological point of view culture is more than just material remains. Illustrating this, Clark (1957) published a diagram showing the various aspects of culture and their interaction (Fig. 3.1). A period of theoretical discussion on those issues followed, led by Binford (1965) and Clarke (1968).[15] In particular, Clarke's polytheti c model of archaeological cultures expressed that only a small region shows certain uniformity and the periphery is interacting with other areas. Thereby several culture-defining aspects may 'act independently' (Clarke 1968, 246).

Reviewing the literature in order to find an appropriate definition of the concept of *archaeological culture*, it seems that a definition has been generally abandoned in Anglo-Saxon research. In accordance with Jones, although the problems of the concept of culture in archaeology had been comprehensively argued, *archaeological culture* has not been suitably defined, either by processual or by postprocessual archaeologists (Brather 2001, 448; Jones 2007, 122 f.).

Finds and features are the primary source of prehistoric research. We do require a model as a tool to classify and compare archaeological material in order to facilitate further interpretation. Despite all the discussion on culture in archaeology, we have to define how this study deals with the concept of culture, because we shall be examining what influence copper had on what, in central European research, are still called *archaeological cultures*.

In order to avoid interpretative problems in this study the term *archaeological culture* will be replaced by *archaeological group*. Here that term shall be used in the least interpretative sense as possible, in order not to influence any preliminary explanations. So a definition needs to be found which does not necessarily contain a social, religious, economic etc. reading. Therefore, we should go back to Childe's definition. This seems to be the most useful one, because whenever we try to explain the prehistoric past, we finally refer to finds and features as the primary source of every archaeological study.

[15] The ideas and concepts of culture of the 'New Archaeology' have been discussed in various handbooks. Thus, we can refer to them and do not have to present them again (cf. Erickson 1998; Johnson 1999).

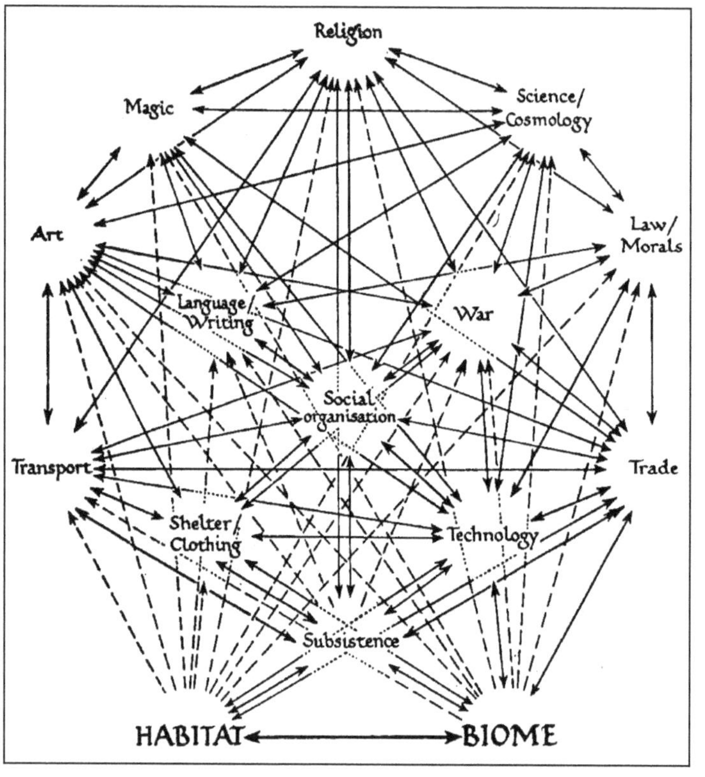

Fig. 3.1 Interrelations of various aspects of culture (Clark 1957, Fig. 25).

Furthermore, his definition is practical for this research, because reconsidering the history of research this concept has been chiefly used and prehistory has always been described, classified and interpreted by reference to similarities of the material remains. Childe's definition has lately been reduced to a minimum and more abstract one by Johnson:

> "An archaeological culture is a repeatedly recurring assemblage of traits – pottery and house forms, burial practices – seen over a discrete time and space. It may or may not relate to a human culture"
> (Johnson 1999, 189).

This definition is an ideal foundation for further research, because it summarises similar archaeological traits – finds and features. It is important to keep in mind that this definition is only used for technical reasons, as a tool, to ease the analysis of archaeological assemblages. As from a present perspective cultural breaks in prehistory can only be recognised by changes in the material remains, the taxonomy of artefacts and other remains has to be relied upon. Certainly, if we interpret these remains we have to be aware that pots or other features are not equal to people. An archaeological group is just an abstract construct in order to categorise archaeological material spatially and temporally and to ease comparisons. In addition to artefacts, other aspects such as economic, social or ritual patterns characterise culture as well, but in our definition those rather interpretative aspects will not be considered to define an archaeological group.

Often an *archaeological group* describes a more local and chronologically restricted facies of a larger, more widespread *archaeological complex*. Of course, archaeological cultures are an expression and interaction of several aspects of cultural activities as it is shown in Figure 3.1 above. This, however, is the result of our analyses of finds and features mostly by evaluating *inter-archaeo-cultural* analyses. As long as we keep in mind that the concept of archaeological groups only delineates typological similarities among the archaeological traits, that concept can be used in the following study to compare archaeological remains of the large research area covered in the current study.

4. Overview of Copper Using Groups and the Regional and Chronological Development of Copper Metallurgy within central Europe

This chapter outlines the development of early copper metallurgy in the study area. This area includes roughly the Carpathian Basin and the northlpine region, as defined as central Europe in Chapter 2.1. It will be shown below that in different regions of central Europe copper use appears at different points in time. It is necessary to introduce and describe the copper-using archaeological groups of the research area and to review the metallurgical evidence, and its relationship to the cultural, economic and social fabric. This will constitute the archaeological background for the later compilation of data and will provide the basis for checking whether the database constitutes a representative sample of the entire metallurgical evidence in the area of research.

Throughout the 5th and 4th millennia BC we see numerous signs of various inventions and new technologies, which justifies the assumption that economic and social change was underway in Europe. Probably the most important developments were the adoption of ploughs, wagons, horse riding and, finally, copper metallurgy. Ploughing fields and also the secondary exploitation of sheep, goats and cows, for milk and wool, increased food production. Sherratt (1981; 1982) called these innovative developments by the term 'Secondary Products Revolution'[16]. Besides, the settlement pattern changed during that time. Whereas Early and Middle Neolithic communities preferred to inhabit areas of fertile soil (e.g. loess), after c. 4000 BC people were able to cultivate regions with heavier and/or less fertile soils. These factors may have increased food production and, consequently, freed some people to concentrate on producing other goods, including copper (Müller, J. 2001, 437 f.).

[16] Sherratt's idea of the *Secondary Products Revolution* has been discussed several times (cf. Sherratt 1996; Vosteen 1996a; ibid. 1996b). Although the secondary exploitation of animals as an innovative development has not been questioned, a punctuated change affected by impulses from outside Europe (specifically from the Near East) and the chronology of that development has been criticised. From today's perspective, the secondary products revolution was indeed an innovative development, but it was a longer-lasting process which was adopted throughout at least the 4th and 3rd millennium BC and, until now, its European origin is hard to locate. Furthermore, all its elements – ploughing, milking, riding and wheeled vehicles – developed independently. For example, in the Carpathian region domestic animals may have been exploited for milk during the 6th millennium BC (Craig et al. 2005), whereas early evidences of ploughing date to the first half 4th millennium BC (Vosteen 1996b).

While the first metal objects, small beads and awls, appeared in the Near East around the 8th millennium BC (e.g. in Cayönü Tepesi [TR]), the earliest European copper finds have been found sporadically in Neolithic contexts in Rumania. However, copper artefacts emerged in larger amounts around 4500 BC in the Balkans (Pernicka 1995, 31 f.; Ottaway and Roberts 2008, 195 ff.). Dating to this period, the cemetery of Varna in East Bulgaria has produced a quantity of exceptional copper and gold artefacts and the interpretation of the cemetery has finally allowed a detailed interpretation of the local society to be developed, revealing that it functioned at an outstanding technological level[17] (Renfrew 1986). Although this site is important for European prehistory and often discussed in research, we will not consider it in detail, because it is located outside the research area. Also situated outside the study area, but worthy of mention are two famous copper mines, Rudna Glava (SRB) and Aibuna (BG), famous for being probably the earliest evidence for copper mining in Europe. Both have been radiocarbon dated to the 5th millennium BC. It has been argued that these may have been the sources which supplied the earliest metal workers of the Balkan and Carpathian region with copper (cf. Černych 1978; Jovanović and Ottaway 1976; Jovanović 1982; Pernicka et al. 1993).

4.1 Overview of Copper-using Archaeological Groups within the Area of Research

The following section characterises the metal-using archaeological groups within the study area and illustrates the chronological development and spread of metallurgy in that region. It should be noted that there is not scope to discuss in detail each of the many archaeological groups included within the vast geographical and chronological area. The fine chronology of the archaeological groups, chiefly based on artefact typology, has been debated in several studies on which this research relies. Therefore a brief impression of the groups will be given, referring the reader to the relevant literature. The features, distribution

[17] For further information the reader is referred to the recently published article by Higham et al. which includes all the important references (Higham et al. 2007).

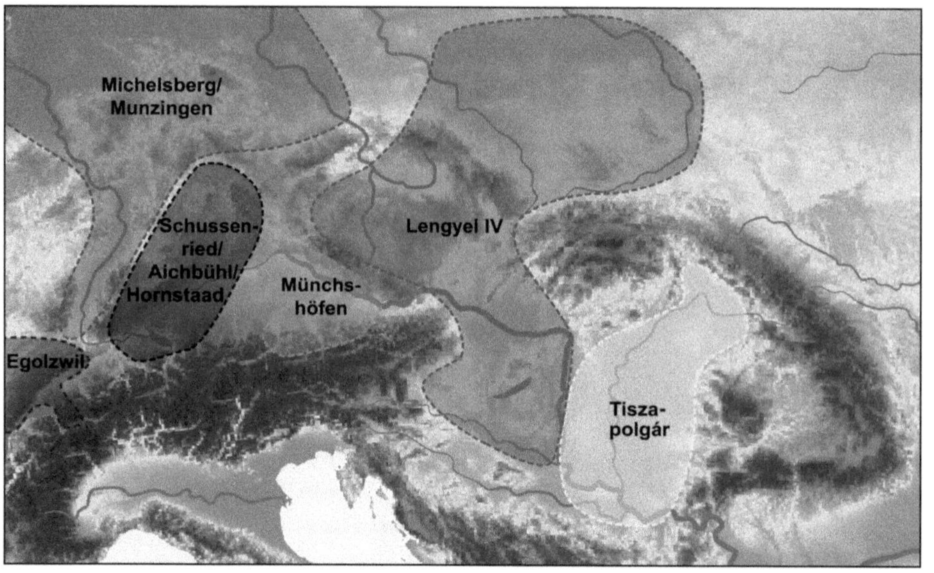

Fig. 4.1 Distribution of the archaeological groups from approx. 4300 – 4000 BC.

and chronology of the groups will be described, but with a specific focus on evidence of metallurgy[18]. That description outlines the general information on metallurgy in the various regions of the study area, so that it can be clarified whether the available trace element analyses are representative for the particular region and period.

The earliest evidence of copper metallurgy within the area of this study was found in the context of the Tisza group of the so-called Tisza-Hérpály-Csőszhalom complex[19] in east Hungary and in the Transdanubian Lengyel group. Beads and other small objects – mostly jewellery – appeared sporadically, e.g. at the cemeteries of Mórágy-Tűzkődomb and Zengővárkony in south-east Hungary (Kalicz 1992, 12 f.; Zalai-Gaál 1996). These archaeological cultures are dated generally to the first half of the 5th millennium BC (Regenye 2001; Visy 2003, 101 & 103).

As many authors assume, during the second half of the 5th millennium BC, a cultural and economic transition occured with the appearance of the eastern Hungarian Tiszapolgár group (Fig. 4.1); this involved changes in settlement structures and mortuary practices (Bognár-Kutzián 1963; Bognár-Kutzián 1972; Parkinson 2006). Although not many settlement sites are known, the houses and settlements became smaller and were more widely distributed than in previous periods. In contrast, the size of the cemeteries grew, and the dead were chiefly buried outside the settlements with burials orientated East-West (Bognár-Kutzián 1972, 150 – 159; Sherratt 1983b, 36). Copper objects have been quite frequently found at sites dated to this period. In particular, heavy copper axes were discovered. Although most of these are single finds, they can be dated typologically by comparing them with axes which were found in grave contexts of the Tiszapolgár and subsequent Bodrogkeresztúr groups[20] (Fig. 4.1) (Patay et al. 1963; Patay 1975; Vulpe 1975; Patay 1984; Schalk 1998). While the hammer-axes were seen as related to the Tiszapolgár group, the shaft-hole axe-adzes are generally connected with the later Bodrogkeresztúr group[21] (Schalk 1998, 33; ibid. 37). Although Boroffka (2009) suggested that axe-adzes were cast with a quite simple technique using open moulds, Kienlin and Pernicka (2009, 270 f.) argued on metallographic analyses that these axes were cast in bivalve moulds. This indicates that, even as early as the first half of the 4th millennium BC, copper metallurgy was complex and fully developed in parts of our area of research.

Occasionally smaller jewellery of copper and gold, such as beads, rings or bracelets have been discovered in graves (Bognár-Kutzián 1972, 138 f.; Patay 1975, 17 f.). Also some evidence of copper work has been discovered in Tiszapolgár contexts. Bognár-Kutzián (1972, 164) mentioned a 'crucible-like' vessel from grave 17/55 at Tibava (H) and one from grave 2 at Tiszapolgár 'Hajdúnánás Road' (H) in which copper lumps were found. In the same paragraph, she talked about "traces of melted metal observed in some Lucska vessels". However, the copper composition has not been analysed.

[18] These archaeological groups have been defined by present scholars. Their chronological and geographical borderlines refer to the current state of research. They cannot always be clearly defined and often overlap with each other. Even the so-called 'absolute chronological' radiocarbon dates only reflect an approximate value (cf. Aitchison et al. 1991; Raetzel-Fabian 2001). Consequently, chronological dates and regional distributions have to be seen as an approximation, based on current knowledge.

[19] Parkinson (2006, 43 – 63) recently published an overview of the Late Neolithic and the Copper Age on the Great Hungarian Plain, containing appropriate references and radiocarbon dating. Another detailed picture of distribution, chronology and material remains including a history of research of the Tizsa group is given by Kalicz and Raczky (1987).

[20] The following maps (Fig. 4.1 – 4.7) show the distribution of the archaeological groups within the area of research. They are based on the literature used in the text and on two prehistoric atlases (Buchvaldek, M. et al. (eds.) (2007), *Archeologický atlas pravěké Evropy*, Praha; and Von Schnurbein, S. (ed.) (2009), *Atlas der Vorgeschichte. Europa von den ersten Menschen bis Christi Geburt*, Stuttgart).

[21] The calibrated radiocarbon dates from Bodrogkeresztur contexts cover predominantly the first half of the 4th millennium BC (Parkinson 2006, Table 4.1).

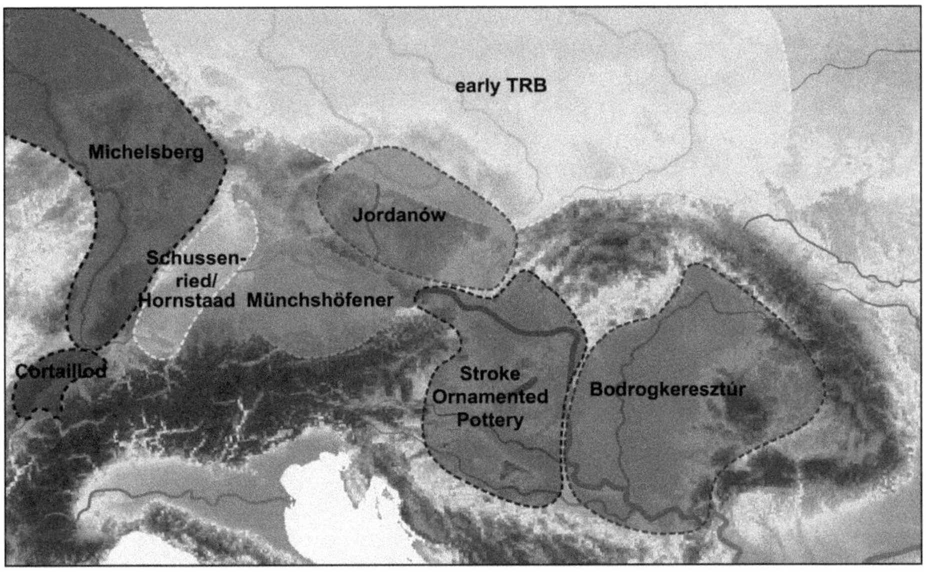

Fig. 4.2 Distribution of the archaeological groups from approx. 4000 – 3800 BC.

The Tiszapolgár and the Bodrogkeresztúr groups can be discussed together, because a strict division between these two phases cannot be defined. They extend beyond each other spatially as well as temporally – sometimes within one cemetery, e.g. at Tiszapolgár-Basatanya (Bognár-Kutzián 1963). This also bears comparison with the radiocarbon dates, which coincide at the first quarter of the 4th millennium BC, while the Bodrogkeresztúr group emerged later and lasted longer (Parkinson 2006, 55 ff.). We do not know much about the settlements of the Bodrogkeresztúr group, but the mortuary practices are quite similar to those of the Tiszapolgár group, although the types of artefacts vary (Patay 1975). Finally, it can be assumed that, with regard to the amount of copper artefacts, their shape and size, and the few well explored mines in the Balkans which may have provided the raw material for the metal workers of the Tiszapolgár and Bodrogkeresztúr group, copper metallurgy was fully developed and was an important part of society at that time (Ottaway 1982, 16 – 20).

Analogues for the eastern Hungarian copper objects have been found over a wider area. On the Transdanubian and Slovakian sites (Fig. 4.1) of the late phase of the Lengyel group (also called Epi-Lengyel or phase IV of the Lengyel group), heavy axes, spectacle spirals and pieces of sheet copper have been discovered (Novotná 1970; Schalk 1998; Schreiner 2007, 109 ff.). However, at the end of the Bodrogkeresztúr group (~ 3800 BC) the number of copper artefacts diminishes in archaeological contexts (Černych 1992, 51 ff.).

Finally, we should briefly mention an archaeological complex called Stroke- Ornamented Pottery (Fig. 4.2) which, separated into several regional groups such as Bajč-Retz or Balaton-Lasinja II-III, extended over an area from Croatia to Moravia, crossing Transdanubia (Kalicz 1991; Kalicz 1995). In addition to the various types of larger copper axe which can be paralleled in the contemporary Bodrogkeresztúr group, hoards including copper and gold discs provide evidence for a developed metallurgy and wider connections (Parzinger 1992; Kalicz 1995, 40 f.; Horváth and Simon 2003, 116 ff.).

Whereas in the Carpathian Basin from around 4500 BC onwards copper metallurgy was already highly developed, the archaeological evidence in the northalpine region indicates that here, metallurgical progress was limited at this time. Only a few, and mostly small, objects are known. Currently the earliest copper finds, which were found in the area north of the Alps, are an awl and a piece of copper wire from Schernau (D); this site is dated to the second half of the 5th millennium BC (Gleser and Schmitz 2001, 366). During the late 5th/early 4th millennium BC, and especially from contexts of the Neolithic Münchshöfen group in south-eastern Germany (Fig. 4.2), copper artefacts only occasionally appeared (Bartelheim et al. 2002, 60 f.) For a long time it was supposed that these were only casual imports from the Carpathian Basin, but now there might be a sign of local copper production. In Brixlegg (A) traces of copper smelting were discovered. Slag was excavated on a slope of the Mariahilfbergle near Brixlegg (Bartelheim et al. 2002; Höppner et al. 2005). The slag has been analysed with the result that it originated from tetrahedrite, a typical fahlore of that region (Höppner et al. 2005, 299). If the archaeological context in which the analysed samples of slag were found has been correctly dated to around 4000 BC, this would be the earliest evidence of sulphide ore smelting in that region. However, it has not yet been finally proven whether these are the remains of a developed metal production, because these early finds remain singular in the northalpine region and mines dating to that period have not been discovered[22]. Up to

[22] Slag finds, a fragment of sheet copper, a bead and two tuyères indicate at least sporadic metallurgical activity on that site. Three ^{14}C-dates ranging from 4500 to 3650 BC from layer SE 6 and a fireplace below that point to early copper production at Brixlegg (Bartelheim 2002, 41 f.). Although it is argued that the features indicate that the copper was more probably worked between 4500 and 3900 BC (because of the Münchshöfen pottery found on-site) (Bartelheim et al. 2002, 60; Höppner et al. 2005, 299), it has to be considered that it could also date to the later phase. Some pieces of slag and two copper artefacts in one

now it has not been possible to presume that the northalpine region had a flourishing metal production centre before around 4000 BC, as is evidenced for the eastern part of the Carpathian Basin and the Balkans. The central European copper objects of that period are most likely to have been imported from the Carpathian Basin. This emphasises finds such as the axe (type Şiria) from Überlingen (D), which has typological parallels in Bodrogkeresztúr contexts (cf. Matuschik 1997a).

However, after 4000 BC copper objects became more frequent in archaeological contexts in the northalpine region. In the horizon of the Cortaillod, Pfyn, Altheim and Mondsee groups (c. 3800 – 3400 BC)[23] indications of copper metallurgy can be located at various sites (Fig. 4.3) (Ottaway 1982; Matuschik 1991, 28 & Fig. 1). Most of our knowledge comes from excavations of the Alpine lake dwelling villages and settlements and enclosures of the Altheim group; very few burials have yet been excavated (Driehaus 1960; Ruttkay 1981; Matuschik 1991; Engelhardt 1994; Ottaway 1999; Schibler 2004). This period features not only a wide range of various copper artefacts (e.g. awls, flat axes, daggers, beads and other jewellery), but also a couple of crucibles and castings (Schlichtherle and Rottländer 1982; Obereder et al. 1993, 7; Matuschik 1998, 209 ff.; Bartelheim et al. 2002, List 2 & 3). Instead of nearly pure copper, which was used earlier on, copper which contains remarkably more arsenic than other impurities was predominantly used during the 4th millennium BC. It has been noted that arsenical copper has various improved mechanical properties and, thus, it was used deliberately over wider area and time[24] (Strahm 1994, 11 f.). Despite the large number of copper artefacts from the Cortaillod-Pfyn-Altheim-Mondsee horizon, evidence of mining related to these groups is still absent.

Eastwards, in Bohemia copper artefacts were also used. Most of the evidence is single finds, which can be typologically dated to the Cortaillod-Pfyn-Altheim-Mondsee horizon. These objects are mainly flat axes, daggers and jewellery (e.g. spectacle spirals) (Dobeš 1990; Matuschik 1996; idem 1997). Partly contemporary with the Cortaillod-Pfyn-Altheim-Mondsee horizon is a large archaeological complex spread over northern Europe; the so-called TRB complex (Fig. 4.3 – 4.4; dated from the late-5th to early-3rd millennium BC). However, it partly overlaps with the area this study examines and is closely related to the contemporary archaeological groups in central Europe. Copper artefacts (e.g. various types of axes, rings, spirals and further jewellery) were quite frequently found in the northern TRB groups and, due to the copper composition, importation from the East Alpine and Carpathian region is postulated, since in Silesia and Bohemia copper is comparably rare (Midgley 1992, 294 – 302; Klassen 2000, 299 f.; Klassen and Stürup 2001). In features of the Middle German Baalberge, Salzmünde and Bernburg group[25], local facies of the TRB complex, copper finds have only sporadically been found (Jacobs 1989, 3 ff.; Müller J. 2001, 411 f.). Additionally, the excavated TRB settlement from Makotřasy in Bohemia yielded crucibles and copper prills, which proves that metal artefacts were produced there[26] (Pleslová-Štiková 1985, 175 ff.). Contemporary with a decrease of metal objects in the Carpathian Basin is a rise of the number and complexity of copper finds in the Cortaillod-Pfyn-Altheim-Mondsee horizon. We can assume that the centre of metal production may have changed from there to the Alpine region; perhaps this was affected by a new metallurgical impulse coming from the West.

After a peak of metal production, metallurgy apparently stagnated in the northalpine region during the following period. Almost no copper appears in features of the northalpine Horgen-Goldberg III-Cham horizon[27] and the Czech Řivnáč and Jevišovice groups (Fig. 4.4), although these archaeological groups have been well explored[28] (Bartelheim 2007, 192 f.). One crucible, however, was recently found at a site of Cham group in Sallmannsberg (D), which points to local metal working in Bavaria at that time[29]. The Horgen, Goldberg III and Cham groups are contemporary from c. 3300 – 2800 BC (Matuschik 1992; Hafner and Suter 2003, 35).

layer (SE 6) and two further tuyères without archaeological context (Bartelheim 2002, 41 f.) are a matter of a few singular finds. Displacement has also to be taken into account, especially because the site is situated on a slope. The exact location of the slag and copper finds is not marked in the drawings, even if it is indicated in the captions (Bartelheim et al. 2002, Fig. 5 – 7). Although this site indicates sporadic early copper usage there, it neither gives sufficient evidence for widespread extensive metallurgy nor for copious mining in the area north of the main chain of the Alps before c. 4000 BC.

[23] All four archaeological groups of the Cortaillod-Pfyn-Altheim-Mondsee horizon can be combined, because they are quite similar concerning the typology of their finds and features (Schlichterle 1990, 222). The dating of the Cortaillod-Pfyn-Altheim-Mondsee horizon refers to Obereder et al. 1993, Lüning 1996, Billamboz 1998, Ottaway 1999 and Hafner and Suter 2003.

[24] From an archaeological point of view, the emergence and distribution of arsenical copper during the 4th millennium BC has been discussed several times (e.g. Sangmeister 1971; Ottaway 1978b; Schubert 1981; Kuna 1989; Matuschik 1998; Klassen and Stürup 2001). In Chapter 7.3.9 we will return to that issue and discuss it from a metallurgical perspective in detail.

[25] Dated between c. 3800 and 2750 BC (Midgley 1992, 225 ff.; Müller J. 1999, 77).

[26] ANo 58024 – 508031, ANo 98902 – 98903.

[27] The Goldberg III group is considered to be a regional facies of the Horgen-Cham horizon (Schichterle 1999).

[28] For further literature see: Itten 1970; Burger 1988; Matuschik 1992; Ottaway 1999, 251 – 268; Schichterle 1999; Hafner and Suter 2003, 33 – 45.

[29] I would like to thank A. Hanöffner for that information (pers. comm., November 2008).

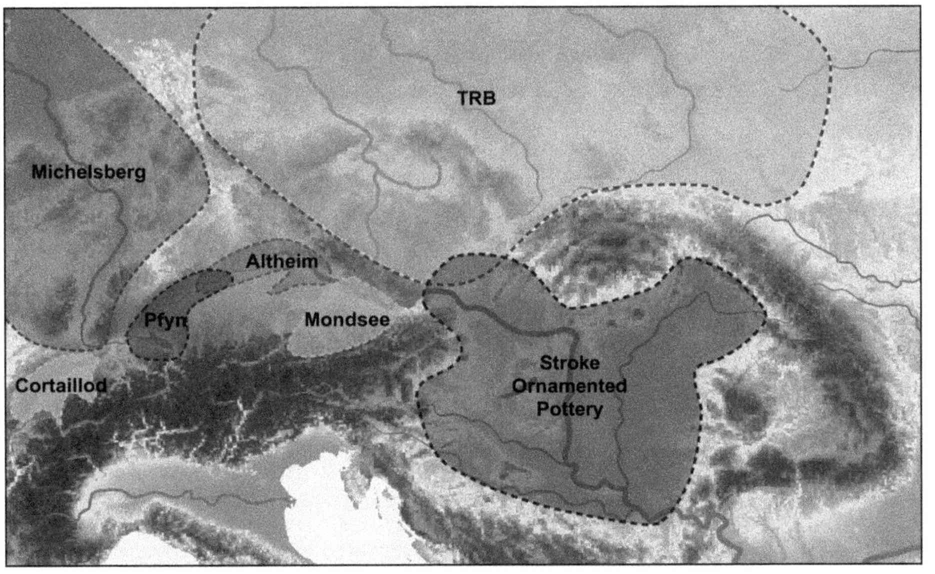

Fig. 4.3 Distribution of the archaeological groups from approx. 3800 – 3400 BC.

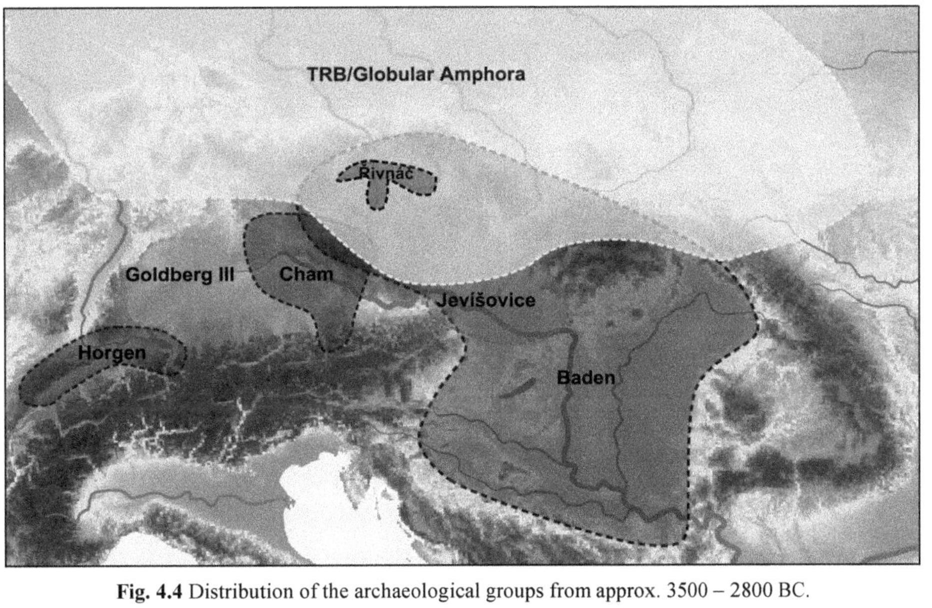

Fig. 4.4 Distribution of the archaeological groups from approx. 3500 – 2800 BC.

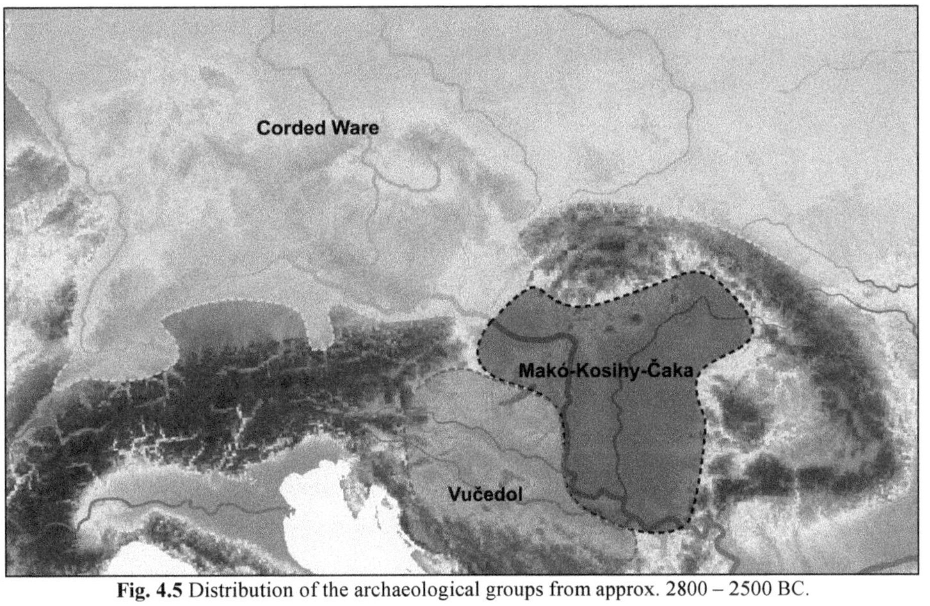

Fig. 4.5 Distribution of the archaeological groups from approx. 2800 – 2500 BC.

A decrease of metallurgical remains can also be observed in the Baden complex. The Baden complex lasted over a period at least from about 3500 to 2800 BC and is divided into three periods: the early Boleráz, the classic Baden and the late Bošaca/Jevišovice B phase (Maran 1998b, 500 ff.; Schreiner 2007, 16 ff.). It covered an area from the Carpathian Basin in the East and to the Czech Republic and north-eastern parts of Austria in the West, so that the spread of the Baden and the TRB complex partly overlapped geographically and chronologically (Fig. 4.4)[30] (Banner 1956; Chropovský 1973; Lenneis et al. 1995, 145 – 177; Endrődi (ed.) 2004).

During the first half of the 4th millennium BC evidence of metallurgy declined rapidly in central and eastern-central Europe (Maran 1998b, 512 f.; Schreiner 2007, 82). If copper objects appear in context, these are from exceptional features, for example the rich graves of Vörs (H) or Lichtenwörth in Lower Austria (Banner 1956, 33 f.; Willvonseder 1937, 18 ff.). Nevertheless a couple of single finds, mainly flat axes, were found and classified as being related to the groups of Horgen, Cham and Baden (Novotná 1970; Dobeš 1990; Říhovský 1992; Pászthory and Meyer 1998; Schalk 1998).

In comparison with these regions, the amount of copper artefacts rose in the Middle-Elbe-Saale region from around 3350 BC. Thus, J. Müller (2001, 412 ff.) proposed a fully developed copper technology for the Globular Amphora and the last phase of the TRB complex.

According to the archaeological evidence, sweeping cultural and economic changes are seen in central Europe during 3rd millennium BC. The widespread Baden complex disappeared in the Carpathian Basin and, from around 2700 BC, a number of smaller, local and contemporary archaeological groups are reflected in the archaeological remains. Probably influenced by the Vučedol complex, located south of the Danube, the Makó-Kosiny-Čaka group spread over the Carpathian Basin (Fig. 4.5) and was subsequently replaced by the archaeological substrate of the Somogyvár-Vinkovci, Pitvaros, Nyírség and early Nagyrév groups (Fig. 4.6)[31] (Tasić 1984; Bóna 1992; Vollmann 1999; Schreiner 2007, 85 – 98). The archaeological records of these groups are fairly similar and they overlap spatially. These Carpathian Basin archaeological groups have been defined, merged, split and redefined several times over during past decades of research (cf. Kalicz 1982). Pottery types, which have been found largely in settlements, distinguish the particular groups (cf. Tasić 1984; Bóna 1992; Visy 2003). Schreiner (2007) recently proposed uniting amongst others, the Makó-Kosiny-Čaka, Somogyvár-Vinkovci, Nyírség and Nagyrév groups as a *spätäneolithischer Kulturkomplex*, because they all were "[...] *kulturell relative ähnlich* [...] (culturally quite similar)" (Schreiner 2007, 85). Except for the ceramics, other finds, above all metal objects, are very rare, but are quite uniform (Visy 2003, 142).

A number of single-bladed shaft-hole axes are a notable characteristic of the Carpathian Basin during the first half of the 3rd millennium BC (Batorá 2003). Although several of these artefacts have been found in graves as well as in settlements, their function is not yet understood. Utilitarian functions, such as use for tools or weapons, have been discussed as well as a ceremonial or prestige usage. A number of bivalve moulds used to cast these axes have been discovered in settlements of, for example, the Makó-Kosiny-Čaka, the Somogyvár-Vinkovci and the Vučedol groups (Batorá 2003, 22). They document a local production of these shaft-hole axes. In terms of Hungarian research this period is already defined as the Early Bronze Age, because linked to contexts of these groups new, more distinctive types of metal artefact and technique appeared and it had been assumed that those were made of tin-bronze. However, copper analyses have shown that these objects were still not being made of alloyed tin-bronze (with more than 4% tin), but were of fahlore-copper (Liversage 1996; Visy 2003, 141).

In contrast to the Carpathian Basin where several partly contemporary and geographically-overlapping groups existed, two widespread archaeological phenomena stretch across most of the rest of our research area during the 3rd millennium BC.

Firstly, the Corded Ware complex extended over a huge area from the today's Baltic States to the Rhine valley (Fig. 4.5), and represents new ideas, religious beliefs and social structures in central Europe. The first impulses of the Corded Ware complex came either from the archaeological cultures of the Black Sea region or from the subsequent TRB complex in central Europe, but there is not enough evidence to demonstrate which of these had been the 'parent' area (Czebreszuk 2004a, 468; Heyd 2007).

The next part of this chapter will concentrate on the Corded Ware complex as it appears within the area of research. There, according to a wide range of ^{14}C- and dendrochronological dates, the Corded Ware complex lasted from 2900 until 2200 BC, though these dates differ from region to region (Furholt 2003). The remains of the Corded Ware complex are predominately known from burials, but the Corded Ware complex in Switzerland is chiefly known from lake dwellings[32]. The mortuary rites

[30] As the issue of the genesis of Baden complex, its development and the relationship to neighbouring groups is too comprehensive to discuss it appropriately at that point, it must be referred to an in-depth work written by Schreiner. He offers a current summary including the most important references on the distribution and characteristics of the Baden complex (Schreiner 2007, 78 – 85).

[31] According to Hungarian terminology, these archaeological groups cover the typo-chronological periods Early Bronze Age FB I and FB II (cf. Maran 1998a, 315). Generally to classify the Hungarian Bronze Age a systematic of six phases – Early Bronze Age I – VI – based on Mozsolics' studies is used (1967). For an overall introduction to the Bronze Age chronology of the Carpathian Basin and an overall discussion about the Hungarian Copper and Bronze Age terminology the reader is referred to Kalicz (1982) and Novotná and Novotný (1989).

[32] The Corded Ware complex is divided into a variety of regional groups which are generally distinguished from each other by pottery styles; these styles may reflect pre-existing ceramic variability. An outline of

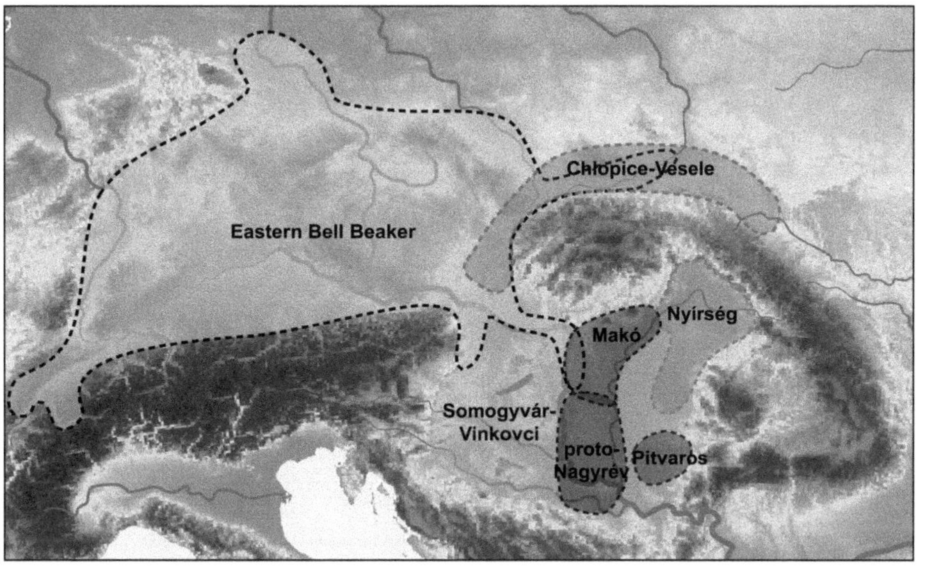

Fig. 4.6 Distribution of the archaeological groups from approx. 2500 – 2200 BC.

of the Corded Ware mark a change in ritual and social habits in central Europe. While the preceding culture appears to have been egalitarian, with villages, enclosures and especially collective graves showing mutuality in life and death, the appearance of the Corded Ware complex marks the near-disappearance of large, long-lived settlements and a shift to the practice of interment in single graves as the dominant practice[33]. The bodies were buried in flat graves or under mounds, most often in a crouched position, in a gender-specific manner, with males laid out West-East and females East-West, both facing the south. The grave goods were also gender specific and comprise stone battle-axes, amphorae, pots with wavy lines and the eponymous beakers adorned with corded or incised line decoration (Fischer 1956; Buchvaldek 1986; Häusler 1991; Wiermann 1998; Turek and Černý 2001).

This gender specific burial rite, the individualised grave goods and a qualitative and quantitative variation in these grave goods allow us to identify ranking of the dead: thus battle-axes, for example, are interpreted as status symbols. Accordingly, the Corded Ware community has often been portrayed as the first central European stratified society (cf. Wiermann 1998; Strahm 2002; Heyd 2007 including further references). However, this claim should be treated with caution as it is based solely on funerary evidence, which does not necessarily mirror the situation in everyday life. The graves may indeed display different status, but this could just be a ritualised expression.

Although the Corded Ware culture is distributed over a large area and many sites have been well excavated, evidence for metalworking has not frequently been found in comparison with the number of sites and features[34]. A number of small copper objects, mainly jewellery (for instance beads, rings and spirals) have been discovered, mostly from grave contexts (cf. Strahm 1971; Ottaway 1992; Šumberová and Hora 1992; Schreiner 2007). Tools, axes, daggers and awls, have been found in Swiss lake dwellings or as single finds. Double axes (of Zabitz type), shaft-hole axes and flat axes have generally been discovered as stray finds and assigned to the Corded Ware period on typological grounds (Kibbert 1980; Batorá 2003). Direct evidence of metallurgy, such as castings or moulds, has not yet been found. Finally, the relatively small amount of copper and its appearance in graves indicates a certain value and shows that copper was used as a status symbol at the period (Heyd 2007. 341 ff.).

While the Corded Ware complex dominated the eastern part of Europe, a second archaeological complex appeared in the area of research – the Bell Beaker phenomenon. As a pan-European phenomenon, its interpretation and the general issue of Bell Beaker metallurgy have already been discussed in Chapter 1.3 and 1.4, here the discussion deals only with the Bell Beaker phenomenon as it pertains to the study area, comparing it with earlier and later groups.

The so-called Eastern Bell Beaker group is defined by relatively homogeneous finds and features and has been radiocarbon dated to 2500 to 2000 BC (Müller and van Willigen 2001; Heyd 2007, 334). Archaeological remains classified as Eastern Bell Beaker group are distributed throughout the northalpine region from the Upper Rhine valley to Budapest (Fig. 4.6).

these groups is given in the proceedings of the *Schnurkeramik-Symposium* (Buchvaldek and Strahm (eds.) 1992).

[33] Except for Switzerland where the Corded Ware complex is known from several lake dwellings, but not from graves (cf. Hafner and Suter 2003).

[34] Bertemes and Heyd, referring to work by Jacobs and by Šumberova and Hora, discuss the quantity of copper objects and comment that copper would not have been rare (Jacobs 1989; Šumberova and Hora 1992; Bertemes and Heyd 2002, 212). Certainly copper was used and passed on to us, but it has to be taken into account that compared with the number of sites and the amount of other materials found in Corded Ware contexts (ceramics, bone, flint and stone) copper is still exceptional. Bartelheim, for example, has recently spoken about a quantity of 5 % of graves containing copper artefacts in the northalpine region (Bartelheim 2007, 193).

Characteristic Bell Beaker pottery and *Begleitkeramik* (common ware)[35], which is typical for the Eastern group, are primarily found in graves, although recently more and more traces of settlements have come to light. Virtually no traces of buildings remain, so that pits are the only evidence for settlements (Endrődi 1998; Turek and Peška 2001; Heyd et al. 2004; Dvořák (ed.) 2006). However, numerous burials have been excavated in cemeteries comprising up to 20 – 30 graves in southern Germany and up to more than 100 burials in Austria, Moravia and Hungary[36] (Heyd 2007, 332). Analysing the mortuary practice anthropologically has shown that within the Bell Beaker East group the dead were buried in single graves, in the foetal position, with a strict bi-polar gender differentiation and mainly in flat graves, but sometimes under a mound. This is similar to the Corded Ware complex, but instead of an East-West orientation the Bell Beaker graves were usually orientated North-South, while men were laid on the right side, with their head facing to the North and women on the left side, head towards the South (Fischer 1953; Havel 1978; Häusler 1991; Engelhardt 1998; Müller A. 2001; Vander Linden 2003 & 2007a).

There are also elements that connect the grave goods found in the Bell Beaker and Corded Ware complexes. In both, a fairly narrowly-defined set of objects were used. Both have beakers – but differing in styles and shapes – and instead of, for example, Corded Ware battle-axes and amphorae, the Bell Beaker set or package consists of beakers, bowls, flint arrowheads, wrist-guards, bow-shaped pendants, V-perforated buttons, finger rings, copper awls and tanged daggers (although in general just a selection of these artefacts were deposited in any individual grave[37]: Shennan 1975b; Fischer 1976; Case 2004). In addition to items from this range, some Bell Beaker graves contained artefacts such as small exquisite sheet gold ornaments and rings, flint blades and stone tools (probably for working metal), so the Bell Beaker assemblage varies from grave to grave. Gold artefacts are especially seen as prestige goods, which indicate a higher status for the buried person (e.g. Heyd 2007, 341 ff.).

The range of Bell Beaker grave goods also shows gender dimorphism: weapons and tools are mostly associated with men and rings, buttons and awls mainly with women. Only a small number of burials deviate from this pattern (Müller A. 1993). Quantitative and qualitative analyses of grave assemblages has led archaeologists to conclude that Bell Beaker society was stratified (Heyd 2000a; Heyd 2007).

This theory is chiefly supported by the idea that metal objects were elite status symbols and that metalworking played a significant role in Bell Beaker communities. As several graves that contained cushion stones and other metal workers' tools (cf. Moucha 1989, 215 f.) are also exceptional in terms of their opulent grave goods and structural features, such as ring ditches or mounds, this has led commentators to argue that metal workers – or the people who dealt with them – were so important that it was appropriate to portray them as 'metallurgists' in their graves (Butler and van der Waals 1966; Bátora 2002; Bertemes and Heyd 2002). The set of metal objects found in Eastern Bell Beaker contexts mainly includes (tanged) copper daggers (Fig. 4.7b, c and h), awls (Fig. 4.7l) and oar-headed pins with an S-profile rolled head (Fig. 4.7d). Recently, several rings and sheets metal made of copper, electrum and gold have been found. Two types of ring appear in contexts of the Eastern Bell Beaker group: firstly, spiral finger rings (so-called *Noppen-* or *Lockenringe*) (Fig. 4.7k) and, secondly, ring-shaped ornaments with a flat hammered terminal, often decorated by small punches (Fig. 4.7g). The same type of decoration is also found on the sheet metal ornaments of copper and gold (Fig. 4.7f) (Metzinger-Schmitz 2004, 240 f.). The appearance of gold artefacts in Bell Beaker graves also indicates the wealth of Bell Beaker users.

In contrast to the Iberian Peninsula where there are several sites in which direct evidence of metal work and Bell Beaker pottery is associated (Rovira and Delibres de Castro 2005, 514), evidence for metalworking is extremely rare on sites of the Eastern Bell Beaker group. Until now, the only direct evidence for metalworking from an Eastern Bell Beaker context is the famous mould for casting a tanged dagger from Luděřov (CZ) (Fig. 4.7a) (Hajek 1966, 214).

As regards the homogeneity of the Eastern Bell Beaker group, there is some difference between the finds and features of the Hungarian Bell Beaker Csepel-group and those of Bell Beaker users in the Rhine-Main and the Middle-Elbe-Saale regions of Germany. The Csepel-group is very local and isolated from other Bell Beaker areas; it is only found around Budapest (Schreiber 1973). Whereas inhumation dominated in the rest of the East group, in Hungary cremation was more common and represents a continuation of an earlier local tradition (Kalicz-Schreiber and Kalicz 2001, 442). In addition to many beakers which are characteristic of the Eastern Bell Beaker group, single-handled jugs are very common there. Because this pottery is analogous to local styles, the Csepel-group is seen as strongly related to the Somogyvár-Vinkovci group and, even more, to the Nagyrév groups (Kalicz-Schreiber and Kalicz 2002). Furthermore, the radiocarbon dates (ranging between 2900 and 2000 cal BC) are, in most cases, older than for

[35] Hájek (1966, 210) defined as *Begleitkeramik* undecorated bowls and handled cups, which often appear in Bell Beaker contexts. In contrast to *Begleitkeramik* which is regarded as typical for the Eastern Bell Beaker group, the common pottery of both East and West groups has recently been classified as Complementary Ware by Strahm (2004; 2008). In general, common pottery indicates close links to local, previous pottery traditions.

[36] Heyd (2007, 335 ff.) sees the growth in the cemeteries as a chronological development from his earliest Bell Beaker phase to the Early Bronze Age. This will probably have to be reviewed when the recently excavated huge cemeteries from Moravia and Hungary are published (per. comm. J. Peška, July 2007).

[37] In other areas of the Bell Beaker phenomenon this set varies slightly. In parts of Western Europe, for example, Pamela points were commonly used.

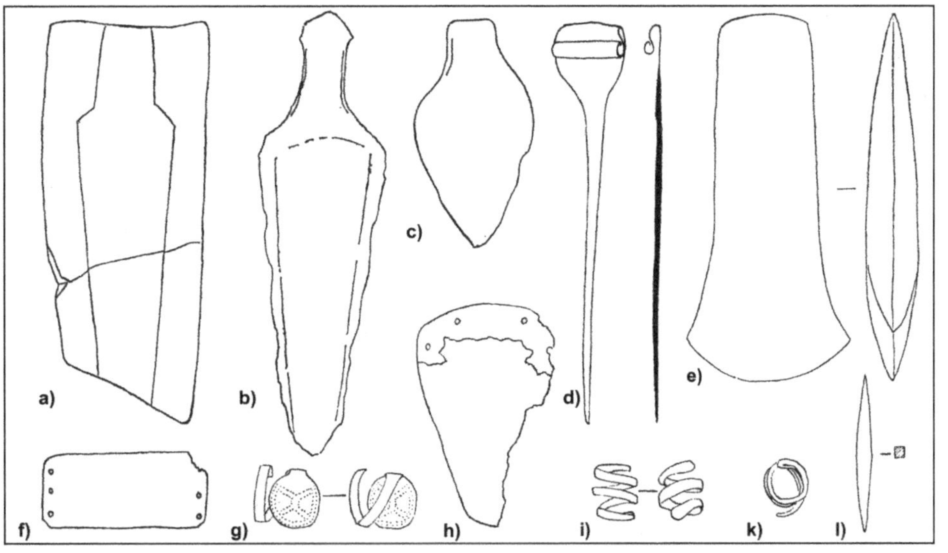

Fig. 4.7 The sand stone mould from Luděřov and a selection of metal finds of the Eastern Bell Beaker group **a)** Luděřov (CZ), **b)** Stehelčeves (CZ), **c)** Thalmassing (D), **d – e)** Budkovice (CZ), **f)** Vřesovice (CZ), **g)** Borkovany (CZ), **h)** Šlapanice (CZ), **i)** Lysolaje (CZ), **k)** Lechovice (CZ) and **l)** Lysolaje (CZ) (after Kuna and Matoušek 1978, Fig. 1 & 2).

the rest of the Eastern Bell Beaker group[38] (Müller and van Willigen 2001). In contrast, the Bell Beaker sites from the northern part of the Upper Rhine valley, Hessen and the Middle-Elbe-Saale region vary chiefly in terms of the styles of the beakers. Whereas these beakers are typologically more similar to the forms seen in the Western Bell Beaker group, their features and dating do not differ very much from the rest of the East group (Sangmeister 1951). By keeping this in mind, we can include both regional groups in the Eastern Bell Beaker complex.

In central Europe, where the Corded Ware and Bell Beaker complexes overlap both spatially and chronologically, both prehistoric phenomena are, when viewed from a more generalised perspective, quite similar in terms of their features: there are virtually no settlements, but many gender-differentiated single graves, in which the dead were buried in a crouched position with a standardised set of grave goods (Heyd 2007, 361 – 368). However, they differ markedly in their details: the graves are differently orientated, and the grave goods feature different types of beaker and a different 'package' of artefacts. It seems that each group distanced itself intentionally from the other one (cf. Fischer 1953; Fischer 1976; Turek and Černý 2001). When the Bell Beaker phenomenon appeared in central Europe, Bell Beaker users may have reacted to the Corded Ware rites that were already present. They probably adopted aspects of the local rites, but also included their own traditions. This can be seen not only in the domestic pottery that follows local traditions, but also in the similarity-but-difference in the gender-differentiated funeral traditions, with individuals buried with beakers and weapons. This arguably indicates a certain interaction between the users of the two different types of pottery. The reason for this variation could be that both shared a similar ideology but expressed it differently because the groups could have been defined around specific aspects of people's lives such as technology, economy, society and/or religion (Benz et al. 1998, 184 f.).

Meanwhile, it is clear that there was chronological overlap not only between the Bell Beaker and Corded Ware complexes, but also with the Earliest Bronze Age groups[39] in central Europe. The appearance of the latter is marked by the appearance of new styles of pottery and an increase in the use of metal objects and is interpreted in terms of socio-economic changes (Bertemes and Heyd 2002, 187 f.). Nevertheless, there was continuity of tradition between Bell Beaker users and various Earliest Bronze Age groups, demonstrating their contact and interaction (e.g. Kadrow and Peška 1999; Bertemes 2000; Krut'ová 2003).

The archaeological features of the Earliest Bronze Age in the region north and north-west of the Carpathian Basin are closely related to the previous Bell Beaker and Corded Ware features. In general, the Earliest Bronze Age

[38] Recent excavations near Budapest may clarify the nature of the Hungarian Bell Beaker group, but they are not yet published. The author had the chance to study some of the finds at the conference "*Bell Beakers along the Riverside*" in Szentendre (H) in May 2009. Initial results from the excavation of over 1000 graves indicate that the Hungarian Bell Beaker graves are, both typologically and chronologically, more closely related to the rest of the Eastern Bell Beaker group than previously suspected. An all-encompassing interpretation of the relationship and a final evaluation has to await the detailed publication of the cemeteries. It may well be that the results of previous analyses, discussed here, may have to be questioned; future research will probably change the current picture.

[39] As will be shown in this section, the term 'Earliest Bronze Age' encompasses several archaeological groups which cannot clearly be classified as either Neolithic or Bronze Age. Remains of these groups overlap typologically and chronologically with both. Whereas the types of metal objects are comparable with typical Early Bronze Age finds, the features tend to be more closely connected with the earlier period. Since, during the history of research, those sites were chiefly arranged as the first phase of the Early Bronze Age, the term 'Earliest Bronze Age' is adequate. In terms of relative chronology it corresponds with Reinecke's Bronze Age phase A1 and the other contemporary phases.

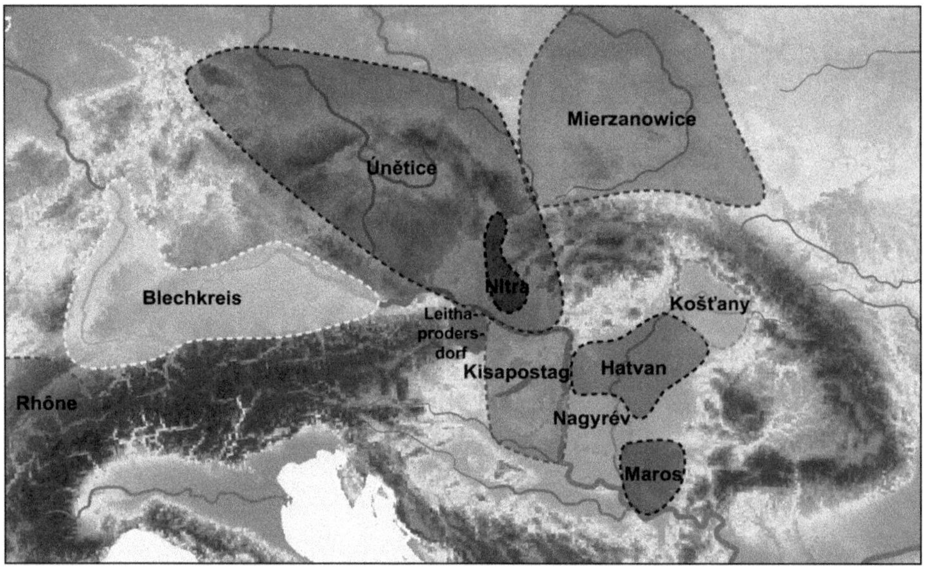

Fig. 4.8 Distribution of the Early Bronze Age groups from approx. 2200 – 2000/1800 BC.

groups between western Slovakia and the Rhine are unique archaeologically. In this area, settlements are still virtually unknown and most of the remains come from graves. Compared with previous cultures, the cemeteries grew to an enormous size, often far exceeding 100 graves (e.g. Singen am Hohentwiel [D], Gemeinlebarn [A] or Výčapy-Opatovce [SK]) Pottery styles changed and metal objects appeared much more frequently in the graves. There was also a larger variety of types of metal artefact, which are almost exclusively made of copper and not tin-alloyed bronze (Junghans et al. 1954). However, the dead were buried in flat single graves, mostly as crouched inhumations, with a gender-specific contrast in orientation. The Earliest Bronze Age sites have been divided into various regional groups (Bertemes and Heyd 2002; 188 ff.). The analyses of several large cemeteries have led to the definition of various regional groups which differ in types of grave goods, pottery styles and burial rites. Although graves were usually orientated on gender-specific lines, the orientation (N-S or W-E) can vary regionally; so can the construction of the grave chamber, with stone cists being used in some cases. Recently archaeologists have reflected on this issue and tried to combine these smaller groups into larger complexes, allowing a better and more comprehensive examination. The following descriptions are mostly based on chronological, regional and typological information as presented in the literature.

Firstly, there are the Earliest Bronze Age groups of the Carpathian Basin of phase FB I and FB II (after Kalicz 1982), such as the Makó-Kosiny-Čaka and Somogyvár-Vinkovci, the Pitvaros, Nyírség and early Nagyrév groups, including the Csepel-group, all of which cover a horizon ^{14}C-dated to between c. 2800 to 2200 BC (Forenbaher 1993, 247 f.).

The Únětice complex extended from Moravia to Middle Germany (Fig. 4.8), appearing around 2300 BC (Rassmann 1996, 204; Müller J. 1999). Thus, this research encompasses the proto- and early Únětice groups (according to Zich [1996]: *Stufe 1* and *2*), lasting until 2000 BC. According to various authors, these have left hardly any evidence of metal, except for a small number of wire ornaments and rings (Zich 1996, 262, 269, 300; Bartelheim 1998, 167). Moucha classifies the proto-Únětice group as Eneolithic and separates it from the classical Únětice group (Moucha 1963, 51; idem 2005, 7). In contrast, J. Müller was able to demonstrate that Bartelheim's and Zich's typological phases of the Únětice complex are not comparable with the absolute chronology (Müller J. 1999, 70 f.). He was also able to prove that people of the Únětice complex were commonly using tin-bronze (Müller J. 2002). Therefore, the finds associated with the classical Únětice group are excluded from this study; the proto-Únětice group, however, is included.

Between the Carpathian Early Bronze Age groups and the Únětice complex another archaeological complex has been located in western Slovakia, Moravia and southern Poland. The so-called *Epi-Schnurkeramische Komplex* (Epi-Corded-Ware complex [ECWC]) was defined on the basis of the analyses of the cemetery of Iwanowice (PL) and circumscribes the early Mierzanowice and the Chłopice-Vesele groups (Kradow 1991; Kadrow and Machnik 1992; Peška and Šebela 1992; Kadow and Peška 1999). The Eastern Slovakian Košt'any group can also be seen as part of the ECWC (Fig. 4.8) (Furmánek et al. 1999, 29 ff.). It is assumed that the ECWC reflects the phase *Reinecke A0* when, 2350 – 2100 BC, the cultural transformation between Chalcolithic and Bronze Age took place (Bertemes 2000; Bertemes and Heyd 2002; Schreiner 2007). Rooted in this complex and strongly linked with it, the west Slovakian Nitra group probably emerged, and went on to influence the development of the Únětice complex[40] (Točìk 1963; Bertemes and Heyd 2002).

[40] According to radiocarbon dates, the Mierzanowice and Nitra groups may have lasted until around 2000/1900 BC (Kadrow 1997, 235).

Finally, the Earliest Bronze Age groups of the region between the Rhine and the Enns River have already been described as the *Blechkreis* by Vogt (1948) (Fig. 4.8). Then, it was divided into the regional groups Adlerberg, Straubing, Neckar, Singen and Unterwölbling[41]. Bertemes (2000) then reunited these as the *Danubian Early Bronze Age*. In the following discussion the term *Blechkreis* will be used for that complex. By typological comparisons and the fact that elements of the Bell Beaker and Corded Ware mortuary practices occur in later features, several archaeologists have argued that the genesis of the *Blechkreis* is a regional variation of a combined Corded Ware and Bell Beaker tradition (cf. Lißner 2004).

Absolute dating evidence is rare. Consequently, chronological studies of the central European Early Bronze Age are mostly based on typological sequences (cf. Ruckdeschel 1978a; Bertemes 1989a; Zich 1996; Bartelheim 1998). The typological phases of the Bronze Age of the northalpine and Carpathian regions have been compared and put into an absolute-chronological context by various authors (e.g. Liversage 1994, 62 f.; Maran 1998, 311 – 354). The correlation of these different systems has led to figure 4.9.

According to the few radiocarbon dates, which are mainly from the cemetery at Singen am Hohentwiel (D) with some from Middle Germany, southern Poland and Moravia, the earliest phase in Austria, the Czech Republic, Germany and southern Poland extends over a period from c. 2200 BC to c. 1800 BC (Rassmann 1996; Kadrow 1997; Furholt 2003, 259 f.). In Switzerland, Bronze Age A1 ended with the re-establishment of the lake dwellings c. 2000 BC (Hafner and Suter 2004b) and in the Carpathian Basin, the corresponding Early Bronze Age phases FB I and FB II lasted from around 2700 until 2200 BC (Maran 1998, 352 f.). The Early Bronze Age groups Kisapostag, Nagyrév, Maros and Hatvan are classified as phase FB III and tin-bronze was regularly used in those contexts, so that they are excluded from this research (Liversage 1994; Maran 1998).

We can conclude that there are four complexes of the Earliest Bronze Age within the area of research:

1. The Earliest Bronze Age groups of the Carpathian Basin.
2. The ECWC and the Nitra group.
3. The proto-Únětice complex.
4. The *Blechkreis*.

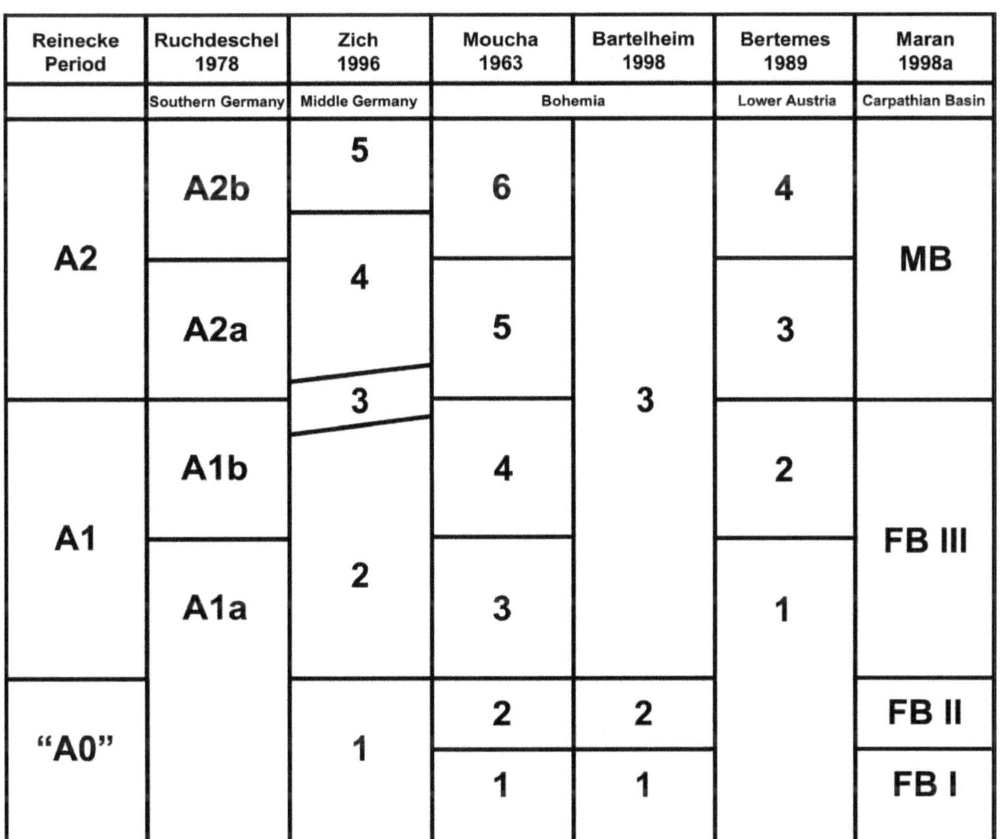

Fig. 4.9 Table comparing the different chronological schemes of the Early Bronze Age periods in central Europe.

[41] Köster 1966; Ruckdeschel 1978a; Gebers 1978 & 1984; Krause 1988; Bertemes 1989a; Möslein 1997.

All of the Earliest Bronze Age groups show elements of continuity from previous traditions. This is shown in the continuing use of the individual mode of burial, with the bodies laid in a crouched position in an unmounded grave and orientated according to gender, recalling Corded Ware and Bell Beaker practice. It is also reflected in material culture: objects such as small pots, daggers and spiral finger rings were still used as grave goods, although their styles and shapes had changed (Zich 1996, 340 ff.; Heyd 2000a, 445 – 456; Kim 2005, 123 – 129). Others were made of a new material: for example the bone V-perforated buttons were replaced by similarly shaped metal tutuli and, furthermore, new artefacts such as oar-headed pins were used[42].

After c. 2000 BC new elements appeared. Settlements, in particular tell-settlements in Hungary, and enormous rich graves such as the famous examples at Leubingen (D), Helmstedt (D) and Łeki Małe (PL) are also found; these demonstrate the establishment of a hierarchical social structure (Bartelheim 2002). Furthermore metal hoards, for example the southern German deposits of ingot torcs (*Ösenhalsringe*)[43] emerged; these had not been deposited during the Corded Ware and Bell Beaker phase and the Earliest Bronze Age. These hoards have been interpreted as metal traders' stores of copper (Krause 2003, 257 ff.; Kristiansen and Larsson 2005, 108 ff.). Metal in general became more common and new casting techniques allowed the production of more complicated objects, such as metal-hilted daggers. Tin-bronze was now regularly used (Kristiansen and Larsson 2005, 112 f.). Today, we call this period the *fully developed Bronze Age* and in the northalpine region the first phase is usually called the *Early Bronze Age* (encompassing Reinecke Bronze Age A2 and B).

This development, however, did not take place in every region of the study area at the same time and not every component appeared everywhere simultaneously. Tin-bronze use, for example, was probably established around 1800 BC in southern Germany, whereas it had already been in use in the Middle-Elbe-Saale region and in the Carpathian Basin around 2200 BC (Müller J. 2002).

In the light of the evidence reviewed above for the appearance of the single grave tradition around 2500 BC and for the variability in the grave goods being deposited, with valuable objects being interpreted as status symbols, several authors have argued that this signified a change to stratified societies in central Europe around this time (Bertemers 2000; Strahm 2002; Heyd 2007). The division of labour is often seen regarded as a concomitant in this process, because it is assumed that demand for the great amount of copper and especially tin will have required a wide-ranging, structured organisation (e.g. Kristiansen and Larsson 2005). Consequently, several authors have argued for a direct connection between the establishment of metallurgy and the introduction of stratified social systems (e.g. Shennan 1975a; Strahm 2002, Kristiansen and Larsson 2005).

By contrast, Bartelheim has recently argued that in central Europe no archaeological evidence emphasising social differences is directly related to metallurgical activities, even for the Early Bronze Age. He argued that Chalcolithic and Early Bronze Age communities settled in areas with fertile soil, such as loess, rather than close to copper ore deposits (Bartelheim 2007, 196 – 207). Whereas Bartelheim claimed that Early Bronze Age cemeteries do not show social hierarchy apart from some rich graves, such as the one from Leubingen (D) (Bartelheim 2007, 197), others have interpreted the cemetery evidence of the Eastern Bell Beaker and Nitra groups in terms of social stratification (Bátora 1991; Heyd 2007).

Cemeteries with single burials are the ideal foundation for analysing hierarchies in prehistoric societies, but as has been shown above, not all periods deliver the same kind of archaeological record. However, settlements can also sometimes be a sign of social diversity. Sites of the 4th millennium BC had already provided evidence suggesting that society was differentiated. A house-specific spectrum of finds in lake villages (e.g. Reute [D]), or hill forts in the Czech Republic (e.g. Makotřasy [CZ]) and other small lowland settlements suggest the existence of social differences within settlement clusters during the 4th millennium BC. The lack of graves of the period, however, does not allow one to analyse social stratification in detail (Müller, J. 2001, 430 ff.). Studying the very few graves dated to that time which have been found in the area of research also indicates social hierarchies, as reflected by the richly-furnished graves such as at Vörs (H) (Banner 1956, 33 f.). Even if the impact of copper metallurgy on Chalcolithic society had been rather small, it is still not clear whether certain communities – above all the Eastern Bell Beaker group – played a particular role in distributing metallurgical technology and expertise.

It could be demonstrated that in the study area evidence of copper appears in various regions at different times. Also the switch from copper use to tin-bronze use occurred at different dates in different areas. While copper occurs in one area only sporadically, in other regions complex artefacts were already in regular use at this time. Interestingly, when copper emerges commonly in archaeological contexts, gold also often appears – for instance, the gold discs found in relation with the Tiszapolgár and Bodrogkeresztúr groups (Parzinger 1992). At the northalpine region gold appears regularly not until the Bell Beaker period. In Bell Beaker contexts several golden and even silver finger rings and quadrangular sheet metals have been discovered (Metzinger-Schmitz 2004, 240 f.).

[42] Over the last decades the issue of continuity and discontinuity between the Late Neolithic and Early Bronze Age has been discussed in detail in a number of articles. Therefore the reader is directed to the proceedings of several conferences focusing on this topic (Aspes and Fasani (eds.) 1982; Richter et al. (eds.) 1989; Müller, J. (ed.) 2002).

[43] These objects were chiefly interpreted as copper ingots, although they were also found in graves as neck rings (Junk 2003, 11 f.). They may, of course, have served this dual role, with wearing them being a convenient way of carrying around one's wealth.

In general, gold is much less frequent than copper. The reason why both metals emerged more or less at the same time could be that the two metals are shaped with similar techniques (e.g. by hammering) – although copper has to be smelted from ores instead of native appearing gold. Silver, on the other hand, has only been used from the 3rd millennium BC. In principle, the appearance of gold chronologically varies in the same way than copper.

There is also a varying relationship between the different sources of archaeological evidence for metal using, such as graves, settlement and single finds. This complicates the comprehensive and diachronic analyses of the social value of metal. Whereas massive, heavy axes of the late 5th and early 4th millennium BC have usually been found as single finds, copper evidence from between c. 4000 and 3000 BC has been almost exclusively found in settlements, and that of the 3rd millennium BC, in grave contexts. Accordingly, it is hard to say whether we are dealing with comparable data in terms of the metal's utility value, social value and religious value. While a copper axe found in a house context indicates that it is likely to have been used mainly as a tool, the tanged dagger in a Bell Beaker grave had definitely been deposited with religious intentions. Both objects, however, may have had a certain material value and may have expressed a certain status of the former owner.

This variation has to be considered when metal objects over such a large area of research are studied. Consequently, the present study focuses on the technical innovation and processes of transferring technological expertise rather than on the social values of metallurgy. Nonetheless, the impact of metal technology cannot be neglected, but the main focus will be on the question of the level and the spread of certain techniques of copper production. This can also be analysed using information from single stray finds, because the production process probably did not depend on the later use and deposition.

4.2 THE CHALCOLITHIC OF CENTRAL EUROPE

In the last section we encountered the problem that the period covered by the term 'Chalcolithic' is not standardised over the area of research. Reviewing the international literature discussing the period between the emergence of copper and the final establishment of tin-bronze technology in central Europe, archaeologists of various regions use different terms: *Late Neolithic*, *Final Neolithic*, *Eneolithic*, *Chalcolithic*, *Copper Age* and even *Earliest Bronze Age*. These terms have been used synonymously to describe a historical, chronological and typological period as well as a particular stage of cultural or economic development. Thus, it is necessary to offer some commentary on terminology.

A general grouping according to the Three-Age system (after C. J. Thomsen) is based on material which is typical for certain periods. The Stone, Bronze and Iron Ages are defined on the basis of the emergence of the particular material (Thomsen 1836). Soon after the introduction of that model it was obvious that a fourth material – copper – was also frequently used and, according to Thomsen's system, requires its own period. The Copper Age was situated between the Stone and Bronze Ages. Nevertheless the idea of a Four-Age system did not take root and the original Three-Age system has still been used (cf. Lichardus 1991a; Klassen 2000, 17 ff.).

The idea of there being a separate period between the Neolithic and Bronze Age, however, was never abandoned, because between the Neolithic (i.e. the period of establishing a productive economic system by egalitarian societies) and the Bronze Age (i.e. the phase of 'industrialised' usage of tin-bronze, the hierarchising of communities and, probably, the division of labour) we can observe cultural, ritual and economic changes in archaeological contexts. During the 5th and 4th millennia BC there is evidence for various inventions and new technologies which justify the assumption that economic and social change occurred in Europe. Probably the most important developments were the adoption of ploughs, wheeled vehicles, horse riding and, at least, copper metallurgy (Sherratt 1981; ibid.1982). Sherratt (1981) summarised these innovative developments as the *Secondary Products Revolution*[44]. An intensified agriculture, ploughing, the exploitation of sheep, goats and cows for milk and wool, and the association of smaller unities to villages may have facilitated the emergence of an infrastructure which benefited copper metallurgy. Metallurgy – prospecting, mining, processing and trading – needs a sophisticated infrastructure and manpower (Ottaway 2001). The time between c. 4000 and 2000 BC is a period when several transformations occur in the archaeological record. It is important to mention that the indications of these changes are not continuous within the area of research and have left traces only sporadically. Thus they are not a linear and complete area-wide development which would fit into a pure evolutionary model. Taking this into account, it seems difficult to find an appropriate label that characterises both the region and the period of this research as a whole.

In archaeology, a number of concepts labelling a certain period or phase are used. Which concept is the most appropriate for this study? Probably the most famous attempt to define the period in question was made at a conference hosted by Lichardus in 1988. In the conclusion of this conference, Lichardus gave a list of attributes which ought to characterise the *Kupferzeit als historische Epoche* (Lichardus 1991b, 787 f.). Comparing that catalogue with the archaeological background of our area of research it is obvious that these characteristics do not appear in every part of the area at every time. Some characteristics, a stratified organisation of society for instance, are nowadays seen as an attribute of the Bronze Age (cf. Bertemes 2000; Strahm 2002; Kristiansen and Larsson 2005). A second problem is that within the whole

[44] He ascribed them to a narrower date range than we now know to be the case.

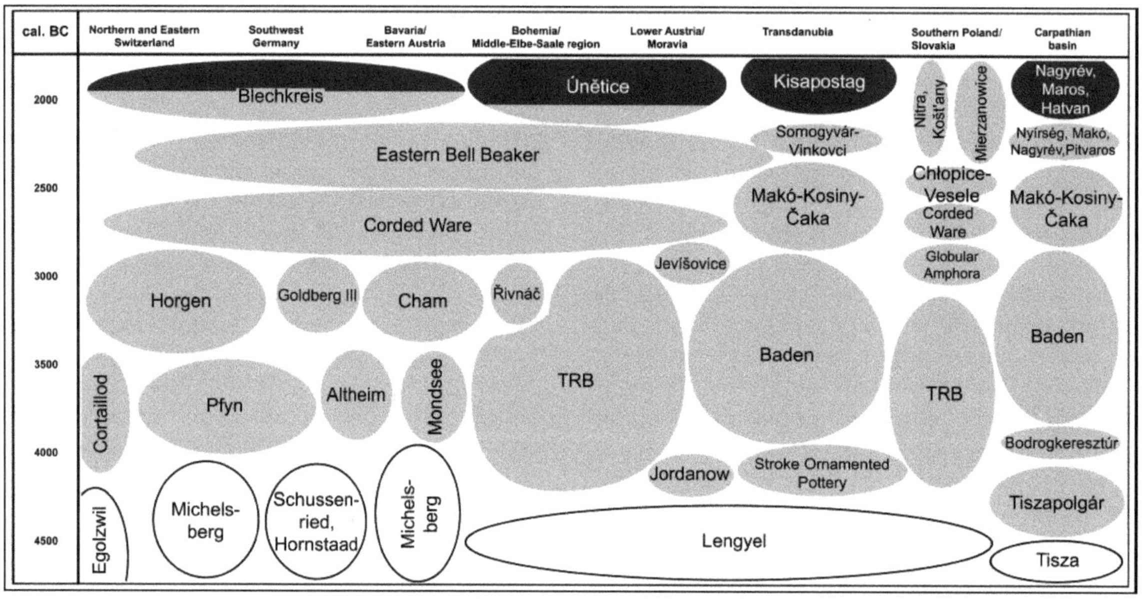

Fig. 4.10 Chronological table showing the archaeological groups of the area of research; the white groups are defined as Neolithic; grey, Chalcolithic and dark grey, Bronze Age.

area of this study there are no unified standards in terminology. The best example is Hungary compared with the northalpine region. In Hungarian terminology 'Bronze Age' is used to describe archaeological groups from around 2800 BC onwards, because from an archaeological point of view, there is a break between the Baden complex and the following Somogyvár-Vinkovci and Makó-Kosiny-Čaka groups. Also from then on metal artefacts become more common in archaeological contexts, although these objects are not made of tin-bronze (Kalicz 1982, 132 f.). Accordingly, in Hungarian research the Bell Beaker phenomenon is seen as part of the Early Bronze Age, whereas in Czech or German terminology it is assigned to the *End-* or *Eneolithic* or *Copper Age* (cf. Visy 2003, 141 ff.; Czebreszuk 2004b).

Thus we are facing various concepts of periodisation which sometimes overlap[45]. First of all, a concept of archaeological similarities defines a chronological horizon. This would reflect the concept of material cultures as described above, but interpreted as a chronological phase. Lüning, for instance, used this model (Lüning 1996). Secondly, one might define chronological phases which are framed by relative or absolute dating. A stylistic or typological model defining a horizon of similar artefacts, for example Reinecke's chronological system for the Bronze Age, is a third method to characterise a prehistoric period. The fourth idea is a division by the appearance of a new, commonly used material in the corresponding phase, according to Thomson's Three-Age system. Finally, a classification by social and economic criteria: this is the most complicated taxonomy, because the factors are mainly interpretative and not always obvious in every area. Childe was the first scholar who developed that periodisation with aid of Marx's idea of cultural evolution (McNairn 1980, 79 f.). The theories are not on the same level, and consequently they are not mutually exclusive, as they focus different topics. These various concepts are overlapping, and they are often used in parallel in prehistoric studies, so a definition has to be chosen with regard to the research question in hand.

Due to the large size of our research area and the variability of the archaeological evidence, it is useful to create a concise and consistent definition for the phase covered in this study. Thus, because this research investigates copper metallurgy, we will use a definition that is based on metallurgy. Therefore the term *Chalcolithic* seems appropriate, because it does not cause confusion with terms such as Neolithic, which describes a larger social and economic development. *Chalcolithic* denotes a phase of technological development which comprises various archaeological groups that either produced artefacts from native copper or from copper ore, or left behind so many copper objects that we can assume that these people were able to smelt and work copper, although they have not bequeathed any actual traces of metal-working. From a socio-cultural perspective Chalcolithic societies are still on a Neolithic level (Strahm 1982, 20 ff.). Because this is a minimalist definition, we say that every group using copper regularly as a substance is Chalcolithic, and the term *Chalcolithic* does not characterise a closed chronological period. According to this definition, during the Neolithic copper was merely used sporadically and the Bronze Age is characterised by an extensive usage of tin-bronze. The start and finish of the Chalcolithic varies from region to region and has to be delimited for every area separately (Fig. 4.10). Bell Beaker using communities are situated at the end of this development and it is often unclear whether they are still on a Neolithic or already on a Bronze Age 'level'.

To sum up, the term *Chalcolithic* does not define a continuous period and a linear development during which copper metallurgy, secondary products, animal traction,

[45] The following paragraph is based on Strahm's list of concepts of periodisation and its discussion (Strahm 1981).

megalithic or other collective buildings, villages etc were established. In different regions and periods between c. 4000 and 2000 BC, the archaeological background is quite varied in matters of structures, artefacts and typology. At some periods the archaeological record may be dominated by settlement evidence, at other times burials are the predominant source of information. The dominant element is copper, whose use emerges in waves all over the research area throughout that time.

4.3 DISTRIBUTION OF COPPER METALLURGY IN CENTRAL EUROPE WITH PARTICULAR REGARD TO TRACE ELEMENT ANALYSES

The following section will present an overview of various research projects which have tried to clarify the development of metallurgy by undertaking metallographic analyses of copper objects from central Europe. Prehistoric metal objects have been analysed for over 50 years, not only by means of the conventional interpretation of finds and their contexts, but also through scientific methods, and in particular through analysing their trace element composition. The fundamental idea of this method is that prehistoric metal workers were either not able to clean raw copper completely, or else they intentionally retained certain trace elements within the metal, so that traces of impurities from the ore appear in certain amounts in the smelted, melted and cast artefacts. It is assumed that analysis of the composition of metal items allows one to identify the origin of the metal and also manufacturing techniques. Thus the most important works on trace element analyses within the area of this study will be reviewed. By comparing the metallurgical and archaeological results[46], the chronological and chorological development of Chalcolithic metallurgy will be outlined.

The first systematic work involving the archaeological interpretation of trace element analyses of copper and bronze objects was undertaken by Otto and Witter during the 1930s and 1940s. They published their results in the *Handbuch der ältesten vorgeschichtlichen Metallurgie in Mitteleuropa* in 1952 (Otto and Witter 1952). In order to clarify the origin of ores from which the Middle German copper artefacts were made, they analysed around 1400 objects including axes, daggers, torcs, rings and other jewellery from late Chalcolithic and Bronze Age contexts.

According to the trace element configuration Otto and Witter classified their dataset into six groups that they called *Leitlegierung* (characteristic alloys) and localised their provenance (Tab. 4.1). In a further step, Otto and Witter (1952, 31 ff.) tried to locate the provenance of the ore from which the artefacts were produced. Therefore they acted on the assumption that the composition of products reflects the original ore. Finally they postulated diverse orebodies from where the metal had come and, by means of find distributions, they retraced prehistoric trade routes (Otto and Witter 1952, 48 ff.).

Soon after Otto and Witter's work a group led by Pittioni from Vienna presented their investigations on primary Austrian early copper technology (Pittioni 1957). This group was also looking at the relationships between the product and resource. In a similar way to Otto and Witter, Pittioni (1957, 4) and his colleagues tried to shed light on the provenance of copper. They, however, operated with semi-quantitative trace element analyses[47], because they assumed that the exact proportion of trace element impurities was not characteristic for the copper-ore, but that the ratio of the various elements provides relative amounts when compared to each other. They assumed that the smelting and melting processes had an impact on the precise percentage of trace elements, but the general

Leitlegierung	Composition	Provenance
Pure Copper	99 – 100% Cu	n/a
Crude Copper	< 99% Cu	mainly Frankenberg/Eder, Vogtland
Arsenical Copper	< 8% As; remainder < 0.5%	Zwickau
Fahlore	4 – 6% Ag, As, Ni, Sb in total	Saalfelder Erzrevier
Other Metals including Ni, As and/or Ag	< 4% Ni, As or Ag	Alpine region
Tin-alloyed Bronze	Sn < 14%	„Bronze", i.e. not an ore

Tab. 4.1 Composition of div. *Leitlegierung* and their provenance (according to Otto and Witter 1952, 32 f.).

[46] The relevant studies' methods of comparison and (statistical) grouping of trace element analyses will be explained and discussed in detailed in Chapter 8.1.

[47] Instead of a fully quantitative analysis which produces exact values of percent by weight (cf. Otto and Witter 1952; SAM I & II), semi-quantitative analyses deal with scales which cover a certain span of concentration. A particular key figure codifies a definite range of a trace element concentration (cf. Pittioni 1957, 6, Table 1; Patay et al. 1963, 42). Semi-quantitative trace element analyses cannot be considered in this research, as they are not necessarily comparable on a one-to-one basis with fully quantitative analyses and, moreover, because semi-quantitative analyses cannot be included in numerical statistical evaluations (Ottaway 1982, 76; Rau and Willing 1991, 360).

proportion of trace elements does not change (Pittioni 1957, 3 f.). Finally, it was proposed that the origin of the material could be determined by comparing the objects' trace element signature with various ore deposits, especially in the eastern part of the Alps (Pittioni 1957, 64 ff.).

A project applying the same methods and asking similar questions was run by Patay, but he and his colleagues were looking at early metal finds from Hungary (Patay et al. 1963). Therefore the problems of this research were the same as those facing Pittioni's project.

The most extensive project which has investigated early metallurgy by trace element analyses in central Europe has been the *Studien zu den Anfängen der Metallurgie* (SAM) by Junghans, Klein, Sangmeister and Schröder (SAM I; SAM II). It is not necessary to describe the procedure of the whole project, the problems with which they were confronted and their cooperation during the compilation and evaluation of the analyses, because Schröder has already described this in detail (Schröder 1991). Thus the results of the SAM-project can be summarised briefly.

From 1952 to 1970 they assembled around 12000 trace element analyses of prehistoric copper and bronze artefacts from all over Europe (SAM II/1, 11). The project aimed to clarify the origin and spread of copper and bronze technologies. Based on this huge number of analyses Klein developed a systematic grouping method: the so-called *Stammbaum* (genealogy) (Junghans et al. 1954; SAM II/1, 13 ff.)[48]. This genealogy arranges the copper objects according to the similarity of their configuration in 29 hierarchical groups. These groups differ from each other by a limited value of the five trace elements arsenic, antimony, bismuth, nickel and silver[49]. In a further step Junghans and his colleagues tried to ascertain how these statistical groups can be interpreted historically. For that purpose they compared the different types of metal with their regional and chronological milieu, because it was assumed that – as with ceramics – prehistoric craftsmen were using a particular resource and technique for a long period. Consequently, it was postulated that there was a coherence between a certain type of copper and a particular region and time period (SAM II/1, 16).

Their whole area of research was therefore divided into 67 areas according to the spatial and archaeological characteristic of each region. Then, they examined what kind of copper occurred in each area (SAM I, 59 ff.; SAM II/1, 20 ff.; ibid., Map 81). The SAM-group found out that certain types of copper predominated in specific regions and over specific times (SAM II/1, 25 ff.).

[48] Later, the *Stammbaum* was slightly adapted by Sangmeister (Sangmeister and Strahm 1973, 198 – 201).

[49] In Chapter 8.1.1 this systematic grouping method will be discussed in detail.

Later projects were mostly based on the results of the SAM-project; these will be outlined briefly below.

In 1982 a significant piece of research concerning the methodological investigation of early metallurgy was published by Ottaway (1982)[50]. Her area of research ranged from the Alps to the Danube and from the Swiss Jura mountains to Vienna. Chronologically Ottaway looked at early copper artefacts from around 3400 BC up to 2000 BC (Ottaway 1982, 11 f.). After outlining the archaeological evidence of her research (Ottaway 1982, 29 – 68), she studied the metallurgical composition of copper artefacts (Ottaway 1982, 69 – 185).

Ottaway adopted a new method for evaluating trace element analyses using cluster analysis, grouping her database according to the similarities of trace element concentration in order to make them easier to handle. Each cluster of objects sharing a similar trace element configuration was interpreted as constituting a certain type of copper (Ottaway 1974; Ottaway 1982, 117 ff.). Although cluster analysis had already been used in archaeology by Hodson in order to compare fibulae from the La Tène cemetery of Münsingen-Rain (D) (Hodson 1969b), Ottaway established this methodology in archaeometallurgical research and later other scholars also used it (e.g. Krause 1988; Pernicka 1995; Krause 2003). Her work represents a milestone in archaeometallurgy; thereafter, cluster analysis became the most popular way to analyse trace element compositional data in archaeology as it allows the reduction of a huge amount of data by treating every object equally and making it more manageable without losing information by grouping.

Through analysing the archaeological material Ottaway recognised a chronological development which she divided into three horizons. During her first horizon, including the archaeological groups, Pfyn, Cortaillod, Altheim and Mondsee group, copper metallurgy was established. After a phase of stagnation – during the second horizon including the Boleráz group, Horgen group, the Lüscherz group of the early Corded Ware complex in Switzerland and maybe the Cham group (Ottaway 1982, 65), the number of copper items increased again and also the variety of objects expanded in the third horizon (Baden and Corded Ware complex) (Ottaway 1982, 65 f.). Using the results of the cluster analyses Ottaway was able to assign specific types of copper to each of the three horizons. Arsenical copper dominated during the first horizon. The second and, especially, the third horizon showed a larger variety of copper types. These types of copper contain several impurities such as antimony, bismuth and silver (Ottaway 1982, 181 ff.). Finally Ottaway drew the conclusion that the establishment of copper metallurgy in central Europe was not due to a single archaeological group; rather, the know-how was introduced simultaneously over a more or less contemporary horizon (Ottaway 1982, 195).

[50] Ottaway's book is based on her PhD thesis (Ottaway 1978a).

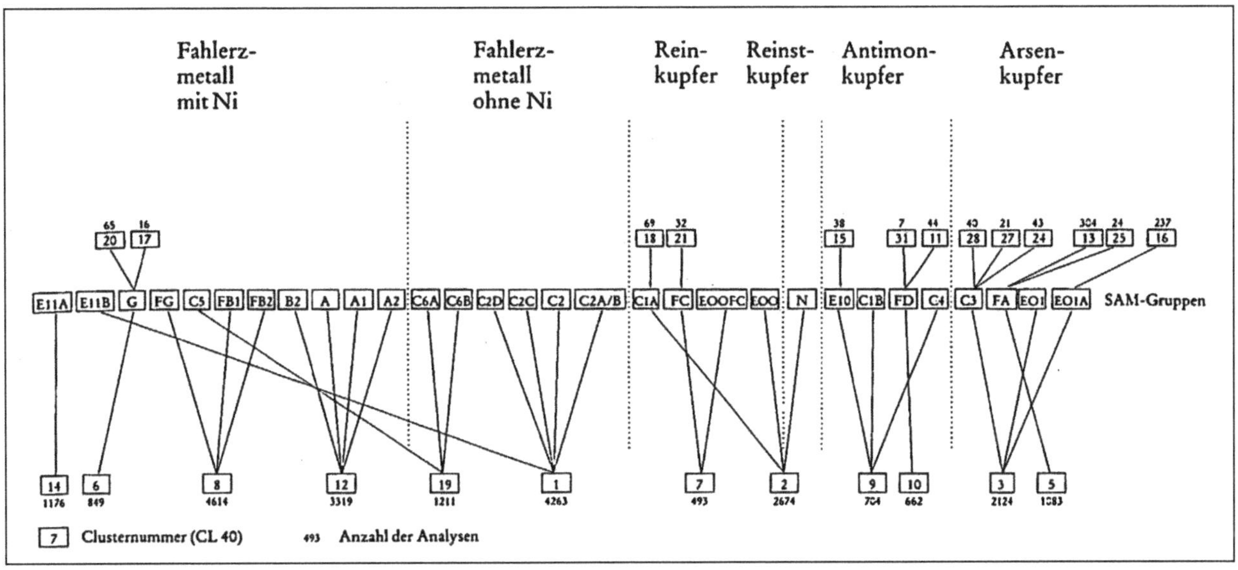

Fig. 4.11 Comparison of the SAM-groups with the results of the SMAP's cluster analysis (after Pernicka 1995, Fig. 43).

During the 1990s two projects continued in the tradition of the SAM-project. The first, the so-called *Stuttgarter Metallanalysenprojekt* (SMAP), compiled a database by collecting trace element analyses from various projects from all over Europe. So the SAM-dataset became part of a much larger dataset comprising over 37500 analyses (Krause and Pernicka 1996, 278). In a further step the whole database was subjected to cluster analyses and these clusters were compared with the results of the *SAM-Stammbaum* and digitally mapped (Rau and Willing 1991; Pernicka 1995, 97 ff.; Krause and Pernicka 1996; Krause 2003, 20 ff.). The distribution and the constitution of the new clusters were comparable to the SAM-group's results (Pernicka 1995, 97 ff.; Krause 2003, 24). Figure 4.11 compares the cluster tree resulted from SMAP with the material groups defined by the SAM-project.

While SMAP did not assay trace element analyses from prehistoric objects, the following FMZM-project (*Frühe Metallurgie im Zentralen Mitteleuropa*) conducted about 2800 new analyses of Neolithic, Chalcolithic and Bronze Age objects from central Europe and especially from the area of the former GDR. This project tried to shed light on the chronological development of copper and bronze metallurgy in that region in comparison to neighbouring areas (Krause 2003, 25 f.).

The results and the archaeological interpretation of both the SMAP- and the FMZM-project were published by Krause (2003). Krause tried to show the chronological development of metallurgy in central Europe, focusing on the northern part of the Únětice complex (Krause 2003). He aimed to clarify the origin, the distribution and, especially, the chronological development of Bronze Age metallurgy based on modern chronological data and trace element analyses (Krause 2003, 10). Therefore, he classified the copper objects by several cluster analyses. After grouping the objects according to their trace elements arsenic, antimony, bismuth, nickel and silver (Krause 2003, 88, note 168; ibid., 132), these clusters were discussed with regard to their chronological and geographical distribution. Similar clusters were interpreted concerning their trace element composition – as pure-, arsenic-, antimony- or fahlore-copper[51] (cf. Krause 2003, Fig. 40).

Firstly, he described a cluster analysis which had been undertaken on a dataset of 26525 trace element analyses by Pernicka (Pernicka 1995, 97 ff.; Krause 2003, 86 ff.). The entire data of his research are published at the *Stuttgarter Metallanalysendatenbank* (Krause 2003: CD-Rom). Then, Krause split the database up into several regional studies and ran cluster analyses on trace element analyses from, firstly, Early Bronze Age contexts north of the Alps (2726 analyses), then from eastern Germany (1653 analyses) and then Bohemian Únětice sites (1812 analyses). One further study contained material from the Neolithic and Bronze Age of the Middle Danube and western Carpathians (4167 analyses). Finally Krause explored Bronze Age ingots from central Europe (4637 analyses). In all of these datasets, the compositional data obtained from the Bronze Age are in the majority (Krause 2003, 87 – 119; ibid., Fig. 35).

Although Sangmeister had proposed a chronological and geographical coherence between copper objects and their compositional configuration, Krause (2003, 120) wanted to verify this using a larger quantity of new data, independent dendro- and radiocarbon dates and advanced

[51] Fahlore-copper is defined after Otto and Witter as a type of copper which contained a remarkable amount of arsenic, antimony, silver and sometimes nickel. In total these trace elements can account for up to 6% by weight (Otto and Witter 1952, 32). The so-called fahlore-coppers originate chiefly from sulphide ores, e.g. tennantite ($Cu_{12}As_4S_{13}$) or tetrahedrite ($Cu_{12}Sb_4S_{13}$) (Moesta 2004, 270), and were used at least since the Early Bronze Age (Junk 2003, 3). Some archaeological evidence, however, is known that indicates a smelting of sulphide copper ores even during the Chalcolithic (Höppner et al. 2005).
The concentration of sulphur in prehistoric copper objects has not comprehensively enough determined to include that element in a broad study. Subsequently, sulphur as a trace element of Chalcolithic copper objects have to be neglected in the present research.

computer-analytical techniques. He mapped the spread of the varying types of copper, and he selected only the larger clusters, because he argued that smaller groups, comprising fewer than 100 objects, would result in a dispersal that was too diffuse (Krause 2003, 121). The spread of twelve larger clusters of the entire database was illustrated. Several maps show various concentrations of particular copper types (Krause 2003, 121 – 132).

Krause then examined the clusters chronologically. To summarise, he assumed that fahlore-copper was used earlier than previously thought. In every region a number of objects made of fahlore-copper can be dated to the 3rd millennium BC or sometimes even earlier (Krause 2003, 145). After a brief excursion regarding Neolithic and Chalcolithic copper objects (Krause 2003, Chapter 6.4), the author focused on the topic of Bronze Age metal work (Krause 2003, 145 – 157). A selection of typical Bronze Age finds such as various pins, rings, halberds, daggers and torcs were analysed in order to test which types of copper were present and how the particular types of copper are geographically distributed. Accordingly the author postulated a spread of similar types of copper from the Carpathian Basin, via Bohemia and central Germany, to southern Scandinavia, including an offshoot towards south-western Germany and Switzerland (Krause 2003, 199 f.).

Krause could then confirm, using a larger database and modern methods of statistical analysis, what the SAM-group had already suggested. There was a regional and chronological distribution of certain types of copper. Finally, he concluded that Bronze Age metallurgy in Europe had developed during a process which can be divided into four horizons. After a consumption of Neolithic and Chalcolithic pure-, arsenic- and antimony-copper, fahlore-copper began to be used around 2500 BC. Then the use of fahlore-copper was fully developed from 2300/2200 BC, and this led into the third step, when tin-bronze was adopted from about 2000 BC. After around 1800 BC tin-bronze technology was fully developed and adopted in the northalpine region (Krause 2003, 263 ff.).

In addition to Krause's book, several regional studies emerged from the SMAP- and the FMZM-projects (e.g. Klassen 2000; Matuschik 2004). These studies chiefly discussed the trace element composition of prehistoric artefacts by focusing on rather smaller geographical, chronological or archaeological areas; for example, a particular region (e.g. Schalk 1998), an archaeological group (e.g. Metzinger-Schmitz 2004; Bernard 2005), a cemetery (e.g. Krause 1988; Bertemes 1989) or a specific type of artefact (e.g. Matuschik 1996; Kienlin et al. 2006). Here, it is not possible to discuss each work individually. In general, most of these studies connect typological and chronological analyses with a metallurgical examination of copper or bronze artefacts. Although a detailed discussion of these studies would go beyond the scope of the chapter, their results are taken into account and will be referred to in this study.

4.4 DEVELOPMENT AND DIFFUSION OF COPPER METALLURGY THROUGHOUT THE AREA OF RESEARCH

By connecting the extensive research which has been done in order to clarify the origin and distribution of metallurgy, it is possible to show the following picture of the development of copper metallurgy throughout the research area.

In the current state of research, the first metal objects appear in the research area on the Great Hungarian Plain during the first half of the 5th millennium BC, possibly due to an impulse from the Balkans. Copper, however, was only found sporadically at sites dating to that period. During the second half of the 5th and early 4th millennium BC many, often heavy copper artefacts produced using complex procedures have provided evidence for a fully developed metallurgy of the Tiszapolgár and Bodrogkeresztúr groups in the Carpathian Basin. In the northalpine region, however, copper objects are still exceptional at that time. The early copper objects were mostly made of pure copper without noticeable impurities.

Before 4000 BC, copper finds were very rare in the area north of the Alps, but after that arsenic-bearing copper was used[52]. From that time onwards copper metallurgy was adopted all over the research area and is documented by numerous objects and metalworkers' tools. It is possible that the impulse came from the Carpathian Basin, because various artefacts show parallels with objects in this area, for example the Şiria axe from Überlingen in southern Germany (Matuschik 1997a).

In the Carpathian Basin, however, the evidence of copper metallurgy was stagnating from around 3800 BC onwards and not many copper artefacts have been found. With a chronological shift, in central Europe the same phenomenon is recognised at sites dated between approx. 3500 and 2800 BC. The copper objects that have been found in the more widespread groups such as the Baden, later TRB, Cham and Horgen groups came from extraordinary contexts, for example rich graves or hoards. Gold has occasionally been found in these contexts.

As demonstrated by the growing number of shaft-hole axes and other objects that have been discovered, copper use increased again in the Carpathian Basin from c. 2900 BC, whereas in the distribution area of the Corded Ware complex metallurgical evidence was still rare. With the arrival of the Bell Beaker phenomenon in central Europe copper and then even gold artefacts became more common. From the mid-3rd millennium BC onwards copper artefacts were made of various types of fahlore-

[52] Hosler, Lechtman and Holm defined different types of arsenic-bearing copper: arsenical copper (< ~0.1% As), copper-arsenic alloy (~ 0.1 – 0.5% As) and arsenic bronze (~ 0.5 – 7% As) (cf. Lechtman 1996, 481). For now, this division will not be considered, as it would anticipate the results of the statistical analyses. Labelling of particular types of copper should be avoided as far as possible until the statistics have been interpreted.

copper. At the turn of the 2nd millennium BC metallurgy was eventually established and fully adopted. New networks may have led to a social change and a hierarchisation of the Bronze Age community. But this development had only started during the Earliest Bronze Age. From around 2000 BC onwards we can see the hierarchical communities of the 'fully-developed Early Bronze Age'; at the typo-chronological Bronze Age phase Reinecke A2, when copper alloyed with tin – typical tin-bronze – was regularly used all over Europe.

Despite the varying date of the appearance of copper and bronze metallurgy in Europe, a certain uniformity in the sequence has been suggested. To conclude, the development of early metallurgy in central Europe is not a clear evolutionary process leading straight from a less to a fully developed stage. There are increasing and decreasing phases which occurred all over the study area at a certain time. Developments and changes occurred in different regions at different times. Thus we cannot define one fixed chronological boundary for the whole area of this research.

In the Carpathian Basin, the Chalcolithic covers a period from around 4700 BC until 2400/2200 BC and in the northalpine region from c. 4200 BC until 1800 BC; in Middle Germany and what is now the Czech Republic, from approx. 3800 – 2000 BC. During the central European Chalcolithic, metallurgy had been introduced in two stages. The first impulses of copper metallurgy from the Balkans and the Carpathian Basin may have arrived in the northalpine region after c. 4000 BC, but this connection seems to break down around 3500 BC. Then, around 2700/2500 BC, with the appearance of the *spätäneolithischer Kulturkomplex* in the Carpathian Basin and the Bell Beaker phenomenon, it appears that a second impulse finally re-established metallurgy in the northalpine region.

Strahm and Hauptmann observed the emergence, expansion and establishment of metallurgy in a five-stage development which, according to them, expanded fairly uniformly in the Near East and in Europe. Despite the chronological variation in different areas, the adoption of metallurgy by prehistoric societies has been described as *Metallurgical Developmental Phases*. The first *preliminary* and *initial stages* of the relatively sporadic appearance of artefacts – chiefly made of native copper minerals – were followed by an *innovative period* when copper production rapidly increased and mainly pure and arsenical copper was used. After a decline of metallurgical evidence in archaeological contexts in Europe between c. 3300 and 2800 BC, metallurgy became re-established during the *Consolidation Phase* and more complicated techniques used to smelt fahlores and other sulphide ores emerged. Finally, accomplished tin-bronze-technology metallurgy was fully established. The last phase, dating to 2000 BC, has been called the *Industrial Phase*, because of the exploding number of metal objects, intensive metalworking, as documented at various copper mines and smelting sites, and a stratification of society – presumably based on the production of, and trade in metal and the exigency of organising this (Strahm 1994; Strahm and Hauptmann 2009). In Strahm's and Hauptmann's opinion this development is founded on a changing ability to exploit different types of ore deposits. Firstly, pure copper from the surface would have been collected. After a phase of exploiting surface-near oxide ores, it is suggested that complex sulphide ores from deeper levels were exploited which were more difficult to mine (Strahm and Hauptmann 2009, 121 ff.). Whereas, from a metallurgical point of view, the early two phases are still Neolithic and the last one is *Metallicum* (after Strahm 1981), the *Innovation Phase* and the *Consolidation Phase* cover the Chalcolithic.

In discussing the changing developments from the first appearance of copper to the extensive establishment of tin-bronze metallurgy, through fluctuating amounts of metallurgical evidence at different periods, it is essential to take into account the variability of the main sources of archaeological evidence during the Chalcolithic. Whilst evidence pertaining to the 4th millennium BC is chiefly represented by settlements, most of the archaeological remains dated to the 3rd millennium BC have been discovered within a grave context. In comparing all periods this study covers it must be borne in mind that settlements provide evidence of everyday life, whereas graves and their furnishings typically idealise and ritualise the buried person. The metallurgical evidence found in both categories differs as well.

In principle, grave goods are finished products, while in settlements, in addition to finished products, the remains of production processes, such as castings, crucibles or moulds, have been found. Thus, the quantity of evidence from a period may depend on the predominant type of context. Between the earlier and the later 4th millennium BC the archaeological record does not vary much, but the number of copper finds declines. Accordingly, it can be suggested that, for example Baden, Horgen and Cham groups, communities in whose contexts copper is less frequently found, had less access to copper and therefore metal was of higher value and treated much more carefully through re-use and recycling of the previously existing copper materials.

It is essential to consider the recycling of copper artefacts for all periods. In contrast to pottery or stone artefacts, metal can be re-used and recycled continually. It has to be considered that archaeological groups from which only a few copper finds are preserved may have re-used their artefacts to a greater extent than others. In some cases the type of archaeological record has to be taken into account as well. Since only finished products were usually used as grave goods, then evidence of metal work, such as slag, crucibles or moulds, is not discovered in burial contexts. Direct evidence of copper production is only found in contexts of archaeological groups of which not only graves, but also settlements are known. Additionally, the preservation of metal artefacts in burial contexts may be further restricted by the particular burial rite employed, or by grave robbing. The depositional and recycling practices during this period may account for our fluctuating data of copper finds throughout the study area (Taylor 1999).

Perhaps a further reason for the decline of copper finds during the late 4th millennium BC can be attributed to the refusal of communities to adopt copper technology. The acceptance of technologies in prehistoric societies is governed by complex cultural processes. Metal acquisition and the transmission of metallurgical expertise require smoothly-operating social networks which may be interrupted by cultural changes. It is assumed that metallurgical knowledge was kept secret and passed on only to a selected few. In turn, these individuals may have had superior social, political and religious status founded on their exclusive skills and their networks (Kristiansen and Larsson 2005, 51 – 60; Roberts 2008b).

Finally, it has to be mentioned that although a huge number of copper artefacts and metalworking tools demonstrate a fluctuating use of copper, Chalcolithic copper mines have not been discovered within the study area. Consequently, our exploration of Chalcolithic and particularly Bell Beaker metallurgy in central Europe is restricted to the analysis of the finished copper products.

5. The Database

To provide an appropriate foundation for investigating the research questions (cf. Chapter 1.1) a compendium of data from across the entire area of research, in the form of a database, will be presented and its content described in detail. Before that, it is essential to discuss how these different analyses have been carried out.

It was necessary to ascertain whether the trace element analyses contained in the database were suitable for statistical analysis and whether they can be considered as representative of the general evidence for copper use in the area of research which is described in Chapter 4.

5.1 The Procedure of carrying out Trace Element Analyses

Because the present study aims to review published trace element analyses and uses these records in order to clarify the specific question of whether there existed a typical Bell Beaker metallurgy, the procedure of analysing the trace element composition of Chalcolithic copper objects will be described. The trace element analyses that have been carried out over the past 70 years were acquired by different methods. In general three methods – optical emission spectroscopy (OES), X-ray fluorescence spectroscopy (XRF) and neutron activation analysis (NAA) – have been employed in order to clarify the composition of copper objects (detailed description: Pollard and Heron 2008, chapter 2).

Since the current research has involved no new compositional analyses, and relies instead on these past analyses, it is essential to address the issues of applicability and comparability of the methods. The statistical comparability of the trace element analyses of this study's dataset will be clarified fully in Chapter 8.5. The methods that were been employed to extract samples and to obtain the compositional data by the respective project have already been explained in their relevant publications (e.g. OES by Klein 1954, 89 – 103; NAA by Ottaway 1982, 79 – 86; XRF by Lutz and Pernicka 1996). The OES method was chiefly used by earlier study-groups (especially by Otto and Witter and the SAM-project). By running an OES, a sample that has been obtained by drilling a small hole (Ø c. 1 mm) is vaporised. Thereby, an element-specific light spectrum is emitted which then can be measured. According to the characteristic spectrum of the examined element and the intensity of the light emitted, the concentration of the impurity can be evaluated. This effect is caused by activating the outer shell electrons which results in exciting them to a higher energy level. If the electrons drop back, they emit light with this element's typical spectrum. This effect is known as a quantum jump. The method enables researchers to determine a wide range of elements to a relatively high degree of sensitivity, such as antimony, nickel etc (cf. Hughes et al. 1976; Härke 1978, 170 – 188; Pollard and Heron 2008, 24 f.).

In principle, the XRF method has a similar basis, but instead of light, the emission of element-specific radioactive radiation (X- and γ-rays) is used to determine the elements present. Although the limit of detection is inferior to the OES, the XRF method allows various elements to be analysed without destroying the sample and, furthermore, analysis time is shorter. Consequently, a great deal of data can be compiled quickly. The disadvantage of the non-destructive measuring is that only the often corroded surfaces of an artefact can be analysed. A solution to this problem is provided by drilling holes and taking samples from inside the object, and analysing those (cf. Lutz and Pernicka 1996; Pollard and Heron 2008, 38 – 45). This method was chiefly used for trace element analyses by Pernicka at the Max-Planck-Institute in Heidelberg (D) – especially for the FMZM-project (Krause 2003, 15).

In contrast, NAA method allows much smaller amounts of a sample to be used than had been the case with the previous methods. Generally, NAA allows not only quantitative measuring, but is also capable of analysing qualitatively more impurities than OES and XRF (Perincka 1984, 517 f.). Even so, bismuth, lead and nickel can, at best, be only unsatisfactorily measured (Ottaway 1982, 87). In brief, NAA works by bombarding a sample with neutrons. The addition of neutrons to the particular elements produces new isotopes of a specific atomic weight and radioactive half-life. The intensity of induced radiation can be measured and thereby the concentration of certain impurities determined. The greatest benefit of NAA is – besides its remarkable accuracy – that artefacts can be analysed less destructively and with relatively small sample materials compared to the other methods. Thus, NAA is ideal for investigating trace element composition of ancient and precious objects (cf. Ottaway,

1982, 81 – 86; Pollard and Heron 2008, 50 – 56). Even so, this method has not yet been applied often in the projects that are used as a base for this study. As far as the author knows, the only projects which use NAA data are those run by Ottaway (1982) and those carried out on the finds from Makotřasy (CZ) (Pleslová-Štiková 1985).

All these techniques have both advantages and disadvantages, and the decision for or against the usage of certain methods depends on issues such as how much time and money is available for carrying out trace element analyses, or on the artefact itself, its size, preservation and the available material. Because researchers have had to deal with different situations in each case and due to improvements in measuring techniques, several methods have been used by various projects. Accordingly, it needs to be asked whether the results of these diverse analyses can be combined in one study and whether they are inter-comparable.

5.2 THE DATABASE OF CHALCOLITHIC TRACE ELEMENT ANALYSES

According to the previous results, it is possible to select data from the numerous published trace element analyses by means of several geographical, chronological, metallurgical and archaeological criteria[53]. This section will, firstly, outline the different criteria that have been applied for compiling the database on which the present research is based and, secondly, will describe the information that is contained in the database and the archaeological background that is listed in the attached catalogue of the entire database (cf. CD-Rom).

5.2.1 CRITERIA USED IN COMPILING THE DATABASE

The foundation of the present research is the published trace element analyses from different projects, and the main aim is to re-evaluate them statistically, to provide new ideas for interpreting Chalcolithic metalwork and to compare these analyses in order to clarify whether there was a specific Bell Beaker metallurgy. Most of the trace element analyses used in this study have been carried out by the SAM-project (cf. Chapter 4.3). Krause recently published a collection of over 35000 trace element analyses from copper objects that were carried out by several study-groups, covering nearly the entire Old World and dating chiefly from the emergence of metal to the Late Bronze Age. This huge dataset, also known as the *Stuttgarter Metallanalysen Datenbank*, offers an ideal source for compiling data for new researches (Krause 2003).

In the present study, additional trace element analyses from other projects are also included. In order to be integrated in the database of this research, data must have fulfilled several criteria, which will be discussed now. A hierarchical list of criteria is checked for all trace element analyses that are included in the database and all data and information have been double-checked with other references.

The first criterion which must be fulfilled is that the geographical location where the analysed object was found must be exactly known. All trace element analyses that are included in the present survey have had to be taken from objects found within the geographical area of this study (see Fig. 4.1). Data from outside the area of research, as it has been defined in Chapter 2, are not considered here. For a further regional comparison it is essential that the find place of all analysed objects is known and clarified – either by double-checking at least one reference in literature or by verifying records at the relevant museum. The references are given in the database (cf. database variable: LITERATURE).

The criteria used in compiling the dataset of this study are derived from the geographical and chronological definition of the research area that has been given in Chapter 2. In this regard, the question of deciding what level of tin content constitutes deliberate alloying needed to be addressed. A threshold of 4.5% was adopted, based on the chronological difference between rather lower- and higher-alloyed tin-copper[54]. Whereas artefacts dated in the Reinecke period A1 chiefly contain less than c. 4% of tin, from period A2 onwards, objects are definitely alloyed with more tin. Nevertheless, a tin concentration of around 1% already affects the properties of copper in various ways (e.g. colour, tensile strength and hardness).

The appearance of typical tin-bronze differs chronologically and regionally (cf. Chapter 4.3). Whilst it is rather easy to find the earliest artefacts that are considered for the present research, it is more difficult to set the chronological finish line for the data collection. Because this study examines copper metallurgy and its influences on Bell Beaker metallurgy, all pre-Beaker copper finds will be included as long as they fulfil the other criteria. The end-point of the set is marked by the Earliest Bronze Age, a period that has typologically often been interpreted as part of the Bronze Age, but, as, for instance, Strahm (1982; 1988) has shown, it is historically still pre-Bronze Age – i.e. Neolithic – and, from a metallurgical point of view, the numerous metal artefacts of that phase are predominately not made of bronze with high levels of tin.

[53] The references of the projects that provided trace element analyses to this study are displayed in the *Comments of the Catalogue* (p. 181 – 187). At this stage I would like to express my appreciation to Dr. R. R. Wiermann (Museum Schloss Bernburg) who made unpublished trace element analyses form Middle Germany available to this research (ANo 100088 – 100131). These data are of outstanding importance for a successful evaluation of central European copper finds of the 3rd millennium BC – especially in comparison with the analyses from Drosa (D) and the Eastern Bell Beaker group.

[54] For the detailed discussion on the issue of whether there is a threshold that marks a limit to define intentionally alloyed and on the chronological outline of this study, see Chapters 2.2 and 7.3.4.

The chronological criteria pertaining to whether an Earliest Bronze Age object is included or not are very difficult to set and have to be examined for each object. As radiocarbon dates are still very rare for this area of study, it is only possible to decide this by typological-chronological correlations and by referring to the authors who studied and dated the particular finds and sites. The chronological issues which determine whether a trace element analysis is considered or not need to be described in more detail. For example, as regards the regions (from West to East) that are studied and some larger archaeological complexes, such as cemeteries, the relevance of copper objects was individually ascertained when gathering the data. Therefore, the most significant literature is that which discusses typo-chronological classification in any given part of the study area. Features such as graves that contain objects that are obviously younger, or else copper objects that contain more than 4.5% tin have to be seen as *termini post quos* and are younger than the copper objects in question. So, these cannot be included in the database. Because of the lack of independent dating, the Swiss and southern German Bronze Age chronology is closely connected to both the typo-chronology for southern Bavaria – developed by Ruckdeschel (1978), which is based on earlier works by Reinecke (1930), Holste (1953) and Hundt (1958), and the radiocarbon dated cemetery from Singen a. Hohentwiel (D) (Krause 1988). Compared with other regions dealt with in this study, in Switzerland complexes of the Earliest Bronze Age are rare. Apart from there being a large number of flanged axes, discovered as stray finds, rather few graves have been excavated that date to the late 3rd millennium BC (cf. Hafner and Suter 1997; Hafner 2001; Hochuli 2001). David-Elbiali (2000) developed a chronological system for the Swiss Bronze Age and our dating more or less follows this system. Thus, only objects that date to her phase A1 are included in the database. From comparing radiocarbon dates (in particular from Singen)[55], the re-occupation of the Swiss lake dwellings and typological sequences, the Earliest Bronze Age in Switzerland and southern Germany, started at c. 2200 BC and lasts until around 2000/1800 BC. For the compilation of the database, the dominant copper finds of the earlier Early Bronze Age phase are taken chiefly to be spiral rings, triangular undecorated dagger blades, roll-headed and minor or undecorated disc-headed and oar-headed pins. Furthermore, the slightly flanged axes of type Salez[56] are also dated to that period (Krause 1988, 219; Bill 1997, 253). The famous torc ingots (*Ösenhalsringe*), however, are generally dated later (Junk 2003, 11). This horizon is generally labelled as Early Bronze Age phase A1; in the case of Bavaria, Ruckdeschel (1978a, 297 ff.) separates an individual phase A1a.

[55] A collection of radiocarbon dates from southern German Early Bronze Age sites are published in works of Becker et al. (1989) and Furholt (2003).

[56] Flanged axes of the types Griesheim and Neyruz are typologically very similar to the type Salez. Therefore they are considered to be more or less contemporary, or perhaps slightly younger axes, and are also included in the database (cf. Abels 1972, 16; Krause 1988, 223 ff.).

Regarding the Austrian (especially Lower-Austrian) sites, the chronology there is closely related to that of southern Germany. For the moment, the cemetery at Gemeinlebarn (A) is the best studied and published site. It is both dated to the Earliest and Early Bronze Age and also provides a great number of trace element analyses (Bertemes 1989). Bertemes (1989a, 164 ff.) typologically connected his phases of the cemetery at Gemeinlebarn with the southern German chronology, so that his phase *Gemeinlebarn Stufe 1* is considered to be more or less contemporary with Ruckdeschel's phase A1a[57]. Hence, this study is restricted to the graves of *Gemeinlebarn Stufe 1*.

Finally, in the northalpine region – from Switzerland to southern Germany and Austria – where the archaeological remains that are considered as Earliest and Early Bronze Age have been labelled as so-called *Blechkreis*, all finds can firmly be connected with Ruckdeschel's typology and with the radiocarbon dates from the Singen cemetery.

Looking further, towards eastern Lower Austria and Moravia, this area is – in terms of archaeological remains of the Earliest Bronze Age – closely related to western Slovakia and southern Poland (cf. Chapter 4.1). Recent studies have named the archaeological groups dating to the end of the 3rd millennium BC as the Epi-Corded-Ware complex (ECWC) (Kadrow and Machnik 1992; Kadow and Peška 1999), although it is still not fully ascertained how this complex is connected to the dates concerning the most relevant components, such as Bell Beaker, Corded Ware, early Mierzanowice and Nitra group (cf. Bertemes and Heyd 2002). These uncertainties, however, make it obvious that these archaeological groups are at least overlapping, interacting and possibly contemporary. Copper finds which are classified as ECWC will be included, but in the database they are named according to the particular archaeological group (e.g. Mierzanowice or Nitra group) in order to clarify whether they are similar or different in terms of their metallurgy.

The geographically restricted Loretto-Leithaprodersdorf group is typologically closely related or perhaps even part of the ECWC. It is located in the very eastern part of Austria and roughly synchronized to the phase Bronze Age A1a, and it has provided a few analysed copper objects that have been included in the database. The Loretto-Leithaprodersdorf group, as in the rest of the ECWC, shows several similarities to both the previous Bell Beaker and the later Únětice complexes (Neugebauer 1994, 49 – 56).

North of the regions discussed, the picture of Earliest Bronze Age metallurgy is slightly different. As mentioned before, in archaeological contexts defined as the proto- and early-Únětice complex in the Czech Republic, eastern Germany and Poland, copper finds are relatively rare. By checking the most important studies in that period not a single trace element analysis can be included in the

[57] Metal finds dated younger than phase *Gemeinlebarn 1* contain mainly more than 4% tin. This fact supports excluding graves that date to *Stufe 2* and later (Bertemes 1989a, 168; Liversage 1994, 81).

database (cf. Moucha 1963; Zich 1996; Bartelheim 1998; Lauermann 2003; Moucha 2005)[58], because these metal artefacts either date from after *Stufe 2* (according to Zich 1996) or contain more than the threshold value of 4.5% tin. Several archaeometallurgical studies that have been carried out on Únetice copper objects (in particular in eastern and Middle Germany) found that typical tin-bronze had already been adopted by around 2200 BC (Müller J. 2002; Krause 2003, 267). Independent dates for the Únětice complex are very rare and many ^{14}C-dates still remain unpublished – especially from Czech sites (Müller J. 1999).

In Hungary the situation is very difficult, too – both in Transdanubia and on the Great Hungarian Plain. We have already come across the problem that, in Hungarian terminology, post-Baden archaeological groups are considered as Bronze Age, although compositional analyses have indicated that most of metal objects dating to the 3rd millennium BC contain far less than 4% tin. The terminology is based on the history of Hungarian research. Because of new influences, especially from the more southern Vučedol group and the qualitative and quantitative increase of metal artefacts, the period after 2700 BC is referred to as the Bronze Age (Maran 1998, 312 ff.). Despite the fact that most of these finds are not made of tin-alloyed bronze, but of copper (often impure), the terminology is still used. In consequence, the Hungarian Earlier Bronze Age is, according to our definition, Chalcolithic. For that reason, it is essential to clarify which of the Hungarian archaeological groups are relevant for the present study and of which groups' trace element analyses will be considered here. In order to illuminate this, several studies that shed light on the chronology and archaeocultural interactions in Hungary during the 3rd millennium BC have to be taken into account. But, first of all, we can refer to Liversage (1994) who studied the development of metallurgy in the Carpathian basin. He demonstrated that high-tin bronze (i.e. copper containing more than 4% Sn) was usually not used before the appearance of the Kisapostag, Maros, Ottomány, Hatvan and late Nagyrev groups[59]. Moreover, at around 2200 BC further changes can be recognised from the archaeological remains. The first tell-settlements were occupied on the Great Hungarian Plain by people whom we name today as, for example, the Hatvan, Ottomány or Perjámos groups. This typo-chronological phase is named as Early Bronze Age phase III after Hungarian terminology (Forenbaher 1993, 247 f.; Maran 1998, 315 f.; Visy 2003, 142 ff.). Based on the small number of published radiocarbon dates (Meier-Arendt (ed.) 1992; Forenbaher 1993), only copper objects of Hungarian provenance are included in the dataset, and these are not earlier than approximately 2100 BC or Early Bronze Age phase II. The end of the relative chronological Early Bronze Age phase II in Hungary is chiefly parallel with the beginning of Reinecke phase A1 and the system of the northlpine Bronze Age periods (cf. Maran 1998b, 316). Consequently, the period examined here ends slightly earlier in Hungary than in the rest of the area of research. As these periods are mainly based on a variety of types of metal artefacts, they logically reflect a new spectrum of objects.

It has been shown that, apart from individual knowledge of the objects that have been analysed, e.g. the find place, sometimes the collection of data has made it necessary to decide whether a particular artefact and its archaeological context is to be included in the dataset. The reason for this is that the metallurgical remains of the Chalcolithic period in the different regions within the study area are not uniform. The results of the data collection process will be presented below.

5.2.2 THE CONTENT OF THE DATABASE

The database presented in the catalogue includes, in addition to the trace element analyses, further information which is coded in order to make it more manageable for statistical analyses. These codes refer to the *Handbuch zur Stuttgarter Datenbank* (Krause 2003, 289 – 296), but they had to be changed slightly in order to fit the individual requirements of the present research. Both the decoding and a detailed explanation of the database are described in the *Comments on the Catalogue* placed at the end of this book (p. 181 – 187). The structure of the present database is based on the *Stuttgarter Datenbank* in order to offer good comparability with the main bulk of published data and to enable links for further studies. The content of the database and, in particular the chronological and archaeological information, relates to the characterisation of the research area presented in Chapter 4 and the criteria for the data collection in Chapter 5.2.1.

The total database consists of 1943 trace element analyses of copper objects that have been found throughout the research area and contains the following information: chronological, archaeological and geographical information on those finds which rely on the referenced literature (LITERATURE)[60].

Apart from the actual measured concentrations of copper (CU) and its trace elements (SN, PB, AS, SB, AG, NI, BI, AU, ZN, CO and FE) in weight-by-weight percent, the exact find place (FIND PLACE) of the objects and in which feature they were found (FEATURE) are the most important information of the database for the subsequent investigations. The category 'SITE' gives a more detailed description of the feature in the language of the publication that is referenced. This makes it easier to find the site in the corresponding publication. The general geographical region (AREA), a rough chronological dating (PER) and if possible the archaeological classification (CULT) of the object or site where it was discovered are also displayed.

[58] This relates to Zich's phase 1 and 2, Bartelheim's phase 1 and 2 and Moucha's phases 1 – 4.

[59] This has been randomly double-checked on several samples of these archaeological groups that had been published in SAM II.

[60] The categories of information and variables are addressed in the text in brackets and written in capitals in the same way as they are labelled in the database.

Moreover, the typological classification is also recorded. Firstly, the specific typological (TYPE) and functional classification (TYPE 2) is decoded and secondly the object is named (OBJECT). Although the category 'TYPE' is useful to classify and date finds typologically, it is too detailed for functional analyses in comparison with the properties of copper and for manufacturing processes. For instance, to draw a wire and form a ring it was probably not necessary to take into account whether a simple *ring* or a *spiral finger ring* was being produced. Good malleability may have been important for both artefact types. In consequence, artefact types with functions that can be assumed to be related, and similar shapes, are combined as category 'TYPE 2'. This eases a later interpretation of object type concerning the mechanical characteristics of copper. For example, various types of spiral ring (TYPE 409, 413, 414, 416 and 417) are classified as 'TYPE 2: 410' (cf. *Comments of the Catalogue*).

Furthermore, all trace element analyses have been ordered according to the SAM-project's material groups (MG), as their systematic is not only commonly known but also often applied in other studies. The SAM material groups are defined by fixed thresholds which allow classification to a single analysis (cf. Chapter 8.1.1). As most of the other projects that dealt with classifications of trace element analyses also related their clusters to the SAM-grouping, this will allow comparison of the results of the present study with other investigations.

The total number of 1943 trace element analyses includes 243 double-analyses (DOUBLEANALYSES) of 110 different objects: this encompasses results from where a specific object has been analysed at least twice by the same or another laboratory in order to compare the results (cf. Chapter 7.2.2). Allowing for these, then, composition analyses of 1810 different copper objects from 569 different sites are included the study. Each site, e.g. cemetery or settlement, is labelled with an individual number (PLACENo), so that such sites can provide analysed artefacts from more than one feature. This has to be kept in mind for further study, because we focus on each trace element analysis as a single source of archaeological information. Even if one closed find, a grave pit for instance, contains more than one analysed object it may relate to various metallurgical resources. Thus, all analyses, including the double-analyses, are considered in the further study.

5.2.2.1 THE CONTEXTUAL, CHRONOLOGICAL, GEOGRAPHICAL AND ARCHAEOLOGICAL-CULTURAL CONTENT

The analysed artefacts have mostly been found in graves – closed features that can mostly be classified geographically, chronologically and archaeologically. Thus 976 analyses relate to these closed contexts.

In contrast, the find contexts relating to 429 trace element analyses are harder to categorize. All of these objects are defined as single finds because a clear find context could not be certified by literature review or in museum collections. Usually these objects could only be dated via typological parallels. Their find places, however, are known. Items with unknown or uncertain find places are excluded.

Another large category of features that have produced copper or copper-related artefacts are settlements. Some 304 trace element analyses have been undertaken on copper objects from settlement contexts. Sometimes an archaeological-cultural classification is not possible, because the finds were gathered by unsystematic, early late-19th/early-20th century collectors (Ottaway 1982, 32; Obereder et al. 1993, 6). Nevertheless, due to typological similarities and because the find place is known, it was nearly always possible to put these finds into a rough chronological order.

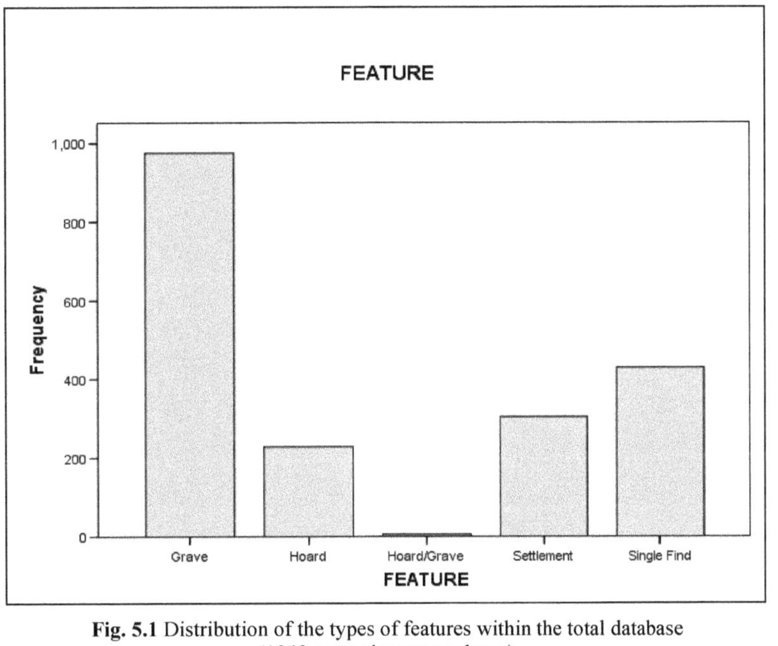

Fig. 5.1 Distribution of the types of features within the total database (1943 trace element analyses).

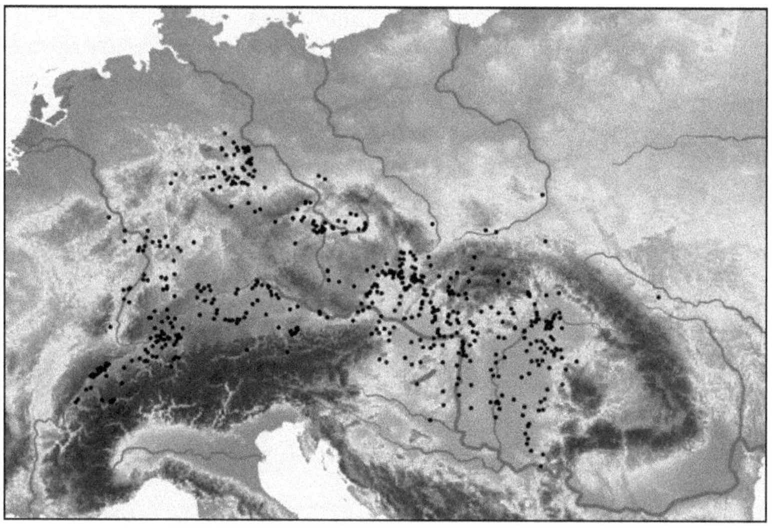

Fig. 5.2 Geographical distribution of the material included in the database.

Finally, 229 analyses come from artefacts which have been discovered in contexts that are regarded as hoards. Hoards are often interpreted as either votive deposition – as gift to the gods – or as 'profane' deposits (especially as regards metal artefacts), buried for safekeeping, etc.; in other words, with such hoards we are dealing with metal that has been hidden because of its value, deposited in order to reuse it at a later time. In our case, hoards are simply treated as an assemblage of objects found in closed features, indicating a deliberate and contemporary deposition; however these features are neither graves nor settlements[61].

With five further objects from Kelsterbach (D; ANo 146 – 150), it is not clear whether they come from a grave or a hoard, but they are most likely to be from a grave; they are definitely from a closed undisturbed context (Wiermann 2004, 259, No. 521). Consequently, the present study covers all of the four basic types of archaeological features: graves, settlement sites, hoards and single finds (Fig. 5.1).

The finds that are examined are distributed throughout the geographical area of research, although there are several regional clusters, with dense concentrations of findspots (Fig. 5.2). This, however, reflects the general distribution of the archaeological remains of the studied period and often the area of operation of local archaeologists. In principle, the distribution of trace element analyses in the database correlates with our current knowledge about Chalcolithic metallurgy in the study region (cf. Chapter 4.1).

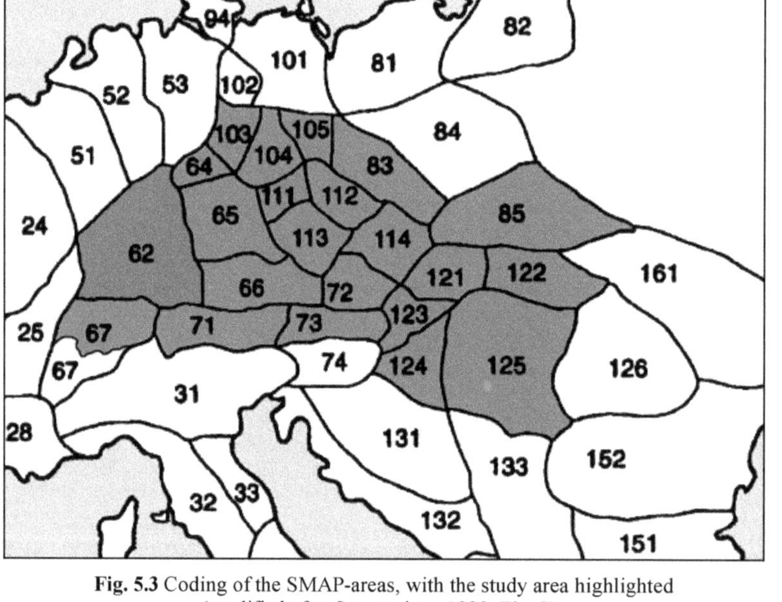

Fig. 5.3 Coding of the SMAP-areas, with the study area highlighted (modified after Sangmeister 1999, Fig. 2).

[61] For detailed argumentation see the discussion on the interpretation of hoards and depositions in Bradley 1990, 5 – 39.

Every trace element analysis has been assigned an area code (AREA) which indicates a geographical region in order to ease subsequent regional evaluations (with statistics). This coding refers to a system which has been already applied by the SAM-group and improved by the SMAP-project (SAM II/1, 20 – 25; Rau and Willing 1991). The area of research encompassed 22 SMAP-areas, whereby AREA 67 is split, because only the region south of the Rhone valley is part of this reaserch (Fig. 5.3). The so-called SMAP-areas, however, do not fit exactly into the area of this study, so that two finds from AREA 51 and two from AREA 53 have been added to the database[62]. All four objects were found very close to AREA 62. In contrast, none of the available trace element analyses from AREA 64 fulfils the criteria of the database. The rest of the study area shows variable frequencies of data (Tab. 5.1).

A period-based categorization, analogous to the regional coding, has also been used. This approach has been adopted because most single finds and many of the other finds have not been independently dated, and are only datable by means of typological sequencing and comparisons[63]. The period coding (PER) gives a rather vague chronological classification which reflects the wide date range within which an artefact may belong. The following six periods have been established[64]:

PER 10 = before c. 4000 BC
PER 21 = c. 4000 – 3500 BC
PER 25 = c. 3500 – 2800 BC
PER 31 = c. 2800 – 2300 BC
PER 35 = c. 2500 – 2000 BC
PER 40 = c. 2200 – 2000 BC

The periods should offer a rough orientation for the dating of the analysed finds. The correlation table (Tab. 5.1) displays approximately the same picture as the frequency of occurrence of metallurgical evidence in the general archaeological contexts in central Europe (cf. Chapter 4). Consequently, it may be concluded that the distribution of the content of this study's dataset reflects the overall distribution of metallurgical evidence and the spread of copper across central Europe during the Chalcolithic. The database of trace element analyses reflects the unevenness of the metallurgical evidence over the study area. This needs to be borne in mind, as well as the fact that an artefact, and a single feature, may provide more than one analysis. This is the case at PER 31, for example: while relatively little metallurgical evidence exists, many trace element analyses are available, because over 200 analyses have been carried out on the beads and reels. These artefacts were parts of necklaces found at the lake dwelling at Vinlez (CH) (ANo 2876 – 2921) and in two graves at Drosa (D) (ANo 34448 – 34610)[65]. What the comparison

	PER 10	PER 21	PER 25	PER 31	PER 35	PER 40	Sum
AREA 51	0	0	0	1	0	1	**2**
AREA 53	0	1	0	1	0	0	**2**
AREA 62	3	10	15	19	3	169	**219**
AREA 65	2	0	1	0	0	1	**4**
AREA 66	1	5	9	3	16	47	**81**
AREA 67	0	70	0	127	0	67	**264**
AREA 71	2	0	0	0	0	2	**4**
AREA 72	0	7	30	1	1	40	**79**
AREA 73	0	65	0	0	0	0	**65**
AREA 83	0	1	0	0	0	1	**2**
AREA 84	0	0	0	0	0	15	**15**
AREA 85	0	0	0	8	0	41	**49**
AREA 103	0	17	20	200	6	0	**243**
AREA 104	0	1	2	3	0	0	**6**
AREA 105	0	0	0	7	0	0	**7**
AREA 111	0	15	0	9	6	2	**32**
AREA 112	6	18	3	2	15	6	**50**
AREA 113	0	0	0	1	0	0	**1**
AREA 114	10	25	18	0	30	1	**84**
AREA 121	15	44	39	5	0	247	**350**
AREA 122	25	58	4	0	0	23	**110**
AREA 123	2	15	1	1	3	16	**38**
AREA 124	6	25	10	9	1	0	**51**
AREA 125	34	109	5	16	1	20	**185**
Sum	**106**	**486**	**157**	**413**	**82**	**699**	**1943**

Tab. 5.1 Correlation of the regional and chronological distribution of the analyses.

[62] ANo 1944, 3954, 22149 and 22151.

[63] The few absolute chronological dendro- and radiocarbon dates that are available to date sites where the analysed artefacts have been found are listed in the database under the category "COMMENT".

[64] The in-depth explanation of the period codes is given in the *Comments of the Catalogue*.

[65] For both the description of the database and the later statistical analysis, it is essential to take that issue into account. Consequently, this issue will be returned to below.

of the regional and chronological spread of the analysed copper objects within the database shows is also confirmed by the relationship between period and feature (Fig. 5.4). Here, the distribution reflects the general archaeological picture described in Chapter 4; evidence for the early 4th millennium BC copper use was predominantly found in settlements and as single finds, while 3rd millennium BC objects were discovered in graves. Overall, this reflects the broad picture of archaeological evidence of the individual periods.

The archaeological-cultural classification of the analysed objects is more complex. As mentioned above, this categorisation relates to the referenced literature and is recorded in the database as the variable 'CULT'. However, not every archaeological group within the study area is represented in the database because some – and particularly the early groups, and those in which few objects had been found – did not provide many data. Nevertheless, the dataset includes a broad variety of archaeological groups, which generally also mirror the relative amount of copper finds that are known from the particular archaeological group. Thus, for instance, archaeological groups of the first half of the 4th millennium BC and the Earliest Bronze Age groups are better represented in the database than others (Tab. 5.2).

What the comparison of the regional and chronological spread of the analysed copper objects within the database shows is also confirmed by the relationship between period and feature (Fig. 5.4). Here, the distribution reflects the general archaeological picture described in Chapter 4; evidence for the early 4th millennium BC copper use was predominantly found in settlements and as single finds, while 3rd millennium BC objects were discovered in graves. Overall, this reflects the broad picture of archaeological evidence of the individual periods.

The archaeological-cultural classification of the analysed objects is more complex. As mentioned above, this categorisation relates to the referenced literature and is recorded in the database as the variable 'CULT'. However, not every archaeological group within the study area is represented in the database because some – and mainly the early groups, and those in which few objects had been found – did not provide many data. Nevertheless, the dataset includes a broad variety of archaeological groups, which generally also mirror the relative amount of copper finds that are known from the particular archaeological group. Thus, for instance, archaeological groups of the first half of the 4th millennium BC and the Earliest Bronze Age groups are better represented in the database than others (Tab. 5.2).

CULT

		Frequency	Percent	Valid Percent	Cumulative Percent
Valid	n/a	501	25.8	25.8	25.8
	ABC	11	.6	.6	26.4
	ALT	5	.3	.3	26.6
	BAA	23	1.2	1.2	27.8
	BAD	50	2.6	2.6	30.4
	BBC	82	4.2	4.2	34.6
	BER	2	.1	.1	34.7
	BK	284	14.6	14.6	49.3
	BOD	66	3.4	3.4	52.7
	CHA	1	.1	.1	52.8
	COR	47	2.4	2.4	55.2
	CWC	325	16.7	16.7	71.9
	GIII	14	.7	.7	72.6
	KST	23	1.2	1.2	73.8
	LEN	11	.6	.6	74.4
	LLG	27	1.4	1.4	75.8
	Mier	56	2.9	2.9	78.6
	MKC	11	.6	.6	79.2
	MON	65	3.3	3.3	82.6
	MUE	8	.4	.4	83.0
	NGY	1	.1	.1	83.0
	NIT	247	12.7	12.7	95.7
	PFY	14	.7	.7	96.4
	PIT	16	.8	.8	97.3
	RIV	3	.2	.2	97.4
	SAL	5	.3	.3	97.7
	SOP	3	.2	.2	97.8
	TC	39	2.0	2.0	99.8
	TZC	3	.2	.2	100.0
	Total	1943	100.0	100.0	

Tab. 5.2 Number of trace element analyses from the archaeological groups (CULT) in the study area (see *Comments of the Catalogue* for codes).

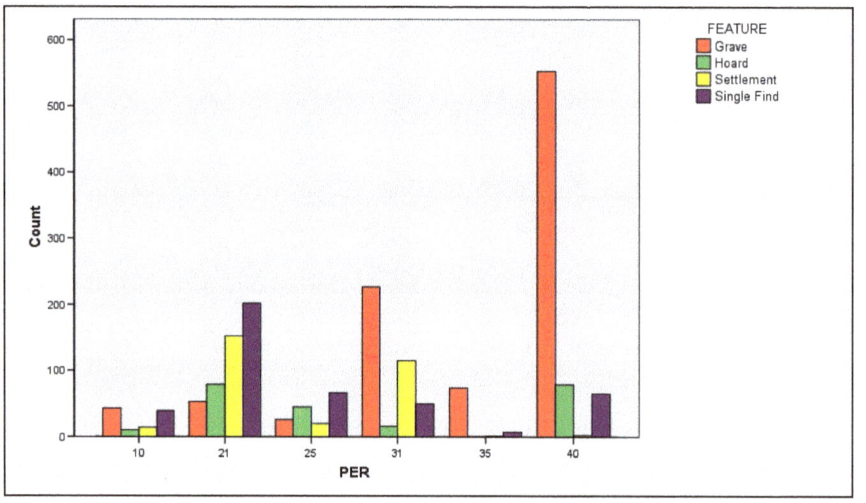

Fig. 5.4 Correlation of period and feature (total database excluding finds from Kelsterbach [D]).

The correlation between the chronological periods and the archaeological groups that have furnished trace element results for the database is listed below (see *Comments of the Catalogue* for the codes):

PER 10 = before c. 4000 BC (LEN; MUE; TC; TZC)
PER 20 = c. 4000 – 3500 BC (ALT; BAA; BOD; COR; MON; PFY; SOP)
PER 25 = c. 3500 – 2800 BC (BAD; BERN; CHA; GIII; RIV; SAL)
PER 31 = c. 2800 – 2300 BC (CWC; MKC)
PER 35 = c. 2500 – 2000 BC objects definitely related to Bell Beaker context (BBC)
PER 40 = c. 2200 – 2000 BC (ABC; BK; KST; LLG; Mier; NGY; NIT; PIT)

It is crucial to mention that these correlations do not always fit exactly into the absolute chronological phases and some archaeological groups may overlap at the borderlines. Therefore, the connection between period and group has to be seen independently and should only offer a guide for later interpretation. The TRB complex for instance, which extended from at least 4000 BC to 2800 BC, is split up into its regional and chronological facies Baalberge, Salzmünde and Bernburg (following Midgley 1992 and Müller J. 1999).

5.2.2.2 THE FINDS

In the following section the variety of artefact types that are included in the database will be described. One issue for the present study has been whether Chalcolithic metal workers were able to choose certain types of copper based on the mechanical properties of the metal, and whether there was a correlation between metal type and artefact function. Therefore, the main focus should be more on the functional classification of the artefacts than on their typology, which is chiefly used for chronological and chorological ordering. Consequently, the relevant information for us now is headlined as TYPE 2 in the database[66]. These types are presented as functional categories, such as axes, daggers etc (cf. Tab. 5.3). It is clear, for instance, that tools, such as axes, have to fulfil other functions than rings or sheet metals. Concerning those functions certain types of copper may have been used to produce these artefacts. In describing the finds we must take into account that copper tends to corrode and can be worn down, so that an object's current shape may not necessarily correspond to its original appearance. This is particularly the case with small objects.

Category	TYPE 2	Object	Frequency
Axes	100	Flat axes	250
	110	Flanged axes	121
	122	Shaft-hole axes	19
	130	Axe-adzes and hammer axes	269
	134	Chisels	29
Daggers/Knives	200	Daggers	21
	212	Riveted daggers	101
	213	Rivets	6
	215	Tanged daggers	71
	300	Knives	28
Rings	400	miscell. Rings	97
	405	Massive rings	49
	410	Spiral rings	213
	420	Willow-leaf-shaped rings	91
Sheets and Wires	500	Sheet metals	65
	510	Wires and small spirals	54
	550	Reels	182
Pins	600	Pins	78
Copper production	700	Castings, crucibles etc	30
Other finds	800	Beads	98
	810	Awls and two fish hooks	71

Tab. 5.3 Number of trace element analyses of objects included in the total database (see *Comments of the Catalogue* for codes).

[66] The following frequency tables and diagrams present the total number of trace element analyses that fulfil the relevant criteria including the double-analyses, because these are all included in the later statistical evaluation. If not mentioned otherwise, the double-analyses do not significantly affect the general proportion displayed in the frequency diagrams.

This chapter concentrates on the various non-Bell-Beaker-finds, but in order to enable a direct comparison of Bell Beaker and non-Bell Beaker evidence, the trace element analyses carried out on Bell Beaker related objects are also included in the subsequent sections. The Bell Beaker related objects that produced trace element analyses for this study are presented and discussed in detailed in an additional section (Chapter 5.2.2.3).

AXES

Axes are probably the most thoroughly examined type of artefact within our research area, having been studied typologically, chronologically and archaeometallurgically. The largest number of trace element analyses has been carried out on axes (TYPE 2: 100 – 134), with 688 analyses undertaken from various types of axes. The typological schemes used to describe these axes have varied in different regions within the study area (e.g. Novotná 1970; Abels 1972; Patay 1984; Dobeš 1989; Říhovský 1992; Schalk 1998; Pászthory and Meyer 1998). The chronological and chorological ordering of the axes represented in the present dataset is based on these studies. The table shows the chronological distribution of the analyses corresponding to the artefact types (Tab. 5.4).

Heavy shaft-hole axe-adzes and axe-hammers are dated to early copper using periods – the so-called *heavy implement horizon* (TYPE: 127, 128 and 131; Fig. 5.5a & 5.5b). These finds appeared mostly as single finds throughout the Carpathian Basin and neighbouring areas. Most of these large axes have not been found in a secure context, but typological comparison with the few parallels discovered in graves has allowed their chronological status to be assessed (Schalk 1998, 25 – 36).

		PER 10	PER 21	PER 25	PER 31	PER 35	PER 40	SUM
TYPE_2	100	3	129	81	30	1	6	250
	110	0	1	3	3	0	114	121
	122	0	3	4	12	0	0	19
	130	67	188	6	8	0	0	269
	134	1	13	4	9	0	2	29
	200	0	6	0	9	1	5	21
	212	0	16	1	32	6	46	101
	213	0	1	0	1	0	4	6
	215	0	9	1	5	56	0	71
	300	1	2	1	2	0	22	28
	400	8	0	7	7	0	75	97
	405	5	2	0	5	0	37	49
	410	12	5	17	28	0	151	213
	420	0	0	0	1	0	90	91
	500	2	8	5	13	2	35	65
	510	1	19	9	4	0	21	54
	550	1	6	4	163	0	8	182
	600	1	0	0	4	2	71	78
	700	0	21	4	4	0	1	30
	800	2	40	0	55	0	1	98
	810	2	17	10	18	14	10	71
SUM		106	486	157	413	82	699	1943

Tab. 5.4 Chronological distribution of the various types of artefacts (TYPE_2).

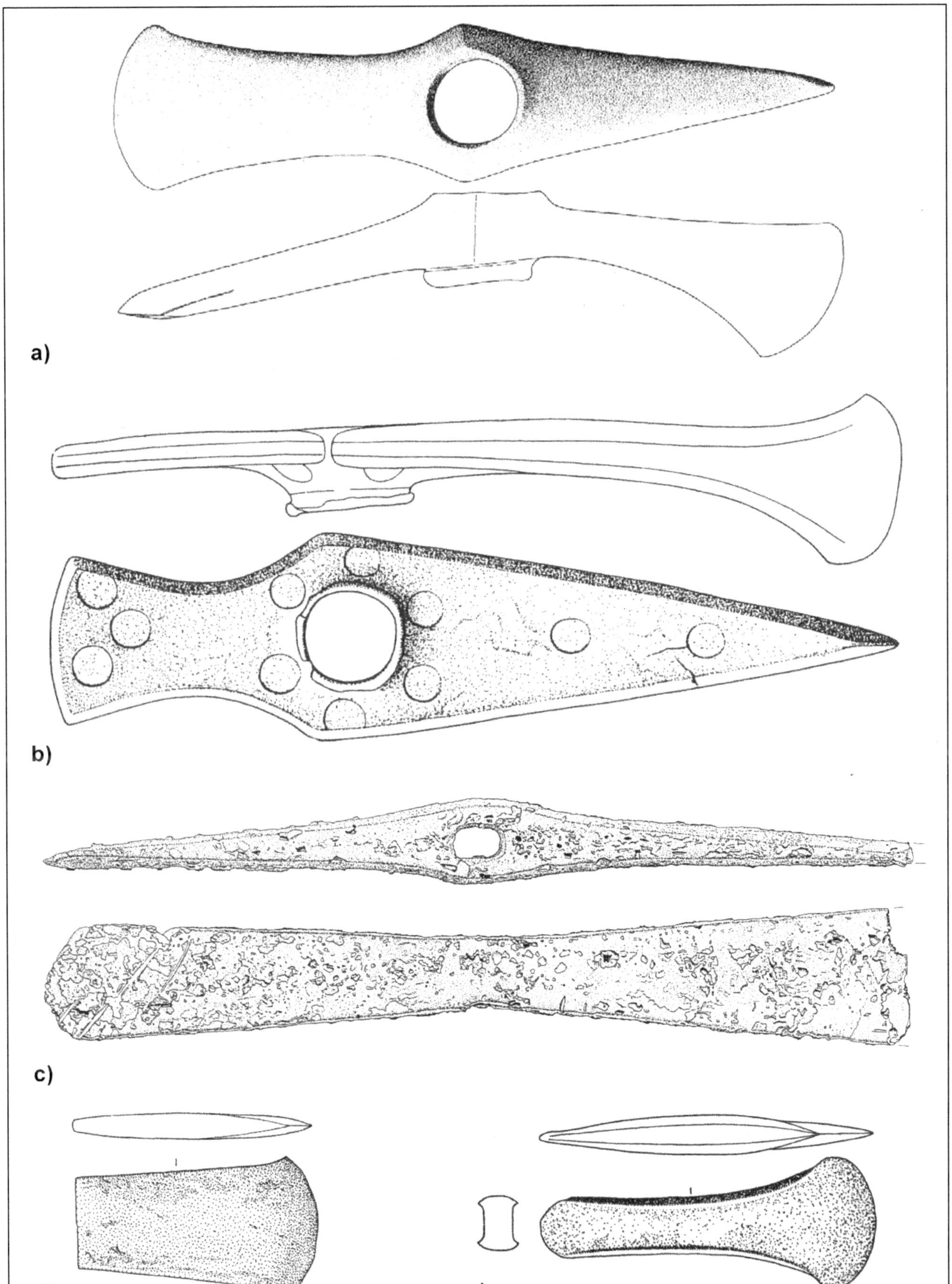

Fig. 5.5 Selection of various types of analysed axes: **a)** shaft-hole axe-adze, type Jászladány, Jászladány, Grave 18 (CZ) (Patay 1984, Plate 39, No. 407) **b)** axe-hammer, type Handlová, Handlová (SK) (Novotná 1970, Plate 5, No. 99) **c)** double-edged axe, type Zabitz, Zimmern unter der Burg (D) (Matuschik 1997b, Fig. 11.5) **d)** flat axe, type Altheim, Unterach (A) (Mayer 1977, Plate 11, No. 137) **e)** flanged axe, type Salez, Salez 'Sennwald' (CH) (Abels 1977, Plate 3, No. 32).

Primarily, but not exclusively in the northalpine region, different types of flat axe have been found at sites that date to the 4th millennium BC (TYPE 2: 100; Fig. 5.5d)[67]. During the late 4th and first half of the 3rd millennia BC single-edged shaft-hole axes were used (Fig. 5.6). These axes are considered more difficult to cast than flat axes and may therefore reflect a change in metal technology (TYPE 2: 122). A corpus of singe-edged shaft-hole axes has recently been compiled and studied by Bátora (2003). Double-edged axes of Zabitz type (Fig. 5.5c) were used at the same time as these (Kibbert 1980, 35 f.).

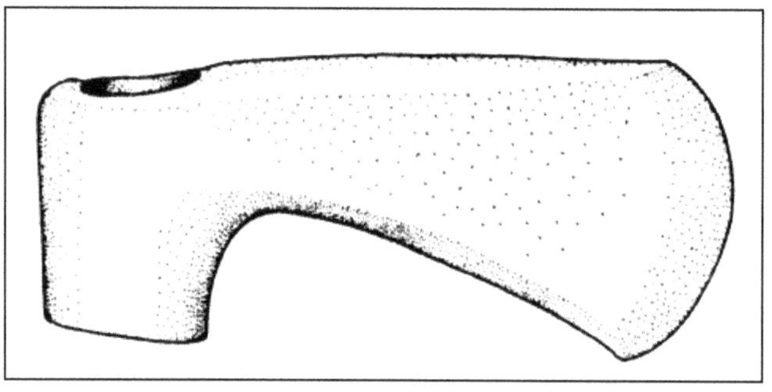

Fig. 5.6 Single-edged shaft-hole axe from Budapest-Óbuda (H) (Kovács 1977, Fig. 14,12).

A new development in the typological sequence of axes is visible at the beginning of the Early Bronze Age when flanged axes were frequently used (TYPE 2: 110). In particular, the so-called Salez axes (Fig. 5.6d) that date to the very end of the period covered by this study have been analysed and discussed several times (Mayer 1977; Bill 1997; Kienlin et al. 2006, 458).

Chisels must also be mentioned (TYPE 2: 134). Although they have not been found as often as other types of axes, they appear in archaeological contexts throughout the 4th and 3rd millennia BC. Generally, they can be compared with the morphology of flat or flanged axes, but they are considerably narrower (cf. Schalk 1998, 59 f.).

DAGGERS

According to archaeological contexts, daggers and knives occur regularly for the first time during the 4th millennium BC (Tab. 5.4). At that time a variety of types of dagger have been used, e.g. triangular daggers sometimes with a tanged hilt, both with and without rivets. However those early copper daggers differ significantly from the tanged daggers from Bell Beaker contexts and from the Early Bronze Age riveted daggers. Typical for the Slovakian Nitra group are the willow-leaf-shaped knives (e.g. Schalk 1998, 66 – 70).

In principle, daggers are typologically distinguished by size and shape of the blade and by the type of hafting. The haft is either round, flattened or trapezoidal, and the hilt plates would have been bound or riveted together (Fig. 5.7). Hilts were made of organic materials, so they usually do not survive. For detailed typologies of early daggers (pre-Bell Beaker period) the reader is referred to the works of Ottaway (1982, 41 – 46), Matuschik (1998) and Schalk (1998, 63 – 66). The daggers from Reute (D), Levice and Malé Levåre (both SK) are dated to the 4th millennium BC (Fig. 5.7a-c), while the example from St. Blaise (CH) dates to the mid-3rd millennium BC (Fig. 5.7d). The characteristics of the Earliest Bronze Age daggers have been comprehensively studied, for example by Ruckdeschel (1978, 55 – 89) and Krause (1988, 49 – 63). The Earliest Bronze Age daggers have rounded sometimes omega-shaped hilts, rivets and usually a flat blade (Fig. 5.7e-f).

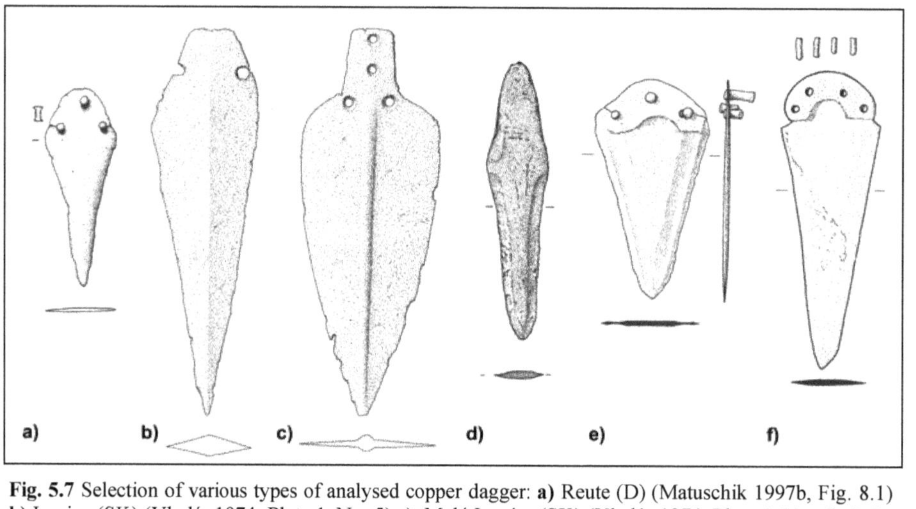

Fig. 5.7 Selection of various types of analysed copper dagger: **a)** Reute (D) (Matuschik 1997b, Fig. 8.1) **b)** Levice (SK) (Vladár 1974, Plate 1, No. 5) **c)** Malé Levåre (SK) (Vladár 1974, Plate 1, No. 6) **d)** St. Blaise (CH) (Strahm 1971, Fig. 31.1) **e)** Singen Grave 7 (D) (Krause 1988, Fig, 13.7) **f)** Singen Grave 63 (D) (Krause 1988, Fig. 14.63).

[67] The axe types and especially the flat axes of the northalpine region have been studied in detail by Ottaway (1982). She defined the types Bevaix, Robenhausen and Thayngen of Swiss flat axes by using cluster analysis and then clarified their archaeological-cultural association. Whereas the Bevaix axes could not be connected with an individual archaeological group, the other two types were mostly discovered in Pfyn contexts (Ottaway 1982, 33 – 40). Other typological classifications can be seen in the literature mentioned above.

RINGS

Beside the great number of rings of different sizes, two types need to be discussed in more detail. The first are the copper spiral finger rings which, in German terminology, are called *Noppen-* or *Lockenringe*. These have been found in many Earliest Bronze Age graves (TYPE 413). They have been discussed by Ruckdeschel who distinguishes between right- and left-twisted spiral finger rings (Ruckdeschel 1978, 142 ff.; ibid., 163 – 168). A few of those rings have been found in graves of the Eastern Bell Beaker group. Spiral finger rings from Bell Beaker contexts, however, are mainly made of gold or electrum[68].

The second interesting type of ring is the willow leaf-shaped ring (TYPE 420) that is typical for the Nitra group (Fig. 5.8). These are characterised by a flat hammered terminal which has the shape of a willow leaf, complete with midrib. The other end is usually pointed (Schalk 1998, 75 ff.).

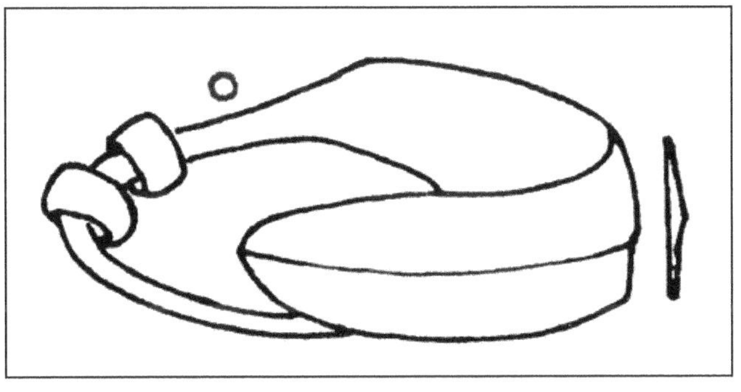

Fig. 5.8 Willow leaf-shaped ring with two bone beads from Výčapy-Opatovce (SK) Grave 6 (Schalk 1998, Plate 14, No. 5).

Rings have been found in archaeological contexts of all Chalcolithic periods (Tab. 5.4). The number of trace element analyses that have been provided by all types of rings reflects the general picture of the use of copper. In contrast to PER 40, between PER 10 and PER 31 many fewer rings have been found. With the growth in the overall number of copper artefacts during PER 40, the number of rings that have provided trace element analyses also increased. This can probably be explained by the fact that there was a general rise in the deposition of metal objects in graves during PER 40; in these graves small rings – commonly used as jewellery – are better preserved and thus easier to spot.

Because of the small size of most of these artefacts, the only contexts that reveal their function or symbolic role are graves. There they appear to have been used as jewellery: for example as finger rings, as arm spirals or as hair decoration.

SHEETS AND WIRES

There are countless suggestions regarding the use of sheet metal objects or of other small finds such as copper wire or other fragmented copper pieces. Again, finds from graves or hoards, together with special, decorated artefacts, indicate that some sheet metals may have been used as jewellery (e.g. as reels or diadems).

The peak seen in PER 31 is deceptive, as the majority of the large quantity of analyses comes from two graves from Drosa[69] (D; grave 8 and 9). Taking this into consideration, the quantity of analyses does not vary drastically over the entire Chalcolithic in eastern-central Europe (Tab. 5.4).

Some analyses carried out on several types of spectacle- and hook-spiral ornaments are included in the database (TYPE: 528 and 529). These types of artefact have been comprehensively studied by Matuschik (1996). He was able to develop a chronological and chorological sequence for these artefacts, which has been applied to classify the finds in our database. These spirals appear in archaeological contexts more or less diachronically from Switzerland to the Carpathian Basin.

Spirals and smaller rings were probably produced by drawing and hammering the metal to a thin wire. The wire objects were then bent in order to achieve the ring or spiral. This would require a certain malleability and elasticity in the copper.

PINS

The greatest variety of pins derives from Earliest Bronze Age contexts. All pin types are combined in 'TYPE 2: 600', but in order to compile the dataset the typo-chronological classification reflected by category TYPE is useful. Apart from the exceptional sample from Jászladány (H) which is a disc-headed pin with a twisted shaft (ANo 14382; Schalk 1998, 111 f.), all analysed pins are dated to after 2800 BC (Tab. 5.4).

The various pins differ by typological characteristics (cf. category 'TYPE'). Whereas 28 of the 78 pins are undecorated disc- or oar-headed pins (TYPE: 603), 31 are of similar shape, but with a decorated head (TYPE: 604). The rest are chiefly roll-headed pins (TYPE: 610). According to these typological attributes, which can be used for dating, the artefacts here discussed have been dated to the Earliest Bronze Age phase – at the latest, Bronze Age phase A1 (cf. Kubach 1977; Ruckdeschel 1978a; Novotná, 1980). According to these, roll-headed pins date generally earlier than disc- or oar-headed pins.

Three so-called 'Cypriot' pins (ANo 10902, 13091 and 13127) from the Carpathian Basin are considered in the

[68] Electrum is an alloy with gold as the main element, bearing natural silver impurities. The colour of electrum is more silver than golden.

[69] In the Corded Ware graves 8 and 9 at Drosa 158 reels have been found and analysed by the FMZM-project (cf. *Metallanalysen Datenbank* published by Krause [2003]).

database. Although in the Middle-Elbe-Saale region Zich (1996, 301 ff.) regards them as dating somewhat later, the relevant samples from the Carpathian Basin were discovered in contexts of the Nitra and Pitvaros group and both groups are part of this study (Bona 1965, 18 ff.; Tocìk 1963, 722).

EVIDENCE OF METAL PRODUCTION

The TYPE2-code 700 summarises all trace element analyses that have been taken from finds that cannot be considered as finished products; for example, unfinished shaped copper pieces or items that relate to smelting, casting and other metal working processes. Nearly all the evidence for metal production that is included in the database for this research has been found in settlement contexts. Taking into account the fact that metalworking evidence predominately appears in settlements, it is hardly surprising that finds such as smelting crucibles or castings considered in this study have been discovered at sites dating to the 4th millennium BC: this is the period where settlements are the principal source of finds. Objects such as moulds or casting cakes that provide direct evidence for metalworking have hardly been found from late Chalcolithic contexts. This is probably because graves are the main source of archaeological finds for the 3rd millennium BC.

OTHER FINDS

Finally, two additional categories of copper artefact need to be discussed: beads and awls. The former have been found in all periods, but most of the examples considered here come from the settlements of Seeberg 'Burgäschisee-Süd'[70] and Vinelz[71], both in Switzerland. In general three types of bead are included in the database: ring-beads, cylindrical beads and biconical beads (TYPE 2: 800). The other type of find is copper awls[72] (TYPE 2: 810). These are mostly single-pointed, and they were commonly used throughout the 4th and 3rd millennia BC. However, their function is not known and several possible uses have been suggested, including the perforation of materials such as animal skins, and even tattooing the body (Husty 2004, 41).

5.2.2.3 THE BELL BEAKER FINDS

This section describes the typology and regional distribution of the objects specifically related to Bell Beakers. The database includes 82 trace element analyses of one flat axe, two pins, two quadrangular copper sheet metals, 14 awls, seven riveted daggers, 55 tanged daggers and one unspecified fragment of a dagger blade[73]. Double analyses exist for two artefacts: the awl from Ludeřov[74] (CZ) and the tanged dagger from Mühlhausen[75] (D). The geographical distribution of the analysed objects more or less corresponds to that of the Eastern Bell Beaker group (Fig. 5.9).

DAGGERS

Although A. Müller (2001, 594) calculated that only 7.21% of all grave finds of the Eastern Bell Beaker group are daggers, these are the most frequent copper artefact within the entire Eastern Bell Beaker group and in the current study's database. Daggers with a pronounced tang are seen as typical of the Bell Beaker package (TYPE 2: 215). The first comprehensive study to focus on the Bell Beaker-related metal objects, and particularly tanged daggers, was carried out by Kuna and Matoušek (1978). The authors developed a typological sequence of four phases for the tanged daggers, reflecting variability in the length, weight, size and shape of the tang (cf. p. 23, Fig. 4.7). Later studies mainly revisit Kuna's and Matoušek's

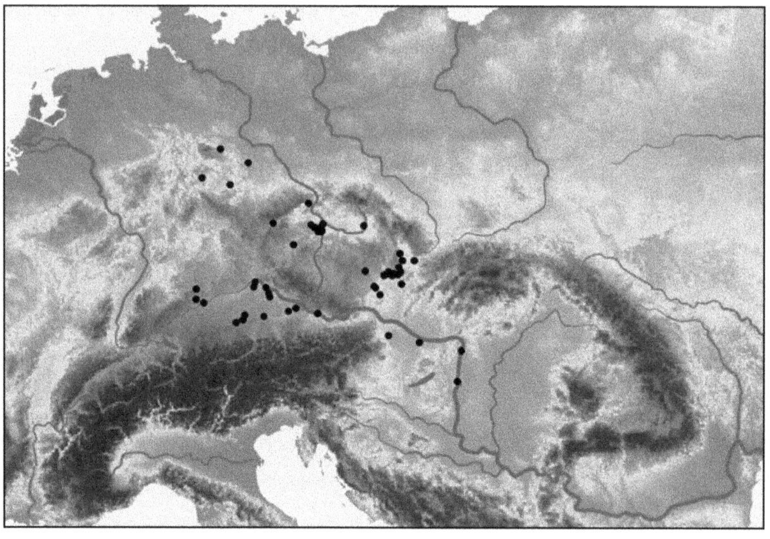

Fig. 5.9 Geographical distribution of Bell Beaker sites with analysed finds.

[70] ANo 2987 – 3026.
[71] ANo 2876 – 2921.
[72] This category contains two fish hooks (ANo 22288 and 100039).
[73] Comparable metal finds from Bell Beaker contexts are displayed in Fig. 4.7 (Chapter 4.1).
[74] ANo 19931 and 29848.
[75] ANo 32228 and 32328.

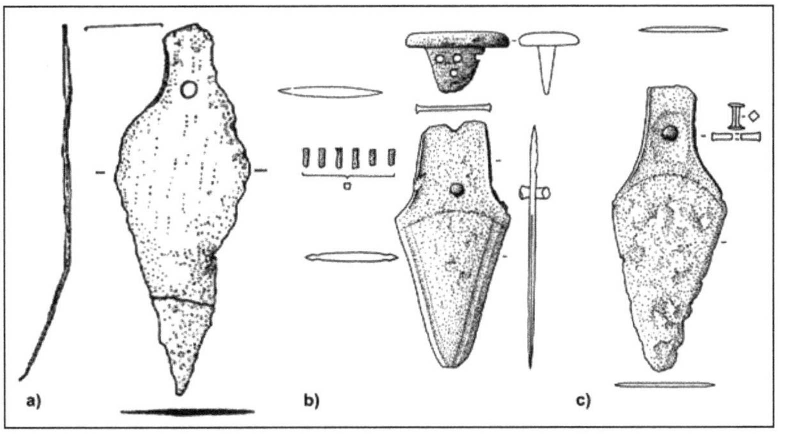

Fig. 5.10 Tanged, single-riveted daggers from **a)** Hülen (D) (Matuschik 2004, Fig. 2.8) **b)** Shrewton (GB) (Gerloff 1974, Plate 1, No. 12) **c)** Sittingbourne (GB) (Gerloff 1974, Plate 1, No. 13).

classification (cf. Heyd 2000a, 269 – 275; Metzinger-Schmitz 2004, 121; Matuschik 2004, 287 ff.). Sometimes the tang has hammered flanges on each side, often with notches which indicate that they were used to fix an organic hilt – perhaps using a cord. A few samples of tanged daggers found in Bell Beaker contexts also feature two rivets. The dagger found in Hülen[76] (D) is unique in central Europe for having just a single rivet. Comparable finds have been discovered in Beaker contexts in Shrewton and Sittingbourne (both GB) (cf. Gerloff 1975, 29 f., No. 12 & 13) (Fig. 5.10). The typological similarity between the dagger from Hülen in southern Germany and the two English examples may point to the often discussed long-distance connections of the Bell Beaker network (cf. Chapter 1.2).

Daggers were usually found in male graves, often near the upper part of the body. Thus, female burials with daggers can be considered as exceptions (Müller A. 2001, 595 f.). Within the area of this research tanged daggers with a tapering hilt are almost exclusively associated with Bell Beaker contexts, so they can be regarded as typical artefacts of the Eastern Bell Beaker group. One exception is the dagger from Egeln 'Bleckendorf'[77] (D), found in a grave associated with an awl and Corded Ware pottery (Matthias 1968, 9 f.). Despite the fact that the dagger is connected to the Bell Beaker set by its typology, the pottery style and the comparably early ^{14}C-date justifies classifying this grave as belonging to the Corded Ware complex[78] (Furholt 2003, 189). Nevertheless, these types of tanged dagger are typical of the Eastern Bell Beaker group, so that the very few stray finds of such daggers, for example those from Győr[79] (H) and Moosinning[80] (D), can be labelled as Bell Beaker finds.

Apart from tanged daggers, some triangular daggers with rounded butt and rivets have been discovered in graves – chiefly with Bell Beaker common ware. For this reason, and because of their typological similarities to Early Bronze Age riveted daggers, these are seen as chronologically younger than tanged daggers (Matuschik 2004, 289). This issue is especially relevant for grave 1 from Safferstetten (D) which included a dagger with five rivets[81], a wristguard and an undecorated handled cup. During the history of research the grave has been classified either as belonging to the Bell Beaker or the Earliest Bronze Age, depending upon the scholar[82].

Although the available radiocarbon dates do not definitely reconfirm a chronological difference between tanged and riveted daggers in central Europe, the stylistic parallels imply a degree of connection, especially since the both archaeological groups are considered to be contemporary to some extent.

AWLS

The awls found in Bell Beaker graves can be discussed more briefly, because of their rather simple shape (TYPE 2: 810). Their length tends not to exceed c. 10 cm; some are single-pointed, and their other end has been hammered flat (Heyd 2000, 278 f.). Usually these awls are quadrangular in section and are either parallel-sided, or they swell in the middle. As discussed above (Chapter 5.2.2.1), the purpose of these 'tools' is not clear. For the Eastern Bell Beaker group it has been noted that awls appear primarily, although not exclusively, in female graves (Müller A. 2001, 597). Some samples include fragments of an organic handle, such as the awl from grave 9 at Künzing 'Bruck' (D) (Schmotz 1992, Fig. 13).

[76] ANo 21597.
[77] ANo 100096.
[78] KIA-162: 2850 - 2820 (10.1%); 2620 – 2570 (51.7%); 2520 – 2500 (6.4%) cal BC; (Müller, J. 1999, 80, No. 18.).
[79] ANo 6533.
[80] ANo 192.
[81] ANo 276.
[82] Whilst Holste (1942, 8 f.) and Dehn (1952, 23) defined grave 1 from Safferstetten as Early Bronze Age, Ruckdeschel (1978a, 77 ff.) and Heyd (2000b, 79) classified it as a Bell Beaker grave.

FLAT AXE

Although the flat axe from grave 1 at Koštov (CZ)[83] is supposedly from a secure Bell Beaker context, Hájek (1968, 44) was not sure whether this was a closed feature. The other axes found in the study area, and which are often considered to be linked to the Bell Beaker phenomenon, cannot be considered as definitely Bell Beaker-related. Comparable finds generally date to the Earliest Bronze Age[84].

PINS

Two pins are included in the dataset – one from Praha-Libeň[85] and one from Vyškov[86] (both CZ). Both items are oar-headed pins with an S-profiled rolled head. Even though pins do not appear very often in Bell Beaker graves, when they are encountered they tend to be of the same type. Metzinger-Schmitz has listed ten samples of these pins found in Eastern Bell Beaker group contexts (Metzinger-Schmitz 2004, 241).

SHEET METALS

Finally the database contains two sheet metal copper objects, found in a Bell Beaker grave in Westerhausen (D)[87]. They are rectangular and made from very thin copper sheets. A few comparable objects have been found sporadically throughout the study area, particularly in Bohemia and Moravia (Husty 1999, 106 ff.). Apart from copper, similar sheet metal objects have been found in gold, and the silverish colour of some of them indicates the use of a silver-rich gold, also known as electrum. They are usually around 7 by 4 cm in size and punched and perforated at their narrow edges. Thus, it has been suggested that they had been fixed to leather or fabric (Husty 2004, 49 ff.). As some of the samples have been excavated lying near to the skull, it has been assumed that these sheets were worn as hair ornaments or diadems (Lehrberger et al. 1997, 151).

Although copper, gold and electrum spirals, rings and other jewellery have been sporadically discovered in sites of the Eastern Bell Beaker group, trace element analyses of those finds could not be compiled for this study. Nevertheless, this compilation of trace element analyses can be considered as a representative cross-section of the Bell Beaker copper finds.

5.3 THE REPRESENTATIVENESS OF THE DATA

Finally, it can be concluded that the content of this database clearly reflects the current status of research in terms of the general distribution of metallurgical evidence and the numerical spread of copper relating to the individual periods, areas and archaeological groups within the study area (cf. Chapter 4). Consequently, the database is considered to be representative for the entire area of research and, in comparison with the results of Chapter 4, is generally reflective of the metallurgical evidence and knowledge of the central European Chalcolithic. With the exception of rings, the database represents all types of Bell Beaker metal object that have been discovered in central Europe.

Nonetheless, it is essential to take into account that the number of trace element analyses does not always reflect the archaeological picture of the metallurgical evidence. This is illustrated if only the sites that have individual PLACENo's are considered. Especially in the case of PER 31 this picture belies the prehistoric truth. Also the large number of single finds of PER 21 contrasts with the cemeteries of the Earliest Bronze Age that contribute a much larger number of copper artefacts (Fig. 5.11). Bearing this in mind, the database, covering 1943 trace element analyses and a wide chronological and geographical area, does offer enough potential for a comprehensive statistical analysis and interpretation. The database includes a sufficient set of trace element analyses in terms of their geographical, chronological and archaeological background knowledge to enable the specific aims of this research to be addressed.

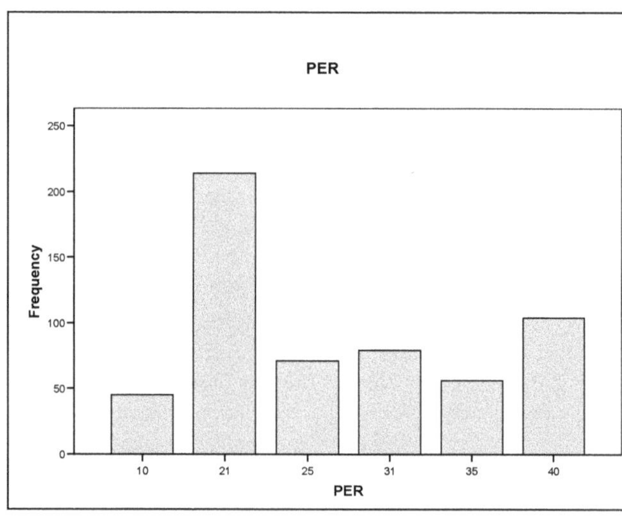

Fig. 5.11 Distribution,by period, of the sites that provided trace element data (only considering PLACENo).

[83] ANo 7546.
[84] Metzinger-Schmitz (2004, 241) argues that all axes previously defined as being Bell Beaker-related have to be seen as dating to the Early Bronze Age. The grave from Koštov shows at least similarities with both archaeological groups, thus the situation is comparable to the discussion regarding the archaeological-cultural classification of grave 1 at Safferstetten and the chronological status of riveted daggers in Bell Beaker contexts.
[85] ANo 3343.
[86] ANo 3471.
[87] ANo 100127 and 100128.

6. From Ore to Metal: The Effects of Ore Formation and Smelting Processes on the Trace Element Composition of Copper Objects

As we have seen in Chapter 5, a number of trace element analyses of Chalcolithic copper artefacts from central Europe have been carried out. These have chiefly focused on two themes: firstly, researchers such as Otto and Witter (1952) attempted to clarify the provenance of the worked copper and, secondly, other researchers tried to describe the chronological, spatial or cultural development of metallurgy (e.g. SAM I & II; Ottaway 1982; Krause 2003). This chapter will review the various ways in which the trace element data have been interpreted archaeologically, will reflect on those and will try to open up new perspectives for understanding trace element analyses. In doing so, it is essential to demonstrate the reasons for accumulation of impurities within workable copper. There are two occasions when both enrichment and loss of impurities is possible: either during natural ore formation or during metalworking processes.

The earliest comprehensive studies which analysed the trace element impurities of copper artefacts focused chiefly on the issue of the provenance of prehistoric ores and on whether special routes of distribution could be located (e.g. Otto and Witter 1952; Pittioni 1957). Whereas Otto and Witter compared quantitative spectral-analyses of copper artefacts with ore compositions from various Alpine and Middle German deposits, Pittioni and his colleagues made semi-quantitative analyses. Pittioni considered that exact quantitative figures of trace element composition did not reflect directly the original concentration as it existed in the ore-body, but suspected that the ratio of the various elements within the ore and the artefact is the same. In other words, if for example the arsenic concentration was dominant within the particular ore, arsenic will also be the main impurity in an artefact made using that ore. They argued that it was not acceptable to draw a direct conclusion from the artefact and ore compositions by comparing absolute quantitative percentages of impurities, since their concentrations vary to some extent within a deposit. Consequently, semi-quantitative analyses compensate for the variability of trace element configuration within one and the same ore deposit (Pittioni 1957, 3 ff.)[88]. However, it is difficult to derive conclusions from semi-quantitative analyses of the ore-body without also taking into account the effect of smelting and other metallurgical preparation processes on the composition of copper impurities.

We need to consider those smelting processes and the ways in which prehistoric, and in particular Chalcolithic, metallurgists worked: how did they smelt, and how effectively did they smelt? Were they able to produce pure copper? As virtually no furnaces, slag and other direct remains of metalwork have been found in archaeological contexts within our chronological and chorological framework of research, and since experiments pertaining to Chalcolithic smelting practices have rarely been conducted in our study area, this chapter must rely on examples drawn from other areas and periods. Other regions, such as Britain and Ireland, Spain and the Near East, provide outstanding archaeological remains for Chalcolithic copper metallurgical studies. Although it is crucial to bear in mind that these examples do not necessarily allow us to compare the archaeological remains of those regions with our study area, the referenced studies may suggest some further perspectives for interpreting prehistoric copper metallurgy in the study area.

A brief explanation is required here of why trace elements accumulate in copper ores and why they appear in copper products, bearing in mind that the process of smelting aims to separate the gangue from copper bearing ore and to obtain metallic copper from its ore. It may be that not all the impurities can be removed through this process; conversely, there may be deliberate trace element enrichment of the metal. Enrichment of trace elements may occur not only with the purpose of intentional alloying, but also because of mixing or recycling ores and/or metals, both deliberately and accidentally.

The following sections will describe ore formation and deposition processes and illustrate operations of prehistoric ore-processing. I will attempt to shed light on the process of accumulating impurities in prehistoric copper artefacts. In order to understand how trace elements react during these operations, it is necessary to take some – at least generalised – geochemical and thermodynamic processes into account.

[88] In principle, the same accounts also for the semi-quantitative trace element analyses of Chalcolithic artefacts from the Carpathian Basin, which Patay et al. (1963) published.

6.1 THE THERMODYNAMICS OF ORE FORMATION AND THE ENRICHMENT OF TRACE ELEMENTS

In simple terms, ore formation can be described as the process whereby a certain mineral accumulates, crystallises and is deposited. Thereby, metals and ions undergo chemical and physical transformation in a way which will be outlined briefly in order to explain why trace elements appear in the ore.

The following paragraphs can only briefly describe ore formation and provide a general picture of copper ores' genesis; abundant technical literature exists to provide an in-depth explanation of ore-forming, geochemical processes and deposition, (Mason 1966; Evans 1980; Robb 2005 with references). Ore formation and deposition are closely linked processes. Ore genesis, the formation of minerals by compounds of ions, results from the crystallisation, precipitation and accumulation of minerals. Accordingly the following description does not distinguish specifically between these processes.

In general, three processes are involved in the formation of ores and their deposits[89]. Firstly, magmatic and hydrothermal processes: metal ions, which through heat and high pressure within the Earth's interior are in solution in magma or hot water, crystallise and precipitate if these circumstances change (e.g. the solution cools down). These igneous processes form primary ores (cf. Robb 2005, part 1 & part 2).

The second encompasses sedimentary processes (including so-called supergene processes), under which minerals are relocated and accumulate under particular conditions, in a particular chemical milieu or in sedimentary traps. Due to mechanical or chemical weathering (e.g. erosion by wind and water or alteration of the chemical milieu) metal ions are dissolved and relocated. Then, minerals with similar chemical and physical characters (e.g. specific weight and solubility) are deposited in the same place which 'attracts' these minerals by its characteristics. Copper ore genesis accounts for supergene enrichment, by which oxide and sulphide ores mineralise, caused by descending solutions (Maynard 1983, 63; Robb 2005, 238 ff.).

Finally, metamorphic processes transform the chemical compounds of primary ores. These are converted into new, secondary minerals, because of changes to their chemical and physical milieu (cf. Evans 1980, 42 ff.). In particular, copper fahlores are created by metamorphic ore formation (Petrascheck and Pohl 1982, 44).

In conclusion, ore formation and deposition is influenced by the chemical and physical milieu. From a chemical point of view, a redox reaction between a metal, in our case copper, and an oxidising agent forms a chemical bond with a partner[90] (e.g. oxides, hydroxides, oxy- or sulfo-salts, or sulfides) during ore genesis (Robb 2005, 8 ff.). This can be generalised with the following redox reaction (here using the example of copper-(II)oxide [tenorite]):

$$2Cu + O_2 \rightleftharpoons 2CuO \qquad (6.1)$$

The energy which is necessary to run a redox reaction is called oxidation or redox potential *Eh* and is defined by the *Nernst equation* (equation 6.2):

$$Eh = E^0 - \frac{R \cdot T}{n \cdot F} \ln K \qquad (6.2)$$

where

Eh:	Redox potential (V)
E^0:	Standard electrode potential (V)
R:	Universal gas constant (R = 8,314 J/mol·K)
n:	Amount of substance (mol)
F:	Faraday constant (F = 9,64853·10^{-4} C/mol)
K:	Equilibrium constant[91] $\left(K = \frac{c(CuO)^2}{c(Cu)^2 \cdot c(O_2)} \right)$

According to the *Nernst equation*, the redox potential *Eh* is proportional to the temperature and quotient of the concentrations of reaction's educts and products, the equilibrium constant. Subsequently, under certain conditions, or more precisely under the influence of a particular pH-value, temperature and pressure, different redox reactions can occur and a particular type of ore becomes stable. However, pressure, temperature and pH-value are interdependent. To sum up, in an area in which a special redox potential and a pH-value exists, a certain (copper) ore accumulates and is deposited (Fig. 6.1). This area, circumscribed by so-called geothermical boundaries, is specific for an individual type of ore (Mason 1966, 67 – 74 & 166 – 173).

[89] It needs to be kept in mind that these types are a simplified classification. The division may vary from author to author and can be subclassified into various more detailed processes; furthermore, the categories are considerably overlapping (cf. Robb 2005, 2 ff.). For present purposes, however, it is sufficient to deal with an overview of the procedures in order to obtain a general idea of ore formation and to understand trace element enrichment.

[90] For an extensive introduction to redox reactions the book "Thermodynamics of Natural Systems" by Anderson can be highly recommended. There, he published a comprehensive overview of thermodynamic processes in nature (Anderson 2005, 335 – 365).

[91] Here the calculation of the equilibrium constant is demonstrated using the example of the oxidation of Cu to CuO.

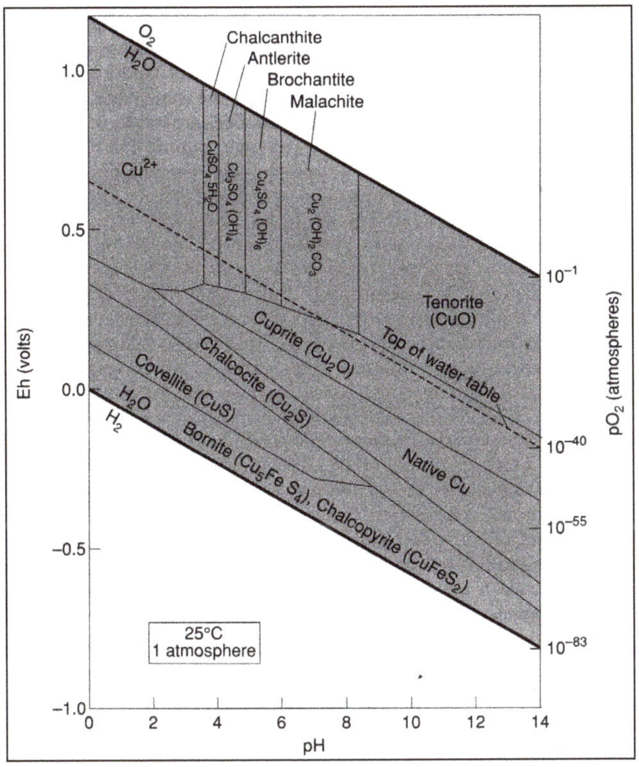

Fig. 6.1 Eh–pH-diagram for stable copper ores (Robb 2005, Fig. 4.15).

Many copper deposits originate not only as a result of primary hydrothermal processes, but also by secondary enrichment. Supergene processes form many of these secondary copper sulphide deposits. We need to take a closer look at the supergene enrichment of copper sulphide deposits since, as mentioned above, the shift to the use of copper sulphide ores is often reported as a change in prehistoric metallurgy (see Chapter 4.4). During supergene copper enrichment iron firstly oxidises and crystallises on the surface at the so-called gossan, because iron ions usually have higher redox potential than copper (Fig. 6.2). Copper ions, soluble in rain water, subside through the *leached zone* into the ground. Nevertheless, native copper (almost pure elementary copper) may also appear near to the surface and was apparently the first kind of copper to have been used (Coghlan and Willows 1962, 59). Below the gossan, in the *oxidation zone*, oxidised ores – copper oxides, hydroxides and carbonates, such as azurite, cuprite or malachite – precipitate and accumulate. Copper sulphates, however, do not react and sink downwards, because of their lower redox potential. They are separated by a barrier (e.g. water table or rock buffer), which prevents oxygen and carbon dioxide from seeping into the *reduction zone*, where sulphates and copper react and precipitate as copper sulphides and other copper sulphosalts. Usually, sulphur-bearing copper ores form at some depth in the ground, so that digging and mining these ores requires a greater deal of energy, logistics and know-how than open-cast digging, where light- and air-supply do not have to be organised. Whereas these reactions do continue in near surface layers, at greater depths where no supergene enrichment occurs, we find primary, mostly hyperthermal copper ores such as chalcopyrite (Anderson 1955; Evans 1980, 184 – 187; Maynard 1983, 63 – 70; Robb 2005, 238 – 245).

^{1}H 2,1																	^{2}He -
^{3}Li 1,0	^{4}Bi 1,5											^{5}B 2,0	^{6}C 2,5	^{7}N 3,0	^{8}O 3,5	^{9}F 4,0	^{10}Ne -
^{11}Na 0,9	^{12}Mg 1,2											^{13}Al 1,5	^{14}Si 1,8	^{15}P 2,1	^{16}S 2,5	^{17}Cl 3,0	^{18}Ar -
^{19}K 0,9	^{20}Ca 1,0	^{21}Sc 1,3	^{22}T 1,5	^{23}V 1,6	^{24}Cr 1,6	^{25}Mn 1,5	^{26}Fe 1,8	^{27}Co 1,8	^{28}Ni 1,8	^{29}Cu 1,9	^{30}Zn 1,6	^{31}Ga 1,6	^{32}Ge 1,8	^{33}As 2,0	^{34}Se 2,4	^{35}Br 2,8	^{36}Kr -
^{37}Rb 0,8	^{38}Sr 1,0	^{39}Y 1,2	^{40}Zr 1,4	^{41}Nb 1,6	^{42}Mo 1,8	^{43}Tc 1,9	^{44}Ru 2,2	^{45}Rh 2,2	^{46}Pd 2,2	^{47}Ag 1,9	^{48}Cd 1,7	^{49}In 1,7	^{50}Sn 1,8	^{51}Sb 1,9	^{52}Te 2,1	^{53}I 2,5	^{54}Xe -
^{55}Cs 0,7	^{56}Ba 0,9	$^{57-71}$La-Lu 1,1-1,2	^{72}Hf 1,3	^{73}Ta 1,5	^{74}W 1,7	^{75}Re 1,9	^{76}Os 2,2	^{77}Ir 2,2	^{78}Pt 2,2	^{79}Au 2,4	^{80}Hg 1,9	^{81}Tl 1,8	^{82}Pb 1,8	^{83}Bi 1,9	^{84}Po 2,0	^{85}At 2,2	^{86}Rn -
^{87}Fr 0,7	^{88}Ra 0,9	$^{89-}$Ac- 1,1-1,7															

Fig. 6.3 Copper (red) and its most frequent trace elements (yellow) according to electro-negativity (after Kraume 1964, 42; Mortimer 1983, Fig. 13.3; Rosenqvist 1983, 338 f.).

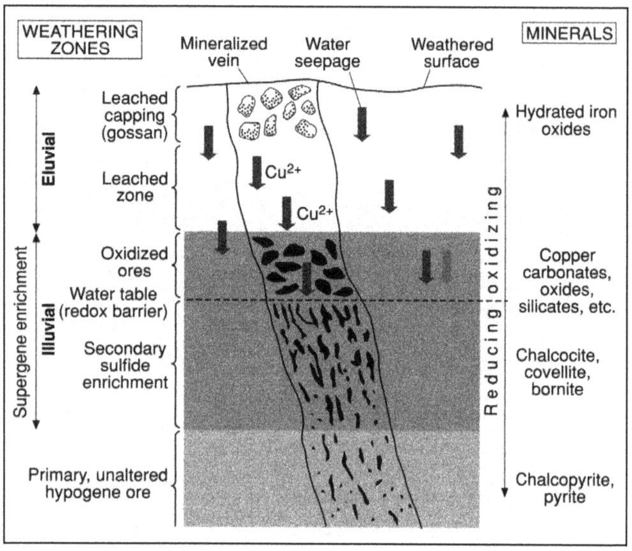

Fig.6.2 Schematic profile of a copper deposit (Robb 2005, Fig. 4.13).

As we have seen, the reason why ore accumulates at a specific spot depends on physical and chemical interaction and on the surrounding milieu. Consequently, under the same conditions ions of trace elements which have analogue characters to copper ions react similarly to copper. This analogy is indicated by similar atomic radius and ionic charge, which are also reflected by the electro-negativity; in other words, ions with similar electro-negativity undergo parallel reactions and form analogous chemical bonds (Mason 1966, 133). Thus, ions of these elements, trace elements, which react in a similar way to copper, can replace copper ions in the crystal lattice of minerals during crystallisation[92] (Mason 1966, 132 ff.; Scharbert 1984, 178 ff.). The most frequent trace elements associated with copper ores are marked in the Periodical Table of Elements in Figure 6.3.

In this very brief description, it has been shown that copper ores are chemical bonds of copper and a salt. Trace elements can react as compounds of the salt and, because of their similar chemical character to copper, they may substitute copper in the crystal lattice. Accordingly, trace elements react to influences such as temperature, pressure or oxidising and reducing conditions in a similar way to copper. As a result, these impurities can be present in copper.

6.2 PREHISTORIC MINING AND SMELTING PROCESSES

Now that the processes of copper ore deposition and formation have been explained, it is necessary to examine where copper ores occur within the study area, and to explore the beneficiation and smelting procedures and their effects on copper composition.

Although a variety of copper-rich areas are distributed throughout Europe (Fig. 6.4), only four major districts within the study area have provided any evidence for prehistoric copper exploitation: the Valais and Grisons in Switzerland, the Greywacke area in the eastern Alps, the Middle German low mountain region and, finally, in the East, the western Carpathian and Slovak Ore mountains. That said, no evidence of pre-Bronze Age exploitation has been found at the famous huge copper ore deposit of the Mitterberg in Austria, nor in the Saxon and Czech Ore Mountains or in the Harz Mountains, and no definite evidence of Chalcolithic mining has yet been found in the area of this study (Eibner 1982; Shennan 1995; Bartelheim et al. 2002; Krause 2003, 30 – 43). In contrast to archaeological remains from central Europe, excavations and experiments from the Iberian Peninsula, southern France, Ireland and the Near East have shed some light on the issue of Chalcolithic metal production (cf. Rothenberg (ed.) 1990; Montero-Ruiz and Rodríguez de la Esperanza 2004; O'Brien 2004; Bourgarit 2007; Hauptmann 2007; Nocete et al. 2008)[93]. Thus, the following paragraphs cannot rely solely on central Europe, but also cite examples from other regions in order to try to create a possible picture of prehistoric copper smelting processes and to clarify the reason for trace element accumulations and loss in copper artefacts.

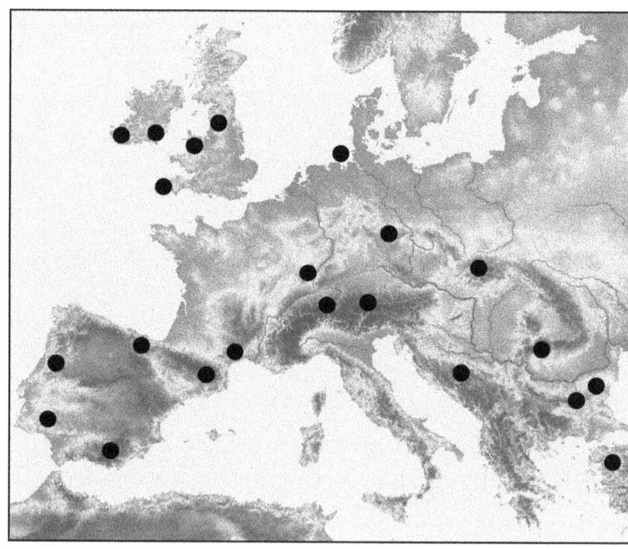

Fig. 6.4 Principal copper ore deposits in Europe.

[92] *Goldschmidt's Rule* presents a detailed explanation of which trace elements are more likely to enter a crystal lattice and replace the primary element. According to Manson, the rule explains that it is possible that (Mason 1966, 132):

1. if ions have the same radius and the same charge, both enter a given crystal lattice with equal facility.
2. if ions have similar radii and the same charge, the smaller ion enters a given crystal lattice more readily.
3. if ions have similar radii and different charge, the more highly charged ion enters a given crystal lattice more readily.

[93] A summary of the most important experiments simulating early copper smelting has been recently provided by Hauptmann (2007, 219 ff.).

First of all, copper deposits suitable for mining had to be located by Chalcolithic metallurgists. Although it is hard to find archaeological evidence for prospection activities, the fact the copper was mined indicates that experienced metallurgists were able to recognise adequate deposits where they existed – probably by spotting significant landmarks or other indicators. For instance, the copper oxide ores were probably identified by their striking colour (cuprite being reddish, malachite green and azurite blue) and by copper's effects on vegetation. Because copper only appears in specific regions, prospectors had to look intentionally for copper deposits that were sometimes difficult to access, and this means that the prospectors must have been experienced. During the earliest stage of copper mining it has been suggested that native copper was collected from the surface. Later, copper ores were exploited from open diggings and were mined underground. For this the copper- bearing material is pecked with stone hammers or picks and carried out of the mines. It can be assumed that, because of evidence from previous Neolithic flint mining, experience in mining may have been passed on to Chalcolithic miners. Fire-setting has been attested in several places and shown experimentally to be an ideal prehistoric method for loosening and breaking solid rock. A fire is set with the purpose of heating the rock, and by cooling the hot matrix with water, the stone breaks up, allowing the material to be collected (cf. Henderson 2000, 214; Ottaway 2001, 90 ff.; O'Brien 2004, 455 ff.; Timberlake 2007, 29; Strahm and Hauptmann 2009, 121 ff.).

As copper is chiefly available as ore-mineral bonded with oxides, carbonates and sulphides, and native copper is extremely rare in nature, copper ores have to be prepared and worked up before an artefact can be produced. These procedures aim to separate pure, workable metal from its compounds – the oxides, sulphides and other impurities. Mined copper contains various impurities and gangue material, mostly silicate rocks. Thus, in the first step, the so-called beneficiation, the cupriferous rock is crushed, ground and sorted (cf. Hauptmann 2007, 217 – 228). Pea-size has been suggested as an ideal volume for smelting by Craddock (1995, 161). The actual ore-rich rocks were manually separated from the gangue matrix – probably immediately after mining and close to the mines, as discoveries of stone hammer and rock spoils indicate on Ross Island (IRL)[94] (O'Brien 2004, 465).

The earliest evidence of smelting comes from Tepe Ghabristan and Tal-i Iblis in Iran – both dated to the 5th millennium BC. According to Chalcolithic and early Bronze Age examples from Ireland (Ross Island) and the Near East (e.g. Timna valley, site 39 [IL], and Faynan area[95], Wadi Fidan 4 [JOR]) these hearths were, to a greater or less extent, vases or pits sometimes bordered by a small stone chamber and most likely filled with charcoal as fuel (Tylecote 1980a, 199; Rothenburg 1990, 4 – 8; O'Brien 2004, 466 ff.; Hauptmann 2007, 217 – 228). This method is called crucible smelting. Since metal and slag were not tapped before the Bronze Age, after the smelting process the crucible contains a mixture of copper prills and slag[96]. The slag needs to be cracked and the enclosed copper slags picked out. In a further step copper prills may have been melted into larger pieces (cf. Levy et al. 2002, 430 ff.; Hauptmann 2007, 223 – 228). For example, the slag pieces found at the Mariahilfbergle (A) have intentionally been broken and within them traces of copper prills were detected (Höppner 2005, 300). This record indicates that crucible smelting may have been undertaken in central Europe even during the first half of the 4th millennium BC.

During the actual smelting, copper ore is reduced with the aid of heat and carbon from the burning charcoal. Copper ore is smelted, and oxygen as well as other impurities are removed. For this, the atmosphere inside the hearth needs to be reducing in order to separate impurities, especially iron and oxygen, from copper (Henderson 2000, 220). The assumed prehistoric smelting temperature is around 1200°C when copper melts and carbon monoxide reduces the ore[97] (Pernicka 1989, 630). The air supply and the fuel are crucial in reaching the required temperatures. Although Chalcolithic evidence of tuyères and pipes is extremely rare, a good air supply is essential for efficient smelting (Batorá 2002; Bougarit 2007, 7 f.). Even with blowpipes or bellows, the reducing conditions within the crucibles were probably quite poor, so that the impurity content of the metal was high (Craddock 1995, 126 ff; Hauptmann 2007, 217 ff.). If in Chalcolithic central Europe crucible smelting was common, 'furnaces' may have been just small pits, under a metre across. The reason why evidence of crucible smelting has been scarcely explored may lie in the construction of these hearths and the fact that the slag was crushed into lumps. Our lack of knowledge concerning the metallurgical technology of the 4th and 3rd millennia BC is probably the consequence of the poor circumstances of preservation of these features (cf. Craddock and Meeks 1987, 190 ff.). However, Mariehilfbergle in Austria presents one of the very few pieces of evidence for crucible smelting in central Europe. At that site, dating to the late 5th/early 4th millennium BC, pieces of intentionally broken slag bearing copper prills have been discovered (Höppner et al. 2005, 300).

It has also to be taken into account that the addition of fluxing agents for smelting more complex ores may have been necessary. Fluxing agents can be siliceous or ferrous metals often containing typical trace elements, such as arsenic, antimony, lead etc. (Tylecote 1980a, 188 f.). According to the similar properties of trace elements it is most likely that these impurities, at least in traces, also

[94] On Ross Island a copper mine has been excavated and extensively studied by O'Brien (2004). That site gave not only evidence of copper mining related to Bell Beaker contexts during the 3rd millennium BC, but also provided indications of mining, beneficiation and smelting.

[95] In literature various spellings are used for that area in south-west Jordan: Faynan, Feynan, Feinan.

[96] The reason why copper concentrates as beads within ferrous silicate slag is because of different densities, the slag has a quite low free-running temperature and copper is slightly soluble (cf. Tylecote 1980a, 184 f.).

[97] The melting point of pure copper is 1083°C.

pass into copper. Concrete evidence of the utilisation of flux and its precise composition has not yet been found on Chalcolithic sites in central Europe.

It has often been suggested that oxide copper-ores (e.g. malachite) were used before sulphide copper ores (e.g. chalcocite), the explanation being that, firstly, oxide copper ores are easier to exploit, being found in the upper oxidised zone (cf. Fig. 6.2), and, secondly, sulphide copper ores are supposedly harder to process into workable copper (e.g. Strahm and Hauptmann 2009, 122 f.). Sulphide-containing ores need firstly to be transformed into oxides. Copper sulphides can be converted into oxides by burning the sulphur in a previous operation, so-called *roasting*[98]. In contrast to scholars such as Bachmann or O'Brien, who advocate a two-step roasting and smelting operation (Bachmann 1993; O'Brien 2004, 470 f.), others have plausibly demonstrated that roasting and smelting can be done simultaneously. Accordingly, a separate roasting procedure is unnecessary in order to turn sulphide into oxide ores (Rostoker et al. 1989; Klein and Lechtman 1999; Lorscheider et al. 2003; Moesta 2004). In particular, sulphide ores mixed with oxide and/or carbonate copper ore allow the smelting temperature to be lowered to around 800°C (Rostoker et al. 1989). In contrast to the often argued thesis that fahlores and other sulphide ores have to be treated by complex smelting techniques, these ores can be smelted in one process even by simple crucible smelting (Moesta 2004, 271). Moreover, it has been proved that co-smelting of oxide ores together with sulphide ores works even under prehistoric conditions. In this way it was also possible to produce an alloy deliberately, if a copper ore contains a further metallic component (e.g. arsenic) (Rostoker and Dvorak 1991; Klein and Lechtman 1999). Furthermore, Ryndina and her colleagues have demonstrated that even as early as c. 4000 BC metallurgists in the north-eastern Balkans were already capable of dealing with sulphide ores (Ryndina et al. 1999). Furthermore, the slag found at the Mariahilfbergle originates from sulphide fahlore – probably the local occurring trahedrite (Höppner et al. 2005, 301 f.).

Consequently, the ability to work with sulphide copper ores does not mark a fundamental change in copper metallurgy as is frequently assumed; however, as mentioned above, these ores were usually more difficult to mine, because they formed deposits at greater depth than copper oxides. Furthermore, it could be argued that it was not the capability to smelt, but rather the concentration of trace elements (e.g. arsenic, antimony or nickel) in sulphide ores, their transition into copper, their plausible influence on artefacts' properties and the skills to control them which constituted the key revolutionary aspects in the development of metallurgy. The following flow-chart sketch summarises the process for converting copper ore to workable copper and finally artefacts. These are commonly named as the metallurgical *chaîne opératoire* (Fig. 6.5).

[98] For an introduction to the thermodynamics of roasting see: Rosenqvist 1983, Chapter 8-3.

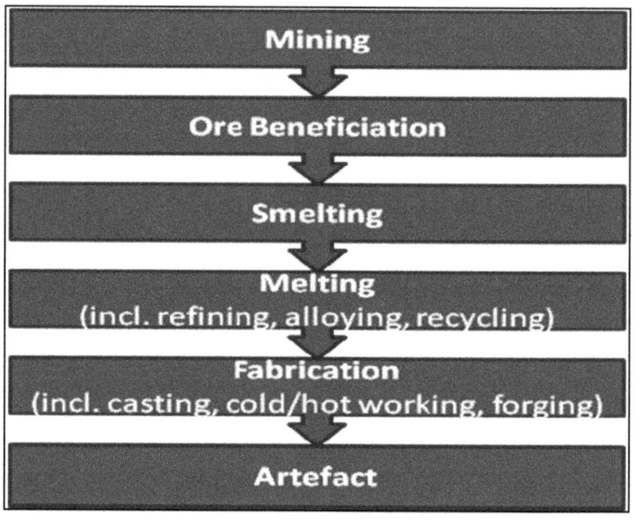

Fig. 6.5 Flow-chart showing simplified processes for converting copper ores to finished artefacts.

Finally, after the metal has been cooled, pure copper should be aggregated. At that stage, the copper charge is not pure enough to work, so that, especially after crucible smelting in Chalcolithic times, it has to undergo an additional refining process. Refining is perceived as a reheating of the smelted copper in order to drive off remaining impurities and enclosed oxides. Impurities oxidise, precipitate and can be detached (Craddock 1995, 202 ff.). Tylecote (1987, 193) noticed that every re-melting process would further refine the copper.

At the beginning of this chapter it was noted that ore deposits are located either outside, or on the periphery of, the study area. Copper artefacts, however, have been discovered all over that area, and so it is necessary to explore where the production sites could have been situated and how copper may have been passed (presumably by some kind of exchange) to consumers. Given the rarity of the evidence for copper mining and production in our study area, it is hard to identify trade routes. Local copper production is proved by crucibles and slag found in settlements of the 4th millennium BC (e.g. in Pfyn, Mondsee and TRB contexts; cf. Chapter 4.1). However, we do not know how the raw material was transported from the copper deposits which are located elsewhere (Fig. 6.4). In the case of the 3rd millennium BC the character of the production sites can only be assumed. We know, from 3rd millennium BC examples from Valencina de la Concepción [E] and Khirbat Hamra Ifdan (Faynan [JOR]), that copper was produced in specific workshops (Levy et al. 2002; Nocete et al. 2008). In each case, the archaeological record covers the full metallurgical *chaîne opératoire* and indicates that we are dealing with specialist workers – whose work may have been coordinated and controlled by élites – and a wider distribution of the copper products[99].

[99] At Khirbat Hamra Ifdan ingots and their associated moulds have been found, so that it is clear that copper was cast in a standardised shape in order to export the metal (Levy et al. 2002, 432 f.). From central European Early Bronze Age contexts Salez axes and *Ösenhalsringe* are also seen standardised ingots (Junk 2003, 11 f.), but these objects are dated to the very end of the studied period or even later.

Even if this model cannot be transferred to central Europe where comparable sites have not been excavated, it is apparent that copper was distributed from the mining areas to the consumers. However, the acquisition of copper over greater distances requires a distribution network. Perhaps individuals or communities – perhaps identifiable as archaeological groups – were specialised in that networking.

6.3 COPPER ORE SMELTING AND ITS EFFECTS ON TRACE ELEMENT COMPOSITION

After looking beyond the study area at some elementary theories of Chalcolithic copper production, the question of how trace elements act and react during smelting, melting and casting processes will now be discussed. This should finally explain why workable copper contains quantities of impurities, in spite of care during preparation.

Both ore genesis and smelting are chemical reactions. It has been shown in Chapter 6.1 that ore-forming processes are oxidation reactions. Smelting aims not only to remove oxides from copper, but also to extract other compounds of more complex ores such as iron, arsenic etc. During smelting with charcoal, oxygen and high temperature impurities would be extracted from the copper ores and, in an ideal case, these impurities agglomerate in slag. However, trace element analyses have demonstrated that prehistoric copper objects were mostly not free from impurities. Consequently, we have to ask why these artefacts are 'polluted'.

In order to purify copper minerals, they were reduced by heat and a reducing agent, namely carbon monoxide. Carbon monoxide, which arises from burning charcoal[100], reduces copper oxides (cf. Tylecote 1987, 107). Subsequently, pure copper and carbon dioxide are produced (shown here on the redox reaction of copper-(I)oxide [cuprite] and carbon monoxide [see equation 6.3]).

$$Cu_2O + CO \rightarrow 2Cu + CO_2 \qquad (6.3)$$

Smelting is the reverse redox reaction of ore genesis, and the chemical bonds between copper and its compounds have to be broken. In order to find out which impurities remain in the copper, we have to understand how the various elements react during these procedures. In terms of thermodynamics, copper is extracted from its energetic stable state inside the crystal lattice. This has to be done by feeding the system with energy in form of heat. Energy must be high enough to break the chemical bonds of the copper ore. This energy can be measured as *standard free energy* $\Delta G°$ (for in-depth information: cf. Mason 1966, 67 ff.; Rosenqvist 1983, chapter 3). Standard free energy can be expressed as:

$$\Delta G° = \Delta H° - T \cdot \Delta S°,$$
$$\Delta G° = -R \cdot T \cdot \ln K \qquad (6.4)$$
$$\text{and } \Delta G° = R \cdot T \cdot \ln pO_2 ,$$

where
$\Delta H°$: Change in enthalpy (J)
T: Reaction temperature (K)
$\Delta S°$: Change of entropy (J/K)
R: Universal gas constant (R = 8,314 J/mol·K)
K: Equilibrium constant (see equation 6.2)
p: Pressure (here of oxygen) (Pa).

According to equation 6.4, $\Delta G°$ is proportional to temperature and pressure of the system, so that with rising temperature and pressure the stability of metal oxides drops. Temperature and pressure could be controlled by prehistoric metal workers by changing the amount of fuel or by increasing or decreasing the air supply (Hauptmann 2007, 221).

It is also considered that the more negative the value of $\Delta G°$, the more difficult an oxide is to reduce (Charles 1980, 156). The exothermic reaction of the redox system (here: $2C + O_2 \rightarrow 2CO$) must provide more energy to split an ore than an endothermic reaction needs to break up a molecular ore from its stable state. If – below the smelting temperature of c. 1200°C which is assumed for copper smelting processes – the standard free energy value of a molecule is higher than the $\Delta G°$ value of carbon monoxide, then it is likely that it can be melted and transformed into the copper during this procedure (Pernicka 1989, 630 f.). The stability of various oxides is illustrated by the *Ellingham diagram*, where the free energy is plotted against temperature (Fig. 6.6). Comparable with the ore formation, elements with similar chemical characters react in a similar way.

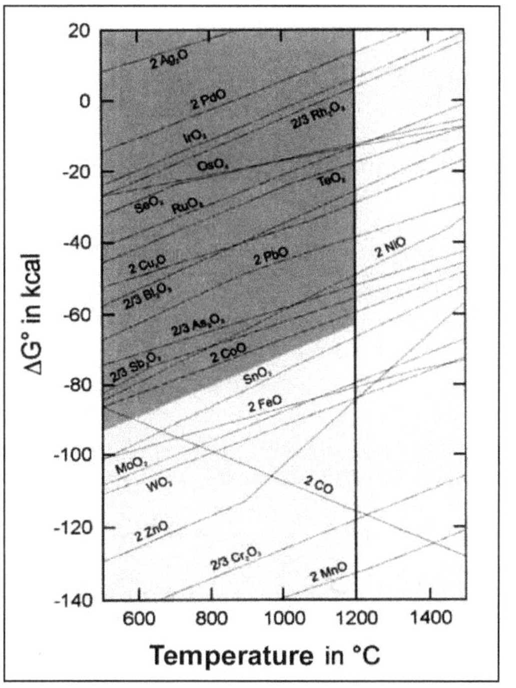

Fig. 6.6 Ellingham diagram showing the change in standard free energy $\Delta G°$ with temperature of various oxides of trace elements (after Prange 2001, Fig. 19).

[100] The burning of charcoal can be illustrated simply as: $2C + O_2 \rightarrow 2CO$.

The Ellingham diagram displays the $\Delta G°$ of various metal oxides and carbon monoxide versus temperature. All oxides within the grey shaded area can be reduced whilst carbon monoxide is oxidised. Thus, at the assumed smelting temperature of c. 1200°C these elements can be transferred to raw copper during smelting and – theoretically – used for provenance studies (Dannatt and Ellingham 1948; Pernicka 1989, 630; Craddock 1995, 198). This, however, occurs only under standardised conditions, with only pure copper oxides (e.g. cuprite) and without producing slag (Charles 1980, 157 f.; Prange 2001, 23).

As prehistoric metal workers could not achieve standardised laboratory conditions, it is difficult to explain the origin of smelted copper by using trace element analyses on their own. McKerrell and Tylecote proved experimentally that a significant loss of impurities occurs during smelting, even of those that were expected to stay in copper, such as antimony, arsenic, bismuth and lead (McKerrell and Tylecote 1972; Tylecote 1980b). By using charcoal, regulating the air supply and using small hearths prehistoric metal workers were able to control the conditions of smelting and of the resulting metal (Bamberger 1985).

6.4 THE ARCHAEOLOGICAL INTERPRETABILITY OF TRACE ELEMENT ANALYSES

For the moment, we cannot determine from the weight percentage of trace elements in finished products either the composition and types of ores used or the source of the particular deposits. Referring to Sperl (1975, 16), various processes affect the trace element analysis of a single artefact. Firstly, there is a variation within the ore deposit itself. Then, during mining and preparation, the workers decide which ore goes into the smelting by choosing those samples which appear to them to be more productive. The smelting process itself concentrates the main metal content and its related trace elements. This is the most influential operation on trace element composition and the loss of components cannot be replicated. Finally, it has to be considered that every measurement varies with statistical error. This, however, can be counteracted by comparing a larger number of analyses. These have to be taken into account if the trace element composition is to be examined.

Direct remains of metal work (slag, ore, crucibles etc) are very rare in our study area, and so we have no idea of hearth size, for example, and can only make assumptions about the temperature, pressure and concentration of oxygen and other reacting agents during smelting. It is not possible to say whether the same percentage of trace elements which were in the original ore is reflected in the products (Friedman et al. 1966). Furthermore, we cannot prove whether, during smelting and/or casting, copper has been recycled or mixed. One can suggest that as soon as people collected and melted solid rock into metal that could be cast, they probably realised that metal could be re-melted and re-cast over and over again.

Even if ones assumes that fluxes had not been involved and if the original source area for a raw material is known, it is still hard to relate the finished products to the ore using just trace element analysis, because the quantity of trace elements accumulated in copper minerals varies within a deposit. Whether the range of impurities tends to concentrate in the copper or within the slag is mainly controlled by the redox potential. A higher redox potential enhances the chance that trace elements oxidise and accumulate in slag, but then copper may also oxidise and tend towards passing into the slag. In contrast, if there is a stage of pure copper present during smelting, then trace elements tend to stay in copper (Yazawa 1980). It is also necessary to keep in mind that several impurities (e.g. arsenic and antimony) are highly volatile in an oxidising atmosphere (McKerrell and Tylecote 1972). Apart from that, trace element compositions can vary within one deposit, and conversely different deposits can feature similar characteristics of impurities (Budd et al. 1996, 168 f.). Due to these factors we must conclude that we cannot use the trace element composition of a finished copper artefact on its own to provenance the basic material and the original ore deposit.

A method that attempts to clarify the provenance of copper is the analysis of lead isotope ratios of artefacts and certain ore deposits. The numerical proportion of lead isotopes is determined by the geological history of the deposit and does not change throughout metallurgical procedures. Given that ore deposits tend to have a characteristic lead isotope profile, it is possible to list the deposits from where the copper may have come, and rule out those from where the copper may not have come. The ore used for an artefact can only come from a deposit that matches both the elemental composition and lead isotope ratios. Hence, a possible provenance of copper can only be determined by combining both trace element and lead isotope analyses (e.g. Pernicka 1995, 99 – 105; idem 1999; idem 2000; Northover et al. 2001).

The method has been critically evaluated by Budd et al. (1996), who concluded that it offers a great opportunity to investigate the provenance of prehistoric copper. Although these authors had already stated that more analyses are necessary for comprehensive studies, unfortunately there are very few lead isotope analyses of objects within the study area and of Bell Beaker related objects in general. Thus, unfortunately, they cannot be taken into account for this research[101].

[101] Several rather local projects have been studying lead isotope analyses in combination of trace element analyses in order to determine the provenance of raw copper. They have focused on Chalcolithic and chiefly Bronze Age copper finds (e. g. in the Aegean: Pernicka 1995, 106 – 116; the Balkans: Pernicka et al. 1993; Bohemia and central Germany: Niederschlag and Pernicka 2002; the Inn valley: Höppner et al. 2005, 301 – 311; the Slovakian ore mountains: Schreiner 2007; and the Valais: Cattin 2008). Nevertheless, in the case of central European Chalcolithic and particularly Bell Beaker sites the published data are not sufficiently numerous to integrate them into the present study.

In contrast to Otto and Witter's assumption, it is impossible to explain the origin of copper ores only by the use of trace element analyses of finished products. The trace element composition of copper artefacts does indeed reflect the configuration of the parent ores, but predominantly it reflects the technique and the conditions by which it was produced; for instance, the chemical and physical conditions during smelting: whether the material was mixed with other types of copper; whether flux had been added. Thus, the main reason why it is not possible to infer the ore source directly from the trace element composition of the copper artefact, even if the source is known, is that smelting, refining and melting massively influence the copper configuration.

Several earlier studies have focused on obtaining trace element analyses and interpreting them archaeologically. These works have already convincingly demonstrated that a regional and chronological diversity existed by grouping trace element analyses of (central) European prehistoric copper artefacts and comparing these clusters with their archaeological background (cf. Chapter 4). They have evaluated trace element analyses with the archaeological contexts. Hence, these researchers were able to define a spatial and chronological range of copper types, based on similarities in trace element content. As a result, a spatial and chronological variation in the use of particular types of copper has been interpreted in terms of a regional and chronological development – a diffusionistic spread of metallurgy – at first from East to West and then, in a second pulse, from the West (cf. SAM I, 90 f.; Strahm 1994; Cevey et al. 2006, 25). Because direct correlation from the composition of a finished copper artefact to a certain ore deposit is not appropriate, most researchers speak of 'workshop-recipes', implying that copper with comparable trace element composition has been made with a similar technique and from a similar ore – using a similar recipe (cf. SAM I & II; Rau and Willing 1991; Strahm 1994). These recipes are typical for certain areas and times. However, the ingredients of these recipes could not be evaluated and it is now necessary to investigate what other reasons may explain the use of different types of copper.

7. The Effects of Trace Elements on Copper Properties and Their Archaeological Interpretation

The aim of this research is to establish whether prehistoric metallurgists (especially those associated with the Bell Beaker phenomenon) deliberately chose a 'recipe' and its 'ingredients'. It has already been shown, in Chapter 4.3, that it is possible to define certain 'workshop-recipes', characterised by distinctive trace element compositions of copper artefacts, in different areas and at different periods during prehistory. In seeking to explain why these recipes were so special for a certain region and time, we have noted that the composition of copper artefacts is influenced both by the configuration of the resource and by the manufacturing processes. It is therefore worth considering whether there is evidence for an intentional selection of types of copper and, if there is, determining what might have been the reasons for the prehistoric metal workers' choices.

It should be remembered that the trace element analyses have primarily been determined from finished artefacts. One might reasonably expect from these results that, if Chalcolithic metal workers intentionally produced certain copper types, they chose recipes in order to control the characteristics of the artefact. One reason for a deliberate selection may have been the chemical, physical and mechanical properties of copper which are affected by the presence and concentration of several trace elements[102]. Positive effects can be improved smelting, casting or working properties, because factors such as ductility, hardness, malleability, tensile strength and solidification temperature are improved by increasing concentrations of certain impurities. A desired change in colour can also be seen as a positive effect. Conversely, some elements can have a negative effect, for example by making copper brittle (and thus difficult to work and use) when present in sufficient quantity.

While prehistoric metal workers would not have based their choice of metal on a knowledge of its chemical composition, they probably realised that copper from a specific source, or of a particular colour, had specific properties and thus they may have sought out particular coppers.

The deliberate adding or alloying of impurities could occur either by selecting ores which naturally contain a particular element in high concentration, such as the arsenic-bearing copper ores olivenite ($Cu_2(AsO_4)(OH)$) or enargite (Cu_3AsS_4), and working them in such a way that the desired (trace) elements accumulate in the copper, or else by adding an ore containing a second element, e.g. arsenic trioxide (As_2O_3), which would produce a copper-arsenic-alloy (cf. Charles 1967, 25; Rapp Jr. 1988, 25). In both cases, we may assume that prehistoric metallurgists would have been following 'recipes' in the selection of these raw materials. This is particularly the case with arsenical copper, whose intentional production has been argued for by various authors (cf. Charles 1967; McKerrell and Tylecote 1972; Northover 1989; Budd 1991a; Lechtman 1996; Junk 2003).

The question of the extent to which deliberate alloying took place is currently hard to answer, because within the study area, the configurations of the original ores are not known. However, it is known that workable copper can be extracted from ores that contain various elements, and that these elements can be transferred into the copper. Thus, in considering the range of observed copper compositions, we have to take into account the possibility that there may have been both non-intentional and deliberate alloying. Although the boundary between the two is hard to distinguish, we shall assume that we can only speak of deliberate addition if a trace element is present at a level of over c. 2% (Rovira and Montero 2003, 15).

Some previous studies have already shown that compounds such as arsenic harden copper to a certain degree (e.g. Budd 1991a; Lechtman 1996). Several other studies have also suggested that particular types of copper were used, and have discussed the influence of impurities on these coppers; however these have mostly focused on a specific type of artefact, a period or a type of trace element (e.g. Charles 1967; Northover 1989; Junk 2003; Kienlin 2008).

Beside these rather recent archaeometallurgical studies, the further sections rely mainly on a number of research papers from the first half of the 20th century (Hofman 1924; Archbutt and Prytherch 1937; Dies 1967). Although those explored the effects of impurities on copper mainly from the perspective of addressing modern, industrial issues, they can suggest ways in which trace elements may have influenced the properties of copper in prehistoric metallurgy.

[102] The various properties of copper that can be influenced by impurities are explained in the glossary.

In 1937 the *British Non-Ferrous Metals Research Association* at the *National Physical Laboratory* conducted a fundamental project on the influence of impurities on the physical and chemical properties of solid copper[103] (Archbutt and Prytherch 1937). The project investigated how trace elements at various concentrations – especially at low concentrations – affect preparation and casting, and how impurities influence working properties, density, tensile strength, hardness, fatigue, softening or melting temperature[104], micro- and macrostructure, solubility of the impurities in solid copper and electrical conductivity (cf. Archbutt and Prytherch 1937, 1 – 18). These metallurgical studies are pertinent for archaeological research as low concentrations of impurities are commonly found in prehistoric artefacts. Moreover, the results of these experiments are applicable to prehistoric material because the graphite moulds which were used in such work have similar casting conditions and cooling rates to those of prehistoric moulds (Budd 1991a, 42). The type of mould is important, since experiments on bronze cast in sand, clay and metal moulds have demonstrated that the type of mould influences the metallography and the properties of cast copper (e.g. Staniaszek and Northover 1982; Wang and Ottaway 2004).

Nevertheless, not all of the characteristics that were examined in these old studies are relevant for archaeometallurgical research. Properties such as electrical conductivity, for instance, were obviously not important for prehistoric people[105]. In contrast, characteristics such as the softening and solidification temperature, fatigue or tensile strength, may have been significant for Chalcolithic tools and their manufacturing process. As will be demonstrated, impurities do not only have negative effects on the properties of copper. We can presume that the types of copper that contained beneficial trace elements were deliberately prepared.

The development of the skills that allowed prehistoric metalworkers to produce specific types of copper could explain their chronological and chorological distribution in Chalcolithic times – from the initial use of pure, arsenical and fahlore copper to the final production of tin-bronze. Attention will be focused on the trace elements bismuth, iron, tin, lead, nickel, silver, antimony and arsenic, as they are the commonest to be found, and they have the greatest technological impact on prehistoric copper artefacts. Other trace elements will not be disregarded, however, since they too can exert a significant influence on properties of copper. In ascertaining the properties of the impurities, Archbutt and Prytherch (1937, 2) undertook their tests using cast, cold- and hot- rolled and annealed samples. The results of these investigations will be reviewed briefly in order to discuss their applicability to archaeological research. It should be noted that rolling is a modern technique, used to shape and to reduce the thickness of metal and, thereby, to strengthen it. Rolling metal is an indicator of the material's strength. Thus, metal that still can be rolled with increasing concentrations of impurities has increased strength. Prehistoric metal workers hammered the copper to achieve similar results, but with much more effort.

7.1 Metallurgical Properties and their Use by Chalcolithic Metallurgists

This section focuses on the archaeological interpretability of trace elements, on the impurities' influences on the properties of copper, and on the metallurgical knowledge of Chalcolithic people. It therefore discusses the chemical, physical and mechanical effects of particular copper impurities as determined by various metallurgical experiments conducted by several laboratories. Therefore, it will concentrate on the results and on the conclusions of these experiments and their relevance for archaeological research.

The experiments carried out by Archbutt and Prytherch (1937, 1) and others featured tests of density, fatigue, hardness, notched-bar impact, reverse bend tests and tensile tests[106]. They also assessed the effects of differing composition on casting, cold- and hot-working, melting and solubility. The experiments were carried out using copper samples containing a range of trace element concentrations and most of the test objects were subjected to cold- and hot-working in order to explore the influence of trace element concentrations on prepared metal. All treatments, both cold- and hot-working, change the properties of copper. Cold-working involves handling the metal below the re-crystallisation temperature, e.g. by hammering or rolling, in order to shape it. Coghlan and Willows (1962) and Kienlin (2008, 262 ff.) reported that

[103] Archbutt and Prytherch (1937) have summarised in their book various studies that had investigated the effects of trace elements (e.g. arsenic, bismuth, iron, oxygen) on copper. In several volumes of the *Journal of Metals*, the results of the single studies have been published in-depth by researchers who had been involved in that project. Those primary studies will also be taken into account and referenced at this chapter.

[104] It is necessary to define two significant temperatures in a brief notice. Both are named in different studies differently. The softening temperature (also freezing temperature) is the characteristic temperature from which a substance (in our case a copper-alloy) starts to soften, but it is not fully melted, yet. In phase diagrams this temperature is marked by the solidus line. Below the solidus, a metal is solid and stabile.
The temperature above that a substance is completely melted is called solidification (melting) temperature; below it begins to solidify. This marks the liquidus line. Between liquidus and solidus line substances are generally soft and crossing several partly soluble and metastabile phases.
The re-crystallisation temperature is comparable hard to define. It marks a temperature when fluid crystallises by forming new crystals. This process is, however, time and temperature depending. So it is estimated that the re-crystallisation temperature is around "the 50 % softening temperature, taken from the hardness/temperature graph derived from isochronal annealing (annealing for a fixed period)" (Budd 1991a, 54 f.).

[105] Just on a side note, most of the here studied trace elements lowers the electrical conductivity, significantly in oxygen-free copper and especially arsenic and antimony (Archbutt and Prytherch 1937, 121 f.; Smart 1970b, 410 ff.).

[106] Overviews of the experiments to measure the effect of trace elements on the various properties are also listed in the glossary of this study. The detailed set-ups of the tests are described in the referenced literature of the following section.

this improves the hardness and resistance of native, pure copper by imposing pressure on the copper minerals, which are dislocated and agglomerated. However, beyond a certain point, continued cold-working can make an object brittle. In contrast, hot-worked metal has been heated above the re-crystallisation temperature. Annealing, however, is a form of heat treatment at temperatures around the respective re-crystallisation temperature. Since, during cold-working, dislocations within the metal lattice enlarge and cracks can appear, annealing softens the metal and counteracts that by re-crystallisation. After annealing, the metal is finally cold-hammered again. Usually, these procedures aim to improve hardness and strength of copper. Re-crystallisation, however, often degrades the strength of metals (Junk 2003, 23 ff.). Annealing is attested from the early metallurgical periods onwards. In central Europe, metallographic analyses have shown that cold-smithing after annealing, in order to improve hardness, was not commonly used before the first half of the 4th millennium BC (Kienlin and Pernicka 2009, 262 – 265).

As discussed above, the melting, softening and re-crystallisation temperatures affect the annealing, hot-working and casting characteristics of metals. A lower softening temperature allows the metal to be kept soft for longer before it fully crystallises and, thus, it can be cast better. On the other side, a higher melting temperature means that more heat needs to be generated to melt the copper. These properties have always been essential for metal workers.

Tensile tests provide information about the strength of a metal object by showing how ductile a material is, that is, the extent to which it can be deformed without fracture. This is important in order to conclude whether an object had been cold- or hot-worked. The properties of ductility and malleability are crucial when shaping artefacts by hammering thin sheet metals or drawing wires without cracking. Lechtman (1996, 497 f.) reports that good tensile strength is necessary for producing axes, hammers or other tools that have to cope with stress. Another essential test to clarify the influence of impurities on copper's properties is the notched-bar test. This examines the resistance of materials in reaction to a sudden, concentrated impact. The energy required to deform the sample can be measured and its resistivity determined. It also provides an ideal indication on hardness and brittleness (Archbutt and Prytherch 1937, 9; Junk 2003, 43). Brittleness, ductility and cold-working properties can also be investigated by reverse bend tests, which measure how much strain a sample can endure before it breaks. Hardness and strength are particularly significant properties for the use of a finished product, while ductility and elongation are more essential for the manufacturing processes, such as casting or forging. The behaviour of copper at different temperatures is crucial to the fabrication process. Critical temperatures such as those of the melting, freezing and re-crystallisation of the metal are very important for casting, annealing and hot-working.

Colour can be assumed to have been an important characteristic for prehistoric metalworkers and metal users, not only because of its aesthetic appeal but also because it would have aided the identification of different coppers. While the effect of trace element concentrations on colour has been determined for modern specimens, it is hard to assess the original colour of ancient copper artefacts, because many are heavily corroded, having laid in the ground for many years, and in other cases the artefacts have been tinted, during the conservation process, to the colour that conservators estimate was original[107].

In conclusion, it is clear that the mechanical properties of artefacts are influenced by the way they have been cold- or hot-worked after casting. As this study investigates whether particular sorts of copper were chosen because of their technical characteristics, the effects of working have to be regarded – but only secondarily. In contrast to Budd (1991a), Junk (2003) or Kinlien (2008), the primary topic of the present research is not an examination of the qualities of the artefacts. Rather, this survey wants to clarify whether there is chronological or regional patterning in the use of different types of copper, and if there is, whether it varies from archaeological group to archaeological group. If so, the question is whether these types were used because of the different properties of the material, or for some other reason.

The next section will summarise the results of a range of studies undertaken to assess the influence of several impurities in copper. The focus will be on those trace elements that both appear frequently in prehistoric copper objects[108] and have an impact on the properties of copper in a way that may have been relevant for Chalcolithic metal workers. First of all the characteristics of pure copper will be described.

7.2 Chemical, Physical and Mechanical Properties of Pure Copper

In general, the most significant property of metals is their capacity for plastic deformation. In other words, up to a certain point metals can be shaped by stretching, hammering or casting without perishing or breaking. Thus, copper, like every metal, can be formed into all kinds of shapes, from thin sheets to massive bars. This is determined largely by the structure of the metal's lattice (for details of which, see Henderson 2000, 208 – 210).

[107] This issue was recognised during a study trip to several museums across the area of research in summer 2007. Sometimes, if the drill holes from the trace element analyses were not filled, a change in colour was observable. Often, however, these holes have been filled and varnished, so the number of samples was not enough for a comprehensive examination.
[108] According to SAM and other projects (cf. Chapter 4.4).

According to Otto and Witter (1952, 32), pure copper consists of over 99% copper, originating either from deposits of native copper or from clean, fully oxidised ores. For prehistoric metal workers the important property of pure copper was that it is relatively easy to shape by cold hammering. The melting point of pure copper is around 1084°C (and its boiling point is 2595°C). Fluid copper can be cast into a desired shape before it resolidifies at a temperature of 1065°C (Smart 1970a, Table 17-2). Its re-crystallisation temperature is estimated at 250 – 300°C (Budd 1991a, 55). The hardness of pure copper is around 59 HV[109], but can be increased to 115 HV by cold-hammering (Charles 1980, 167). By cold-working (e.g. hammering) copper is compressed, thereby raising its tensile strength and reducing its elongation (Smart 1970a, 366).

7.3 Chemical, Physical and Mechanical Effects of Trace Elements

The effects of individual trace elements, present either as natural impurities or as deliberate additions, will now be reviewed. In general, if a trace element is present as an insoluble impurity in copper, it appears as an oxide and usually causes brittleness, whereas if a trace element is soluble, its negative effects are less marked (and in some cases it can have a positive effect, as detailed below). Iron, in particular, can be controlled and eliminated by smelting and melting techniques, because it precipitates as both oxides and sulphides (Hofman 1924, 16 – 26; Smart 1970b, 410). Usually, however, soluble trace elements improve copper's ability to be annealed and hot- and cold-worked, because they lower the softening temperature and thereby decrease the amount of heat necessary to soften copper (Archbutt and Prytherch 1937, 44). The behaviour of binary metals with varying compositions, and their critical temperatures, can best be presented by phase diagrams. As will become clear, the behaviour during the transition from the solid to liquid phase is crucial for the effect of impurities on copper.

7.3.1 Oxygen

Oxygen (O) has not been commonly analysed in archaeometallurgical studies, and will thus play a minor role in this study. Nevertheless, it appears regularly in natural copper and thus in prehistoric copper as well.
In solid copper uncombined oxygen is only soluble in very low concentrations. Usually oxygen is present as copper oxide (Fig. 6.1) (Dies 1967, 679 f.; Subramanian et al. (eds.) 1994, 288). In contact with oxygen, for instance during melting and casting, copper reacts and forms cuprite (Cu_2O) which is insoluble in solid copper; but cuprite can be reduced by hydrogen. In this case, oxygen leaves the copper melt as steam, thereby increasing the porosity of the copper. Insoluble Cu_2O causes marked brittleness (Budd 1991a, 48).

The casting properties are more affected by density than directly by oxygen percentage, but density increases in proportion to the amount of oxygen present. Thus, oxygen impairs the casting properties of copper (Archbutt

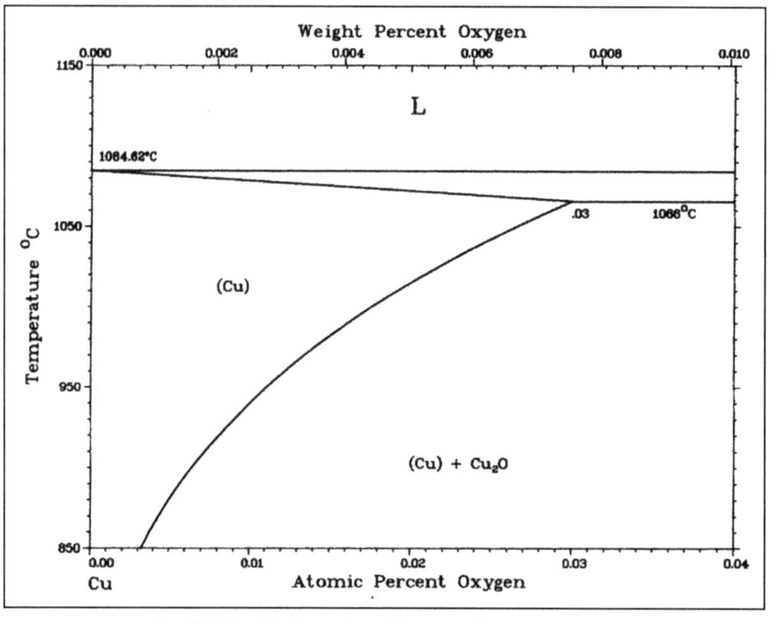

Fig. 7.1 Equilibrium phase-diagram of copper-oxygen (Subramanian et al. (eds.) 1994, 289, Fig. 3).

[109] The Vickers Hardness (HV) is a commonly used method in order to measure and name hardness of metals. It is equivalent to the often used Diamond Pyramid Hardness (DPH). Also the Brinell Hardness (HB) and the Vickers Hardness are both identical on the continuum, at least up to 300 HV, respectively HB (Charles 1980, 167; Budd 1991a, 53; Junk 2003, 40 – 42).

and Prytherch 1937, 28 f.). Both cold- and hot-shortness[110] increases relative to the amount of oxygen (Archbutt and Prytherch 1937, 25). In spite of these drawbacks (such as reduction of ductility and flexibility and higher brittleness), oxygen-bearing copper is considerably harder and has a higher tensile strength than pure copper (Dies 1967, 683). The concentration of oxygen, however, can be counteracted by either melting under reducing conditions or by adding deoxidising agents, such as iron, which reduces oxygen by forming iron oxide (Archbutt and Prytherch 1937, 28 ff.). As most of the trace elements have a higher affinity to oxygen than copper, oxygen tends to react more readily with these impurities during smelting and melting processes. Consequently, when arsenic, antimony or iron are present, for instance, it is rare for copper oxides to appear after solidification (Budd 1991a, 83 ff.).

In conclusion, oxygen is usually an undesirable impurity in copper and generally imparts undesirable physical properties, but for mechanical issues, when acting with other trace elements such as arsenic, it can counteract the negative effects of other trace elements (such as bismuth) and can improve the properties of copper. Overpoling[111], however, works against this and oxygen is required in very small amounts (Archbutt and Prytherch 1937, 34). A high oxygen content causes gas bubbles and porosity that significantly impair the properties of the metal. Thus, in the tests discussed below, the oxygen content has been kept low (chiefly < 0.02%)[112].

7.3.2 BISMUTH

Overall, bismuth (Bi)[113] reduces the strength of copper. Bismuth appears quite commonly in copper, but only in trace amounts. Even in such low amounts, however, bismuth affects copper and it is hard to eliminate through working (Archbutt and Prytherch 1937, 46).

Although bismuth is soluble in melted copper, it is insoluble in solid copper, as shown by the equilibrium phase-diagram (Fig. 7.2), forming a film at the grain boundaries when it cools. This film is reported as the reason of the extreme brittleness of copper/bismuth-metals (Smart 1970b, 415). If copper contains over 0.001% bismuth, the microstructure shows particles of impurities irrespective of whether the metal has been heated or not (Hanson and Ford 1927, 173; Dies 1967, 676 f.).

Hanson and Ford (1927, 172 f.) examined copper that had been alloyed with between 0.001 and 0.1% bismuth and tried not to exceed the amount of 0.015% oxygen. At a content of just 0.02%, bismuth already makes copper hot-short, and at 0.05%, cold-short; at 0.1% it raises brittleness so much as to render the copper hardly workable.

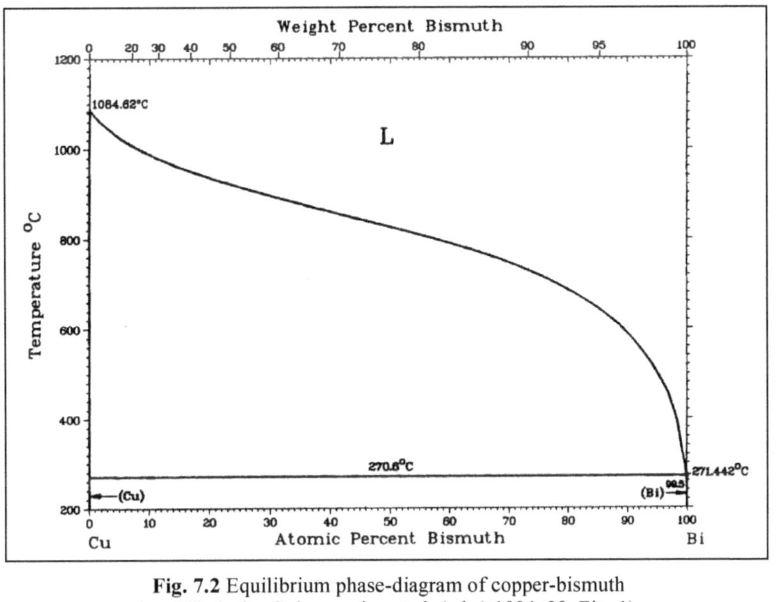

Fig. 7.2 Equilibrium phase-diagram of copper-bismuth (modified after Subramanian et al. (eds.) 1994, 99, Fig. 1).

[110] *Cold-shortness* implies that a cold material can crack, e.g. when it is hammered or rolled, and *hot-shortness* indicates that it can become fracture if it is worked under heated conditions.

[111] *Overpoling* means an elimination of all the oxygen by poling too long. This makes copper more brittle. Poling is a technique to reduce oxygen from copper by stirring the molten metal with a green-wood pole.

[112] In the following all percentage figures refer to the weight-by-weight percentage of the analysed object, unless otherwise noted.

[113] Object of the database do generally not exceed a concentration of 0.620% bismuth (two artefacts with 1.0% and 1.212% bismuth; both from Drosa [D]: ANo 34461 and 34486).

However, if antimony and arsenic are also present, the negative effects of bismuth are reduced, so that – particularly in interaction with antimony – a concentration up to 0.7% can be tolerated (Hofman 1924, 19) (With arsenic, however, such a high concentration cannot be tolerated as it would cause hot-shortness: Archbutt and Prytherch 1937, 99). Arsenic also counteracts the brittleness of bismuth-bearing copper. With 0.6% of arsenic copper can be rolled and is ductile, even if it contains up to 0.05% bismuth (Junk 2003, 34). Nevertheless, copper with up to 0.005% of bismuth on its own can be hot-rolled without cracking. Higher amounts of bismuth, however, will cause problems and breakage. In contrast, cold-rolled samples that had not been pre-heated could be worked un-problematically up to a bismuth concentration of 0.047% (Hanson and Ford 1927, 174 f.). At concentrations of 0.001% bismuth causes a reduction in ductility, as demonstrated though the bend test, and at levels over 0.002% the decline becomes even more rapid (Dies 1967, 676). Hardness also decreases with increasing bismuth concentration, but the presence of oxygen slightly counteracts this (Hanson and Ford 1927, Tab. II). Bismuth has almost no effect on the tensile strength of copper, especially when oxygen is also present (ibid., 175 f.). Fatigue tests have shown that, at up to 0.016%, the amount of bismuth does not change the fatigue range (ibid., 177).

To sum up, the trace element bismuth is regarded as the most damaging impurity. Therefore, from a modern, technological perspective, Dies recommended keeping bismuth below a level of 0.001%, or preferably, 0.0005% (Dies 1967, 678). However, if arsenic and/or antimony are also present, to some extent, this neutralises the negative effects of bismuth and bismuth-containing copper is not as brittle. Looking at the Chalcolithic trace element analyses, there is no analysis that indicates that bismuth is either the only or the predominant impurity in an object. Accordingly, the negative effect of bismuth is apparently neutralised by other trace elements.

7.3.3 IRON

Craddock and Meeks (1987, 187) argued that iron (Fe)[114] has usually been neglected in archaeometallurgical studies[115], even though it has been recorded as a regular contaminant present in prehistoric artefacts. Pernicka argued correctly, however, that it is not worth while studying iron, because it cannot be used to deduce what ore had been used from the finished metal. He suggested that it could only indicate whether the sample is refined or raw copper (Pernicka 1989, 634). It is nevertheless useful to consider iron with regard to its role during smelting and, as we will see, to its influence on the properties of copper.

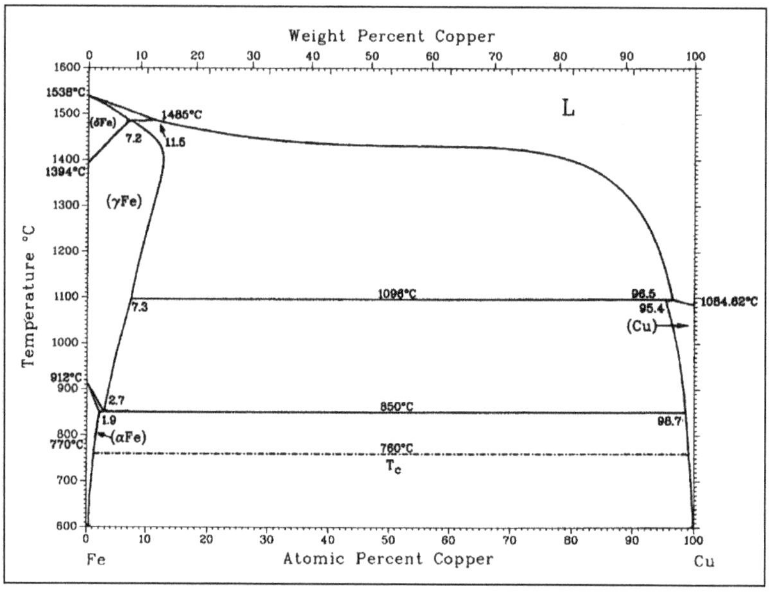

Fig. 7.3 Equilibrium phase-diagram of iron-copper (modified after Subramanian et al. (eds.) 1994, 167, Fig. 1).

[114] Objects of the database do usually not exceed a concentration of 2.0% iron. Only eleven analyses show more than 2.0% of iron (except one artefact; a bead from Gaienhofen 'Hornstaad-Hörnle I' [D] contains 17.6% iron, but only 24.7% copper: ANo 100032).

[115] Although iron could not be measured regularly by SAM (Krause and Pernicka 1996, 279), it may be worthwhile to take this element into account due to the possible interpretations of smelting processes and subsequent analysing projects that are able to measure iron concentrations. Nevertheless, it must be kept in mind for the interpretation afterwards, that iron might cause problems when evaluating the statistics. Analysing the statistics by excluding iron has shown that the general result is not remarkably affected.

As discussed above, iron is very common in copper ores in relatively high amounts, but smelting reduces this and transfers the iron from the ore to the slag, allowing iron-free copper to be produced. Therefore, the presence of iron in a copper object may indicate the smelting technique used: for example in crucible smelting the reducing conditions are not ideal, and so unreduced iron transfers more easily to the copper phase (Craddock 1995, 137). It is also suggested that iron had been used as flux (Henderson 2000, 223); if this was the case, then the impurities may appear in smelted copper as well, at least in trace amounts. Although Craddock argues that the concentration of iron directly reflects the level of copper smelting technology, it must be kept in mind that iron is not only a result of smelting procedure. Iron can appear in smaller traces in smelted copper, because its electrical negativity is similar to that of copper (Chapter 6, Fig. 6.3). Consequently, one cannot automatically infer the smelting technology, or even the successful operation of a smelting, on the basis of iron concentration.

In 83% of the objects in our database the iron contamination is no higher than 0.05%. Iron is soluble in solid copper in small quantities (< 0.2%), but with rising temperatures the solubility increases and in molten copper, at c. 1100°C, over 5% of iron is soluble (Fig. 7.3) (Hanson and Ford 1924, 339; Subramanian et al. [eds.] 1994, 169). Thus, especially during long-lasting smelting operations, some iron may accumulate into copper (Craddock and Meeks 1987, 192). In the presence of oxygen, iron reacts and forms insoluble iron oxides, which precipitate and can be separated from molten metal (cf. smelting procedure) (Smart 1970b, 415). If, however, nickel is present the solubility of iron increases proportional to the concentration of nickel (Dies 1967, 806).

Traces of iron that appear in copper harden the metal (Hanson and Ford 1924, 343). Tensile strength rises with the increase of iron for annealed copper, but not for non-annealed copper (Hanson and Ford 1924, 343 ff.). According to Hanson and Ford (1924, 342 f.) hot- and cold-rolling tests using various iron concentrations were successful, with no material damage. The presence of iron slightly enhances the fatigue range. Furthermore, iron removes oxygen from copper (Archbutt and Prytherch 1937, 38). Also the colour changes from red to gray in proportion to the increasing concentration of iron (Hofman 1924, 20).

Finally, it can be stated that the presence of a small amount of iron does not greatly affect the properties of copper, although iron oxides precipitate and impair the metal. Various experiments, however, have demonstrated that even with simple techniques, such as re-melting or scraping precipitated iron oxides off with a stick, iron can easily be removed from copper (Craddock 1995, 138 f.). Thus, the iron content is quite simple to manipulate and, therefore, an artefact containing a low concentration of iron does not necessarily mean that this object was made of an ore which was (nearly) free of iron.

7.3.4 TIN

The presence of tin (Sn)[116] in copper needs to be treated as a special case, since copper containing around 10% tin, labelled as (tin-)bronze, is often considered to be the first intentionally produced alloy. This, however, does not take into account that other elements, such as arsenic and antimony, may also have been deliberately added to copper. The question of whether a threshold marking the difference between 'natural' and 'intentionally alloyed' tin-bronze can be defined is often discussed in the archaeometallurgical literature (cf. Otto and Witter 1952; SAM, Spindler 1971, Schickler 1981, Pernicka 1995). Some suggest that a concentration of more than 0.1% of tin indicates intentional alloying (Fridman et al. 1966, 1506), while others have argued that even several percent of tin can theoretically appear in copper without being intentionally alloyed, since some copper ores can contain appreciable amounts of tin (Rovira and Montero 2003). Otto and Witter (1952, 46) and the SAM-group (SAM I, 57 f.) acted on the assumption that a tin content exceeding 1% had been caused by intentional alloying, while Liversage (1994, 77 ff.) set the threshold for deliberately alloyed tin-bronze at 4%. As this issue has not yet been resolved, in the current study objects containing over 4.5% tin are excluded in order to ensure that every suggested critical value has been accommodated (and, furthermore, because there is a chronological difference in the use of copper that contains more than 4% tin: cf. Chapter 2.2). Determining whether intentional alloying has taken place – whether it be of tin, or of other trace elements – is very difficult and the author knows of no method to distinguish whether a prehistoric artefact has been alloyed, is a mixture of several ores, or whether its trace element composition reflects the compounds in the parent ore. In the present case, it can only be assumed that the presence of even c. 1% tin in copper results from deliberate alloying, because in central Europe tin-containing copper ores are very rare (Rovira and Montero 2003). However, this Chalcolithic low tin-copper, with less than around 4% of tin, is neither metallurgically nor culturally comparable to the later, classic tin-bronze that contains at least 8% tin. This will be demonstrated below. For present purposes, tin should be treated in the same way as other trace elements, since this study deals with copper metallurgy and chiefly focuses on the effects of trace elements on copper. Nevertheless, throughout this study we shall be returning to the issue of tin-bronze.

Tin is up to c. 22% soluble in solid copper. At higher temperatures and/or concentrations the copper/tin-phases become harder to discern. Since the focus of the current study is on low tin-bearing copper, however, the issues surrounding high tin contents need not concern us (Dies 1967 504 ff.). Even small concentrations of tin lower the melting temperature of copper[117] (Fig. 7.4).

[116] The selection of data is limited to objects with a tin-concentration lower than 4.5% (see Chapter 5.1.2). Nevertheless, just two objects contain more than 4.0% of tin; both contain 4.1 % tin (ANo 319 and ANo 22149).

[117] Copper containing 8% tin has a melting temperature of c. 1000°C and with 13% tin it is around 850°C (Otto and Witter 1952, 41).

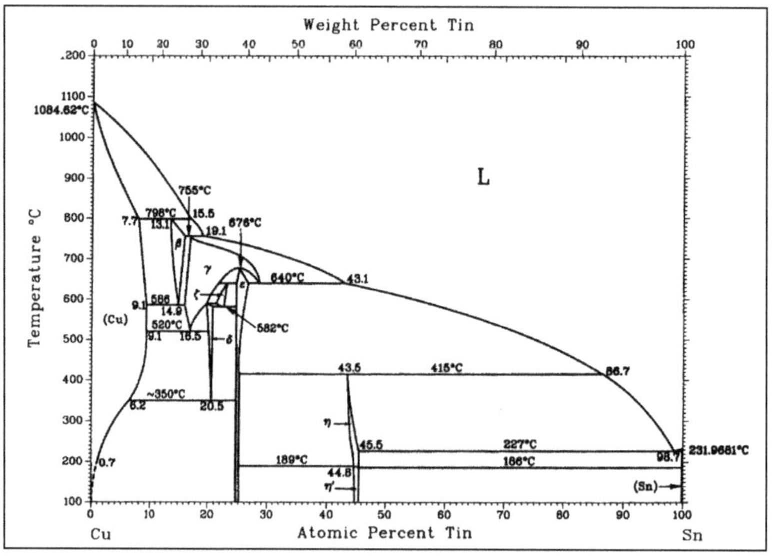

Fig. 7.4 Equilibrium phase-diagram of copper-tin (Subramanian et al. (eds.) 1994, 412, Fig. 1).

Tin can form grains with copper, but the use of especially slow cooling techniques such as sand casting counteracts this and produces solid solutions, especially where the tin content is below c. 6% (cf. Dies 1967, 515). Within the solid phase the elasticity rapidly declines as the tin content increases (Dies 1967, 522). At levels of between 0% and 10% of tin, the hardness and tensile strength of the metal more or less double, but when the tin concentration exceeds 12% the copper starts to become brittle (Lechtman 1996, 488). Malleability increases with increasing tin content, reaching a maximum value at 4 – 6% and decreasing at higher concentrations (Dies 1967, 526). Heat treatment of copper/tin-metals with less than around 5% tin does not affect the metallurgical properties (Dies 1967, 549). Although Lechtman (1996, 501) claimed that copper containing more than 8% tin cannot be hot-worked without brittleness, Wang and Ottaway (2004, 59 – 71) have demonstrated experimentally that copper containing up to 10% tin can indeed be worked after being annealed.

A significant property associated with the presence of tin in copper is its influence on the metal's colour, since the addition of tin changes the colour from the typical copper-orange into a brownish-red at a concentration of c. 3%; above 6% it changes to golden-violet-reddish, and at 10% it turns golden (Dies 1967, 521).

To summarise, it is clear that tin improves the general properties of copper – the hardness and tensile strength improve with increasing tin content and the melting point is lowered, even at a low tin concentration. This improves the casting qualities of copper/tin-metals. Even within a range of c. 1% to 4%, tin improves all these characteristics, while at the classical tin-bronze concentration of 10% significant improvements are observable. A tin content of less than 1% is regarded as an accidental inclusion and does not noticeably affect the properties of copper (Rovira and Montero 2003, 15).

7.3.5 LEAD

In contrast to tin, lead has seldom been discussed as an impurity in Chalcolithic copper, but as we will see below, awareness of its importance has grown during recent years[118]. Lead (Pb)[119] is usually insoluble in solid copper, crystallising and precipitating when the composition contains over 0.04% lead and is not heated to over c. 800°C (Smart 1970b, 415). This indicates that lead can have a negative effect on copper. In a similar way to bismuth, these crystallised precipitations can produce brittleness (Archbutt and Pyrtherch 1937, 51). The presence of oxygen and arsenic in copper/lead-metals generally slightly neutralises the negative effects of lead (Hofman 1924, 19; Archbutt and Prytherch 1937, 52 ff.; Smart 1970b, 416).

Forging tests have demonstrated that, in oxygen-containing copper, cold- and hot-working is possible up to a concentration of c. 0.2% lead; at higher concentrations, the lead makes the metal brittle. A content of over 0.3% lead may cause brittleness during hammering (Archbutt and Prytherch 1937, 52). Nevertheless, if the concentration of lead is below 0.1% it lends the metal higher ductility and malleability (Archbutt and Prytherch 1937, 56). The presence of lead in annealed copper (both with and without oxygen) indicates better cold-working properties. This has been shown by reverse bend tests. From melted copper lead and oxygen react and the lead oxides are volatile. An intentional addition of lead to the melt might be useful for refining copper (Archbutt and Prytherch 1937, 56).

118 There are two reasons why lead deserves to be considered here: first, it has metallurgical effects on copper, and second, earlier studies have suggested that lead may be a characteristic a trace element for Bell Beaker copper finds (cf. Matuschik 2004; Metzinger-Schmitz 2004). Consequently, this issue will be explicitly discussed in Chapter 11 when the specific problem of Bell Beaker metal is discussed.

119 Objects in the database do generally not exceed a concentration of 3.5% lead; those that do exceed this concentration are ANo 34461(6.5% lead), ANo 100025 (9.320%), ANo 100022 (9.9%), ANo 100125 (11.4%) and ANo 100099 (16.2%).

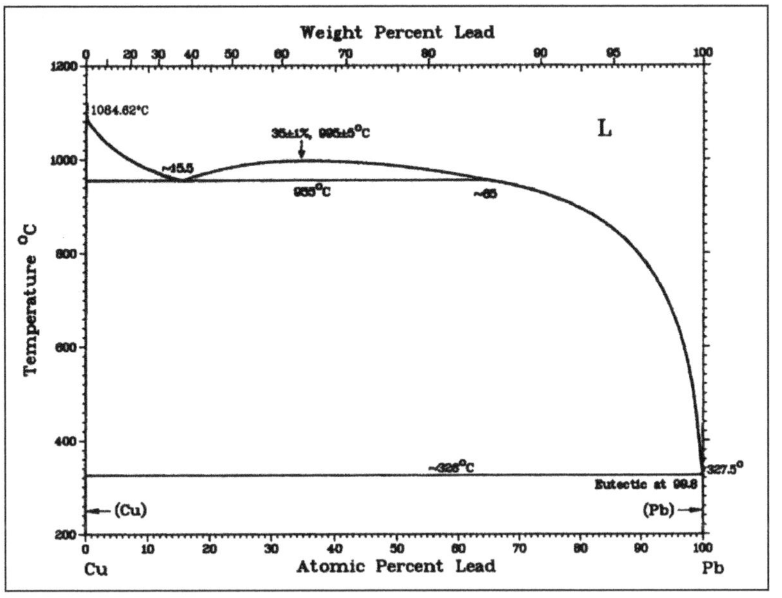

Fig. 7.5 Equilibrium phase-diagram of copper-lead (modified after Subramanian et al. (eds.) 1994, 301, Fig. 1).

Despite its mostly negative characteristics lead has the capability to reduce Cu_2O, and this can be conductive to smelting (Hofman 1924, 19). Even at low concentrations lead lowers the solidification temperature of copper/lead-metals (Dies 1967, 610) (Fig. 7.5), and this considerably enhances their working properties. Thus, one may assume that lead might have been deliberately added to copper in order to improve reduction and softening of copper. Lead can easily be melted out of galena (lead sulphide) (Rapp Jr. 1988, 24). Staniaszek and Northover (1982, 264), who have studied leaded tin-bronze alloys, realised that the amount of lead decreases by 30 – 40% when the metal is heated at the pouring temperature of 1200°C. Thus, the lead content within smelted and cast copper does not reflect the metal's original composition.

Overall, if lead appears as only an impurity of copper, it impairs its strength, with the effect increasing with concentration. Conversely, lead operates as a deoxidiser and in combination with arsenic the negative character is counteracted.

7.3.6 NICKEL

Because copper and nickel (Ni)[120] share a similar atomic radius, electronegativity and lattice constant[121], they also have very similar physical and chemical characteristics (Archbutt and Prytherch 1937, 78; Dies 1967, 783). Nickel shows good solubility in copper, in both solid and liquid phase, and so the two elements act in the same way during melting and other chemical processes. This changes only when the nickel content exceeds c. 20%. Nickel slightly raises the melting temperature of pure copper (Fig. 7.6). In the presence of oxygen, nickel tends to oxidise to NiO which is less soluble in solid copper than cuprite (Ruhrmann 1925, 324). According to Smart, at higher concentrations nickel forms oxides; however, these barely affect copper (Smart 1970b, 415). Bismuth (at > 0.002%), lead (> 0.05%) and sulphur have negative effects on the mechanical properties of copper/nickel-metals (Dies 1967, 792).

In minor amounts (up to 0.3%) nickel has almost no effect on the properties of copper compared with other trace elements. Between 0.1% and 0.9% nickel the hardenss increases only gradually (Ruhrmann 1925, 345). But from c. 2 % onwards, nickel bears more positively on hardness and tensile strength (although an increase in tensile strength is already discernible at a level of 0.1% of nickel: Hofman 1924, 20; Dies 1967, 795). The fatigue limit of various unannealed copper/nickel-metals stays, with some fluctuation, more or less constant; the same applies to malleability (Ruhrmann 1925, 345 ff.). In a smaller range of nickel this impact is enhanced by the additional occurrence of antimony, while in 2 – 3% nickel-copper antimony might interfere (Hofman 1924, 20 f.). Nickel increases the solidification point of copper and, whereas 0.05% nickel scarcely impairs the softening temperature, higher concentrations raise it significantly (Archbutt and Prytherch 1937, 78; Dies 1967, 794). Copper containing nickel is ductile at room temperature (Dies 1967, 795).

Finally, nickel at high concentrations affects the colour. Around 5% of nickel makes the metal much brighter and copper containing 15% is white (Dies 1967, 787). It is reported that a concentration of around 2 % nickel can be easy to recognise as even that amount turns the colour of copper silver-white (Moosleitner and Moesta 1988, 62 f.).

[120] Objects in the database do not exceed a concentration of 6.25% nickel (ANo 68057).
[121] The lattice constant is the distance measured between atoms of a unit cell of a crystal lattice.

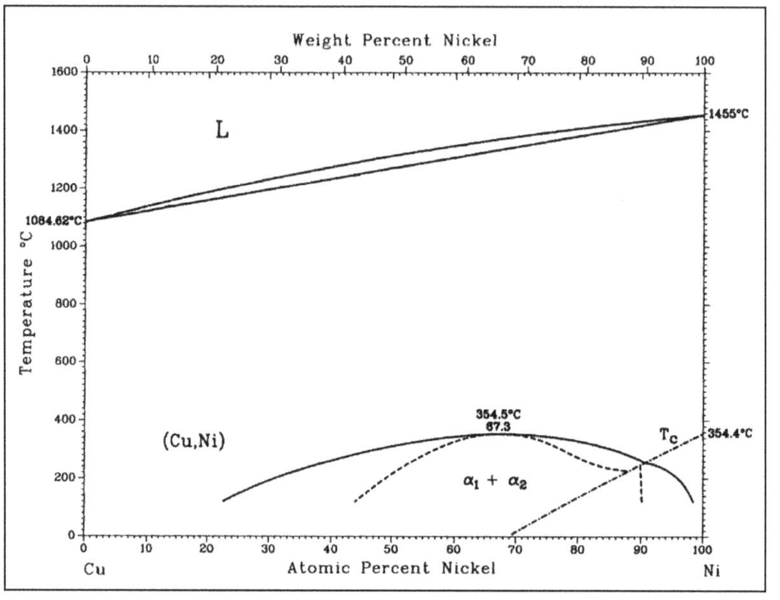

Fig. 7.6 Equilibrium phase-diagram of copper-nickel (modified after Subramanian et al. (eds.) 1994, 276, Fig. 1).

In conclusion, then, as nickel has similar chemical and physical properties to copper and reacts similarly to copper, it does not affect copper very much. It is important, however, that nickel goes easily into the workable copper, even if it is oxidised. The loss of nickel during smelting and heating should be minimal (Tylecote et al. 1977, 329).

7.3.7 SILVER

Like nickel, silver (Ag)[122] has a minor effect on the properties of copper. Both elements have nearly equal atomic radii and, thus, similar chemical, physical and mechanical properties (Dies 1967, 193).

The presence of low amounts of silver in copper raises the re-crystallisation point, from a concentration of

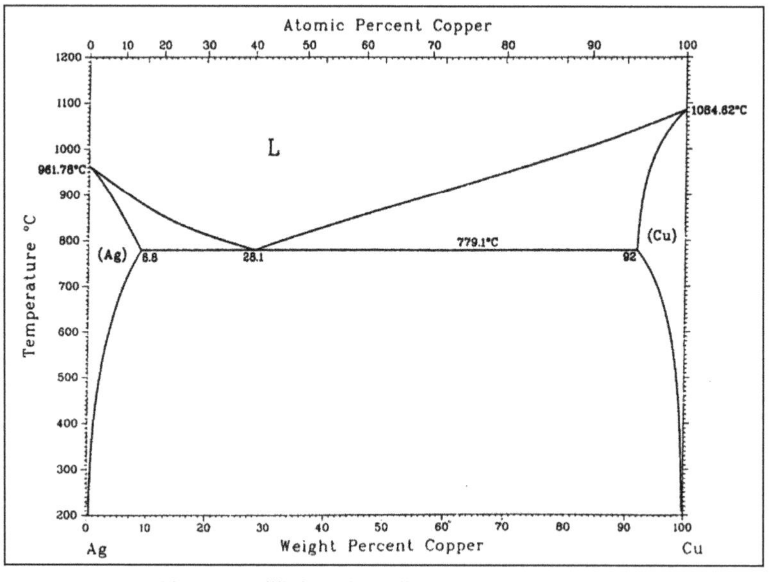

Fig. 7.7 Equilibrium phase-diagram of silver-copper (Subramanian et al. (eds.) 1994, 6, Fig. 1).

[122] Objects of the database do generally not exceed a concentration of 2.7% silver; there is one with 4.5% (ANo 62131), one with 7.3% (ANo 100023) and one with 8.7% (ANo 100026).

0.005% silver upwards. Accordingly, the time necessary to soften copper/silver-metals increases comparative to the quantity of silver present (Dies 1967, 194 f.). Silver is soluble in solid copper and, if the object has been heated to above 200°C, it slightly affects the hardness of copper. It has been shown that at low concentrations (max. 1%) silver raises the softening and crystallisation temperature (Fig. 7.7). If copper contains silver and arsenic, its elastic properties are improved (Archbutt and Prytherch 1937, 83 f.). Within a range of 0.01 – 0.1% silver, however, the melting point does not change compared with pure copper (Dies 1967, 197, Tab. 1).

Copper/silver-metals containing more than 0.5% silver become harder with increasing silver concentrations (Dies 1967, 200 ff.). Fatigue tests have shown that copper containing small amounts of silver is more resistant than pure copper (Dies 1967, 198). Silver, alone and combined with arsenic, improves the elasticity of heat-treated copper (Archbutt and Prytherch 1937, 118).

In general, silver does not have a significant influence on the properties of copper. Comparable with nickel, silver should counteract the negative effects of some impurities. It is also hard to remove by smelting and refining because of its chemical and physical similarity to copper. In addition to a gradual improvement of hardness, the most important effect is the influence on the re-crystallisation point that extends the period during which cold-working is possible.

7.3.8 ANTIMONY

Only in very low concentrations antimony (Sb)[123] forms solid solutions with copper. At 645°C a peak of 5.8 % antimony is in solution; however this drops off rapidly at lower temperatures (Junk 2003, 26). The equilibrium phase diagram shows a fast decline of the softening and solidification temperature with increasing antimony content (Fig. 7.8).

Antimony is volatile especially under oxidising conditions, but the presence of nickel reduces the volatility of antimony (Stahl 1918, 396 f.). In smaller amounts it can vaporise as oxide or be transferred into the slag. As Stahl (1918, 395) reported, a reduction of antimony oxides may occur during the whole metallurgical process. In particular, if an artefact is hot-worked under oxidising conditions (on air) the loss of antimony can be significant[124]. Thus, if for any reason metal workers deliberately want high concentrations of antimony appearing in workable copper, they have to be able to keep antimony within the metal and work it under extreme reducing conditions.

Casting of copper which contains antimony is quite difficult, because it can form crystals within the copper structure which can only be homogenised by annealing. These crystals can cause brittleness (Dies 1967, 672). The biggest problem concerning the properties of copper/antimony-metals is the crystallisation, even with a low concentration of antimony. Even at a concentration of below 0.01% antimony raises the re-crystallisation

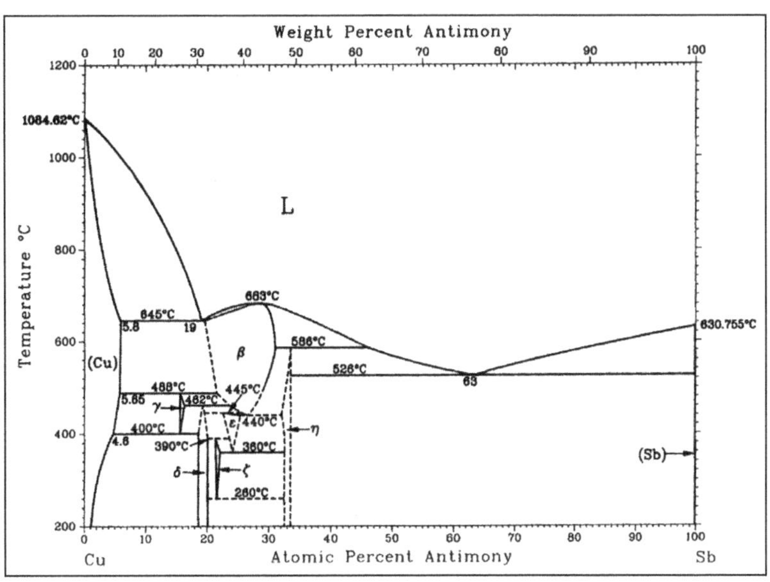

Fig. 7.8 Equilibrium phase-diagram of copper-antimony (modified after Subramanian et al. (eds.) 1994, 372, Fig. 1).

[123] Objects of the database do not contain more than 10% antimony.
[124] The loss of antimony during heating under oxidising conditions has been shown on one sample of antimony/copper-metal by McKerrell and Tylecote. They reported a remarkable loss of antimony after several melting and hot-working processes. Antimony could be reduced from 4.9% to 2.9% by melting and hot working the sample three times. After further operation only 0.9% antimony was left in the copper piece (McKerrell and Tylecote 1972, 211 & Tab. 2).

temperature quickly; consequently copper-bearing antimony can be treated and shaped at higher temperatures without re-crystallising (Dies 1967, 673). It has to be acknowledged that, even at a small concentration, antimony has negative effects on the properties of copper. Nevertheless, antimony both counteracts the even more negative effects of bismuth up to a content of c. 0.2% (Stahl 1918, 395) and in interactions with oxygen and arsenic it even improves the properties of copper (Archbutt and Prytherch 1937, 73).

The most notable effect of antimony is on the tensile strength. This also improves hot and cold-rolled copper and it does not remarkably affect ductility, whereas small amounts of antimony influence the tensile strength significantly. Annealed copper/antimony-metals show an increase of tensile strength with up to 0.2% antimony; these small amounts influence tensile strength even more than the same percentage of arsenic (Archbutt and Prytherch 1931, 272 ff.). Archbutt and Prytherch (1931, 275) supposed that from 0.47% onwards antimony negatively affects the properties of copper. It has already been noted that the hardness of antimony-bearing copper is increased – even more than in arsenical copper. The stated reason for this – compared with arsenic – is that antimony has a larger atomic radius, which "cause[s] higher strain in the lattice, so that the material becomes harder" (Junk 2003, 28).

In order not to impair copper excessively, Hofman (1924, 22 f.) recommended that, if antimony is the only impurity, it should be present at a concentration no higher than 0.05%, because 0.1% of antimony makes copper break under stress and causes edge-cracking during working. Archbutt and Prytherch (1931, 265 – 282) examined various copper samples containing not more than 0.85% of antimony[125]. In contrast to Hofman, they found that copper containing up to 0.47% antimony can be hot-rolled, but more antimony causes problems under standardised conditions and even cold-rolling works perfectly up to 0.85% (Archbutt and Prytherch 1931, 270.). Thus the cold-working properties of antimony-bearing copper are generally better than the hot-working properties.

Antimony and oxygen increase the ductility and malleability of copper and even a small amount of oxygen reduces cold- and hot-shortness of copper when more than around 0.4% antimony is present. Oxygen diminishes elongation of copper/antimony-metals by not affecting the tensile strength (Archbutt and Prytherch 1937, 77).

Overall, in spite of its effect on tensile strength, Smart (1970b, 415) suggested that antimony, if it is the only impurity within copper, should not exceed 0.4% in oxygen-free copper and 0.5% in copper containing oxygen because, otherwise, antimony can precipitate as oxide. Dies (1967, 676) even counted antimony amongst the trace elements which have a negative effect on the properties of copper, although Archbutt and Prytherch (1937, 82) spoke in terms of antimony in combination with arsenic and nickel, of the "beneficial nature of their effects".

7.3.9 ARSENIC

Arsenic (As)[126] is the most frequently examined trace element in ancient copper. Various archaeometallurgical studies have comprehensibly proved that arsenic-containing copper, often called arsenic-bronze, was commonly used in central Europe – chiefly during the first half 4th millennium BC (e.g. Sangmeister 1971; Ottaway 1978, Klassen et al. 1998; Matuschik 1998)[127]. Besides, several authors, such as Charles (1967), Junk (2003), Lechtman (1996) and especially Budd (1991a), to name but a few, have not only discussed the chronological and chorological distribution of arsenic-bearing copper but they have also contemplated the metallurgical influence of arsenic on copper. Prehistoric metallurgists may have recognised arsenic-containing copper by its garlic-like odour (Otto and Witter 1952, 46). This chapter concentrates on the chemical, physical and, chiefly, mechanical effects of arsenic on copper which, at concentrations up to 8% are similar to those of tin on copper (Charles 1967, 24; Lechtman 1996).

In one of the earliest works, Hanson and Marryat (1927) examined systematically, and from a modern, metallurgical perspective, the change of mechanical and physical properties of copper/arsenic-metals by studying various samples containing between 0.053% and 1.04% of arsenic and even concerning the consequence of oxygen. More than 70 years later, during his fundamental PhD-research, Budd (1991a, 59) investigated seven copper samples bearing c. 0.5 – 12% arsenic from an archaeometallurgical perspective.

Similar to antimony-bearing copper, arsenical copper differs from pure copper in colour. Around 4% of arsenic makes it whiter, 8% more silvery (Budd and Ottaway 1991, 134; Moesta 2004, 271). Arsenic is soluble in solid copper up to a concentration of about 8% and solubility does not noticeably vary with temperature rise (Fig. 7.9). The addition of arsenic lowers the softening and solidification temperature remarkably, and density also declines (Dies 1967, 664). This means that arsenical copper can be heated and worked at a wider range of temperatures. Budd (1991b), however, has shown by experiments that the relationship between arsenical content, the re-crystallisation temperature and the alloy structure is quite complicated and is not yet fully explored.

[125] Whilst Archbutt and Prytherch (1931) examined chiefly the effects of antimony on copper that was rolled after casting, it is essential that their results may not count for antimony-bearing copper that was not worked (Junk 2003, 28). Rolling compresses the material and, thereby, the properties are changed.

[126] Objects of the database do not exceed a concentration of 5.159% arsenic.

[127] See also Chapter 4.1 of this study.

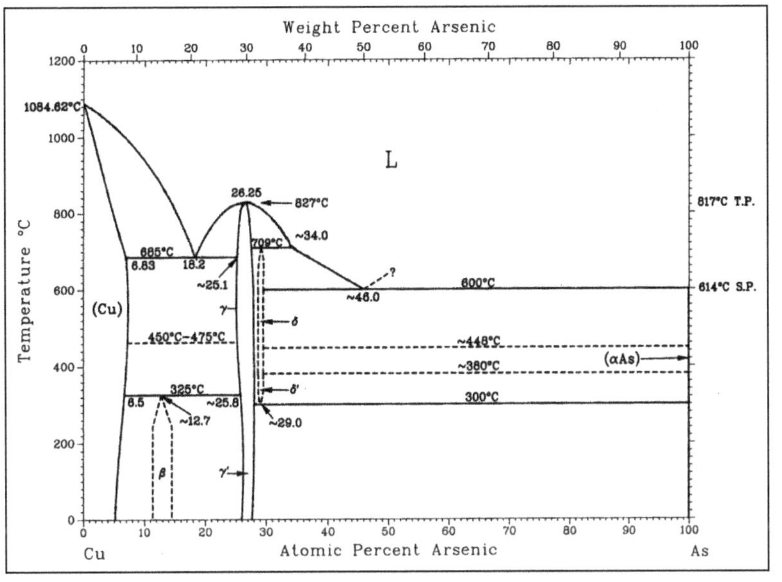

Fig. 7.9 Equilibrium phase-diagram of copper-arsenic (modified after Subramanian et al. (eds.) 1994, 43, Fig. 1).

Hanson and Marryat (1927, 125) mentioned that it would be very difficult to produce sound casts of oxygen-free arsenic-containing copper; however, certain oxygen content eliminates those problems and forms better castings. On the other hand, arsenic supports smelting process, because it helps reducing oxygen from copper ores. Furthermore, as arsenical oxides it separates from liquid copper and, thus, impairing oxides can be skimmed (Charles 1967, 21).

At arsenic concentrations below 0.5%, the properties of copper are not seriously affected, and even copper with 0.8% arsenic can be drawn into thin wires. None of the other studied trace elements improves malleability in the way that arsenic does, and it continues to do so up to a level of around 1% (Hanson and Marryat 1927, 126).

When the concentration exceeds 1% hot-shortness is noticed; and if oxygen is also present, cold-shortness can

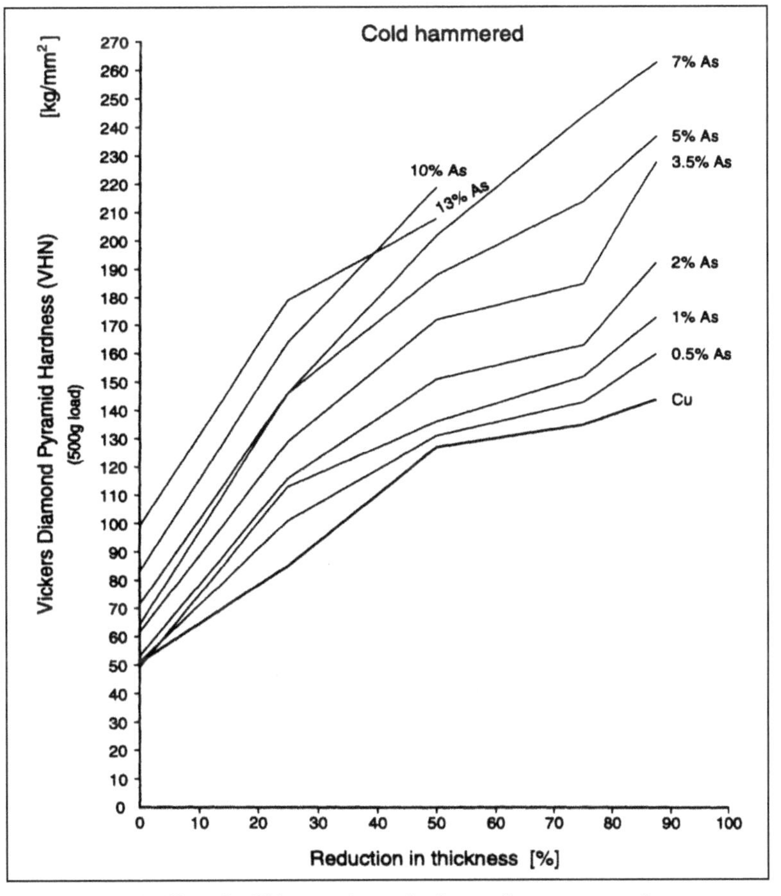

Fig. 7.10 Effect of cold-hammering on hardness of pure copper and copper containing various concentrations of arsenic (Lechtman 1996, Fig. 18).

occur at an arsenic level as low as 0.4%. Copper with between 1% and 1.37% arsenic can still be rolled satisfactorily, although copper with less than 1% can crack during annealing under reducing conditions (Hofman 1924, 21 f.). Experiments by Hanson and Marryat (1927, 142) have shown that arsenical copper with up to around 7% arsenic can be rolled successfully both cold and hot – even into thin sheet, despite the emergence of small edge-notches in some samples. Several other authors have independently confirmed this (Charles 1967; Budd and Ottaway 1991; Lechtman 1996) (Fig. 7.10).

By investigating tensile strength and elongation Hanson and Marryat (1927, 127) assumed that these characteristics depend more on the casting procedure than on the amount of arsenic present. This is not surprising, as the casting techniques and cooling rate generally influence the properties of the object (cf. Kienlin 2008, 255 – 261). Nevertheless, at arsenic concentrations of between 0.1% and 0.86% the tensile strength and density grows; elongation increases up to a concentration of 0.36% and, beyond that level, it decreases (Hanson and Marryat 1927, 127 ff.). The results of tensile tests on rolled metal, however, demonstrate a slight, but constant increase in strength, in particular for cold-rolled metal. In contrast, elongation by and large declines (Hanson and Marryat 1927, 131 ff.). In hardness, cast copper with up to 2% arsenic does not differ greatly from pure copper, but when the arsenic level is 3.5% and over, hardness increases remarkably, so that by a level of 7% it has almost doubled (Lechtman 1996, 488). Several authors have also demonstrated that increasing levels of arsenic have a hardening effect on copper and this can be increased yet further by cold working that reduces the thickness (Hanson and Marryat 1927, 134 f.; Northover 1989; Ottaway and Budd 1991).

Arsenic-containing copper has an improved strength and is highly ductile even at an arsenic concentration of 7%. This explains the good cold- and hot-working properties and the flexibility of arsenic-bearing copper (Lechtman 1996, 499). The tensile strength of cold-worked copper increases by an extraordinary 10 – 30% when the arsenic level is between 0.5% and 1% (Lechtman 1996, 509).

Regarding the fatigue range of arsenic-bearing copper, there is an increase at levels below 0.36%, then a decrease, and then an increase once more with higher concentrations (Hanson and Marryat 1927, 135). As the arsenic concentration increases, the crystal size shrinks in cast copper and is much smaller than for oxygen- or iron-bearing copper. This, consequently, might explain why arsenic does not seem to cause extensive brittleness and why it counteracts the effects of oxygen. However, arsenic does not seem to have a large impact on the grain size of annealed copper (Hanson and Marryat 1927, 139 f.).

The interaction of oxygen in arsenical copper will now be described briefly. As mentioned above, oxygen improves the casting properties of arsenic-containing copper, while arsenic eliminates the negative effects of oxygen (Hanson and Marryat 1927, 162). Whereas hot-working of copper/arsenic-metals is not negatively affected by oxygen, it has been suggested that, for cold-working, metal should contain ten times more arsenic than oxygen. The mechanical properties, in general, and the softening temperature in particular, are not influenced by oxygen (Hanson and Marryat 1927, 162 f.).

To summarise the effects of arsenic on copper, it can be stated that a low concentration of arsenic (below 0.1% As) does not greatly affect the mechanical properties of copper. But a higher content significantly increases hardness, strength and malleability up to around 7%; beyond that point, arsenic starts to cause brittleness. Arsenic/copper-metals have the best working properties and can be cast or hammered into different shapes, even into thin foil.

McKerrell and Tylecote (1972, 210 ff.) proved by several experiments that arsenic is fairly volatile if arsenical copper is hot-treated (melted or annealed) in an oxidising environment, whereas if it is worked under reducing conditions the loss of arsenic is minimised. This was also confirmed by Budd and Ottaway (1991), who found that they could melt arsenical copper under reducing conditions without any loss of arsenic, but during casting from the higher concentrated copper/arsenic-metal, arsenic oxidises to arsenic trioxide (As_2O_3), which is very toxic, and disappears. They, however, consider that the loss is not significant (Budd and Ottaway 1991, 133). By contrast, Lechtman noted that it is very hard to remove all arsenic from arsenic sulphides by oxidation (Lechtman 1996, 481).

Nevertheless, the concentration of arsenic within copper objects relies on the metallurgical procedures used (smelting, roasting etc), and volatile arsenic trioxide gas is highly poisonous. By adding oxygen to the smelting process, for example through blowpipes, prehistoric metal workers would have been able to control the atmosphere within the hearth – making it an oxidising or reducing atmosphere, in other words – and, thereby, to control the arsenic content. This control, as we have seen, meant that the arsenic content could be both raised and lowered deliberately. What is true of arsenic may also be true of other trace elements.

Finally, it can be put on record that arsenic has beneficial effects on the casting and working properties of copper – as long as the amount of arsenic does not exceed around 8%. Arsenic hardens copper; it raises its strength and lowers the softening temperature by not impairing its ductility and workability. Charles (1967, 24 ff.) even reports that the general properties of arsenical copper are equal to, or even better than, those of tin-bronze. The hardness of copper increases with increasing amounts of either tin or arsenic, up to the solid solubility limit of c. 8% in the case of arsenical copper, and beyond that in the case of copper containing tin (Fig. 7.11). Despite this, over the ages tin has been preferred, probably because arsenic is highly volatile and extremely toxic. Nonetheless, arsenical copper was produced over several centuries;

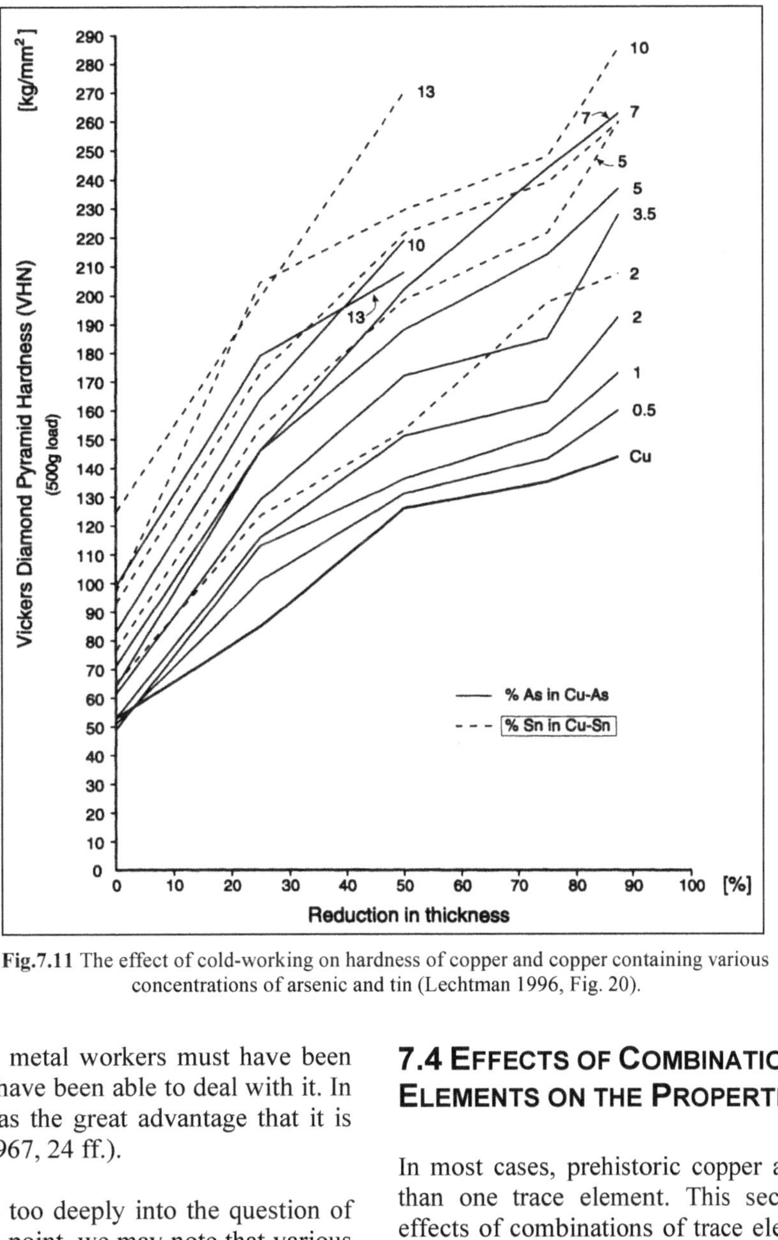

Fig.7.11 The effect of cold-working on hardness of copper and copper containing various concentrations of arsenic and tin (Lechtman 1996, Fig. 20).

consequently, prehistoric metal workers must have been aware of its toxicity and have been able to deal with it. In contrast to arsenic, tin has the great advantage that it is not poisonous (Charles 1967, 24 ff.).

Although we will not go too deeply into the question of deliberate addition at this point, we may note that various researchers have argued that if copper contains more than 1% arsenic, it might have been intentionally added in order to change the properties of pure copper (cf. Ottaway 1982, 131). In prehistory arsenic copper could have been intentionally produced, either by smelting arsenic-rich copper ores, by co-smelting of arsenic sulphides with copper oxides, or by changing smelting conditions (Pollard et al. 1991; Rostoker and Dvorak 1991). Charles stated that by smelting arsenic-bearing copper ores the smelted copper can contain up to around 9% arsenic (Charles 1967, 24), which, however, can be reduced relatively easy via constantly re-melting and/or refining (McKerrell and Tylecote 1972, 209).

Most of the copper which is defined as arsenical copper in archaeological studies contains between 1% and 1.5% of arsenic. However, this has only a small metallurgical advantage in contrast to pure copper (Budd and Ottaway 1991, 139). The greatest, especially cold working, benefits have copper bearing 2 – 6% arsenic (Budd 1991a, 132 f.).

7.4 EFFECTS OF COMBINATIONS OF TRACE ELEMENTS ON THE PROPERTIES OF COPPER

In most cases, prehistoric copper artefacts contain more than one trace element. This section will discuss the effects of combinations of trace elements, particularly as they affect the mechanical properties of copper. Hence, the focus will be on the interaction of the most common and, thus, most important impurities (antimony, arsenic and nickel). The combined effects where other elements are concerned have either been discussed above or will be taken into account below.

Some trace elements – especially the highly soluble ones, such as arsenic, silver and nickel – have a positive effect on the properties of copper when combined with each other. They produce good malleability, raise tensile strength and extend the fatigue range. However, their effect on increasing the hardness of the metal is comparatively small. A high softening temperature, toughness and ductility is reported as desirable properties (Archbutt and Prytherch 1937, 82 ff.).

A combination of antimony and nickel raises tensile strength remarkably (Archbutt and Prytherch 1937, 115 ff.), whereas ductility, toughness and cold-working properties decrease, especially in the presence of greater

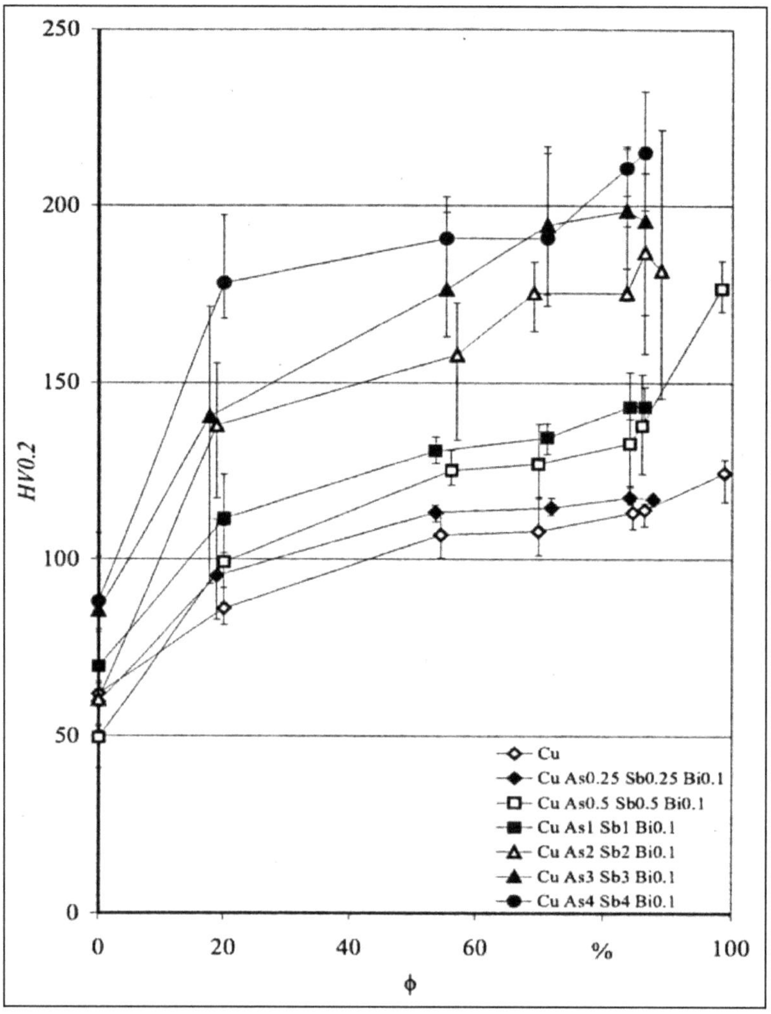

Fig. 7.12 Hardness of various types of fahlore containing varying contents of arsenic, antimony and bismuth. Reduction of thickness by cold working (white symbols: sand cast metal; black symbols: graphite cast metal) (Junk 2003, Fig. 7.26).

amounts of oxygen (c. 0.1%) in copper containing both antimony and nickel (ibid, 92 – 95). In principle, the properties are affected more by the precipitation of oxides during the cooling process than by the actual contents of both elements. Therefore, rapid cooling was desired, and may well have taken place in the casting of small and thin Chalcolithic artefacts (Archbutt and Prytherch 1937, 92 – 95). Although Stahl mentioned that a content of over 2% nickel in antimony-copper has negative effects on the properties (Stahl 1918, 396), both Moesta and Kienlin stated that the high-antimony and high-nickel copper Salez axes have a great hardness and very good working properties (Moesta 2004, 272; Kienlin et al. 2006, 458 ff.); the same is true of the so-called fahlore-coppers that contain lower levels of antimony, nickel and arsenic (Kienlin et al. 2006, 461 f.). Archbutt and Prytherch (1937, 82 – 89) pointed out that ductility, hardness, malleability and tensile strength is significantly improved by increasing concentrations of antimony, arsenic and nickel.

If copper contains two of the three elements antimony, arsenic and nickel, together with just a low content of oxygen (0.01 – 0.02%), then good malleability, high tensile strength, extended fatigue range, ductility and toughness are achieved. The cold and hot working properties are also reported as being good (Archbutt and Prytherch 1937, 82).

Archbutt and Prytherch ran various tests on a range of samples containing between 0.05 and 0.53% of antimony and arsenic. An addition of arsenic and antimony scarcely affects density (Archbutt and Prytherch 1931, 284). Although arsenic and antimony individually increase tensile strength, a mixture of both elements improves this even more – with antimony having a higher impact than arsenic (Archbutt and Prytherch 1931, 284 ff.). The notch-bar impact test that measures toughness does not show particularly good results (Archbutt and Prytherch 1931, 287 ff.). The softening temperature is not significantly influenced by a change in arsenic and/or antimony concentration, but, by contrast, hardness increases with increasing arsenic and antimony content. Here again, arsenic has just a small effect on the hardness of antimony-copper, but antimony increases it (Archbutt and Prytherch 1931, 291 ff.). Junk (2003, 155 ff.) also reports an increase in hardness with increasing antimony and arsenic concentrations. Additionally, the hardness can be improved for all samples by cold-working. The effect is more significant with higher concentrations of impurities (As, Sb > 2%) (Fig. 7.12).

Thus, in combination with arsenic, antimony improves the mechanical properties of the metal and has a bigger influence than arsenic. Both trace elements together improve the mechanical properties of copper to a greater extent than the effect of one element alone.

Overall, it can be concluded that trace elements which individually have a positive effect on the properties of copper (above all arsenic) counteract the influences of negative impurities such as iron or bismuth. In combination with e.g. arsenic or nickel the negative influence of antimony is reversed, so that antimony, when present alongside a trace element that has positive effects in its own right, can significantly improve the properties of copper.

7.5 COPPER IMPURITIES AND THEIR ADVANTAGE FOR CHALCOLITHIC METAL WORKERS

As demonstrated above, various trace elements affect the properties of copper. Because this phenomenon is a key focus of this research, the following section will reconsider those effects in terms of the choices made by Chalcolithic metallurgists. We shall consider not only those metallurgical characteristics relating to finished products, such as brittleness and hardness – important properties as far as tools are concerned, as they are exposed to stress – but also the behaviour of the copper at different temperatures, since this is crucial for melting, cold-working and heat-treatment. Furthermore, the ductility and malleability of a metal determines how easy it can be shaped without cracking or breaking.

Obviously the first impression of an artefact is its colour. Thus, for Chalcolithic users colour may have been a significant factor in the choice of a particular type of copper, especially where jewellery was being produced. As mentioned above, it is quite difficult to recognise a metal object's original colour without damaging the artefact. The colour of ores may have been the key to recognising them. Several copper ores are easy to identify, because of their striking colour – for example azurite is blue, cuprite is red and malachite is green. Impurities, however, can change the colour of copper, in increasing concentrations. Whereas impurities of iron darken copper, antimony, arsenic and nickel brighten it and, above certain concentrations, they tint copper slightly white and silvery. A tin content of between around 3 to 6% turns copper from brownish-red into violet-golden, and at a concentration of c. 10% tin makes it golden in colour.

Brittleness is chiefly caused by trace elements which are either insoluble in solid copper or which form insoluble minerals, such as oxygen, bismuth, lead or even antimony at high concentrations. Tin does not make copper brittle under normal conditions, but after being hot-worked this changes dramatically – particularly if the metal contains more than 8% tin.

If trace elements are soluble at a certain phase, then at that stage copper is (cold- or hot-) workable. Soluble impurities generally increase ductility and malleability of copper. Arsenic and lead affect these characteristics of copper more than all the other examined impurities. Ductility concerns the casting and forming qualities of metal. But also silver – both on its own and combined with arsenic – improves elasticity. Soluble impurities also counteract the negative effects of elements such as bismuth.

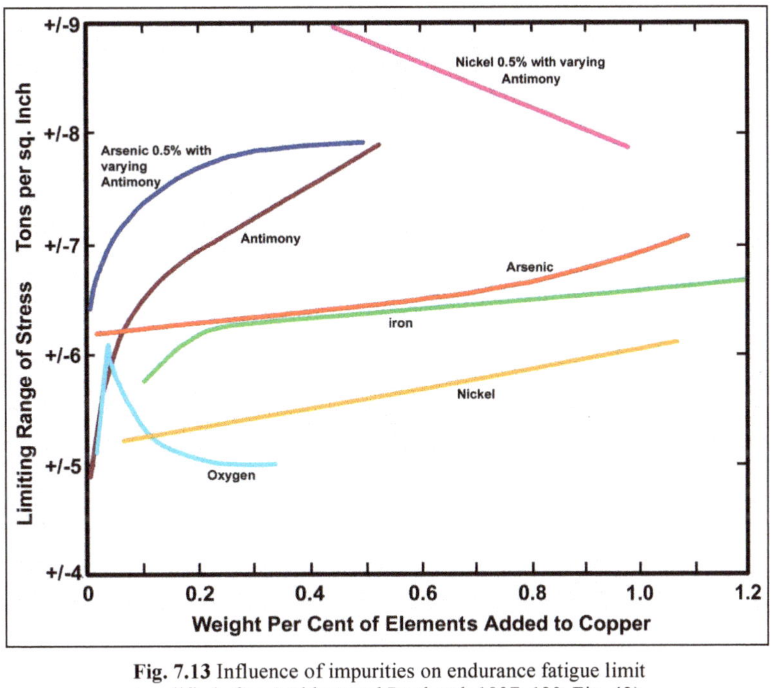

Fig. 7.13 Influence of impurities on endurance fatigue limit (modified after Archbutt and Prytherch 1937, 120, Fig. 42).

All the trace elements – except iron and nickel – lower the melting point of copper. Again, arsenic has the largest effect on solidification and softening temperature of the trace elements studied here. In the case of arsenic and lead, the smallest amount of the particular impurity has the largest effect on the softening temperature.

The results of fatigue tests have shown that particular combinations of trace elements enhance the strain of copper (Fig. 7.13). But here again, impurities which are soluble in solid copper have the best effect (with the exception of nickel). Nickel plus antimony lower stress with rising content. At low concentration oxygen rapidly raises the fatigue limit, while at a concentration of over 0.04% it counteracts it (Archbutt and Prytherch 1937, 119 f.).

For practical reasons, perhaps the most significant properties for prehistoric people were the strength and the hardness of a tool. Whereas some impurities, such as bismuth, decrease the hardness of copper, most of the trace elements increase both hardness and tensile strength. Mainly, however, the increase in of hardness over a certain level also causes brittleness (cf. antimony, iron, lead and oxygen). In particular, antimony (in low concentrations) and antimony plus arsenic and nickel-copper plus arsenic improve these properties; nickel and lead have minimal effect on hardness and tensile strength. The diagram below displays this on the example of annealed copper with varying impurities (Fig. 7.14).

To sum up, all of these properties of copper were important for prehistoric metal workers as well as for the consumers. Trace elements can change the character of finished artefacts, but also the casting and the working properties. The most neutral or beneficial impurities are arsenic, nickel and silver. Antimony can improve the properties of copper, but it needs to occur in combination with arsenic or nickel. If arsenic is the only impurity, it improves working and functional properties, but it can also easily cause unsound casts. The other analysed trace elements make copper better only if these appear in low concentrations or in interaction with other 'positive influencing' partners. In general, combinations of the examined trace elements improve the properties of copper.

As was demonstrated, trace element analyses on their own are not useful for provenance studies (cf. Chapter 6). Nevertheless, Pernicka and other scholars have plausibly shown that by using trace element analysis – but only in conjunction with lead isotope analysis – it is possible to link ore deposits with artefacts (e.g. Begemann et al. 1995; Niederschlag and Pernicka 2002; Höppner et al. 2005). One has to query, however, whether one can draw conclusions by studying exclusively the huge amount of available trace element analyses. We shall focus on finished products and will ask whether their trace element composition can tell us something about the metallurgical skills of the prehistoric metal workers who made them. Since we know that various trace elements, especially in combination, affect the properties of copper, we can ask whether these effects have been utilised.

It is, however, hard to conclude an artefact's physical and mechanical properties directly from its quantitative trace element composition. Factors such as heat-treatment or the cooling rate after casting also influence the characteristics. Therefore analyses, such as micro- and macrostructure tests and hardness tests, have to be undertaken on the objects. In particular, the investigation of the microstructure of finished products has demonstrated that their properties can also be affected by several other factors, such as the cooling rate after casting, the selection of a certain type of mould, or hammering. The microstructure of metal shows whether deformations have taken place caused by, for example, casting techniques or cold- and hot-working. Several

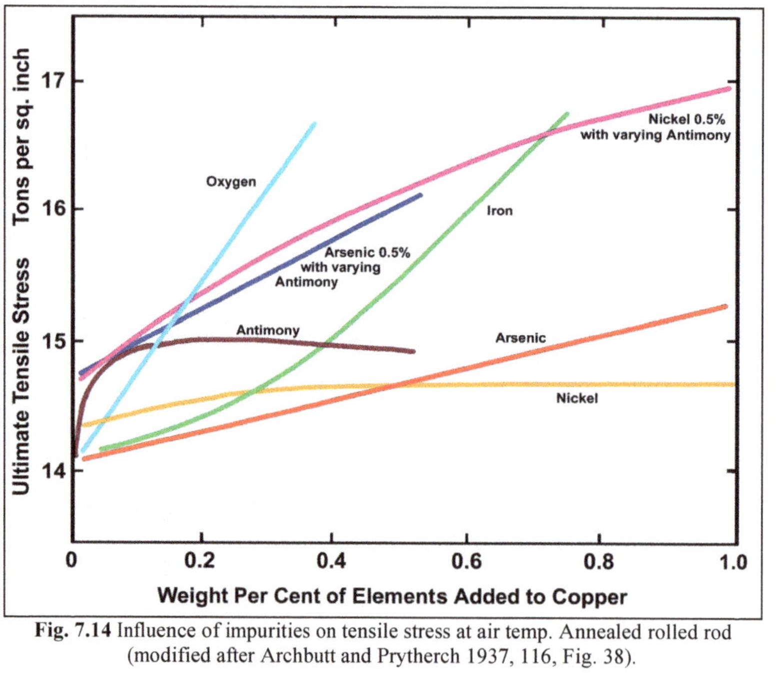

Fig. 7.14 Influence of impurities on tensile stress at air temp. Annealed rolled rod (modified after Archbutt and Prytherch 1937, 116, Fig. 38).

studies have claimed that prehistoric copper artefacts were deliberately worked in order to improve their properties (e.g. Junk 2003, Wang and Ottaway 2004; Kienlin 2008). As Kienlin and his colleagues recently confirmed, these analyses can be destructive to the objects (Kienlin et al. 2006; Kienlin and Pernicka 2009), so probably only a small number of finds can be investigated in this way. However, based on previous research, it is possible to get a general idea of whether impurities improve or impair copper; and by comparing a wide range of artefacts this allows the recognition of copper types with particular properties. By looking at a larger number of trace element analyses singular variations and outliers do not weigh so heavy. A comparison of the metallurgical research on determining the characteristics of these types of copper will give an impression of whether craftsmen at an early stage of metal-technology had the ability to select copper types according to attributes such as workability, hardness, strength etc. The following table presents an overview of the effects of several impurities on the properties of copper that may be the most important ones for Chalcolithic metallurgists (Tab. 7.1). The displayed rating relies on the more detailed descriptions in Chapter 7.3 and 7.4.

The metallurgical properties of various types of copper give some idea as to why they may have been used in order to produce an artefact. Their trace element compositions do not explain the properties of the finished artefact. This can only be examined by analysing the particular object micro- and macroscopically, too (e.g. Budd 191a; Kienlin 2008).

The copper type, however, is the first step in deliberately changing the properties of an object or to improve the smelting, casting and working properties of the metal. Consequently, choosing a certain type of copper influences the entire *chaîne opératoire* and is fundamental for all metal work. The survey presented in the next chapter will clarify whether certain types of copper have been used throughout space and time according to their metallurgical characteristics.

	Sn	Pb	As	Sb	Ag	Ni	Bi	Fe	As/Sb	As/Ni	Sb/Ni
Hardness	+ +	-	+ +	+ +	+	+/-	-	+	+ +	+	+ +
Tensile Strength	+ +	-	+ +	+ +	n/a	+/-	-	+	+ +	+ +	+
Malleability	++	+	+	+	+	+/-	-	n/a	+	+	+
Ductility	n/a	+	+	+	+	+/-	-	n/a	+	+	-
Softening/Solidification Temperature	+	+ +	+ +	+	+	-	+	+	+/-	+/-	+/-
Brittleness	-	- -	+/-	- -	+/-	+/-	- -	- -	+/-	+/-	+/-
Cold working	+	n/a	+	+	+/-	+/-	-	-	+	+	-

Tab. 7.1 Summary of the effects of various trace elements on copper
(+ + : very positive; + : positive; +/- : no remarkable effect; - : negative; - - : very negative).

8. Methods of Statistical Classification in Archaeometallurgy

In Chapter 5.2, the criteria that were applied to compile the dataset of trace element analyses were set out and the database presented. The current chapter presents an overview of some statistical methods that have been used in archaeometallurgy and covers the theoretical basics of statistical analysis used in the present study. The aim of this analysis was to detect metallurgical clusters on the basis of shared trace element composition, which were then interpreted as 'types of copper'[128].

It must be emphasised at the outset that statistical classification can only be a means to an end, used to structure a dataset and to ease its manageability. However, it can also support the archaeological interpretations and, in the current research, it serves to define certain copper types. The process of clustering on its own is archaeologically meaningless; the resultant groups must be interpreted in the light of their archaeological background if they are to make culture-historical sense. It is essential to keep in mind that statistics is an objective method to approach pure figures. The culture-historical interpretation is not necessarily the *result* of the statistical evaluation; rather, statistics can *support* an archaeological explanation by making data structures more clear. Taking this into account, the analysis will compare and combine the results of several statistical methods.

Since the earliest archaeological studies of copper composition, researchers have attempted to find an ideal way to reduce the vast amount of analytical data, particularly by using statistics. As will be shown below (Chapter 8.1), computer-based statistical methods that group, compare and interpret trace element analyses in archaeometallurgy have improved over the decades. In particular, two statistical methods (*two-step cluster analysis* and *principal component analysis*) are presented and their characteristics compared[129]. The reasons why these two methods were considered to be ideal for interrogating the current database will be set out. Because this study is aimed at prehistoric archaeologists rather than statisticians, the following sections chiefly rely on and refer to the general introductions to statistics that have been explicitly written for the field of archaeology.

8.1 Previous Statistical Analyses of Trace Element Compositions: A Methodological Overview

In order to facilitate the archaeological interpretation of large quantities of trace element analytical data, scholars have searched for adequate techniques to classify and compare those data. Various techniques of cluster analysis were found to improve the manageability and interpretability of large datasets without loss of information.

Early studies relied on manual methods of classifying trace element compositional data, but the advent of computer-based statistical methods generally allow greater objectivity by evaluating a much larger dataset. They ease chronological and chorological comparisons, while offering less complicated and clearer visualisations of the results, e.g. by plotting or mapping. The following section will discuss the most valuable techniques that have been used for classifying and interpreting trace element analyses of prehistoric metal finds.

8.1.1 The SAM-Group's *Stammbaum*

The earliest studies that interpreted trace element compositions of prehistoric copper artefacts grouped finds manually to define particular types of copper. This is how, for instance, Otto and Witter defined their six types of copper, the *Leitlegierungen* (cf. Chapter 4.3), and used these to arrange further analyses of finds with critical values of key trace elements (Otto and Witter 1952, 60 – 82). . The specific values that characterise the individual types of copper were appraised after technological and ore-specific considerations, as judged significant in Otto and Witter's opinion (1952, 31). As these types reproduce copper ores, the authors have postulated that artefacts grouped to one of these *Leitlegierungen* would directly reflect the use of a particular ore. In comparing the quantitative trace element compositional data by hand, Otto and Witter

[128] I am indebted to the statistician Frances Provan (Convenor of the Statistical Systems Support team at the University of Edinburgh) for her outstanding support, advice and the time she has spent in order to assist me with statistical and computer issues. Without her assistance and ideas, the following sections could not have been completed satisfactorily.

[129] All statistical analyses discussed in this study were performed using *SPSS Version 15.0 for Windows* software. Guidelines to run statistics with SPSS are provided by various handbooks (e.g. Griffith 2007; Norušis 2008).

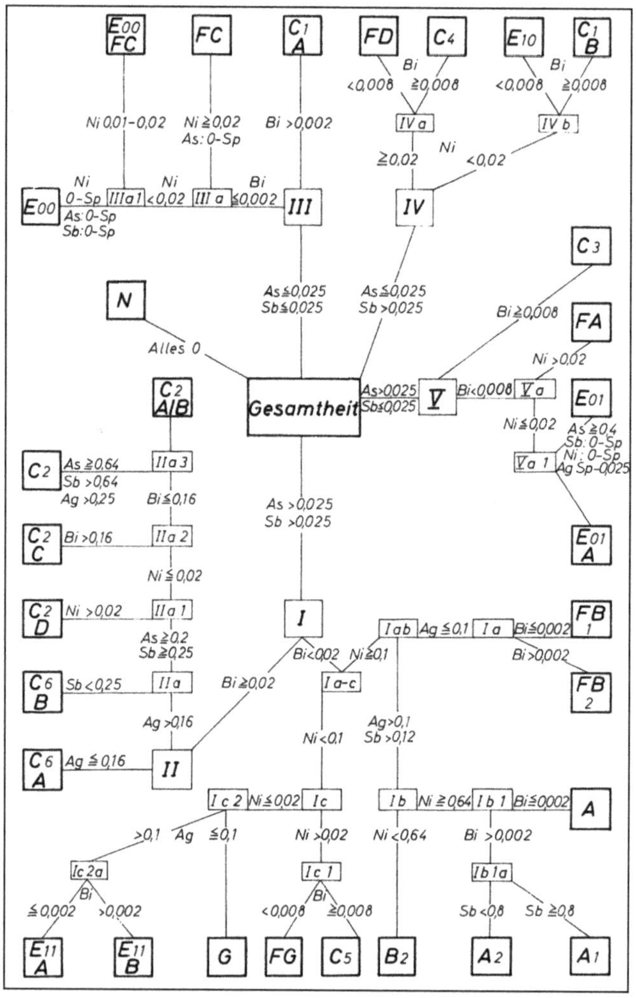

Fig. 8.1 Classification after the SAM-*Stammbaum* (SAM II/3, Diagram 1).

were using their intuition to establish criteria and quantitative thresholds. Compared with later studies, their grouping of data was quite arbitrary and the results were not regarded as being definite. Other studies have applied a similar methodology (Pittioni 1957; Patay et al. 1963), but have instead used semi-quantitative trace element analyses to ascertain the provenance of the prehistoric ores (see also Chapter 4.3).

Further approaches were subsequently made in order to characterise specific types of copper, with attempts being made to move away from Otto and Witter's subjective definition of critical values and to employ larger databases of trace element analyses. In particular, soon after the Otto and Witter publication, scholars aimed to cluster trace element data using objective statistical methods. A comprehensive and long-lasting approach to the statistical grouping of trace element data was made by the so-called SAM-group of scholars, whose key team members were Junghans, Sangmeister and Schröder (SAM I and II; Chapter 4.3).

The SAM-group entrusted the statistician Klein with the task of developing statistical clusters to define copper types (Junghans et al. 1954, 103 – 114). Klein identified certain critical values that distinguish the types of copper, based on relative frequency distribution and the discovery that compositions of objects appear as normal distributions. These thresholds were calculated by graphical frequency analysis, meaning that the frequencies of the elements in question[130] were plotted on probability paper. An indication of the thresholds was given by the maxima and minima of the probability density function (Junghans et al. 1954, 104 ff.). Finally, Klein was able to define four initial groups based on c. 400 analyses. The thresholds of these clusters were marked by an appearance of between 10 and 90% of a particular trace element within the group (Junghans et al. 1954, 109 ff.).

With the increasing number of analyses available and by refining that method, Klein could firstly define twelve (based on 2302 trace element analyses) and later 29 compositional groups (based on c. 12000 analyses) (SAM I, 57; SAM II/1, 13 ff.). The final 29 *Materialgruppen* (material groups) were published as a *Stammbaum* – a two-dimensional cluster tree (Fig. 8.1). After cluster N was defined as copper devoid of any impurities, the first major groups (Group I, III, IV and V) were then characterised by plotting the antimony and arsenic

[130] Firstly, antimony, arsenic, bismuth, lead, nickel, silver and tin have been considered (Junghans et al. 1954, 109). However, further studies of the SAM-project focused only the five elements antimony, arsenic, bismuth, nickel and silver, because they showed the most significant differentiations (SAM I, 57).

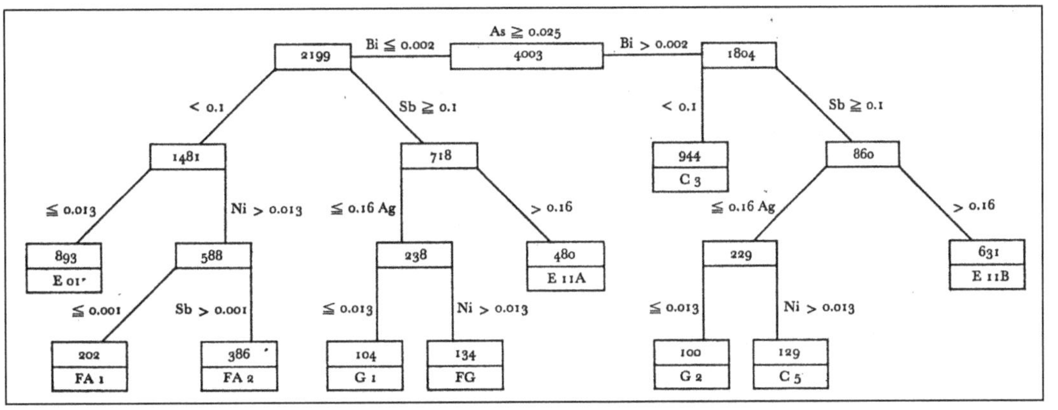

Fig. 8.2 New classification of the SAM-groups C_3, C_5, E_{01}, $E_{01}A$, $E_{11}A$, $E_{11}B$, FA, FG and G (Sangmeister and Strahm 1973, Tab. 6).

concentrations on logarithmic paper – the minima accumulations defining the thresholds[131]. The four groups' were called arsenical resp. antimonial copper. These four groups were then subdivided using typical values for bismuth, nickel and silver. Consequently, every material group (A – G and N) was characterised by a specific configuration of the trace elements arsenic, antimony, bismuth, nickel and silver and thus every analysed archaeological find could be classified into one group[132]. Every analysis could be categorised by considering one threshold after another. This splits the entire *Stammbaum* up into branches (SAM II/1, 14 ff.). The selection of the key trace elements (arsenic, antimony, nickel, silver and bismuth) emerged from the statistics (SAM I, 57).

Some years later, the published 1968 *Stammbaum* (cf. SAM II) was modified again by Sangmeister (Sangmeister and Strahm 1973). The thresholds of some arsenic-richer clusters (C_3, C_5, E_{01}, $E_{01}A$, $E_{11}A$, $E_{11}B$, FA, FG and G) were re-defined (Sangmeister and Strahm 1973, 198 – 201; Fig. 8.2). The database of the present study was grouped according to the old *Stammbaum*, and also according to the modified version, in order both to review Sangmeister's new groups and to allow comparisons of the results of this study's statistics later on. It turned out that the new SAM-groups are, up to a certain level, less detailed than the old and four new groups were detected[133].

Consequently, the database used in the current research has therefore been clustered according to Sangmeister (cf. Fig. 8.2) and also in accordance with SAM II. This results with only a few hitherto unclassificable objects being allocated to a type. However, another problem of the *Stammbaum* was encountered. In the SAM project, copper types with high impurities of bismuth were not classified and types of copper containing only small amounts of arsenic and antimony could not be distinguished on the basis of their silver content.

SAM's grouping system has met with some criticism and subsequently several scholars have tried to improve the methods for clustering and interpreting trace element analyses. Both groups, Waterbolk and Butler (1965) and Slater and Charles (1970), argued that the SAM-*Stammbaum* was grouped only on mathematical criteria without considering the archaeological background. Therefore, Waterbolk and Butler (1965) introduced a clustering method that relates not only to the trace element composition but also to archaeological considerations. Another criticism of the SAM systematic was that it was based on a monocausal decision regarding thresholds of particular elements, based on statistics and without considering the metallurgical significance of these impurities (Slater and Charles 1970).

Furthermore, doubts have been cast regarding the comparability of the analytical results themselves and their archaeological usability (e.g. Lichardus 1991a, 27 f.). These issues are valid, but for the present research one can disregard this, as will be shown in Chapter 8.3. In contrast, the great advantage of the *Stammbaum* is that a single analysis can be compared with material groups according to the thresholds. These thresholds, however, make both the Otto/Witter- and the SAM-grouping unchangeable and, as we will see through the whole evaluation, the results of the clustering changes with the database. Although statistical grouping always depends on the original dataset, several studies using various statistical methods for clustering have not only demonstrated that their groups are similar (Chapter 4.3, Fig. 4.11), but also are related to archaeological contexts and to 'traditional' archaeological interpretations.

[131] Group II is separated from group I by its high concentration of bismuth.

[132] According to the SAM, however, four statistically calculated complexes could not be determined as special groups, as those contain only a few objects, have heterogeneous compositions and could not be interpreted historically (SAM II/1, 15).

[133] The problem appears that some of the 'old' and 'new' SAM-groups are overlapping. At the newer classification (after Sangmeister and Strahm 1973), for example, the $E_{11}A$ also partly includes the groups A and B_2. Group E_{01} now also includes $E_{01}A$ and C_2 is now a part of $E_{11}B$. A combination of both cluster trees still separates the previous groups. FA is split up into FA_1 and FA_2, and, beside group G, G_1 and G_2 became apparent.

8.1.2 HIERARCHICAL CLUSTER ANALYSIS

The major disadvantage of the SAM-*Stammbaum* classification discussed above is the grouping based on fixed thresholds of specific trace elements and the step-by-step evaluation. This issue can be neutralised by other clustering methods that account for an equal observation of several trace elements. Ottaway (1974; 1978a) can be credited with having made the first extensive, and crucial, study which established the use of cluster analysis in archaeometallurgy[134].

In brief, cluster analysis is a multiple variable analysis that offers the possibility to group various objects or cases (here metal objects) by considering the entire range of requested variables (here trace elements) as of equivalent weighting. Cluster analysis examines all variables synchronously, in contrast to the *Stammbaum* that relied on successive, and step-wise examination of individual variables, and to the graphical grouping of data, such as Walterbolk's and Butler's method (1965), which is restricted to the maximum three-dimensional space.

Cluster analysis allows n-dimensional calculations of n numbers of variables. Generally, the similarity between variables in an object is evaluated by measuring and comparing numerical distances between each variable. The objects are then grouped according to their statistical likeness. The *hierarchical cluster analysis* that is most frequently used in archaeometallurgy combines cases according to declining similarity. Then, the distances are related to their corresponding objects and the objects are grouped hierarchically by their mathematical resemblance. The smaller the sum of the distances between objects, the more similar the objects are. The distance between two cases, *a* and *b*, with *n* variables is calculated by *Euclidian distance* d_{ab}, defined by equation 8.1 (cf. Hodson 1969a, 92; Shennan 1997, 223 f.).

$$d_{ab} = \sqrt{\left[\sum_{k=1}^{n} (x_{ak} - x_{bk})^2\right]} \qquad (8.1)$$

An ideal way to display the relationships of the objects is by a cluster tree where similar objects are arranged together more closely than others (cf. Shennan 1997, 234 – 260). The closer the distances between the different variables are, the more similar the objects and consequently, the objects can also be arranged in ascending order according to their distance/similarity. The relationship of the objects can be seen on the distance scale of a cluster tree. One advantage of hierarchical cluster analysis and the deduced cluster trees is that individual cases and smaller clusters are developed from larger clusters. Thus, the next nearest groups can easily be retraced on the cluster tree. This is illustrated in Figure 8.3 with cases D and E. Cases B and C are similar to each other, while case A is the most divergent object. Cases B and C, however, are more similar to A than to cases D and E.

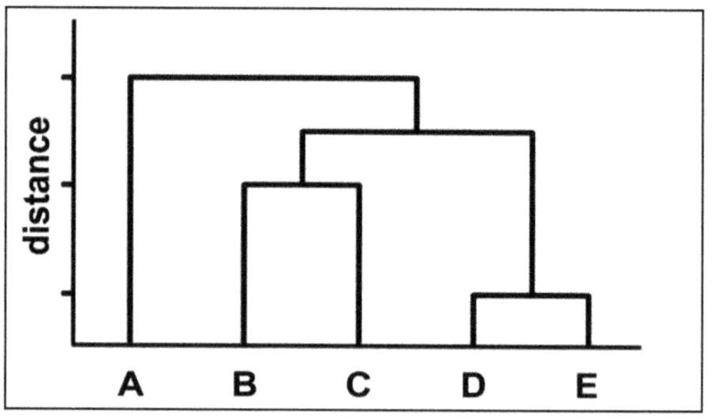

Fig. 8.3 Example of a cluster tree with five cases.

Since Ottaway's publications (1974; 1982), hierarchical cluster analyses have become the key method of studying trace element analyses. Several algorithms can be used to group by means of hierarchical cluster analyses (cf. Shennan 1997, 236 – 245). Ottaway (1978a, 194) used in her thesis a method that was comparable to Ward's method. Christoforidis and Pernicka (1988, 252 ff.) determined that the Group-Average or more known as Average-Linkage method (labelled Between-Groups Linkage method in SPSS) was the most sufficient algorithm for trace element studies. The Average-Link method identifies the arithmetic average among all previously evaluated distances of all variables of the cases. In other words, the entire number of possible distances between all members of the groups are detected, summed up and divided by the number of distances. The result is the average linkage between cases. Finally, the cases are sorted concerning their distances (Shennan 1997, 239 f.). The biggest disadvantage of all methods of hierarchical cluster analyses is that the statistical method itself only arranges the objects after similarities and does not automatically determine the number of clusters or group the objects. Although calculations, such as the Agglomeration-Distance-Plot (Prange 2001, 62 f.; Schreiner 2007, 157), assist the selection of the cluster number, the final decision has to be made by the researcher.

The first applications of cluster analyses in archaeology were not performed to sort out metal groups, but rather to classify artefacts typologically (cf. Hodson 1969a; Ottaway 1982, 33 – 40). Archaeometallurgical studies applied hierarchical cluster analyses chiefly to copper objects from specific archaeological complexes, such as cemeteries or archaeological groups (e.g. Boomert 1975; Krause 1988; Metziger-Schmitz 2004). The broadest study using hierarchical cluster analysis to examine trace element analyses was made by Pernicka (1995) in the course of the SMAP-project when he clustered around 26500 analyses and found that his groups were closely linked with the material groups of the SAM-*Stammbaum* (cf. Chapter 4.3).

[134] On the basis of a very small number of data, Hodson (1969a; 1969b) primarily published first attempts of arranging copper compositions by cluster analysis in 1969. These trials had no significant impact on archaeometallurgy, as his aim was not a metallurgical examination, but a general introduction to cluster analysis using various archaeological examples.

8.2 EXPLAINING THE TWO STATISTICAL METHODS USED IN THE CURRENT STUDY (TWO-STEP CLUSTER ANALYSIS AND PRINCIPAL COMPONENTS ANALYSIS)

Various methods of hierarchical cluster analysis have been used in archaeometallurgy for decades in order to compare and group trace element analyses. The particular statistical method to be used has to be made in the light of the individual research issue. In the current case, it is necessary to choose a technique to group the trace element analyses according to similarities in order to ease the manageability and interpretability of the huge amount of 1943 analyses. The statistics must also permit the possibility to compare both the produced clusters and the single artefacts and to clarify which impurities could be the most significant ones for the particular groups. Hence, two statistical methods come into play, which have not often been used in archaeological research; firstly, the two-step cluster analysis and, secondly, principal components analysis.

8.2.1 TWO-STEP CLUSTER ANALYSIS

In order to detect strictly defined groups of similar cases, it is practical to reduce a larger set of multivariate data by cluster analysis. The method of hierarchical cluster analysis has already been discussed, and it could be demonstrated that the disadvantage of this statistical technique is the subjective choice of clusters. This difficulty can be neutralised by a novel type of cluster analysis, the two-step cluster analysis (TSC). The TSC is a relatively newly-developed technique for clustering and, as far as the author is aware, has not been used in archaeometallurgical studies to date. General information and a brief theoretical overview of the TSC method are discussed below, followed by its applicability in archaeometry using an illustrated example.

The TSC encompasses three advantages over previous clustering methods. Firstly, it can manage much larger amounts of data, while operating at a minor capacity. It can also group not only quantitative data, but also categorical. Bacher et al. (2004, 21), however, noticed that the results of TSC are only correct if only quantitative variables are considered. The present study only incorporates quantitative data, so therefore this issue can be neglected. The third advantage is the most important for the purposes of the current research. TSC is capable of automatically determining the number of clusters by algorithm.

In general, TSC is able to cluster a huge amount of data in two steps, and so it can cope with a more comprehensive quantity of data than the diverse methods of hierarchical cluster analyses. The first step requires a large number of objects to be grouped into small sub-clusters regarding their similar variables. Then, in the second step, these sub-clusters resulting from step one are again clustered into the final clusters (Chiu et al. 2001, 263 f.; Bacher et al. 2004, 3 f.). TSC is methodologically based on the so-called BIRCH method (Balanced Iterative Reducing and Clustering used Hierarchies) developed by Zhang, Ramakrishnan and Livny, which pre-clusters a large database by sorting the data into a so-called Cluster Feature tree (CF-tree) (cf. Zhang et al. 1996; ibid. 1997).

The methodology of the TSC was developed and published in 2001 by Chiu, Chen, Fang, Jeris and Wang (Chiu et al. 2001). The first step consists of building up a CF-tree in order to reduce the volume of data , which then can be clustered easily in the second step. Starting from the original set of data the algorithm gradually configures a cluster tree. The distance measurements between objects determines whether the next object following the previous ones establishes a new cluster (Chiu et al. 2001, 265). In a similar way to the previously discussed methods of cluster analyses, the distances between cases are measured and compared. In order to evaluate the distances (= similarities) between cases and clusters, either the Euclidian or Log-Likelihood distance is used. In general, however, the Log-Likelihood distance (see equations 8.2 – 8.5) which is able to deal not only with quantitative, but also categorical data, is more common[135] (Chiu 2001, 264 f.; Bacher et al. 2004, 4). The Log-Likelihood distance *d* between two objects/ clusters *j* and *s* is as follows (equation 8.2):

$$d(j,s) = \xi_j + \xi_s - \xi_{\langle j,s \rangle} \quad (8.2)$$

where

$$\xi_j = -N_j \left(\sum_{i=1}^{p} \frac{1}{2} \log(\sigma_{ji}^2 + \sigma_i^2) - \sum_{i=1}^{q} \sum_{l=1}^{m_i} \pi_{jil} \log(\pi_{jil}) \right) \quad (8.3)$$

$$\xi_s = -N_s \left(\sum_{i=1}^{p} \frac{1}{2} \log(\sigma_{si}^2 + \sigma_i^2) - \sum_{i=1}^{q} \sum_{l=1}^{m_i} \pi_{sil} \log(\pi_{sil}) \right) \quad (8.4)$$

$$\xi_{\langle j,s \rangle} = -N_{\langle j,s \rangle} \left(\sum_{i=1}^{p} \frac{1}{2} \log(\sigma_{\langle j,s \rangle i}^2 + \sigma_i^2) - \sum_{i=1}^{q} \sum_{l=1}^{m_i} \pi_{\langle j,s \rangle il} \log(\pi_{\langle j,s \rangle il}) \right) \quad (8.5)$$

$d(j,s)$: Distance between cluster *j* and *s*
$\langle j,s \rangle$: Final cluster formed by combination of cluster *j* and *s*
p,q: Number of variables in cluster *j* and *s*
m_l: Number of categories *l*
ξ: Variance within a cluster
N: Number of data records in cluster *j* and *s*, or pre-cluster $\langle j,s \rangle$
σ_{ji}^2: Variance of the *i*-th variable in cluster *j*
σ_i^2: Variance of the *i*-th variable in the entire dataset
π: Probability of the category

[135] The algorithms presented here and used by statistics programs such as SPSS are shown and discussed in more detailed in the referenced literature. SPSS, for instance, offers a command which opens a window that displays all algorithms used by the program. The *Help → Algorithm* command in the task menu of SPSS guides the user to that window. By employing the search function, it is possible to look for the particular equations. The equations used to explain the statistical procedures in the current work are also locatable by using that command. Detailed mathematical derivations of all equations, however, are not essential in order to understand the evaluation and interpretation of the output of the statistics. As this study is using statistics as a tool to ease the archaeological interpretation of copper metallurgy, it is sufficient to make clear how the statistics are to be interpreted and how they correlate data, but not crucial to explain their detailed higher mathematical mechanisms. Thus, these chapters will only cover a general introduction focusing on the practicability of statistical methods for archaeometallurgical research.

It is essential for TSC that the construction of the CF-tree considers cases in sequential order of their input into the database. Objects from the same find place, area or period are usually listed together on the original data sheet. After clustering by TSC, these objects are more likely to be grouped together, even if they differ metallurgically. To avoid this bias, randomising the arrangements of the objects should be done in advance[136] (Norušis 2004, 381). The second step is a hierarchical cluster analysis of the pre-cluster. This second step does not handle the actual number of cases and their variables, but instead processes a smaller amount of pre-clusters consisting of all original cases; thus there is much less calculating effort required. For distance measurement the same algorithms from the first step are used. The pre-clusters are now handled as cases and the distances are determined by using their mean values. Without presenting the output of the first step, SPSS groups those pre-clusters again, considering them as separate cases, and displays the final grouping (Chiu et al. 2001, 266; Bacher et al. 2004, 4; Norušis 2004, 382 ff.).

The chief advantage of the TSC is that the number of clusters that are finally outputted is determined automatically. Therefore, Chiu et al. established a method to determine the ideal number of clusters in two steps (Chiu et al. 2001, 266). Firstly by calculating either the BIC (Bayesian Information Criterion) or the AIC (Akaike's Information Criterion) the maximum number of clusters is projected. For the final cluster J, BIC and AIC are calculated from various amounts of the estimated sub-cluster j by:

$$(8.6) \qquad BIC(J) = -2\sum_{i=1}^{1} \xi_j + m_J \log(N)$$

$$(8.7) \qquad AIC(J) = -2\sum_{i=1}^{1} \xi_j + 2m_J$$

where

N: Number of data records in total

m_J: $m_J = J\left\{2K^A + \sum_{k=1}^{K^B}(L_k - 1)\right\}$

K^A: Total number of data that are used in the TSC

K^B: Total number of quantitative variables that are used in the TSC

L_k: Number of categories for the k-th quantitative variable

The maximum number of clusters is reached when the BIC or AIC grows by an increasing number of clusters (with the increasing number of clusters the BIC/AIC decreases to a certain low point and then climbs)[137]. The second step is based on the change of ratios of distances. Starting from the 'maximum number of clusters' peak estimated in the first step the ratio between close clusters is evaluated. The programme looks for a significant increase of ratio by changing the number clusters. This is based in the assumption that a big change in ratio marks the point when cases are merged that should actually be separated. Thus, the minimum number of clusters is distinguished (Chiu et al. 2001, 266).

The automatic assignation of clusters comes at the expense of the accuracy of arranging objects in clusters. There is the possibility that the TSC separates objects that are actually of similar composition, but had been wrongly classified in an early stage of the analysis, and because of randomising, this may cause accidental variation in grouping. This misclassification is called in statistics 'error of the second kind'. As is confirmed below, the incorrectly grouped cases will emerge by subsequent clustering. Hence, it is essential to cross-check the results of the TSC with an independent statistical method.

8.2.2 PRINCIPAL COMPONENT ANALYSIS

In contrast to cluster analyses, principal component analysis (PCA) does not strictly classify and group cases. PCA is a technique of multivariate analysis that arranges according to similarities and evaluates variables according to their character. It assumes that each case is characterised by the n-number of variables that can be illustrated by standardised vectors in n-dimensional space. These vectors directly reflect the information in the cases. The method should reduce complex multivariate data to two- or three-dimensional pictures, which can be plotted as a scatter plot.

Although PCA has often been described in books on statistics in archaeology, it has not been applied frequently to archaeometallurgical studies – even if this method is beneficial for this particular field[138] (cf. Baxter 1995; Shennan 1997).

The mathematical background of PCA is the linear combination and the relationship of vectors in n-dimensional space. The original n variables of every case are expressed as vectors in n-dimensional space according to the value of the variables. By rearranging the vectors the dimensionality can be reduced. The variation between the variables and the new vectors, however, should be minimised. That process can be described by a linear transformation of the n variables $x_1, x_2, \ldots, x_n$ into new uncorrelated vectors $Z_1, Z_2, \ldots, Z_m$ – the *eigenfactors* (equation 8.8; cf. Baxter 1994, 40 – 42 & 49 f.).

$$(8.8) \qquad \begin{aligned} Z_1 &= a_{11}x_1 + a_{12}x_2 + \ldots + a_{1n}x_n \\ Z_2 &= a_{21}x_1 + a_{22}x_2 + \ldots + a_{2n}x_n \\ &\vdots \\ Z_m &= a_{m1}x_1 + a_{m2}x_2 + \ldots + a_{mn}x_n \end{aligned}$$

where

a: coefficient of the specific variable (In this case, the variables are the trace elements [Sn, Pb, As etc] and the coefficients are their concentrations in percentage)

[136] SPSS offers a command to randomise the data list (Transform → Compute variable → Random Numbers → Nv.Normal). Then, the data input must be sorted according to the randomised numbering.

[137] In this study, the BIC method is used.

[138] Just a few archaeometallurgical studies have employed PCA in order to compare the composition of prehistoric metal finds. For instance, Rassmann (2005) investigated hoards of the Únetiče complex, Cerny and his colleagues (2006) studied Swiss copper artefacts, and Corsi et al. (2005) analysed copper daggers and an axe from Tuscany dating to the second half of the 3rd millennium BC.

The following discussion is a general introduction, as provided by several authors, which should suffice to enable us to understand the interpretation of the PCA results (e.g. Baxter 1994, 48 – 99; Shennan 1997, 265 – 307; Baxter 2003, 73 – 83). Shennan, in particular, simplified the calculations of correlation matrices and vector geometry and illustrates the method of PCA and its practicability very clearly[139].

Each object is a point in space whose location is positioned by its values on the variables. If variables are represented as vectors in n-dimensional space, the angle between them corresponds to differences of the variables' variability. Subsequently, the angles are compared. As an angle between vectors is directly associated with the variation of at least two variables, this is seen as a data reduction. It is useful to express the correlation of variables as cosine functions of vectors' angles. If the value of two variables is the same, the vectors are congruent and the cosine is 1.0. If the vectors are correlated inversely, the angle between these vectors is 180° and the cosine -1.0; uncorrelated variables with an angle of 90° have cosine 0. The angles are defined by the matrix of the variable of one particular case. The correlation matrix displays the relations of cosines of the angles between the respective variables, whereby the closer the value is to 1.0 the more consistent the correlations are (Shennan 1997, 270 ff.; Baxter 2003, 78).

In a comparable manner to cluster analysis, PCA measures similarity as the distance between variables – in this case, the difference between cosines. The general variance of one case can be expressed by a new vector defined by the average value of all variables. The correlation of variables is outlined by the sum of cosines between each variable. Thereby, the highest sum represents the lowest co-variation. This relationship is displayed as a correlation matrix.

The size of the correlation matrix is the square of the number of variables. The total sum of correlations is the addition of the individual eigenfactors. As the total sum is built up by the entire number of components of the matrix, i.e. by the square of the number of variables, the single variance is the reverse calculation. The correlations of one variable with all others divided by the square root of the total sum is the *component loading L* of that particular variable. This is the average variable, the *first component,* which is closest to all variables in terms of the distance of angles. According to Shennan (1997, 279), squaring the loading L and multiplying it by 100 gives the percentage of how much the new component accounts for the variation in a variable. Summarising all the squared loadings leads to the *eigenvalue* λ of a component (equation 8.9).

$$\lambda_i = \sum_{j=1}^{n} L_{ij}^{\,2} \qquad (8.9)$$

[139] An in-depth description of the mathematical procedures is presented in Baxter 1994.

The eigenvalue λ_i for component i is the sum of all loadings L of variable j of n numbers of variables. This in turn, brings the question of by what percentage the new component replaces the variation of the original variables. This can be expressed by the quotient of eigenvalue λ by n number of variables and then multiplied by 100 (Shennan 1997, 280):

$$\frac{\lambda}{n} \times 100 = \text{variation of the original variables} \qquad (8.10)$$

The total variation is the sum of variations of all eigenvalues. The PCA maximises the variation for the first eigenvalue, then for the second and so on. With the increase of eigenfactors, the variation changes only slightly and therefore the last components can be dropped without losing significant variation.

The variables are reduced to one component. This component expresses a certain percentage of the variation of a case. In order to get the next components representing the next lower percentage of variation a further eigenvector has to be defined. It is, therefore, useful to identify a component which is as unrelated as possible. Considering that uncorrelated vectors are orthogonal to each other, it is suggested to define the second component as the vector at a 90° angle to the first component (Shennan 1997, 281). The two components for one case are calculated and then define further components according to the same scheme. The components of every case can be plotted against each other in a scatter-plot. The sum of squared loadings on each eigenvalue reflects the variation of all variables of one case per percent.

In the next step, the original values of the variables are replaced. These new values are called *factor scores.* A factor score can be expressed by the equation 8.11. It is the sum of the standardised values of all observations x_{ij} multiplied by the loadings L of a variable on the particular component k.

$$S_{ik} = \sum_{j=1}^{n} x_{ij} L_{ik} \qquad (8.11)$$

where

S_{ik}:	Factor score for case i of component k
x_{ij}:	Value of variable j for case i
L_{ik}:	Loading of variable j on component k
n:	Number of variables

The factor scores S for every component of each case that can be plotted against other factor scores, i.e. at least against another two in three dimensional space. The result is usually drawn as a scatter plot where every dot represents a case. Hence, the closer the correlated objects are, the more clustered they are.

Finally, a *rotation of the loadings* is suggested in order to improve the interpretability of the analyses. The most frequently used method is an orthogonal rotation, i.e. the *varimax rotation* (Kinnear and Gray 2006, 501 f.). This simply means that by methods of linear combination the components are rotated. Simplified, the unrotated component loadings are mirrored, for instance at the eigenfactor (Shennan 1997, 301 ff.; Baxter 2003, 80 ff.).

PCA combines these dimensions and detects new components that define a new space by combining co-variant variables to one factor (component). The first component reduces most variability of the cases to one axis, the second component the next greater and so on (Lock 1991, 86 ff.). In other words, factors and especially the angles between them reflect the variance of objects – in the current research, the trace element composition of copper artefacts. Consequently, new coordinates for each case are established in relation to the newly defined components. The components plot, however, is only an approximation of variance of data. This approximation may affect the reading of the statistics and should be double-checked with reference to other statistical methods.

The great advantage of PCA is that by using the new coordinates of each case the statistical comparison can be shown as comprehensible scatter-plots. Furthermore, objects are not grouped in definite clusters and the intersections are more obvious. This more closely approximates to historical reality where strict boundaries are unlikely to have existed. The scatter-plots can show that clustered objects are more similar and the axes show which variables are more significant on the comparison of the cases. This is possible considering the direction of the component vectors is determined by the most dominant variables. Plotting the component vectors in association with archaeological information, such as dating or context record, visualises the connection between a specific case and its archaeological background and facilitates its historical interpretation.

A combination of TSC and PCA addresses both of the requirements set out at the beginning of Chapter 8. These include minimising the 'error of the second kind' caused by the misclassification by TSC, the ability of PCA to display intersections of the database without actually grouping them, and the comparison of multivariable data in order to find similar data structures. Despite overcoming these issues, all outliers might not be eliminated; however, these very small variances will not change the statistical results significantly. This theoretical background knowledge will be illustrated further in the next section, using a particular example.

8.3 THE COMPARABILITY OF TRACE ELEMENT ANALYSES

After the first archaeometallurgical projects had published the results of trace element analyses, their practical application was often queried. Firstly, it was often asked whether compositional analyses from different projects and laboratories employing different methods of analysing metal were comparable. Besides, it has been pointed out that the concentrations of impurities may vary between different analyses of same artefacts. In addition, their general relevance for archaeological research has been questioned; and in that context, the significance of (statistical) comparison and grouping was also discussed [140].

8.3.1 THE GENERAL COMPARABILITY OF TRACE ELEMENT ANALYSES

As the database used in the present study contains trace element analyses that have been carried out by different methods, it is essential to ascertain whether the results from them are comparable. Although the comparability of trace element analyses has been repeatedly questioned, very few studies have been undertaken to address this issue. Pernicka (1984; id. 1995, 83 ff.), for example, compared and published the results from various laboratories. He used various analyses from the same artefact, called 'double-analyses', and compared their results in terms of possible variations. Except for slight aberrations in antimony, he was able to offer reassurance that results for the relevant trace elements analysed by a variety of laboratories were indeed comparable. Only the data compiled by Otto and Witter show significant variations beyond any statistic tolerance[141] (Krause 2003, 18). Furthermore Ottaway (1982, 92 f.) compared her own data obtained from NAA analysis with the results from the same artefacts produced by SAM using OES, and concluded that the outcomes of both methods show good comparability. An evaluation of all three methods (OES, XRF and NAA), carried out by Lutz and Pernicka (1996, 114 ff.), proved that these are comparable.

Besides the issue inter-laboratory comparability, there were also doubts as to whether a single small sample may reflect the composition of an entire object. Within a copper object, impurities may not be evenly distributed, particularly because different elements can precipitate or accumulate differently as the cooling rate varies inside a metal. This is especially true for elements which are not soluble in melted or solid copper (e.g. bismuth) and which could accumulate in one area of the object (cf. Slater and Charles 1970, 208 ff.). The argument that bismuth accumulation starts particularly at the wall of the cast can be countered if the samples are taken from inside the artefacts by drilling (Pernicka 1989, 627). The same argument invalidates the interpretation of the content of arsenic. Here, according to McKerrell and Tylecote

[140] Various authors have offered critiques of the comparability, classification, interpretability and significance of projects that have analysed and evaluated trace element composition in archaeological research (e.g. Waterbolk and Butler 1965; Slater and Charles 1970; Lichardus 1991a). These criticisms have been outlined and discussed by Härke (1978), Pernicka (1989) and Krause (2003).

[141] During the research for the author's Master's dissertation the deviation of the Otto/Witter-analyses from the norm was confirmed on the small dataset of trace element analyses of Bavarian Bell Beaker artefacts. It was noticed that, apart from the silver concentrations, all considered impurities – especially antimony and arsenic – are below the concentrations found for the same objects when analysed by the SAM-project (Merkl 2005, 97).

(1972, 211) arsenic-rich artefacts may have higher arsenic concentrations on the surface than inside the object.

The doubts regarding whether one sample is representative of the entire object have already been discussed and eliminated by the SAM-project and again by Pernicka (Junghans et al. 1954, 100 ff.; Pernicka 1984, 523 f.). Pernicka was able to demonstrate that within one artefact the composition varies only slightly and well below the error margin of the methods of the SAM-project. Several analyses obtained from one artefact have shown that, in general, these are similar and show only small variations. Apart from a few outliers, the aberrations are within the statistical standard deviations and the standard error.

To conclude, most of the trace element analyses that are available from different study-groups can be consulted for comparative investigations – apart from the semi-quantitative data analysed by, for example, Pittioni (1957) and Patay et al. (1963). As mentioned briefly in Chapter 4.4, converting these types of trace element analyses into fully-quantitative date causes a crucial problem during statistical evaluations (cf. Ottaway 1982, 76; Rau and Willing 1991, 360). This, however, does not diminish the content of the dataset, because all the objects that were analysed by them have also been analysed by the SAM-group, so applicable trace element analyses of these finds are available. Concerning the question of the general applicability of trace element analyses, research over the last few decades has shown that the attitude towards trace element and scientific analyses has generally changed, and the fundamental refusal with which the SAM-project originally met has disappeared. Today, material analyses are commonly used to investigate archaeological finds, but must always be considered in the context of other typological, theoretical and chronological approaches.

8.3.2 The Comparability of the Trace Element Analyses within this Database

Although it should be clear that trace element analyses compiled by different researchers can be compared, it is worth looking at the degree of comparability of the data used in this study. In order to clarify the similarity of several analyses of one object, double-analyses are included in the database of the present study. Within the database of this study 243 double-analyses could be compiled which have been carried out on 111 different objects. Those include the results of eight different projects and laboratories (cf. Chapter 5.2.2). If the objects have been analysed several times by one project, then the double-analyses are labelled by "#", whereas, if the double-analyses have been carried out on objects by different projects or laboratories, they are labelled "*". In order to compare the double-analyses, a PCA has been run, at first, on the double-analyses carried out on one object by one project and then on an object by different projects (Appendix A). The result of the comparison is shown in a scatter plot which arranges more similar cases closer together. The first scatter plot shows the correlation analyses of artefacts that have been analysed several times by one project (Fig. 8.4).

Each spot on the scatter plot (Fig. 8.4) displays a single compositional analysis. The numbers indicate which analyses have been taken from one artefact; the same figure marks the same object. So, if analyses with the same figure are clustered close together on the scatter plot, then the analyses are representative for the artefact's composition. Since this indeed seems to be the case here (with small outliers), it can be concluded that the analyses are representative for the composition of a copper object.

Now, the trace element analyses, which were carried out by different laboratories on a single object, will be compared using PCA. Again, the outcome is presented as a scatter plot and analogue numbers show which analyses result from the same object (Fig. 8.5). In this case, once more, the scatter plot shows that the vast majority of cases that are clustered together are those labelled with the same number. However, the distances between single cases are larger, because different laboratories and instruments always result in a certain variation which has to be minimised, but taken into account for every chemical or physical analysis. Therefore, a certain standard deviation is to be considered. In our case, however, no analysis project is separately clustered, so we must take into account that these outliers are caused by individual errors of measurement, e.g. the high concentrations of tin within analyses ANo 62160 and ANo 62161 compared with ANo 22238 and ANo 22268 respectively[142].

Thus, the following comparison has to consider whether outliers may be produced by different projects. Although the plot indicate comparability of trace element analyses carried out by different projects, the PCA also depicts a variation in terms of the bismuth values (Appendix A: Rotated Component Matrix). However, none of the projects shows a significant aberration, although a few double-analyses do not correspond with each other.

Loadings of outliers that are caused by errors in measurement or by metallographic variations are neutralised by increasing the number of cases for the statistical evaluation because, when mean values are calculated from a larger dataset, the standard deviation is consequently smaller. This fact applies to a large number of data, so that the variability of single analysis is counteracted. A greater tolerance provides a greater possibility that the results of statistical grouping reflect the historical truth, but only when those clusters are related to archaeological analyses.

[142] The trace element analyses ANo 62160 and 22238 refer to a dagger and ANo 62161 and 22268 to a flat axe, both from St. Blaise (CH). Double-analyses of the dagger are labelled by *2253 and the one of the flataxe by *2269 in Figure 8.5.

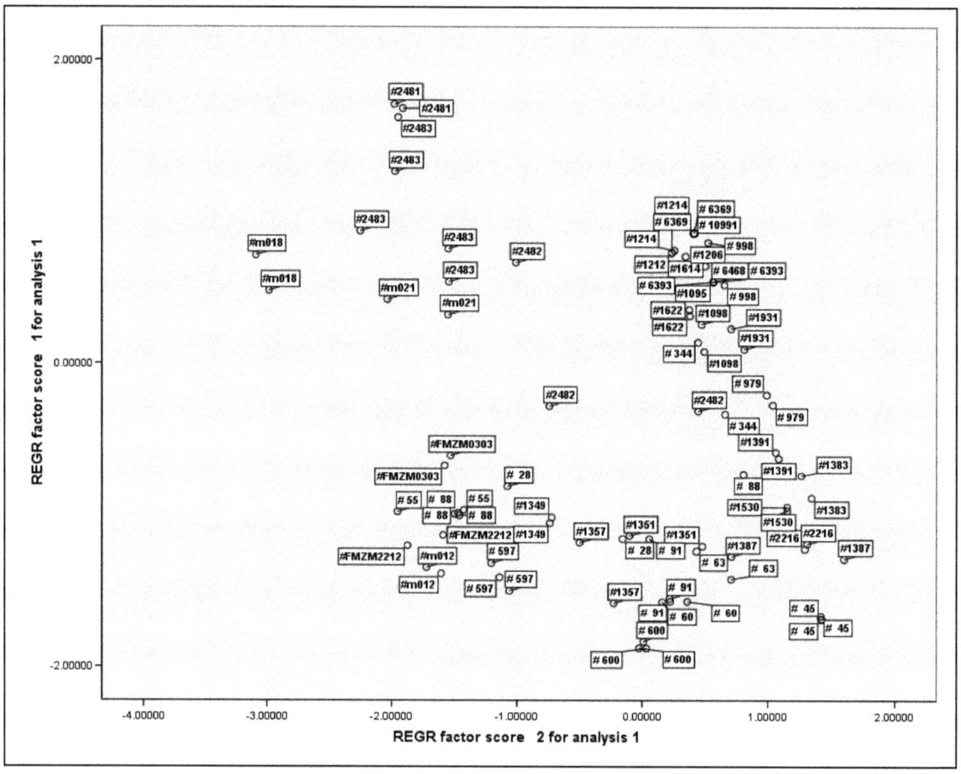

Fig. 8.4 Scatter plot of the principle component analysis concerning the double-analyses of objects each analysed by the same laboratory. Each number indicates an artefact that has been analysed several times; the figures refer to category "DOUBLEANALYSES" at the catalogue (cf. Comments of the Catalogue).

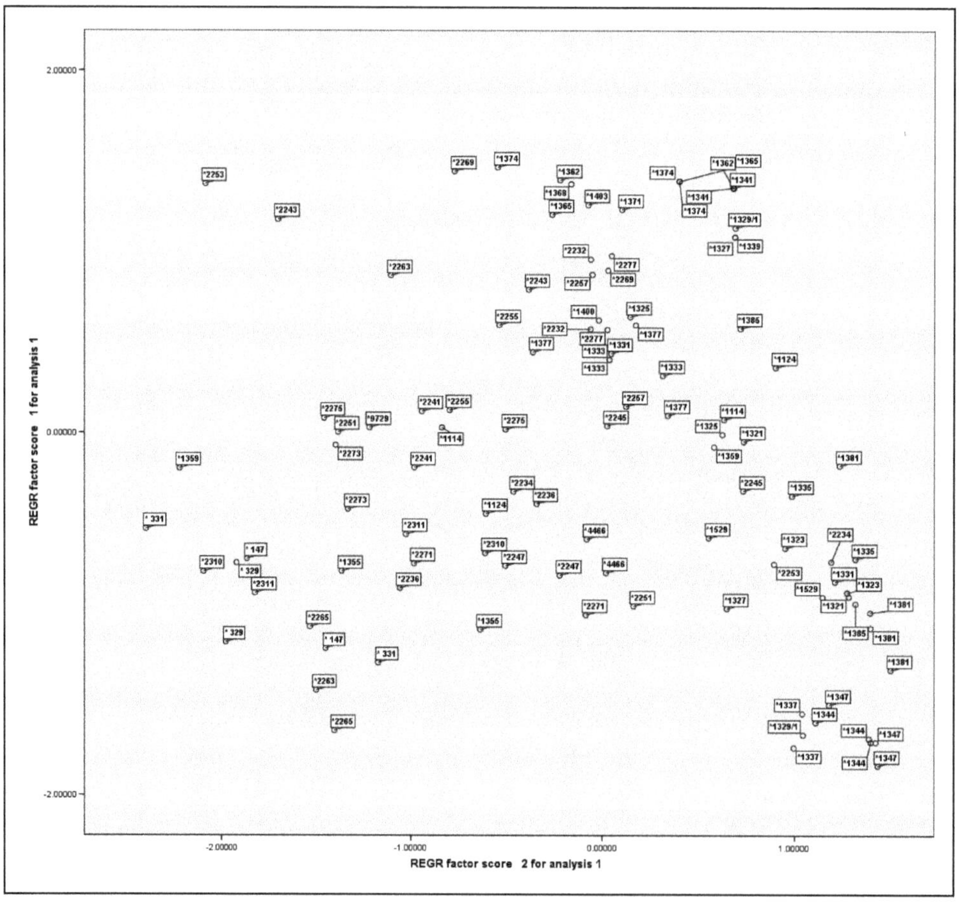

Fig. 8.5 PCA scatter plot for the double-analyses undertaken by various projects on individual objects. Each number indicates an artefact that has been analysed several times; the figures refer to category "DOUBLEANALYSES" of database.

9. Two-Step Cluster Analysis and Principal Component Analysis on the entire Database

As the theoretical description of two-step cluster and principal component analysis presented in Chapter 8 is abstract, a more detailed discussion of these methods and their interpretations is offered in this chapter, using the database of the current research. Firstly, some general comments on running statistical analyses and preliminary preparations are highlighted. Then the process of both statistical methods that are used in the present research and their outcome in SPSS are presented, using the entire database including all the 1943 trace element analyses (cf. Chapter 5). The chapter shows how the output of the statistical analysis is to be interpreted. This should make the results of the other analyses easier to understand by associating then with the SPSS outputs as they are displayed in the relevant appendices[143].

9.1 General Preparations for the Statistical Analysis

Before running any statistical analysis, it is essential to transform the data from weight-by-weight percentage into logarithmic values. As the linear distance between different percentages of trace element concentrations – e.g. between 1% and 2% – is larger than between 0.01% and 0.001%, this would reflect significantly in the statistical results. This great numerical divergence needs to be standardised and with standardised logarithmic values this issue is minimised (Baxter 2003, 75). Various methods can be chosen for standardisation and the negative decade logarithm is used in the current research for practical reasons[144]. Consideration is given to whether a trace element could not be measured – i.e. its value is zero – in which case it cannot be logarithmised (since the logarithm of zero does not exist). For the statistical calculations in the current research, the missing value of 'log 0' is replaced by the constant 3.5, because this is only slightly, but considerably more than the lowest considered values (-log 0.001 = 3.0)[145].

The choice of variables for the statistical examination must be made carefully. As Shennan (1997, 218) pointed out, it is crucial not to evaluate every possible variable, but instead to decide which variable is most appropriate to addressing the specific research question. For instance, Krause and Pernicka (1996, 279) argued that several analysed trace elements (e.g. cobalt, gold, iron, lead and tin) cannot be included within provenance studies and therefore excluded them. However, it is still useful to examine some of these impurities when studying the metallurgical effect of trace elements on copper. Consequently, this study evaluates not only the commonly used elements of arsenic, antimony, bismuth, nickel and silver (cf. SAM; Krause 2003), but also iron, lead and tin. The decision to include the eight elements arsenic, antimony, bismuth, iron, lead, nickel, silver and tin in the present study has been made because these impurities probably provide the best indicator of the intentional use of a specific copper type[146] (cf. Chapter 7).

9.2 The Use of Box Plots to visualise the Statistical Results

Before grouping the database and discussing the results, it is necessary to discuss the methods used to present the output of the statistical analysis. Maximum and minimum scores delimit the size of the dispersion of the groups, but these are extreme values which do not reflect the overall structure of the particular group. The mean value combined with the standard deviation is therefore more useful, since both are defined by the all values of the considered variables, and these give a general impression of the cluster's structure. Maximum and minimum scores,

[143] For further information on running TSC and PCA with SPSS the reader is referred to Kinnear and Gray's user guide (2006).

[144] As most of the impurity content is below 1%, the negative sign turns the logarithmised figures into positive ones. The decade logarithm has been chosen, because the results are smaller than those of the natural logarithm and converts 0.1, 0.01 and 0.001 to whole numbers.

[145] Since the largest dataset consulted for this analysis provides data measured to no more than three decimal points, other measured values are rounded to three decimal digits.

[146] To cross-check the selection of variables, the statistical analyses were also run using the five elements considered in previous studies (Ag, As, Bi, Ni and Sb). The results vary only slightly from the example including eight trace elements. All eight are therefore included, considering that Fe, Pb and Sn may have technological and cultural-historical validity.

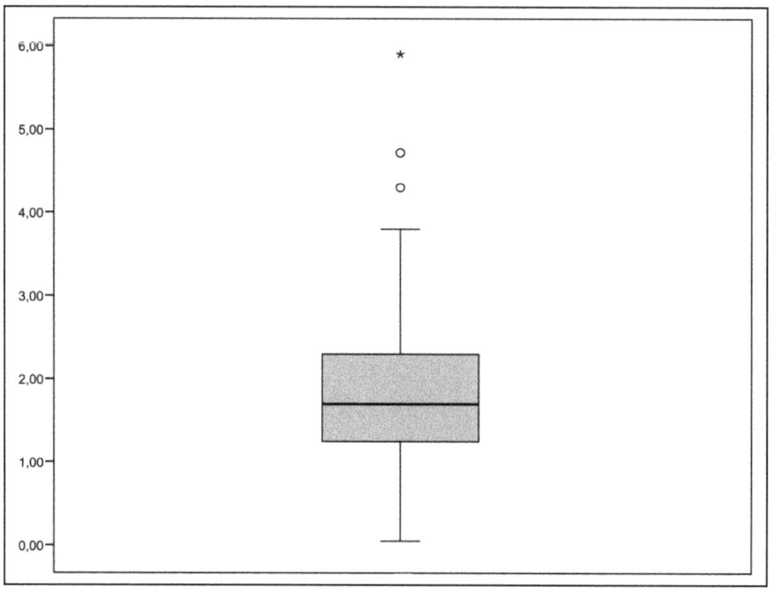

Fig. 9.1 Example of a box plot showing the statistical variation of a dataset.

mean values and standard deviation of the variables within the clusters are presented in the tables in the subsequent chapters. Nevertheless, it has to be borne in mind that the mean value and standard deviation are influenced by even a few extreme outliers. Box plots, however, are an ideal way to visualise the content of the clusters and, at a later stage, the configuration of the copper types that will be defined on the results of the statistical grouping. Since all statistically evaluated groups contain extreme outliers, it is essential to find a method which displays both the average variation of the trace element compositions within a certain group and the distribution of the outlier. For instance, these requirements are fulfilled by the box plots as used in Appendix E.

Box plots visualise the data distribution in four sectors (see Fig. 9.1). The box displays the mid-50% of all data and the *median* (shown as a line within the box) – the score of which the same amount of cases fall below and above. The section including 25% of data below the median is called the *first quartile*; the section above it is called the *third quartile*. The lines above and below the box symbolise the general distribution of each with c. 25% of the dataset beyond the first and third quartile. These two sections are limited by a small bar – the *lower* and *upper hinge*. Using SPSS, circles mark the outliers, and asterisks, the extremes[147]. In Figure 9.1, the box is between approximately 1.25 and 2.30, the lower hinge at 0.00 and the upper hinge at c. 3.75.

[147] For further information: Shennan 1997, 44 – 46.

9.3 THE OUTPUT OF THE TWO-STEP CLUSTER ANALYSIS

Before running a TSC with SPSS, the list of cases needs to be randomised, as explained above in Chapter 8.2.1[148]. After randomisation, the statistical computation is then carried out for the eight variables using the total database including all 1943 trace element analyses (and using the negative logarithmic value of the trace element levels for tin, lead, arsenic, antimony, silver, nickel, bismuth and iron in weight by weight percent). The output directly displays the overall distribution of objects per automatically defined cluster (Fig. 9.2). It also shows the mean values and standard deviation for the logarithmised values of all the impurities considered (see Appendix B: Cluster Profiles). These functions are initialised with the SPSS default.

[148] The author has tested several analyses with different randomisations. These have shown that the variables (here: the percentages of trace elements) account for the clustering much more than the sequence of the data, and the overall contents did not significantly vary, although the total amount of cases per cluster varies. For now, it is stated that the entire database (1943 trace element analyses) is always grouped into two clusters by several TSCs; but the size of these clusters varies slightly. The variation averages out around 5% – the larger cluster includes between c. 1050 and 1100, and the smaller one c. 850 – 900 objects. This variation is caused by the effect of different randomised listing of the data input. As a random listing is used to precede the TSC, this process cannot be reproduced with exactly the same result, apart from using the same list of data, in other words, the data must be listed in the same arrangement.
The general archaeological and metallurgical background of the clusters that are arranged in these groups with slightly varying size tends to be analogous, although a small variation and statistical outliers have to be taken into account. This issue can be considered and reduced by PCA which shows the mistaken arrangement and controls it. The whole issue cannot be described in detail here as it would require too much space at this point, but later on, we will return to the information value of the clusters in comparison with other statistical methods.

Because the output does not automatically show the membership of individual objects within their respective clusters, the command '*create cluster membership variable*' must be ticked at the output-window in order to give the cluster memberships in the data catalogue (CLUSTER_No_I – IV). Otherwise, defining which objects are a part of which clusters is impossible. The output should also display the cluster profile that presents the mean value and the standard deviation of every variable in all clusters.

Cluster Distribution

		N	% of Combined	% of Total
Cluster	1	840	43,2%	43,2%
	2	1103	56,8%	56,8%
	Combined	1943	100,0%	100,0%
Total		1943		100,0%

Fig. 9.2 Distribution of 1943 trace element analyses into two clusters by TSC[149].

The dataset of 1943 objects is divided into two clusters containing 840 and 1103 trace element analyses respectively (Fig. 9.2). As Chapter 10 will illustrate in detail, each of these two groups will be clustered by the TSC in subsequent stages again. This step-by-step clustering will continue for each new developed cluster, as long as the TSC automatically splits a previous cluster into at least two further groups. Each stage of clustering is identified as a certain cluster level, so that Cluster 1 and Cluster 2 resulting from the first TSC are summarised in Cluster Level I[150].

This procedure aims to define types of copper that unite objects that are similar in terms of their trace element compositions (and which thereby may share certain metallurgical characteristics). These can be compared and discussed in archaeometallurgical terms. In other words, the clustering defines various types of copper that differ in characteristic compositions of impurities: e.g. one has significant tin concentration, while another cluster includes arsenic-rich copper objects, etc. The metallurgical content and the variation within a cluster can be visualised using box plots (Appendix E.1). Concerning the box plots resulting from the TSC, the average concentrations, marked by the grey boxes, are more dispersed in Cluster 2 than in Cluster 1. In Cluster 2 the variation of trace elements is larger and, on first impression, the cluster mainly contains objects that are rich in antimony, nickel and arsenic (either singly or in association). The objects of Cluster 1 are generally of a purer copper than those of Cluster 2.

The TSC is currently the most appropriate method to group similar multivariate objects as it chooses the number of clusters automatically and avoids subjective clustering. However, grouping with TSC can 'miss-guided' certain cases at an early stage during the process of running several TSCs. Thus, double-checking and reviewing the results using a second statistical method of multivariate analysis is suggested – either by another type of cluster analysis or the more preferred method, factor analysis (e.g. PCA). Applying PCA is adequate, because all methods of cluster analysis group cases into clearly defined clusters, and within a cluster all cases are handled equally. This makes it impossible to determine boundary areas and outliers. The PCA, however, clearly displays a certain object and its position in relation to other cases within the group. Thus, it offers a way to display intersections and aberrations by scatter plots.

9.4 INTERPRETING THE OUTPUT OF PRINCIPAL COMPONENT ANALYSIS OF THE TOTAL DATABASE

The results of the PCA carried out on all 1943 trace element analyses in the dataset are discussed below and output of the PCA is reported in Appendix C. The calculation results of the correlation matrix of that database for all eight variables are listed in Figure 9.3. This table illustrates the correlation between each variable as worked out by SPSS.

Correlation Matrix

		ng_LogSN	ng_LogPB	ng_LogAS	ng_LogSB	ng_LogAG	ng_LogNI	ng_LogBI	ng_LogFE
Correlation	ng_LogSN	1,000	,286	,348	,418	,437	,452	,090	,339
	ng_LogPB	,286	1,000	,407	,305	,336	,380	,481	,516
	ng_LogAS	,348	,407	1,000	,601	,572	,544	,300	,334
	ng_LogSB	,418	,305	,601	1,000	,840	,628	,362	,195
	ng_LogAG	,437	,336	,572	,840	1,000	,563	,378	,241
	ng_LogNI	,452	,380	,544	,628	,563	1,000	,074	,262
	ng_LogBI	,090	,481	,300	,362	,378	,074	1,000	,292
	ng_LogFE	,339	,516	,334	,195	,241	,262	,292	1,000

Fig. 9.3 Correlation matrix of the PCA based on the 1943 trace element analyses.

[149] This table is copied from the SPSS output. This applies to all tables and graphs concerning the results of the statistics in the present study, unless referenced otherwise.
[150] The detailed description of the contents of the Clusters 1 and 2 is presented in Chapter 10.1.

Total Variance Explained

Component	Initial Eigenvalues Total	% of Variance	Cumulative %	Extraction Sums of Squared Loadings Total	% of Variance	Cumulative %	Rotation Sums of Squared Loadings Total	% of Variance	Cumulative %
1	3,833	47,913	47,913	3,833	47,913	47,913	3,119	38,983	38,983
2	1,257	15,712	63,624	1,257	15,712	63,624	1,971	24,641	63,624
3	,958	11,975	75,600						
4	,614	7,672	83,272						
5	,484	6,052	89,324						
6	,417	5,213	94,537						
7	,288	3,603	98,140						
8	,149	1,860	100,000						

Extraction Method: Principal Component Analysis.

Fig. 9.5 Table of eigenvalues, loadings and rotated loadings of eight components calculated by the eight considered trace elements (Sn, Pb, As, Sb, Ag, Ni, Bi and Fe).

As discussed above, the closer a correlation value is to 1.000, the more closely the two variables are correlated. For example, a relationship seems to be reflected in the results for antimony and nickel and antimony and arsenic, in contrast to bismuth and tin or nickel and tin. However, the other combinations of trace elements seem to illustrate very low correlations to each other.

Communality, which shows how much variance is reflected by the new component, is illustrated in Figure 9.4. For instance, c. 40% of the variance of tin is accounted by the new component and so on.

Communalities

	Initial	Extraction
ng_LogSN	1,000	,399
ng_LogPB	1,000	,723
ng_LogAS	1,000	,598
ng_LogSB	1,000	,805
ng_LogAG	1,000	,762
ng_LogNI	1,000	,667
ng_LogBI	1,000	,553
ng_LogFE	1,000	,585

Extraction Method: Principal Component Analysis.

Fig. 9.4 Communalities.

Figure 9.5 demonstrates crucial information for the further interpretation of the PCA by defining the number of components. The first three columns headlined by *Initial Eigenvalue* consider the sum of particular eigenvalues, i.e. their percentage of variance as they reflect the variables. According to Kaiser's criterion, components that have an eigenvalue of less than 1.0 are not included for further evaluation (Baxter 2003, 81). This is illustrated in a scree plot (Fig. 9.6). The elbow of the graph is significant, because it marks the number of components. The difference between each eigenvalue decline is shown by the curve levelling (Kinnear and Gray 2006; Baxter 2003, 80 f.).

In Figure 9.5, the section headlined *Extraction Sum of Squared Loadings* tabulates the information of the first part again, albeit only for the two considered components. The last section (*Rotation Sums of Squared Loadings*) displays the output for rotated components.
The two tables below show either the component matrix of the un-rotated or the rotated two principal components (Fig. 9.7). For rotation, the method of *varimax with Kaiser's normalisation* is employed.

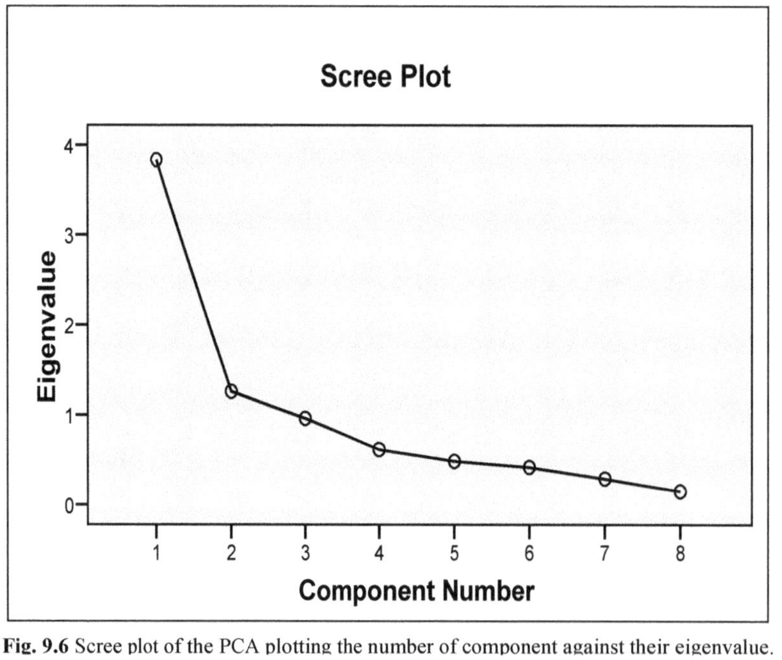

Fig. 9.6 Scree plot of the PCA plotting the number of component against their eigenvalue.

Component Matrix[a]

	Component	
	1	2
ng_LogSN	,608	-,172
ng_LogPB	,640	,559
ng_LogAS	,768	-,087
ng_LogSB	,832	-,334
ng_LogAG	,831	-,268
ng_LogNI	,742	-,341
ng_LogBI	,500	,550
ng_LogFE	,529	,552

Extraction Method: Principal Component Analysis.
a. 2 components extracted.

Rotated Component Matrix[a]

	Component	
	1	2
ng_LogSN	,607	,174
ng_LogPB	,250	,812
ng_LogAS	,699	,330
ng_LogSB	,884	,154
ng_LogAG	,847	,210
ng_LogNI	,810	,101
ng_LogBI	,136	,731
ng_LogFE	,158	,748

Extraction Method: Principal Component Analysis.
Rotation Method: Varimax with Kaiser Normalization.
a. Rotation converged in 3 iterations.

Fig. 9.7 Unrotated and rotated component matrix.

The rotated component matrix demonstrates that the first component is quantified particularly by the values of antimony, silver and nickel, whereas the second component is characterised by lead, iron and bismuth. This is shown by the difference between the loadings of the respective trace elements – the higher the loading, the greater the effect on the composition. The purpose of rotating the components' loading is to change their position. This eases interpretability of the variables as the actual number changes. In other words, the deviation of the variables is now greater and their correlation or disconnection is more obvious. The *Component Transformation Matrix* gives the relationship before and after the rotation (Fig. 9.8) and the *Component Plot in Rotated Space* visualises the correlation of the variables of the two principal components (Fig. 9.9).

Component Transformation Matrix

Component	1	2
1	,850	,527
2	-,527	,850

Extraction Method: Principal Component Analysis.
Rotation Method: Varimax with Kaiser Normalization.

Fig. 9.8 Component transformation matrix.

The *component plot in rotated space* shows two results: firstly, variables with similar loadings are grouped together; secondly, it visualises the results of Figure 9.7. Therefore, there seems to be a connection between bismuth, iron and lead and the subsequent interpretation tries to clarify whether the reason for this is archaeological or metallurgical (cf. Baxter 2003, 78 f.). In considering the results from Chapter 7, the preliminary

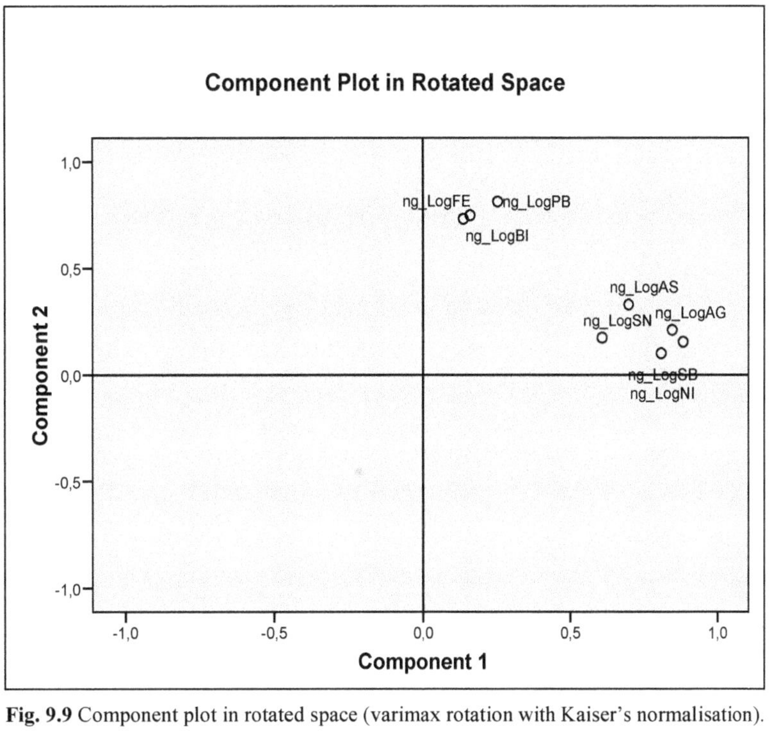

Fig. 9.9 Component plot in rotated space (varimax rotation with Kaiser's normalisation).

impression is that impurities with a more negative effect on the mechanical properties are distinct from the beneficial trace elements, with the exception of antimony. Inter-laboratory variability can be excluded here as an explanatory factor, because higher contents of lead and bismuth have been measured by all projects. Although the content of iron appears in analyses made by the Heidelberg laboratory, inter-laboratory variability can also be excluded in this case as iron has been measured by other laboratories as well. Controlling this by running a PCA without iron delivers a similar result, even if the scatter plot is more condensed.

Finally, a scatter plot displays the regression factor scores of the two principle components (PCA_factor_1 and PCA_factor_2) (Fig. 9.10). According to the *component plot in rotated space*, the X-axis (PCA_factor_2) reflects the lead, iron and bismuth concentrations and the Y-axis (PCA_factor_1), the antimony, silver and nickel content. The concentration of tin and, chiefly, arsenic is not so clearly indicated, since they are located rather between component 1 and 2 at the component plot (Fig. 9.9). The negative decade logarithm is used for the statistical calculations; however, the regression factor scores are reciprocal. Consequently, if their value is higher the concentration of the trace element is low and vice versa. In the current research, this means that objects plotted in the left half of the coordinate system contain comparably more lead, bismuth and iron and cases located in the lower half are rich on antimony, silver and nickel. Moreover, the objects clustered top right consist of pure copper.

To sum up the case of the PCA covering 1943 analyses, the factor scores of two principle components are evaluated to mirror the eight variables. The two components cover a variance of 63.624% of the original variance of the data (with component 1 on its own accounting for 47.913%). Although Baxter (2003, 80) mentions that even around 50% of variance often provides information, this relatively small variance makes it clear why the use of PCA alone will not explain the diversity of copper composition. With consideration of a comparison of PCA and TSC, this uncertainty is smoothed out. As will be demonstrated in detail later on, a comparison of various statistical methods shows related tendencies and analogous archaeometallurgical interpretations.

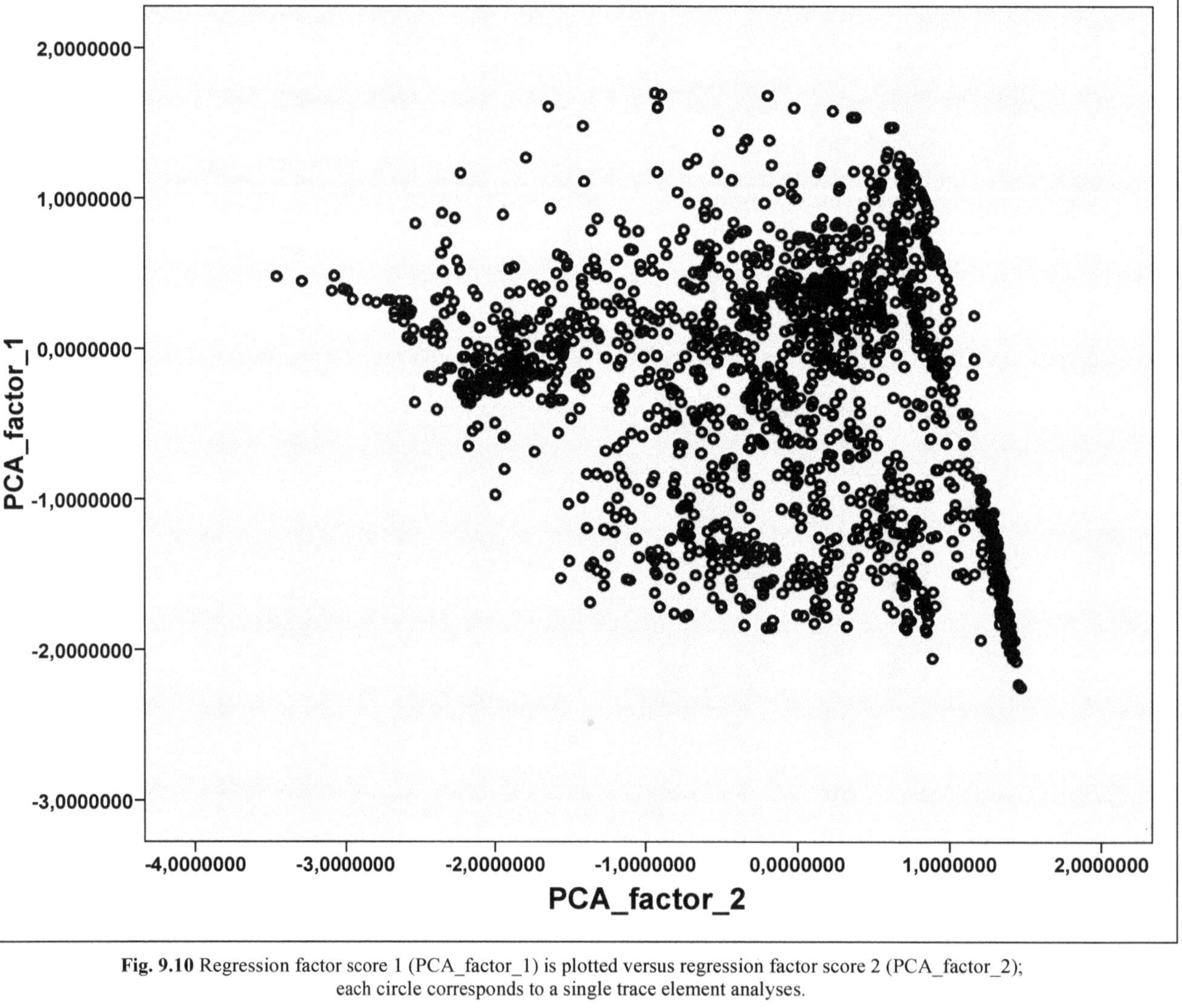

Fig. 9.10 Regression factor score 1 (PCA_factor_1) is plotted versus regression factor score 2 (PCA_factor_2); each circle corresponds to a single trace element analyses.

9.5 ADVANTAGES AND DISADVANTAGES OF TWO-STEP CLUSTER AND PRINCIPAL COMPONENT ANALYSIS IN COMPARISON WITH OTHER METHODS

To conclude the methodology of the two statistical analyses (TSC and PCA) that are employed in the present study, these methods will be compared with the previously popular hierarchical cluster analysis. Although some differences between PCA, TSC and other methods have already been discussed briefly, it is necessary to summarise and discuss the matter in depth, so that future researchers can be aware of the issues involved in selecting an appropriate statistical method(s).

The most significant disadvantage of hierarchical as well as K-means cluster analysis is that the number of groups can only be selected subjectively. Although the cluster tree displays exactly which cases are more similar to each other, the objective discussion of clusters carries the danger of choosing boundaries influenced by archaeological criteria. If these clusters are then interpreted archaeologically, one must be wary of possibly causing a circular argument as a consequence. Christoforidis and Pernicka (1988, 253 f.) have already criticised this procedure and tried to minimise this issue by using the 'Cubic Clustering Criterion' method, which finds the most adequate number of clusters by an approximation. A comparable method is the Agglomeration-Distance-Plot. Both methods calculate values according to various numbers of clusters and, once plotted, the best range of clusters can be identified. Consequently, the decision of which quantity of clusters reflects the expected (archaeological) information is still objective, (Christoforidis and Pernicka 1988, 254; Rassmann 2005, 466). The two-step cluster analysis has the great advantage that it automatically determines the number of clusters and groups the objects solely according to the original data.

The groupings defined by all types of cluster analyses and the *Stammbaum* are definite and the thresholds often appear to be arbitrary. It is difficult to discern whether an object is located in the centre or on the edge of the cluster distribution, or whether it tends to be associated with a particular group but had been classified in a different one due to minor deviations. These issues are minimised by using PCA.

As we have seen PCA offers the opportunity to interpret a larger volume of multivariate data, as in this study. The data are thereby transformed, rearranged and compared according to their similarities in order to determine the correlation between variables. The PCA, particularly the scatter plot of the results, offers the possibility to see how closely certain objects are related to each other. Naturally, this shows overlapping areas and approximates more closely to the prehistoric reality, rather than showing strict clusters. Statistics can never directly represent the true history, but distinct clustering is necessary for the management of large numbers of data. These outcomes are an approach that may highlight tendencies which can then be used as a guideline or directional indicator in order to understand the archaeological record – in particular by comparing results of various statistical methods. As outliers always have to be taken into account in statistical analyses, a larger database and a comparison of several statistical methods may address those issues and reflect a more plausible prehistoric truth.

A combination of the TSC and PCA methods allows both for the objective grouping that defines certain types of copper, and considers individual objects that are not part of a specific cluster in terms of their metallurgy as well as the archaeological background. Statistics in archaeological research are only employed as a tool to ease the interpretation of data and the archaeological interpretation of the statistics is incumbent upon the researcher. The TSC of the present study aims to structure the huge number of this database, but it does not necessarily create general groups that can be compared with results of other cluster analyses.

10. THE RESULTS OF THE STATISTICAL ANALYSES AND THEIR EVALUATION

Throughout the previous sections the two methods – two-step cluster analysis (TSC) and primary components analysis (PCA) – have been identified as the best statistical methods for exploring copper compositions. This chapter will present the results of these statistical analyses and will interpret them as they relate to the question of the uses of different types of copper during the Chalcolithic. The methodology used was as follows: firstly, a TSC was run in order to group the dataset into clusters according to the similarity in trace element composition of the objects. Each of the developed groups was clustered again up to the point when clusters cannot be split by TSC again. Then, the character of the clusters was displayed and presented in various graphs and tables. Next, these clusters were controlled and evaluated by PCA, in order to compare the results of the TSC and to verify possible outliers. Finally, based on the results of these two methods the types of copper were defined. The analyses were run and the results are presented as shown in Chapter 9 and appendices (Appendices B – E). In Appendix E (E.1 – E.102) box plots display the concentrations of the considered elements within a cluster and bar charts show the archaeological, chronological and geographical background of the copper groups[151]. However, before the analyses could be run, some preparation, e.g. randomising the data, had to be made, as discussed in Chapter 9.1 and 9.2.

The results of the sequence of TSC can be illustrated by a cluster tree[152] (Fig. 10.1). That cluster tree is built up by clustering the total database step-by-step, until TSC does not produce any more groups[153]. All in all 71 clusters are

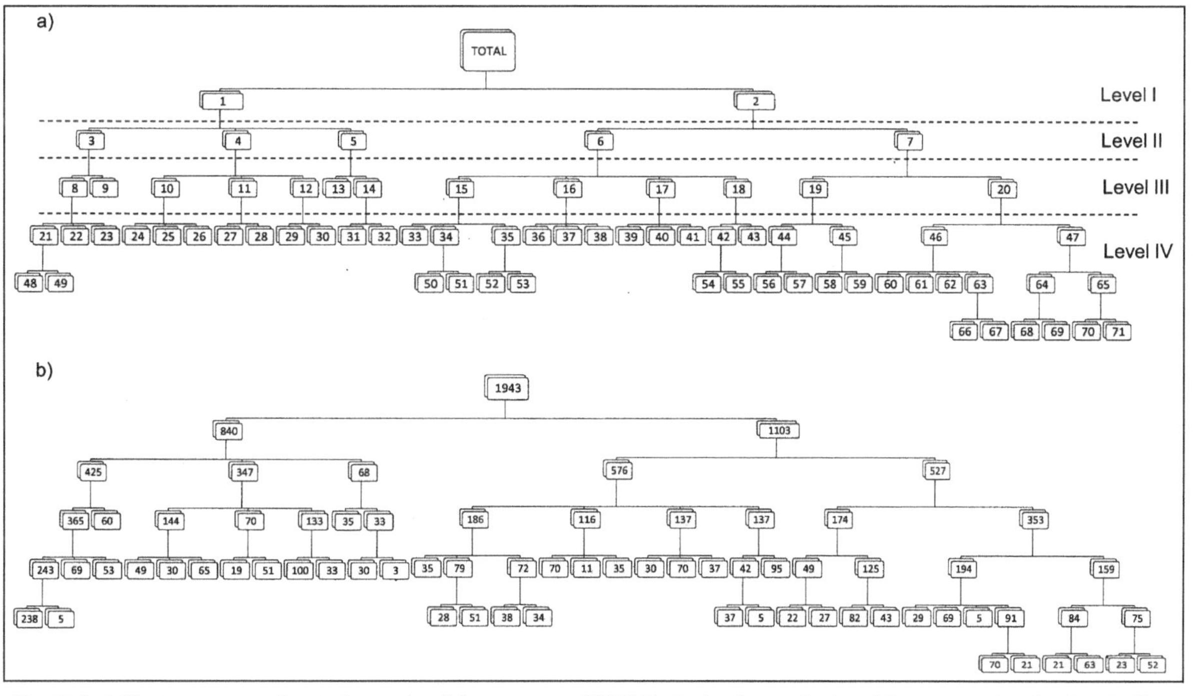

Fig. 10.1 a) Cluster tree according to the results of the sequence of TSC displaying the numbering of the groups; also showing the four levels of the TSC sequence. **b)** Cluster tree according to the results of the TSCs displaying the number of members within each group.

[151] The coding of the database is used as described in Chapter 5.2.2 and in the *Comments on the Catalogue* (p. 181 – 187).

[152] The output of the TSC is added in the Appendix B and the output of the PCA in Appendix C.

[153] Usually, a group of cases cannot be clustered further by TSC either because of the homogeneity of the included objects or due to the small number of cases.

	N	Minimum	Maximum	Mean	Std. Deviation
SN	840	.000	.150	.00047	.007357
PB	840	.000	.290	.00169	.013937
AS	840	.000	4.400	.18704	.461683
SB	840	.000	2.700	.06843	.191489
AG	840	.000	1.750	.04602	.117133
NI	840	.000	.540	.01638	.056326
BI	840	.000	.280	.00584	.021051
FE	840	.000	3.600	.01370	.188209
Valid N (listwise)	840				

Tab. 10.1 Descriptive statistics of Cluster 1.

developed which can be segmented into four cluster levels (Fig. 10.1a). For example, Cluster 1 containing 840 analyses is clustered again. This produces three further clusters (Cluster 3 – 5) in Cluster Level II and those three clusters emerge in a further seven groups in Cluster Level III etc. Cluster Level IV comprises all clusters that cannot be split again; this includes Clusters 9 and 13 which cannot be clustered after Level III. In Figure 10.1b the results of the sequence of TSCs is pictured as well, but here the particular number of members of each cluster is displayed. An analysis belonging to a certain cluster of the respective Cluster Level is recorded in the columns 'CLUSTER_No_X' in the database (cf. Catalogue), where 'X' denotes the relevant Cluster Level (I – IV).

On the basis of various tables and plots, the next section will present all clusters level by level, explain their configurations and review each level by PCA. These clusters are an idealised reflection, summarising objects with similar compositions, and varying outliers have to be considered. However, the clusters reflect very well the structure of each case within its group – especially because of the comparison of several statistical methods. Finally, the groups resulting from the statistical analyses will be compared and summarised as so-called metal groups that reflect the actual types of copper. Tables displaying the correlation between the analysis-number (ANo), the main archaeological information (FIND_PLACE, AREA, CULT, PER, TYPE 2), the cluster numbers of the respective Cluster Level (CLUSTER_No_I – IV) and the interpreted metal group (Metal_groups) are attached at the end of this book.

10.1 CLUSTER LEVEL I

Although the clusters of the Cluster Levels I and II (including Clusters 1 – 7) describe the initial tendencies of the statistical evaluation, the variation is too large to define types of copper. The following tables describe the trace element concentration within each cluster (Tables 8.1 – 8.59). The tables show, at first, the N-number of cases of a cluster; secondly, the minimum and maximum concentration in percent of the objects; thirdly, the mean content of each trace element within the group; and, finally, the standard deviation from the mean value. A high standard deviation (> 0.5) indicates that maxima and minima values have large variance against the mean value. Whether the average value is closer to the maximum or minimum amount has to be checked in order to see which value scatters more. The following descriptions and comparisons are guided by the mean values of the clusters, as the statistically derived clusters include certain random errors that may need to be discussed.

Cluster 1 contains 840 cases. It has, when compared with the other impurities, a significant arsenic concentration, which averages around 0.2%, although some outliers contain up to 4% of arsenic. In average, however, the objects in this cluster show only low levels of impurities (Tab. 10.1; Appendix E.1).

On first examination, the objects of Cluster 1 seem to date to the first half of the 4th millennium BC (corresponding to the temporal code PER 21 applied here). A definite geographical dispersal cannot be recognised, because the two peaks reflect the general distributionn of analyses of that period in the database. The artefacts are chiefly single finds of early dated axe-adzes, flat and hammer axes and some beads. The rest of the artefacts seem to be dispersed regularly (Appendix E.2). This, however, needs to be reviewed in detail after further clustering.

Cluster 2, however, varies much more and includes much higher amounts of all the trace elements than Cluster 1 (Appendix E.1). Not only are the overall element concentrations higher, but also the cluster is bigger (1103 analyses). All sorts of trace elements that are considered here are included in this cluster, but the concentrations vary from small traces to high percentages (Tab. 10.2). These types of copper are generally called fahlores or, more accurately, fahlore-coppers (as we are not dealing with actual ores, but with types of copper).

The objects are generally younger in date than the samples of Cluster 1. Most of the trace element analyses

	N	Minimum	Maximum	Mean	Std. Deviation
SN	1103	.000	4.100	.15911	.562661
PB	1103	.000	16.200	.20901	.836406
AS	1103	.000	5.159	.86601	.834067
SB	1103	.000	9.999	1.16130	1.654830
AG	1103	.000	8.700	.46996	.567845
NI	1103	.000	6.250	.78671	1.102571
BI	1103	.000	1.212	.02593	.079239
FE	1103	.000	17.600	.12874	.669370
Valid N (listwise)	1103				

Tab. 10.2 Descriptive statistics of Cluster 2.

come from objects that date to the 3rd millennium BC (Appendix E.3). A bead from Gaienhofen (Hornstaad-Hörnle I) in southern Germany is noteworthy, because it contains a significant concentration of iron (17.6%)[154], and, therefore, it always appears in the statistics as an outlier. This also accounts for a few lead-rich objects[155] as well. Tests have shown that these singular finds do not change the results of the statistical analysis, because the rather similar concentrations of the other elements in question counteract the effect of those outliers.

A comparison with the scatter plot produced by the regression factor scores of the two principle components of the PCA of the entire database shows that the Clusters 1 and 2 differ considerably (Appendix E.4).

10.2 CLUSTER LEVEL II

In Cluster Level II, Cluster 1 is split up into Clusters 3, 4 and 5; and Cluster 2 into Cluster 6 and 7 (Fig. 10.1). While Cluster 3 consists of very pure copper objects, Cluster 5 indicates a typical arsenical copper – containing arsenic, but not much of other impurities. Cluster 4 is contaminated by more elements than just arsenic (Appendix E.5).

Cluster 3 consists of 425 objects that are free of tin and lead; bismuth and iron are only present in traces. Some artefacts contain smaller amounts of arsenic, antimony and silver, but not more than 0.28% respectively. The largest variation can be seen in the case of nickel (Tab. 10.3).
The objects of Cluster 3 – chiefly axe-adzes and hammer axes, but also some other types – are dated to PER 10, 21 and some to PER 31 and were found in Switzerland, western Slovakia and on the Great Hungarian Plain (Appendix E.6).

Cluster 4 contains 347 cases with rather higher arsenic, antimony and silver contents (Tab. 10.4). In contrast, the nickel concentration is lower. Objects dated to PER 21 dominate this cluster, but younger, even PER 40, finds are included (Appendix E.7).

Cluster 5 contains only 68 trace element analyses. Whereas most of the elements are below 0.1%, iron and

	N	Minimum	Maximum	Mean	Std. Deviation
SN	425	.000	.000	.00000	.000000
PB	425	.000	.000	.00000	.000000
AS	425	.000	.240	.00299	.022080
SB	425	.000	.100	.00082	.005576
AG	425	.000	.280	.00514	.023468
NI	425	.000	.540	.02328	.072263
BI	425	.000	.002	.00004	.000229
FE	425	.000	.001	.00000	.000049
Valid N (listwise)	425				

Tab. 10.3 Descriptive statistics of Cluster 3.

[154] ANo 100032. This significantly high concentration of iron and low content of copper (24.7%) can be explained by the fact that the bead does not consist of smelted copper metal. This bead from Hornstaad-Hörnle I (D) was probably made from an oxide copper ore (Höppner et al. 2005, 294).

[155] ANo 100025, 100099, 100122 and 100125.

	N	Minimum	Maximum	Mean	Std. Deviation
SN	347	.000	.000	.00000	.000000
PB	347	.000	.000	.00000	.000000
AS	347	.000	4.400	.36273	.578535
SB	347	.000	2.700	.16110	.271541
AG	347	.000	1.750	.09873	.164818
NI	347	.000	.300	.00738	.024375
BI	347	.000	.280	.01319	.030746
FE	347	.000	.001	.00000	.000054
Valid N (listwise)	347				

Tab. 10.4 Descriptive statistics of Cluster 4.

	N	Minimum	Maximum	Mean	Std. Deviation
SN	68	.000	.150	.00582	.025418
PB	68	.000	.290	.02084	.045024
AS	68	.000	2.500	.44087	.705330
SB	68	.000	.290	.01809	.046402
AG	68	.000	.210	.03251	.053508
NI	68	.001	.320	.01912	.053710
BI	68	.000	.100	.00454	.013123
FE	68	.000	3.600	.16925	.645646
Valid N (listwise)	68				

Tab. 10.5 Descriptive statistics of Cluster 5.

especially arsenic are dominant in that cluster (Tab. 10.5). The artefacts' origin is located in Moravia and Slovakia and in the Alpine region. Most of the objects are connected to the Mondsee group. The general chronological picture looks similar to Cluster 4 (Appendix E.8).

The TSC of Cluster 2 produced two nearly commensurate clusters (Clusters 6 and 7; Appendix E.9).

Cluster 6 encompasses 576 mostly arsenic-rich trace element analyses. The other typical 'fahlore-impurities', antimony, silver and nickel, also appear frequently. The fairly high average value of lead is also remarkable (Tab. 10.6).

The whole of Cluster 6 is dominated by the analysed finds from the Corded Ware graves of Drosa in Saxony-Anhalt (Appendix E.10). Currently, therefore, we can say that Cluster 6 is marked by finds from contexts dated to the 3rd millennium BC. Additional cluster analyses will clarify whether a distinct 'Drosa copper' can be separated from examples from other contexts.

Cluster 7 can – already at this stage of evaluation – be preliminarily defined as 'Earliest Bronze Age copper', because 494 out of 527 trace element analyses are from objects that are classified as Earliest Bronze Age (21 further analyses are PER 31). Copper objects with higher antimony and nickel concentrations mark this cluster.

	N	Minimum	Maximum	Mean	Std. Deviation
SN	576	.000	4.100	.09214	.463999
PB	576	.000	16.200	.31583	1.112385
AS	576	.000	5.159	.88002	.947327
SB	576	.000	6.000	.23846	.468105
AG	576	.000	8.700	.26761	.604905
NI	576	.000	4.000	.12430	.351139
BI	576	.000	1.212	.04407	.105190
FE	576	.000	17.600	.17383	.876311
Valid N (listwise)	576				

Tab. 10.6 Descriptive statistics of Cluster 6.

	N	Minimum	Maximum	Mean	Std. Deviation
SN	527	.000	4.000	.23230	.646249
PB	527	.000	2.700	.09225	.294662
AS	527	.005	3.600	.85070	.689916
SB	527	.030	9.999	2.16994	1.883037
AG	527	.049	2.700	.69113	.426267
NI	527	.001	6.250	1.51071	1.185929
BI	527	.000	.190	.00610	.017456
FE	527	.000	2.600	.07946	.307551
Valid N (listwise)	527				

Tab. 10.7 Descriptive statistics of Cluster 7.

Also the tin content is significantly higher compared with Cluster 6. The arsenic concentration, however, is comparable with Cluster 6. In principle, the cluster contains objects with the highest levels of impurities of the database (Tab. 10.7).

The regional distribution of the objects in Cluster 7 reflects more or less the distribution of the analyses of PER 40 (Chapter 5.2.2.1). In other words, the finds centre on Southern Germany and the area between Lower Austria and the region along the river Nitra (Appendix E.11).

The result of the PCA indicates that Clusters 3, 4, 6 and 7 in particular can be comparatively well differentiated, while it is impossible to distinguish Cluster 5 primarily from Cluster 4, but also from Cluster 6 (Appendix E.12). Whether there is actually a metallurgical difference between Cluster 5 and the rest will be explained by the next stage of clustering. For now, it seems that Cluster 3 is defined by a type of pure copper, and that Clusters 4, 5 and 6 are defined by a type which has dominant concentrations of arsenic, although Clusters 4 and 6 also contain other trace elements, such as antimony, lead and silver. Cluster 7 includes the objects with the highest concentrations of impurities. The results of Level II clustering, displaying the corresponding mean values of each cluster, are set out in Figure 10.2.

10.3 CLUSTER LEVEL III

If we follow the cluster tree (Fig. 10.1a) at Cluster Level III from left to right, the first clusters to describe are Clusters 8 and 9 that divide up Cluster 3. The box plots show that both clusters are types of relatively pure copper, but the content of Cluster 9 varies slightly more, which is also reflected by the variation of SAM-groups (cf. Appendix E.13). Cluster 8 can be partitioned into three clusters more at the next level, whereas Cluster 9 cannot be further separated by TSC.

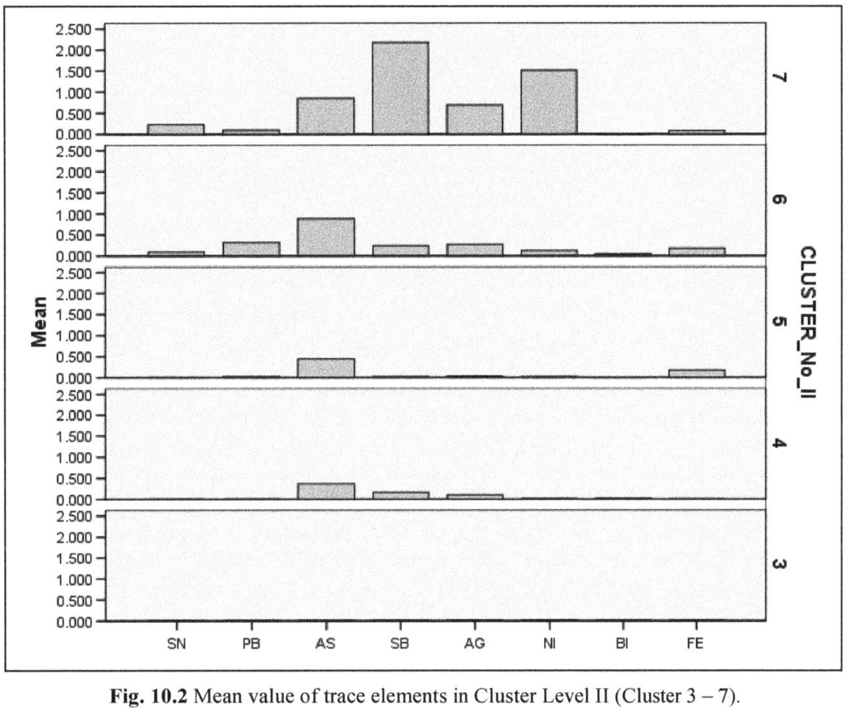

Fig. 10.2 Mean value of trace elements in Cluster Level II (Cluster 3 – 7).

	N	Minimum	Maximum	Mean	Std. Deviation
SN	365	.000	.000	.00000	.000000
PB	365	.000	.000	.00000	.000000
AS	365	.000	.000	.00000	.000000
SB	365	.000	.000	.00000	.000000
AG	365	.000	.280	.00434	.022341
NI	365	.000	.540	.02387	.073116
BI	365	.000	.000	.00000	.000000
FE	365	.000	.000	.00000	.000000
Valid N (listwise)	365				

Tab. 10.8 Descriptive statistics of Cluster 8.

Cluster 8 (365 trace element analyses), which seems to comprise very pure copper objects (Tab. 10.8), will be discussed in detail at Level IV. In general, Cluster 8 primarily includes earlier dated finds, which mainly come from the eastern area of the study area. Most of these are axe-adzes and hammer axes found without contexts (Appendix E.14).

Cluster 9 (60 analyses) includes almost pure artefacts, as well as some with quantities of arsenic, antimony and silver and a wider scattering range of nickel. The concentrations of tin, lead, bismuth and iron are nearly consistently zero (Tab. 10.9). TSC does not split Cluster 9 up again, thus it is already included in Cluster Level IV. The artefacts combined in Cluster 9 are chiefly single finds of axes that are dated to PER 10, 21 and 25, but younger periods are represented as well – including five finds from Czech Bell Beaker graves. In terms of archaeological groups, Cluster 9 shows a certain uniform spread of Baden, Bell Beaker and Corded Ware finds with a slight peak of artefacts from Bodrogkeresztúr contexts – other archaeological groups are less frequent. The objects come predominantly from the eastern part of the study area (Appendix E.15).

	N	Minimum	Maximum	Mean	Std. Deviation
SN	60	.000	.000	.00000	.000000
PB	60	.000	.000	.00000	.000000
AS	60	.000	.240	.02118	.055781
SB	60	.000	.100	.00583	.013918
AG	60	.000	.210	.00998	.029164
NI	60	.000	.430	.01965	.067304
BI	60	.000	.002	.00032	.000537
FE	60	.000	.001	.00002	.000129
Valid N (listwise)	60				

Tab. 10.9 Descriptive statistics of Cluster 9.

	N	Minimum	Maximum	Mean	Std. Deviation
SN	144	.000	.000	.00000	.000000
PB	144	.000	.000	.00000	.000000
AS	144	.000	1.600	.09228	.243586
SB	144	.000	2.700	.33491	.339577
AG	144	.006	1.750	.17672	.211753
NI	144	.000	.001	.00099	.000083
BI	144	.000	.280	.02668	.042428
FE	144	.000	.000	.00000	.000000
Valid N (listwise)	144				

Tab. 10.10 Descriptive statistics of Cluster 10.

	N	Minimum	Maximum	Mean	Std. Deviation
SN	70	.000	.000	.00000	.000000
PB	70	.000	.000	.00000	.000000
AS	70	.000	3.400	.57587	.710366
SB	70	.000	.660	.04716	.086204
AG	70	.000	.680	.07769	.133450
NI	70	.001	.300	.03256	.046612
BI	70	.000	.083	.00523	.014178
FE	70	.000	.001	.00001	.000120
Valid N (listwise)	70				

Tab. 10.11 Descriptive statistics of Cluster 11.

	N	Minimum	Maximum	Mean	Std. Deviation
SN	133	.000	.000	.00000	.000000
PB	133	.000	.000	.00000	.000000
AS	133	.000	4.400	.54336	.642272
SB	133	.000	.820	.03289	.089794
AG	133	.000	.170	.02537	.032038
NI	133	.000	.010	.00105	.000791
BI	133	.000	.078	.00277	.009013
FE	133	.000	.000	.00000	.000000
Valid N (listwise)	133				

Tab. 10.12 Descriptive statistics of Cluster 12.

Three groups can be separated from Cluster 4. Whereas Cluster 10 (144 trace element analyses) is characterised by higher antimony concentrations, followed by silver and arsenic (Tab. 10.10), Clusters 11 (70 cases) and 12 (133 analyses), both contain arsenic-rich cases (mainly E_{01}A-copper after SAM) (Appendix E.16).

Cluster 10 especially includes objects from the Czech Republic and Slovakia, and, in terms of their archaeological contexts, most of these were found at sites of the Baden and Corded Ware complexes and in Bell Beaker graves. The larger amount of single finds of this cluster tends to be from the 4th millennium BC (Appendix E.17).

Cluster 11 is defined as an arsenic-rich copper with lower levels of the impurities antimony and silver (Tab. 10.11). It includes artefacts discovered in contexts of the western Slovak Nitra group. Other objects are from all over the research area, especially from various early Chalcolithic contexts (Appendix E.18).

Cluster 12 consists chiefly of single finds (flat axes) and copper beads from Seeberg (Burgäschisee-Süd) in Switzerland. In general, finds of this cluster date to PER 21, even after considering the 38 analysed beads from the one location at Burgäschisee-Süd[156]. PER 25, 31 and 40 also appear frequently, and the finds of that group are distributed throughout the study area (Appendix E.19). The finds of Cluster 12 are chiefly made of arsenic-rich copper (Tab. 10.12).

The re-splitting of Cluster 5 by TSC results in two further groups, where Cluster 13 marks an end of the tree and Cluster 14 can be split again into two clusters at the next level. Both – Cluster 13 and 14 – are also arsenic-rich copper types, but Cluster 13 differs with regard to its nickel and iron concentration. Cluster 14, in comparison to the previously discussed groups, has some lead content (Appendix E.20).

Cluster 13 is characterised by impurities of especially arsenic and also traces of nickel and iron (Tab. 10.13). Due to its small size (35 trace element analyses), a definite tendency in its configuration is hard to identify. Half of the analyses are dated to PER 21; the rest cover the other periods. The finds were found in the northalpine region and in the area of the western Carpathians (Appendix E.21). Cluster 13 cannot be split again by TSC.

[156] In total, 41 trace element analyses have been carried out on copper finds from Burgäschisee-Süd. The other three analyses appear in Cluster 14.

	N	Minimum	Maximum	Mean	Std. Deviation
SN	35	.000	.150	.01103	.034826
PB	35	.000	.094	.00300	.015858
AS	35	.000	2.500	.36366	.718343
SB	35	.000	.110	.00609	.019021
AG	35	.000	.110	.01186	.021634
NI	35	.001	.320	.03060	.072979
BI	35	.000	.012	.00151	.003364
FE	35	.000	3.600	.32826	.876254
Valid N (listwise)	35				

Tab. 10.13 Descriptive statistics of Cluster 13.

	N	Minimum	Maximum	Mean	Std. Deviation
SN	33	.000	.010	.00030	.001741
PB	33	.000	.290	.03976	.057097
AS	33	.000	2.500	.52276	.692761
SB	33	.000	.290	.03082	.061636
AG	33	.000	.210	.05442	.067346
NI	33	.001	.040	.00694	.009165
BI	33	.000	.100	.00776	.018107
FE	33	.000	.010	.00061	.002423
Valid N (listwise)	33				

Tab. 10.14 Descriptive statistics of Cluster 14.

Cluster 14 is on average richer in arsenic and silver (Tab. 10.14). It covers objects dispersed throughout the entire period of this study. In terms of its regional and chronological distribution no distinct archaeological interpretation can be made (Appendix E.22).

For Clusters 8 – 14 (that were developed out of Clusters 3 – 5), it can be stated that Clusters 8 and 9 are chiefly types of pure or almost pure copper, Cluster 10 can be identified as an antimony-copper, and Clusters 11 – 14 are arsenic-coppers. This is also reflected in the average values of each of these clusters (Fig. 10.3).

In Cluster Level III the TSC develops four groups out of Cluster 6 and two out of Cluster 7. Clusters 15 – 18 emerged from Cluster 6 and those are generally purer than Clusters 19 and 20. The latter two are typical antimony- and nickel-dominated varieties of fahlore-copper (Appendix E.23).

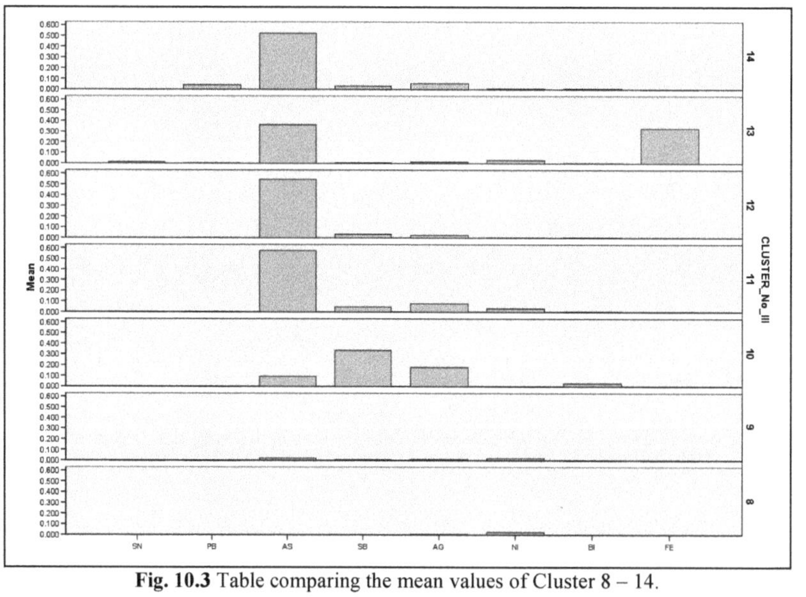

Fig. 10.3 Table comparing the mean values of Cluster 8 – 14.

	N	Minimum	Maximum	Mean	Std. Deviation
SN	186	.000	.200	.00480	.014661
PB	186	.009	6.500	.49702	.735713
AS	186	.060	5.159	1.10061	.827371
SB	186	.008	.946	.10355	.138352
AG	186	.008	.967	.12678	.114003
NI	186	.010	2.600	.17484	.242114
BI	186	.000	1.212	.07073	.143189
FE	186	.020	1.840	.12669	.172765
Valid N (listwise)	186				

Tab. 10.15 Descriptive statistics of Cluster 15.

	N	Minimum	Maximum	Mean	Std. Deviation
SN	116	.000	3.300	.05881	.376983
PB	116	.000	9.320	.10278	.864885
AS	116	.000	5.000	1.16407	1.275785
SB	116	.000	.780	.03434	.091877
AG	116	.000	8.700	.09521	.806504
NI	116	.000	4.000	.18389	.634006
BI	116	.000	.061	.00482	.009708
FE	116	.000	5.600	.31015	.795072
Valid N (listwise)	116				

Tab. 10.16 Descriptive statistics of Cluster 16.

Cluster 15 consists of 186 trace element analyses. 156 of the 186 analyses are from the famous Corded Ware graves of Drosa (D), where in total 163 analyses of copper reels and tubular sheet metal beads have been made. Although this group is dominated by the Drosa finds, the other finds are contemporaneous and geographically similar. Taking into account that the Drosa analyses are from two graves, 19 analyses are considered as contemporaneous to the Corded Ware complex, 6 are Bell Beaker related and 5 objects are classified as Earliest Bronze Age. Moreover, most of the analyses are located in Middle and Southern Germany and Switzerland (Appendix E.24). In terms of their metallurgical composition, the objects of Cluster 15 are considerably rich in arsenic. They contain the typical elements of fahlore-coppers – antimony, nickel and silver – but what is significant is a relatively high concentration of lead (on average 0.497%) and some iron (Tab. 10.15).

Cluster 16 (116 analyses) is similar to Cluster 15 in terms of its metallurgical configuration, except for the concentration of lead, which is much smaller, although with one outlier. A tiny copper point from Tousen (CZ) contains 9.32% of lead[157] and another one from the same site contains 8.7% silver[158] (Tab. 10.16).

The chronological and geographical distribution of Cluster 16 is more diverse than Cluster 15. Most of the analysed artefacts were found in contexts of the Pfyn-Altheim-Mondsee horizon. Nevertheless samples from all the other periods – in particular PER 31 – are represented (Appendix E.25).

Cluster 17 (137 analyses) predominantly contains silver impurities, and then antimony. Arsenic, lead and tin concentration are up to a maximum of around 2% and 4% respectively. With some outliers, the contents of rest of the considered trace elements are relatively low (Tab. 10.17). Nearly two thirds of the analyses within Cluster 17 come from objects that are dated to the 3rd millennium BC. The finds, chiefly jewellery, are spread over the entire study area, except the Carpathian basin (Appendix E.26).

Cluster 18 includes 137 trace element analyses that have a similar chronological spread to Cluster 17 – however with a regional centre of distribution in Slovakia and Hungary (Appendix E.27). From the metallurgical perspective, Cluster 18 is arsenic- and antimony-rich, and contains all the other measured trace elements (Tab. 10.18).

[157] ANo 100025.
[158] ANo 100026.

	N	Minimum	Maximum	Mean	Std. Deviation
SN	137	.000	4.100	.32879	.844877
PB	137	.000	1.500	.15133	.287887
AS	137	.000	2.020	.16824	.328810
SB	137	.000	2.500	.32191	.394573
AG	137	.000	7.300	.61120	.832378
NI	137	.000	.970	.05164	.130571
BI	137	.000	.268	.02022	.042649
FE	137	.000	17.600	.29580	1.614755
Valid N (listwise)	137				

Tab. 10.17 Descriptive statistics of Cluster 17.

	N	Minimum	Maximum	Mean	Std. Deviation
SN	137	.000	.047	.00228	.006521
PB	137	.000	16.200	.41473	1.912914
AS	137	.005	3.800	1.05181	.838588
SB	137	.000	6.000	.51101	.771556
AG	137	.000	1.950	.26118	.339207
NI	137	.001	2.500	.07790	.265717
BI	137	.000	.620	.06495	.117024
FE	137	.000	.011	.00045	.002090
Valid N (listwise)	137				

Tab. 10.18 Descriptive statistics of Cluster 18.

Clusters 19 and 20 (that both emerged from Cluster 7) are remarkably similar in terms of the ratio of impurities. Both Cluster 19 and 20 contain chiefly antimony and nickel, and lower concentrations of arsenic and silver. Other impurities, e.g. tin and lead, are also common, but on average in much lower concentrations. Cluster 19 has a generally higher concentration of trace elements than Cluster 20 – but both are the clusters with highest concentrations of impurities in comparison with the other clusters (Appendix E.23).

Cluster 19 consists of 174 analyses which have higher concentrations of trace elements than the copper objects in Cluster 20 (Tab. 10.19). Virtually no trace element analyses of periods other than PER 40 are included in Cluster 19, and by far most of them came from southern Germany and Switzerland. Nevertheless, it has to be taken into account that in particular the Swiss finds are flanged axes of the Salez type. These are closely related to the *Blechkreis*, although they are often found in hoards or single finds. This cluster also includes most of the

	N	Minimum	Maximum	Mean	Std. Deviation
SN	174	.000	3.200	.22623	.560409
PB	174	.000	2.700	.24210	.439013
AS	174	.080	3.600	1.35311	.782106
SB	174	.250	9.999	3.61725	2.432549
AG	174	.057	2.700	.98105	.508167
NI	174	.050	6.250	2.50042	1.295546
BI	174	.000	.179	.00882	.020534
FE	174	.000	2.600	.22139	.496488
Valid N (listwise)	174				

Tab. 10.19 Descriptive statistics of Cluster 19.

	N	Minimum	Maximum	Mean	Std. Deviation
SN	353	.000	4.000	.23529	.685349
PB	353	.000	2.400	.01839	.135836
AS	353	.005	3.100	.60305	.473667
SB	353	.030	5.000	1.45653	.918981
AG	353	.049	2.600	.54822	.287365
NI	353	.001	4.300	1.02286	.744224
BI	353	.000	.190	.00476	.015578
FE	353	.000	1.120	.00950	.073006
Valid N (listwise)	353				

Tab. 10.20 Descriptive statistics of Cluster 20.

analyses from the famous Earliest Bronze Age cemetery of Singen (D) on Lake Constance (Appendix E.28).

Cluster 20 (353 analyses) is identified as a group encompassing Earliest Bronze Age copper objects from the Nitra region in western Slovakia, even if artefacts from the northalpine region are also included (Appendix E.29). This cluster also consists of objects containing higher amounts of antimony, nickel and arsenic (Tab. 10.20).

In conclusion, Clusters 15 – 20 consist of copper objects that comprise higher concentrations of all considered trace elements than Clusters 8 – 14. On average, Clusters 15 and 16 can be defined as arsenic-coppers, while the finds of Cluster 15 also contain considerable amounts of lead. The remaining Clusters (17 – 20) are types of fahlore-copper containing not only high amounts of antimony, arsenic, nickel and silver, but also relatively high concentrations of iron, lead and tin (Fig. 10.4).

If the results of the TSC of Cluster Level III (Fig. 10.4) are compared with the results of the PCA as displayed in Appendix E.30, it is obvious that the groups with fewer impurities (Clusters 9 – 14) have an extremely overlapping spread, whereas Clusters 15, 19 and 20 are separated from each other and the rest. The scatter plot shows that Clusters 16, 17 and 18 cover a section overlapping the 'purer' Clusters 9 – 14 and the 'dirtier' Clusters 15, 19 and 20 (Appendix E.30). For instance, Cluster 17 is not very significant in the scatter plot of the PCA, as that plot is especially orientated according to the lead, bismuth and iron content (PCA_factor_2), and to the antimony, silver and nickel concentration (PCA_factor_1). The orientations of the two components do not reflect the concentrations of arsenic and tin, so clusters that are defined by arsenic and tin cannot be clearly distinguished by PCA. In these cases it has to be concentrated on the results of the TSC and double-checked by hand.

At this point, some clusters can be grouped into three larger groups and indicate some tendencies for later interpretation. Firstly, the types of almost pure copper (Cluster 8 and 9); secondly, diverse types of arsenical copper which defines Clusters 11 – 16, whereby the concentration in percent is higher in Clusters 15 and 16; and thirdly, the remaining Clusters (17 – 20) are considered as copper types that by previous studies have often be termed 'fahlore-coppers' which are merged with various trace elements in higher concentrations. The next cluster step will shed more light on that issue.

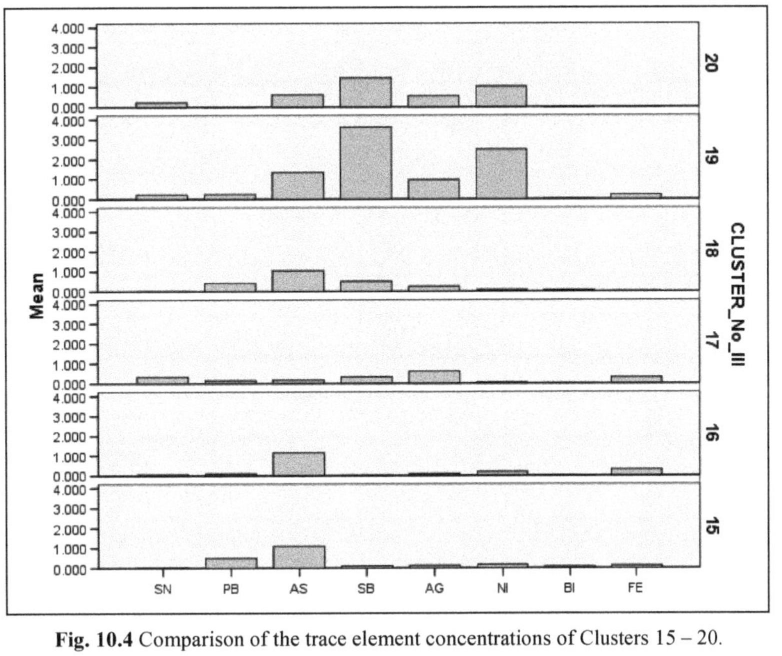

Fig. 10.4 Comparison of the trace element concentrations of Clusters 15 – 20.

10.4 CLUSTER LEVEL IV

This section focuses on the bulk of final groups of the cluster tree. In the following chapter the last branches of the cluster tree will be discussed by referring to the previous clusters from which these emerged. Therefore, several clusters need to be described by summarising them by previous groups, as it is useful to consider their 'origin'. Again, the cluster tree will be portrayed step-by-step from left to right (Fig. 10.1a). Cluster Level IV comprises 40 groups plus Clusters 9 and 13 from Level III, which have already been illustrated in the last section. As mentioned above, Cluster 8 splits into Clusters 21, 22 and 23, whereas 21 can then be divided up again into Clusters 48 and 49. In general, all the clusters that emerged from Cluster 8 contain hardly any trace elements. This is also reflected by the SAM-classification that does not differentiate between Clusters 48, 49 and 23 and typifies those as E_{00}-copper. The reason is that the SAM-classification is not as precise as the TSC. However, Cluster 23 in particular varies in its silver concentration, which is a type of almost pure copper not considered by SAM. In contrast to other purer groups Cluster 22 has remarkable concentrations of nickel (FC-copper after SAM-classification) (Appendix E.31).

Cluster 48 comprises 238 trace element analyses. All of them are free of impurities, except for traces of nickel (only 0.001%; Tab. 10.21). This cluster includes nearly half of the earliest copper finds of the entire database (45 of 106) and many other objects date to PER 21. Consequently, this cluster can be treated as an early copper, even if some later artefacts are included as well. The typical finds of that time, hammer axes and axe-adzes, provided the most of the analyses (Appendix E.32).

Cluster 49 is composed of five objects and all of them were found as a hoard in Stollhof 'Lange Wand' (A) which consists of 26 copper and two gold finds and is identified with the Baden complex (Parzinger 1992). The rest of the 20 analyses of the Stollhof hoard are spread over the relatively pure copper comprising Clusters 9, 23 and 48, and Cluster 30, which is of arsenic-copper. Cluster 49 differs from Cluster 48, as these finds are free from nickel but have traces of silver (Tab. 10.22).

Cluster 22 is defined as nickel-copper (69 trace element analyses). Except for a small amount of objects that contain traces of silver, the artefacts of that group have about 0.1% nickel, but no other impurities (Tab. 10.23). Cluster 22 primarily contains beads from the lake dwelling of Vinelz in Switzerland (Strahm 1971) and other roughly contemporary finds from western Switzerland. Nevertheless, younger as well as older finds are also grouped in Cluster 22 (Appendix E.33).

	N	Minimum	Maximum	Mean	Std. Deviation
SN	238	.000	.000	.00000	.000000
PB	238	.000	.000	.00000	.000000
AS	238	.000	.000	.00000	.000000
SB	238	.000	.000	.00000	.000000
AG	238	.000	.000	.00000	.000000
NI	238	.001	.001	.00100	.000000
BI	238	.000	.000	.00000	.000000
FE	238	.000	.000	.00000	.000000
Valid N (listwise)	238				

Tab. 10.21 Descriptive statistics of Cluster 48.

	N	Minimum	Maximum	Mean	Std. Deviation
SN	5	.000	.000	.00000	.000000
PB	5	.000	.000	.00000	.000000
AS	5	.000	.000	.00000	.000000
SB	5	.000	.000	.00000	.000000
AG	5	.001	.001	.00100	.000000
NI	5	.000	.000	.00000	.000000
BI	5	.000	.000	.00000	.000000
FE	5	.000	.000	.00000	.000000
Valid N (listwise)	5				

Tab. 10.22 Descriptive statistics of Cluster 49.

	N	Minimum	Maximum	Mean	Std. Deviation
SN	69	.000	.000	.00000	.000000
PB	69	.000	.000	.00000	.000000
AS	69	.000	.000	.00000	.000000
SB	69	.000	.000	.00000	.000000
AG	69	.000	.040	.00235	.006303
NI	69	.010	.540	.12203	.128688
BI	69	.000	.000	.00000	.000000
FE	69	.000	.000	.00000	.000000
Valid N (listwise)	69				

Tab. 10.23 Descriptive statistics of Cluster 22.

	N	Minimum	Maximum	Mean	Std. Deviation
SN	53	.000	.000	.00000	.000000
PB	53	.000	.000	.00000	.000000
AS	53	.000	.000	.00000	.000000
SB	53	.000	.000	.00000	.000000
AG	53	.002	.280	.02675	.053267
NI	53	.000	.010	.00106	.001292
BI	53	.000	.000	.00000	.000000
FE	53	.000	.000	.00000	.000000
Valid N (listwise)	53				

Tab. 10.24 Descriptive statistics of Cluster 23.

	N	Minimum	Maximum	Mean	Std. Deviation
SN	49	.000	.000	.00000	.000000
PB	49	.000	.000	.00000	.000000
AS	49	.000	.210	.00733	.032351
SB	49	.000	.440	.08996	.096579
AG	49	.010	1.750	.14245	.269649
NI	49	.001	.001	.00100	.000000
BI	49	.000	.033	.00582	.007230
FE	49	.000	.000	.00000	.000000
Valid N (listwise)	49				

Tab. 10.25 Descriptive statistics of Cluster 24.

Cluster 23, however, is characterised as almost pure copper (53 analyses). It contains objects that have only some silver and traces of nickel as impurities (Tab. 10.24). The finds, generally found in the area in and around the Carpathian Basin, date chiefly to PER 21 and 25. The archaeological classification indicates connections with the Baden culture, but this cluster covers the entire Chalcolithic (Appendix E.34).

Cluster 10 is split into three clusters which cannot be grouped further (Clusters 24 – 26). Whereas Clusters 24 and 26 are antimony-silver copper, Cluster 25 is richer in antimony, arsenic and nickel (Appendix E.35).

Cluster 24 (49 analyses) includes mainly finds of pre-Bronze Age periods and a small peak is marked by Bell Beaker finds. The objects are mainly located in the Czech Republic, but also in Hungary and the northalpine region. In broad terms, these finds are distributed over an area compassing eastern Austria, Moravia and western Slovakia (Appendix E.36). The objects of this cluster are characterised by arsenic, antimony, silver, nickel and bismuth, generally with silver as the dominant impurity, followed by the antimony and arsenic concentrations. Nickel and bismuth are just present in traces, although bismuth concentrations are slightly higher (Tab. 10.25).

	N	Minimum	Maximum	Mean	Std. Deviation
SN	30	.000	.000	.00000	.000000
PB	30	.000	.000	.00000	.000000
AS	30	.030	1.600	.43100	.375338
SB	30	.019	1.550	.50763	.352152
AG	30	.006	1.300	.20000	.277471
NI	30	.000	.001	.00097	.000183
BI	30	.000	.280	.05473	.073802
FE	30	.000	.000	.00000	.000000
Valid N (listwise)	30				

Tab. 10.26 Descriptive statistics of Cluster 25.

	N	Minimum	Maximum	Mean	Std. Deviation
SN	65	.000	.000	.00000	.000000
PB	65	.000	.000	.00000	.000000
AS	65	.000	.000	.00000	.000000
SB	65	.080	2.700	.43985	.351223
AG	65	.030	.660	.19182	.096872
NI	65	.001	.001	.00100	.000000
BI	65	.004	.150	.02946	.028372
FE	65	.000	.000	.00000	.000000
Valid N (listwise)	65				

Tab.10.27 Descriptive statistics of Cluster 26.

	N	Minimum	Maximum	Mean	Std. Deviation
SN	19	.000	.000	.00000	.000000
PB	19	.000	.000	.00000	.000000
AS	19	.000	.090	.02058	.029466
SB	19	.000	.200	.08737	.047004
AG	19	.000	.680	.20000	.209523
NI	19	.001	.300	.07321	.074411
BI	19	.000	.016	.00563	.006130
FE	19	.000	.001	.00005	.000229
Valid N (listwise)	19				

Tab. 10.28 Descriptive statistics of Cluster 27.

Cluster 25 (30 analyses) consists of the same types of impurities as the previously described cluster; however the average concentrations and the ratio of these differ. The predominate trace element is antimony, then arsenic and silver. The concentrations of antimony, arsenic and silver are higher than in Cluster 24, but nickel is hardly present (Tab. 10.26). The regional distribution is similar to Cluster 24, but the dating of the objects point to earlier periods. Most of the objects are single finds of axes from PER 10, 21 and 25 but Earliest Bronze Age finds are also present (Appendix E.37).

Cluster 26 (65 analyses) is defined as antimony-copper (C_1B-copper after SAM). By far the main trace element is antimony with the presence of silver, bismuth and small amounts of nickel (Tab. 10.27). The artefacts of Cluster 26 are mainly single finds – and primarily axes – dated to the 4th millennium BC and including three Bell Beaker finds (Appendix E.38).

Clusters 27 and 28 do not show a significant difference compared with Cluster 11. The artefacts were discovered in contexts of the western Slovakian Nitra group, but both clusters differ metallurgically (Appendices E.39, E.40 and E.41).

Cluster 27 contains 19 trace element analyses and is similar to Cluster 24, but generally purer. Although silver is the predominant trace element, and percentages of antimony, nickel and arsenic are also found, this type of copper does not contain high levels of impurities (Tab. 10.28).

Cluster 28 (51 analyses) differs, due to its higher content of arsenic in relation to the other impurities (Tab. 10.29). Consequently, this cluster is defined as an arsenical copper.

Two groups are developed by TSC out of Cluster 12: Clusters 29 and 30. Both are arsenical coppers, and this is also reflected by the SAM-classification (Appendix E.42).

Cluster 29 (100 analyses), however, has slightly more arsenic and silver on average, than **Cluster 30** (33 analyses) which has, in turn, slightly more antimony (Tab. 10.30 and 10.31). Whereas Cluster 29 consists chiefly of analyses from Burgäschisee, Cluster 30 mostly contains analyses from single finds. It also has to be considered that Cluster 30 is much smaller than Cluster 29. Cluster 30 covers the same geographical and chronological area as Cluster 29, except that it does not include any Earliest Bronze Age finds. Due to their similarity, their later metallurgical interpretation can be related to Cluster 12 in general (Appendices E.43 and E.44).

	N	Minimum	Maximum	Mean	Std. Deviation
SN	51	.000	.000	.00000	.000000
PB	51	.000	.000	.00000	.000000
AS	51	.050	3.400	.78275	.731598
SB	51	.000	.660	.03218	.092824
AG	51	.000	.170	.03212	.031129
NI	51	.009	.050	.01741	.011960
BI	51	.000	.083	.00508	.016242
FE	51	.000	.000	.00000	.000000
Valid N (listwise)	51				

Tab. 10.29 Descriptive statistics of Cluster 28.

	N	Minimum	Maximum	Mean	Std. Deviation
SN	100	.000	.000	.00000	.000000
PB	100	.000	.000	.00000	.000000
AS	100	.005	4.400	.54892	.661361
SB	100	.000	.270	.02243	.039093
AG	100	.010	.170	.03363	.033011
NI	100	.001	.001	.00100	.000000
BI	100	.000	.078	.00265	.009593
FE	100	.000	.000	.00000	.000000
Valid N (listwise)	100				

Tab. 10.30 Descriptive Statistics of Cluster 29

	N	Minimum	Maximum	Mean	Std. Deviation
SN	33	.000	.000	.00000	.000000
PB	33	.000	.000	.00000	.000000
AS	33	.000	2.200	.52652	.589940
SB	33	.000	.820	.06458	.164786
AG	33	.000	.010	.00033	.001744
NI	33	.000	.010	.00121	.001596
BI	33	.000	.039	.00315	.007085
FE	33	.000	.000	.00000	.000000
Valid N (listwise)	33				

Tab. 10.31 Descriptive statistics of Cluster 30.

	N	Minimum	Maximum	Mean	Std. Deviation
SN	30	.000	.000	.00000	.000000
PB	30	.010	.290	.04373	.058459
AS	30	.000	2.500	.57503	.706022
SB	30	.000	.210	.01957	.039708
AG	30	.000	.210	.04987	.064723
NI	30	.001	.040	.00753	.009417
BI	30	.000	.032	.00407	.006913
FE	30	.000	.000	.00000	.000000
Valid N (listwise)	30				

Tab. 10.32 Descriptive statistics of Cluster 31.

	N	Minimum	Maximum	Mean	Std. Deviation
SN	3	.000	.010	.00333	.005774
PB	3	.000	.000	.00000	.000000
AS	3	.000	.000	.00000	.000000
SB	3	.040	.290	.14333	.130512
AG	3	.020	.200	.10000	.091652
NI	3	.001	.001	.00100	.000000
BI	3	.014	.100	.04467	.048014
FE	3	.000	.010	.00667	.005774
Valid N (listwise)	3				

Tab. 10.33 Descriptive statistics of Cluster 32.

All clusters, from Clusters 24 – 30, contain on average the whole range of considered impurities apart from tin, bismuth and iron.

Clusters 31 and 32 are produced by using the TSC on Cluster 14, and the dispersal of the trace element concentrations of these two clusters is displayed in Appendix E.45.

Cluster 31 (30 analyses) is characterised by the arsenic content, but it also contains other impurities (Tab. 10.32). Objects of this group cover, chronologically and geographically, almost the entire study area. It consists chiefly of axes (Appendix E.46).

Cluster 32, the smaller group of only 3 analyses, consists only of three objects and is considered to be a sample that is false as grouped by TSC. It is dominated by antimony and silver contents and, thus, is more similar to Cluster 26 (Tab. 10.33). Cluster 32 includes one flat axe from central Germany[159] (PER 31) and two axes from Slovakia[160] (PER 25).

As discussed above, Cluster 15 contains only analyses of objects from 3rd millennium BC contexts. Cluster 15 is divided into three groups (Clusters 33, 34 and 35) by TSC, and Clusters 34 and 35 can then be separated into two further clusters. All five clusters (33, 50, 51, 52 and 53) need to be discussed together, since all of them include analyses from Drosa (D) (cf. Cluster 15). The metallurgical differences of the clusters that originated from Cluster 15 are shown as box plots in Appendix E.47.

Cluster 33 (35 trace element analyses) shows the largest variety of archaeological contexts of these five groups. The dominance of Drosa finds aside, the contents of this group extends over the northalpine region from Switzerland to western Slovakia, including Middle Germany. This area matches the distribution of the Eastern Bell Beaker group. Nevertheless, Cluster 33 contains steady amounts of objects mainly from contexts of the 3rd millennium BC, and thus it cannot be understood as a typical Bell Beaker copper (Appendix E.48). From a metallurgical point of view, Cluster 33 contains arsenic and antimony, and also some silver and nickel. Traces of bismuth and tin have also been measured in some objects. The relatively high concentration of iron and lead is notable when compared with the previously discussed types of copper (Tab. 10.34).

Cluster 50 (28 analyses) includes just one find that is not from Drosa. This is a ring from a Corded Ware grave in Sömmerda-Orlishausen (D) which is about 120 km away from Drosa (Appendix E.49). This find differs according to the SAM-classification, because of its slightly higher

159 ANo 225.

160 ANo 58002 and 58009.

	N	Minimum	Maximum	Mean	Std. Deviation
SN	35	.000	.200	.01040	.033467
PB	35	.009	1.940	.36357	.485362
AS	35	.060	2.700	.63169	.692310
SB	35	.008	.946	.28460	.242438
AG	35	.012	.967	.19263	.202603
NI	35	.015	2.600	.19163	.470469
BI	35	.000	.196	.02717	.035858
FE	35	.020	1.840	.24671	.353457
Valid N (listwise)	35				

Tab. 10.34 Descriptive statistics of Cluster 33.

	N	Minimum	Maximum	Mean	Std. Deviation
SN	28	.003	.004	.00314	.000356
PB	28	.680	6.500	1.67964	1.150985
AS	28	.516	3.016	1.17061	.547842
SB	28	.026	.070	.04750	.009724
AG	28	.023	.210	.05432	.037011
NI	28	.022	.139	.05789	.030111
BI	28	.042	1.212	.25468	.303089
FE	28	.050	.390	.11607	.074254
Valid N (listwise)	28				

Tab. 10.35 Descriptive statistics of Cluster 50.

	N	Minimum	Maximum	Mean	Std. Deviation
SN	51	.003	.010	.00367	.001669
PB	51	.040	2.360	.35608	.397734
AS	51	.074	5.159	.62569	.856894
SB	51	.024	.176	.07755	.034243
AG	51	.008	.249	.07061	.049289
NI	51	.010	.258	.08055	.057646
BI	51	.003	.062	.01943	.013828
FE	51	.050	.150	.06745	.023138
Valid N (listwise)	51				

Tab. 10.36 Descriptive statistics of Cluster 51.

	N	Minimum	Maximum	Mean	Std. Deviation
SN	38	.003	.014	.00395	.002105
PB	38	.090	.830	.29079	.136290
AS	38	.294	3.492	1.58218	.660656
SB	38	.040	.147	.07250	.024683
AG	38	.110	.442	.19774	.069294
NI	38	.078	.687	.36161	.148478
BI	38	.022	.190	.08174	.034028
FE	38	.050	.250	.11421	.057640
Valid N (listwise)	38				

Tab. 10.37 Descriptive statistics of Cluster 52.

	N	Minimum	Maximum	Mean	Std. Deviation
SN	34	.003	.004	.00303	.000171
PB	34	.040	.220	.10235	.039394
AS	34	.929	3.254	1.69982	.522848
SB	34	.025	.060	.03706	.008399
AG	34	.085	.193	.12365	.025410
NI	34	.078	.312	.18656	.064247
BI	34	.009	.049	.02874	.011057
FE	34	.050	.400	.11471	.091857
Valid N (listwise)	34				

Tab. 10.38 Descriptive statistics of Cluster 53.

content of silver[161]. The rest of the 28 analyses are from Drosa (12 analyses from grave 8 and 15 from grave 9). In comparison, the objects of Cluster 50 have very high concentrations of lead and bismuth. Arsenic is quite frequent, too, whereas tin and other impurities are only present in small traces (Tab. 10.35).

Cluster 51 comprises 50 trace element analyses. Apart from the Drosa material (which mostly comes only from grave 8), the objects date to PER 31. One additional tanged dagger was found in a Bell Beaker grave in Landau/Isar (D) (Appendix E.50). This type of copper is characterised by a significant amount of lead and arsenic. The remaining impurities appear only in minor quantities (Tab. 10.36).

Cluster 52 is characterised by 37 trace element analyses from Drosa (all from grave 9) and one Bell Beaker dagger from Buttelstedt (D). Thus, all objects were found in AREA 103 and they are more or less contemporaneous (Appendix E.51).

The metallurgical configurations of Cluster 52 and **Cluster 53** are very similar. Both clusters have high proportions of arsenic while the composition of the other trace elements is considerably low. However, Cluster 52 differs, because it generally contains higher concentrations of impurities (Tab. 10.37 and 10.38). The 34 analysed artefacts of Cluster 53 are all copper reels from the Corded Ware grave 9 of Drosa (Appendix E.52).
Although the clusters that evolved out of Cluster 15 are archaeologically quite uniform, the trace element compositions vary. Clusters 52 and 53 are typical types of arsenical copper. Despite predominately arsenic and lead concentrations, and a general comparable ratio of elements with Clusters 33, 50 and 51, these differ in terms of the total amount of impurities – the concentrations of which are lower than in Clusters 52 and 53.

Cluster 16 is divided into three subgroups by TSC (Clusters 36 – 38). While all of these are typified by their predominant, but varying, arsenic content, Cluster 37 has higher concentrations of nickel (between 0.14 and 3.0%; Appendix E.53).

Cluster 36 (70 analyses) is archaeologically quite incoherent. It covers all periods, many regions (except the Carpathian Basin) and a wide range of artefacts (Appendix E.54). It is defined by arsenic-rich copper that has other impurities, but only in traces. Significant, however, is the relatively high amount of iron, and some outlying objects that have even several percent concentrations of lead, nickel, silver and iron. In contrast, the antimony content is relatively low (Tab. 10.39).

Cluster 37 is a prime example of the difficulties caused by TSC. The cluster contains only 11 analyses and half of the objects (mostly small copper points) are dated to the 4th millennium BC, whereas the other half are finished artefacts of the 3rd millennium BC. The older objects are considered as lower concentrated arsenic-nickel-copper, and the younger finds consists of higher arsenic and nickel concentrations (Tab. 10.40a and 10.40b[162]).

In spite of its small size of 11 analyses, Cluster 37 consists particularly of analyses from sites of the Swiss Corded Ware and of the Upper Austrian Mondsee group (Appendix E.55).

Cluster 38 (35 analyses) comprises cases that were chiefly found in contexts of the Mondsee group and other contemporary archaeological groups in the northalpine region (Appendix E.56). The composition of that cluster defines it as an arsenical copper without many further impurities (Tab. 10.41). According to the SAM-systematic, Cluster 38 contains mainly group $E_{01}A$ and FA_2, both arsenic-rich, but also contains antimony, nickel and bismuth. An axe from Egolzwil[163] (CH) stands out, as it contains a little more antimony (SAM-group G_1). The

[161] Because the ANo 32335 contains more than 0.16% of silver (0.21%), the trace element analysis is classified as C_6B-copper instead of C_6A-copper.

[162] Sometimes metal groups with the same impurity combination differ in the concentration of these elements. Thus, the particular types of copper have to be characterised. Because they vary in terms of high or low concentrations of defining trace elements, this is additionally noted in brackets (e.g. arsenic-nickel-copper [low] resp. arsenic-nickel-copper [high]).

[163] ANo 100076.

	N	Minimum	Maximum	Mean	Std. Deviation
SN	70	.000	3.300	.09464	.483271
PB	70	.000	9.320	.16994	1.111379
AS	70	.000	5.000	1.03894	1.275258
SB	70	.000	.780	.03216	.110374
AG	70	.000	8.700	.13830	1.038325
NI	70	.000	4.000	.09370	.480193
BI	70	.000	.061	.00770	.011590
FE	70	.000	5.600	.43714	.987656
Valid N (listwise)	70				

Tab. 10.39 Descriptive statistics of Cluster 36.

	N	Minimum	Maximum	Mean	Std. Deviation
SN	6	.000	.000	.00000	.000000
PB	6	.000	.023	.00383	.009390
AS	6	.440	1.470	1.00667	.429729
SB	6	.000	.210	.09833	.079352
AG	6	.000	.001	.00017	.000408
NI	6	.140	.570	.32167	.163636
BI	6	.000	.000	.00000	.000000
FE	6	.035	.260	.11417	.106133
Valid N (listwise)	6				

Tab. 10.40a Descriptive statistics of Cluster 37 (arsenic-nickel-copper (low)).

	N	Minimum	Maximum	Mean	Std. Deviation
SN	5	.000	.000	.00000	.000000
PB	5	.000	.004	.00080	.001789
AS	5	1.700	3.400	2.74000	.716240
SB	5	.030	.230	.09220	.080680
AG	5	.000	.000	.00000	.000000
NI	5	1.200	3.000	2.48000	.725948
BI	5	.000	.007	.00380	.003114
FE	5	.000	.050	.01000	.022361
Valid N (listwise)	5				

Tab. 10.40b Descriptive statistics of Cluster 37 (arsenic-nickel-copper (high)).

	N	Minimum	Maximum	Mean	Std. Deviation
SN	35	.000	.030	.00563	.006504
PB	35	.000	.000	.00000	.000000
AS	35	.036	4.900	1.21617	1.308689
SB	35	.000	.150	.01946	.026156
AG	35	.004	.250	.03891	.045709
NI	35	.001	.044	.01263	.009610
BI	35	.000	.001	.00003	.000169
FE	35	.008	1.290	.13263	.266668
Valid N (listwise)	35				

Tab. 10.41 Descriptive statistics of Cluster 38.

difference between group $E_{01}A$ and FA_2 is that the second one is defined by a content of more than 0.02 % of nickel.

As previously mentioned, Cluster 17 also breaks up into three further groups (Cluster 39 – 41). The box plot shows the ratio of the impurities (Appendix E.57), but this Appendix does not consider ANo 100032 (a bead from Hornstaad-Hörne I [D]), which is significant, as its extraordinarily high content of iron (17.6%). In the further diagrams and descriptions (Appendix E.59 and Tab. 10.42) the bead from 'Hornstaad-Hörnle I' is included again.

Cluster 39 (30 analyses) summarises not only many analyses from the Polish Mierzanowice group (PER 40), but also other, even early Chalcolithic finds, from the western part of the area of research (Appendix E.58). This type of copper mainly contains impurities of silver, arsenic, antimony and iron and traces of tin (Tab. 10.42).

Cluster 40 (70 analyses) is also characterised by dominant silver concentrations, followed by antimony (Tab. 10.43). Arsenic and iron concentrations are lower than the amounts in Cluster 39, while the other trace elements – especially lead, nickel and bismuth – are generally higher. This copper type consists of different

	N	Minimum	Maximum	Mean	Std. Deviation
SN	30	.000	.700	.09243	.153466
PB	30	.000	1.040	.03963	.189466
AS	30	.000	2.020	.37543	.448797
SB	30	.000	1.190	.40460	.350555
AG	30	.090	7.300	1.01603	1.254281
NI	30	.000	.050	.00640	.011649
BI	30	.000	.030	.00113	.005482
FE	30	.001	17.600	.84713	3.176206
Valid N (listwise)	30				

Tab. 10.42 Descriptive statistics of Cluster 39.

	N	Minimum	Maximum	Mean	Std. Deviation
SN	70	.000	3.900	.10796	.529874
PB	70	.000	1.500	.21080	.326488
AS	70	.000	2.000	.09666	.272868
SB	70	.006	1.845	.34567	.373769
AG	70	.060	4.500	.68719	.702410
NI	70	.001	.970	.05310	.141427
BI	70	.000	.268	.03654	.054713
FE	70	.000	7.000	.19299	.834979
Valid N (listwise)	70				

Tab. 10.43 Descriptive statistics of Cluster 40.

	N	Minimum	Maximum	Mean	Std. Deviation
SN	37	.000	4.100	.93822	1.271393
PB	37	.000	1.300	.12938	.249210
AS	37	.000	.920	.13568	.238006
SB	37	.000	2.500	.20989	.449115
AG	37	.000	.730	.13922	.166519
NI	37	.001	.790	.08554	.151889
BI	37	.000	.027	.00481	.006637
FE	37	.000	1.330	.04327	.218971
Valid N (listwise)	37				

Tab. 10.44 Descriptive statistics of Cluster 41.

	N	Minimum	Maximum	Mean	Std. Deviation
SN	37	.000	.000	.00000	.000000
PB	37	.000	.000	.00000	.000000
AS	37	.540	2.900	1.38297	.564687
SB	37	.360	6.000	1.37027	.940684
AG	37	.050	1.950	.58784	.427227
NI	37	.001	.020	.00224	.003897
BI	37	.008	.500	.13378	.130192
FE	37	.000	.010	.00027	.001644
Valid N (listwise)	37				

Tab. 10.45 Descriptive statistics of Cluster 54.

	N	Minimum	Maximum	Mean	Std. Deviation
SN	5	.000	.000	.00000	.000000
PB	5	.000	.030	.00600	.013416
AS	5	.790	2.000	1.52800	.568480
SB	5	.070	.800	.31600	.289879
AG	5	.070	.400	.19000	.147139
NI	5	.001	.060	.01280	.026386
BI	5	.000	.260	.08900	.119394
FE	5	.010	.011	.01020	.000447
Valid N (listwise)	5				

Tab. 10.46 Descriptive statistics of Cluster 55.

types of artefacts from Czech and Middle German Corded Ware complexes, and also finds from other contexts – primarily from the 3rd millennium BC (Appendix E.59).

Cluster 41 includes 37 analyses. It is compared to Clusters 39 and 40 of purer copper, but characterised by a remarkable variation in tin content. Tin varies between traces of 0.004 and 4.1%; only one Bell Beaker dagger is free of tin[164]. The comparison between this cluster and the SAM-groups shows a great dissimilarity, which can be explained by the fact that SAM did not consider tin in its classification (Tab. 10.44). The archaeological composition of this Cluster reflects finds that date to the 3rd millennium BC. Bell Beaker and Earliest Bronze Age complexes are particularly well represented. The geographical distribution shows a general spread over the entire study area – with a slight cluster in Switzerland and the Czech Republic (Appendix E.60).

The aforementioned Cluster 18 chiefly contains analyses of PER 31, 35 and 40. Through TSC analysis, Cluster 18

	N	Minimum	Maximum	Mean	Std. Deviation
SN	95	.000	.047	.00328	.007628
PB	95	.000	16.200	.59777	2.276788
AS	95	.005	3.800	.89777	.895743
SB	95	.000	1.650	.18661	.346149
AG	95	.000	.980	.13771	.194856
NI	95	.001	2.500	.11079	.313920
BI	95	.000	.620	.03687	.100114
FE	95	.000	.000	.00000	.000000
Valid N (listwise)	95				

Tab. 10.47 Descriptive statistics of Cluster 43.

164 ANo 100028.

is divided into Cluster 42 and 43. In turn, Cluster 42 can be separated into two further clusters (Clusters 54 and 55). They all can be discussed together, as Cluster 55 only consists of five cases (Appendix E.61). TSC divides **Cluster 42** in Cluster 54 and 55.

Cluster 54 is characterised as the antimony- and silver-rich copper. Most of the 37 analyses are classified as C_2-copper after SAM (Tab. 10.45 and Appendix E.62). Most of the artefacts included in this group are rings and other jewellery from southern Germany, Austria and Slovakia. The finds date, with exceptions, to PER 40 (Appendix E.63).

Cluster 55 with its five analyses contains four earlier Chalcolithic axes (all single finds) from Slovakia and an arm spiral discovered in a grave at Beizkofen (D) which relates to the *Blechkreis* (Appendix E.64). All of these objects contain between 0.79 and 2.0% of arsenic, some antimony, silver, bismuth and iron, and – except for the arm spiral[165] – all are free of tin, lead and nickel (Tab. 10.46). Cluster 55 principally differs from Cluster 54, because of its low antimony concentration.

Cluster 43 (95 analyses) is an arsenical copper that also contains other impurities, but no iron (Tab. 10.47). All these artefacts that contain high levels of lead impurities were discovered in Corded Ware graves in the Middle-Elbe-Saale region in Germany (Appendix E.62). In summary, Cluster 43 reflects the results of Cluster 18 archaeologically as well as metallurgically.

As explained during the discussion of Cluster Level III, Cluster 19 is chiefly composed by finds of the *Blechkreis*. The objects of this cluster generally consist of fahlore-copper (see above) and this is characterised by quantitatively and qualitatively greater varieties of impurities than the previously discussed copper types. Running a TSC on Cluster 19 separates two further clusters (Clusters 44 and 45), which themselves can be split up again. Thus we can compare the Clusters 56 and 57 – originating from Cluster 44 – and Clusters 58 and 59 – originating from Cluster 45 (cf. Fig. 8.14). All Clusters that emerge from Cluster 19 are dominated by antimony and nickel contents, whilst the other trace elements are less concentrated (Appendix E.65).

Cluster 56 (22 trace element analyses) is primarily characterised by the presence of antimony and nickel, but arsenic and silver are also present. Other trace elements are present in lower concentrations and bismuth is absent (Tab. 10.48). This cluster is formed only by flanged axes – 21 of the hoard from Salez and one from Auvernier (both CH) (Appendix E.66).

Cluster 57 also contains 27 finds only classified as Earliest Bronze Age, and with one exception, only

	N	Minimum	Maximum	Mean	Std. Deviation
SN	22	.000	.150	.01636	.036586
PB	22	.100	2.700	.48727	.643526
AS	22	1.700	2.900	2.27045	.275879
SB	22	6.500	9.999	8.39314	1.036667
AG	22	.910	1.550	1.17455	.136930
NI	22	2.700	6.250	4.56136	.961637
BI	22	.000	.000	.00000	.000000
FE	22	.000	.350	.04150	.073668
Valid N (listwise)	22				

Tab. 10.48 Descriptive statistics of Cluster 56.

	N	Minimum	Maximum	Mean	Std. Deviation
SN	27	.000	.040	.00630	.011485
PB	27	.040	1.700	.49556	.436528
AS	27	1.600	3.600	2.37963	.470323
SB	27	3.500	9.000	5.46296	1.291436
AG	27	.400	1.700	1.15963	.374592
NI	27	1.200	5.500	3.08704	.939168
BI	27	.000	.044	.01044	.008868
FE	27	.000	.030	.00515	.008659
Valid N (listwise)	27				

Tab. 10.49 Descriptive statistics of Cluster 57.

[165] ANo 30.

	N	Minimum	Maximum	Mean	Std. Deviation
SN	82	.000	1.800	.13166	.291884
PB	82	.003	2.500	.21137	.420757
AS	82	.080	3.000	.96543	.564791
SB	82	.250	5.600	2.08988	1.124005
AG	82	.057	2.700	.80077	.583713
NI	82	.050	3.700	1.73973	.808895
BI	82	.000	.179	.01296	.028417
FE	82	.000	1.500	.07804	.203096
Valid N (listwise)	82				

Tab. 10.50 Descriptive statistics of Cluster 58.

	N	Minimum	Maximum	Mean	Std. Deviation
SN	43	.006	3.200	.65205	.933434
PB	43	.000	.120	.01612	.022531
AS	43	.440	2.400	.97851	.417240
SB	43	1.360	8.000	2.92751	1.089947
AG	43	.300	2.200	1.11370	.434452
NI	43	.930	5.900	2.52826	1.032793
BI	43	.000	.028	.00440	.005901
FE	43	.010	2.600	.72256	.767302
Valid N (listwise)	43				

Tab. 10.51 Descriptive statistics of Cluster 59.

flanged axes. Most belong to the famous Salez- and the Hindelwangen-hoards. Besides one pin from Bavaria[166] these finds are also axes of the Salez type (Appendix E.67). From the metallurgical point of view, Clusters 56 and 57 are very similar; however Cluster 57 contains small traces of bismuth and, on average, less antimony (Tab. 10.49).

In conclusion, Clusters 56 and 57, can both be analysed together in Cluster 44 as typical for Salez axes, because both clusters vary only slightly in terms of antimony and bismuth concentrations.

Cluster 58 that includes 82 analyses is also dominated by the concentrations of antimony and nickel, but impurity concentrations are less than in the cases of Clusters 56 and 57 (Tab. 10.50).
The archaeological composition of Cluster 58 shows a range of Earliest Bronze Age artefacts from various sites chiefly from Switzerland, southern Germany, the eastern part of Austria, and Slovakia (Appendix E.68).

Cluster 59 (43 analyses) is metallurgically very similar to Cluster 58, but overall contains slightly more impurities and includes greater varieties of tin and iron concentrations (Tab. 10.51). The analysed objects, however, draw a uniform picture. Nearly all finds were found in Earliest Bronze Age contexts in southwestern Germany. Above all, finds from the cemetery of Singen/Hohentwiel (D) dominate the cluster (Appendix E.69).

TSC separates two further Clusters from Cluster 20 and each of these can be divided into four groups. Those eight clusters are not further separated. From Cluster 46, Clusters 60 – 62, 66 and 67 are separated (Appendix E.70), and Clusters 68 – 71 are derived from Cluster 47 (Appendix E.76). Whilst the smallest cluster (Cluster 62) includes five examples, the biggest (Cluster 66) includes 70 objects. These clusters are characterised by a preponderance of antimony and nickel and by generally smaller percentages of the remaining trace elements.

Cluster 60 (29 analyses) covers a wide range of antimony and nickel concentrations, whereas arsenic stretches across c. 0.5 to 1.5% and silver from c. 0.4 to 0.8%. The rest of the trace elements are either absent or present in low concentrations (Tab. 10.52). The cases of Cluster 60 spread more or less over the whole geographical study area with the exception of the Carpathian Basin and typically relate to the Earliest Bronze Age, while one dagger is classified as early 3rd millennium BC (Appendix E.71).

Cluster 61 consists of 69 trace element analyses: mostly rings, pins and other jewellery which were generally

[166] ANo 9280.

discovered in Earliest Bronze Age graves of the northalpine region, in eastern Austria and western Slovakia. These archaeological complexes can be connected with the *Blechkreis*, Nitra and the Loretto-Leithaprodersdorf group. Also significant in this cluster are the finds from the famous and large Nitra cemetery of Výcapy-Opatovce (SK) (Appendix E.72). Although this copper type is outstandingly similar to Cluster 60, the geographical and chronological variation of the objects is smaller and the content of tin higher than in the previous cluster (Tab. 10.53).

Cluster 62 (5 analyses) shows, on average, the highest antimony proportion, ranging between 1.55 and 3.9 %. The concentrations of the other trace elements are comparable to Clusters 60 and 61 (Tab. 10.54). Although it has to be taken into account that this cluster is only formed by five artefacts, the variation within each impurity is not different from other clusters. Broader archaeological trends cannot be identified from this small number of finds, but it does not vary remarkably from the previous two groups (Appendix E.73). As the analyses were not carried out by different laboratories, but all by the SAM-project, the separation of Cluster 62 is to be explained by reasons of the artefact compositions.

Cluster 66 includes 70 trace element analyses that are nearly exclusively provided by finds from the western

	N	Minimum	Maximum	Mean	Std. Deviation
SN	29	.000	1.120	.06103	.212037
PB	29	.000	.000	.00000	.000000
AS	29	.310	1.850	.90103	.419041
SB	29	.730	5.000	2.58724	1.021067
AG	29	.340	.920	.59483	.165500
NI	29	.750	4.000	1.61310	.768626
BI	29	.002	.017	.00645	.003236
FE	29	.000	.000	.00000	.000000
Valid N (listwise)	29				

Tab. 10.52 Descriptive statistics of Cluster 60.

	N	Minimum	Maximum	Mean	Std. Deviation
SN	69	.000	3.800	.16580	.502609
PB	69	.000	.000	.00000	.000000
AS	69	.230	1.900	.90348	.408677
SB	69	1.150	4.100	2.18406	.582306
AG	69	.310	2.600	.68536	.356138
NI	69	.660	4.300	1.74217	.707173
BI	69	.000	.001	.00001	.000120
FE	69	.000	.000	.00000	.000000
Valid N (listwise)	69				

Tab. 10.53 Descriptive statistics of Cluster 61.

	N	Minimum	Maximum	Mean	Std. Deviation
SN	5	.010	.080	.03000	.029155
PB	5	.000	.010	.00600	.005477
AS	5	.420	1.600	.93000	.499700
SB	5	1.550	3.900	2.85000	.886002
AG	5	.430	1.200	.78800	.337520
NI	5	1.020	2.100	1.52400	.430848
BI	5	.000	.000	.00000	.000000
FE	5	.000	.010	.00320	.004604
Valid N (listwise)	5				

Tab. 10.54 Descriptive statistics of Cluster 62.

	N	Minimum	Maximum	Mean	Std. Deviation
SN	70	.000	2.350	.12714	.375334
PB	70	.000	.000	.00000	.000000
AS	70	.070	1.400	.38657	.204719
SB	70	.650	2.600	1.40157	.417736
AG	70	.180	.720	.39329	.098830
NI	70	.510	2.000	1.00086	.255415
BI	70	.000	.000	.00000	.000000
FE	70	.000	.000	.00000	.000000
Valid N (listwise)	70				

Tab. 10.55 Descriptive statistics of Cluster 66.

	N	Minimum	Maximum	Mean	Std. Deviation
SN	21	.000	.810	.09000	.227002
PB	21	.000	.000	.00000	.000000
AS	21	.080	1.200	.48095	.298511
SB	21	.290	2.200	.93143	.578518
AG	21	.300	1.100	.67810	.222904
NI	21	.100	1.800	.80381	.505376
BI	21	.000	.004	.00057	.000978
FE	21	.000	.000	.00000	.000000
Valid N (listwise)	21				

Tab. 10.56 Descriptive statistics of Cluster 67.

Slovakian Nitra group and especially from the cemetery of Výcapy-Opatovce (SK) (Appendix E.74) and, again, the copper is similar to the last types; however this one is free of lead, bismuth and iron, and contains a range of concentration of tin. Overall, the trace element concentration is below the level of Clusters 60 – 62 (Tab. 10.55).

Cluster 67 (21 trace element analyses) also falls into the same scheme, but, in contrast, the content of the four frequent trace elements – arsenic, antimony, silver and nickel – is more balanced and, in general, slightly lower (Tab. 10.56).

Cluster 67 does not only consist of Earliest Bronze Age finds from Nitra contexts, but also objects from southern Germany and Austria and even a few artefacts contemporary with the Corded Ware complex (Appendix E.75).

Clusters 68 – 71 are derived from Cluster 47, and, apart from Cluster 68, these are antimony- and nickel-rich coppers (Appendix E.76).

Cluster 68 is an antimony- and arsenic-rich copper with silver and nearly free of other impurities, while antimony is dominant here again (Tab. 10.57). The finds are spread

	N	Minimum	Maximum	Mean	Std. Deviation
SN	21	.000	3.900	.30405	.892790
PB	21	.000	.043	.00395	.011227
AS	21	.350	3.100	1.00190	.639356
SB	21	.410	3.700	1.19190	.777995
AG	21	.370	1.300	.68619	.227716
NI	21	.001	.120	.02767	.026764
BI	21	.000	.190	.04286	.043015
FE	21	.000	.001	.00005	.000218
Valid N (listwise)	21				

Tab. 10.57 Descriptive statistics of Cluster 68.

throughout the area of research, with a small centre in Lower Austria. Chronologically the cluster focuses on PER 40 with two outliers dating to PER 21 (Appendix E.77).

Cluster 69 is formed by 63 analyses dating to the 3rd millennium BC. Most of the objects are connected with Earliest Bronze Age contexts – especially Nitra and Carpathian Earliest Bronze Age groups. The artefacts of Cluster 69 are both tools and jewellery (Appendix E.78). Apart from lead, bismuth and iron, the impurities vary up to 2 % for nearly all trace elements (Tab. 10.58).

Cluster 70 and Cluster 71 are comparable in terms of their impurities, in particular, both significant levels of antimony and nickel.

Cluster 70 (23 analyses), however, has a slightly higher range of nickel and contains slightly more silver and iron (Table 10.59). Concerning the archaeological composition, the cluster contains mostly southwest-German finds (PER 35 and PER 40), but also some – more or less contemporary – outliers (Appendix E.79).

Cluster 71 (52 analyses) also includes southern German, Swiss and Slovakian finds. With one exception – a ring with remarkable 4% of tin from a grave from Deszk (H) that is related to the Neolithic/earliest Chalcolithic Tisza group – these objects date to the 3rd millennium BC (Appendix E.80). In contrast to Cluster 70, the copper composition of that cluster contains less nickel and iron, but more lead (Tab. 10.60).

	N	Minimum	Maximum	Mean	Std. Deviation
SN	63	.000	3.800	.48381	.988373
PB	63	.000	.630	.01579	.082522
AS	63	.005	2.500	.40973	.557571
SB	63	.030	2.900	.53530	.505439
AG	63	.049	1.500	.41268	.280281
NI	63	.020	2.100	.49524	.492210
BI	63	.000	.050	.00410	.007075
FE	63	.000	.000	.00000	.000000
Valid N (listwise)	63				

Tab. 10.58 Descriptive statistics of Cluster 69.

	N	Minimum	Maximum	Mean	Std. Deviation
SN	23	.000	2.800	.22313	.623352
PB	23	.000	.020	.00104	.004161
AS	23	.070	1.889	.48648	.402950
SB	23	.050	3.600	1.37339	1.010214
AG	23	.330	1.600	.77057	.318082
NI	23	.022	2.500	1.06161	.742558
BI	23	.000	.013	.00122	.003029
FE	23	.000	1.120	.13930	.257378
Valid N (listwise)	23				

Tab. 10.59 Descriptive statistics of Cluster 70.

	N	Minimum	Maximum	Mean	Std. Deviation
SN	52	.000	4.000	.32523	.928703
PB	52	.010	2.400	.10308	.331966
AS	52	.040	1.350	.47219	.348015
SB	52	.096	3.600	1.27237	.721993
AG	52	.140	1.200	.48346	.235321
NI	52	.040	2.600	.83313	.603109
BI	52	.000	.100	.00569	.014176
FE	52	.000	.050	.00258	.008662
Valid N (listwise)	52				

Tab. 10.60 Descriptive statistics of Cluster 71.

Compared with the SAM-classification, all sub-clusters of Cluster 44, 45 and 46 predominantly encompass A-, A_1- and B_2-copper. In other words, the clusters 56 – 63 and 66 – 71 (except Cluster 68) are an antimony- and nickel-rich fahlore-copper that is often labelled 'Singen copper'[167]. By far, the analysed objects are from Earliest Bronze Age contexts.

The clusters of Level IV marked on the scatter plot of the PCA shows that these small groups overlap considerably and a detailed structure is difficult to identify (Appendix E.81). This indicates that some clusters either may be of similar structure, but differ in terms of their absolute concentrations of impurities, or else they contain some 'miss-guided' cases. Subsequently, it is essential to contrast these clusters thoroughly and compare them with the results of the PCA. Combining them with reference to the output of the PCA will enhance the interpretability of these clusters. As the scatter plot and the bar charts of the mean values of all groups of Cluster Level IV show some similar tendencies between two or more clusters (Appendices E.81 and E.82), a summary of several clusters will ease the archaeological and metallurgical interpretation, and no specific information will be lost. For instance, Clusters 28 – 31 seem to be very similar, so that the statistical clustering developed probably over-detailed groups with only small variations. Now we must establish which clusters can be interpreted in combination and, above all, how their trace element compositions affect the properties of the copper artefacts that have been made of those types of copper.

In our case, the TCS developed clusters of generally uniform configurations, and thus the clusters indeed show that the grouping is not based on the listing of the original dataset, but on the trace element composition. The next section will not only try to eliminate outliers, but it also will connect individual clusters if they are similar by comparing the results of the TSC with the outcome of the PCA. Finally, the metallurgical properties of these new combined metal groups and their intentional use by Chalcolithic metalworkers will be discussed.

10.5 SUMMARY AND COMPARISON OF THE INDIVIDUAL CLUSTERS

In the previous section, the database of 1943 Chalcolithic trace element analyses was been split up into often very small groups by several TSCs. It was possible to break the dataset into 42 clusters – 40 clusters of Cluster Level IV and two of Level III (Clusters 9 and 13). As some clusters of Level IV are very small and consist of analyses reflecting only a certain site, context or even artefact, a comprehensive diachronological examination is difficult. Furthermore, we have to deal with the disadvantage of TSC that objects might have been grouped incorrectly, since to some extent the method considers the listing of the data sheet (cf. Chapter 8.2.1). Hence, it is crucial to evaluate each of the 42 clusters again in order to clarify whether some of them are similar (and can thus be treated together), or whether a cluster includes significant aberrations which perhaps have to be rearranged. In considering the results of the PCA, the following sections aim to re-combine clusters according to their metallurgical compositions and define finally the types of metal which will be interpreted archaeologically later on. Very small numeric variations cause different statistical clusters. Prehistoric metal workers were probably not able to influence these variations deliberately, so that clusters which vary at a small scale can be combined in order to define metal groups, which differ in terms of their metallurgical properties. Throughout the re-combination of clusters, it will be borne in mind whether it is possible to interpret these types of copper in terms of the effects of their compositions on the properties of copper, as outlined and discussed in Chapter 7.

In order to check the TSC results, these must be compared with the distribution of the clusters on the PCA scatter plot (cf. Chapter 9.5) and with their actual configuration, as displayed in the tables (Tab. 10.1 – 10.60). For metallurgical interpretation it is useful to focus on the mean trace element values of the clusters. However, it is essential to keep in mind that for statistical reasons, outliers – which are distinguished both by higher and by lower levels than the mean value – always appear in these clusters, and the larger the groups the more outliers are present. Nevertheless it must be borne in mind that statistical clustering of archaeological data is a tool for facilitating data management and it does not necessarily reflect historic facts. Therefore, based on the statistical results, the clusters will be controlled and rearranged.

Subsequently, the clusters of Chapter 10.1 – 10.4 will be summarised according to their configuration and discussed in terms of their archaeological context and their metallurgical properties. By examining the composition of each cluster, and noting that the first component of the PCA indicated a significance in the antimony, nickel, silver and arsenic concentrations (PCA_factor_1), four major metal groups have been identified and labelled, based on their principal impurity: (almost) pure copper; silver copper; arsenical copper; and antimony copper. Furthermore, these major metal groups can be subdivided into the actual types of copper which will be used to draw the final conclusions[168]. The mean values of impurities of each metal group are displayed in a table at the end of Chapter 10.5.6 (Tab. 10.62).

[167] Antimony-nickel copper has often been called 'Singen metal' or 'Singen copper' (cf. Walterbolk and Butler 1965; Butler and van der Waals 1966; Needham 2002). Here the practice of labelling a type of copper by a site name should be avoided, as this may indicate either a local resource or at least a centre of metallurgical activity. Since no evidence of metallurgical work has been found at the site of the Singen cemetery, this only reflects a find concentration.

[168] The classification of every trace element analysis as the respective final copper type is recorded as 'Metal_group'.

10.5.1 PURE AND ALMOST PURE COPPER

The first main group to be outlined summarises types of copper which are virtually free of trace elements, and whose properties, therefore, are unaffected. It is necessary to subdivide this group between pure copper objects that have only small traces of impurities (not more than 0.001%); almost pure copper with several trace elements present in low concentrations; and types of copper which are pure, but contain either nickel or silver. No cluster appeared that featured the presence of significant amounts of trace elements, such as bismuth or iron, that have a negative effect on the copper. When these impurities are present, they are found in combination with other neutralising elements (e.g. arsenic).

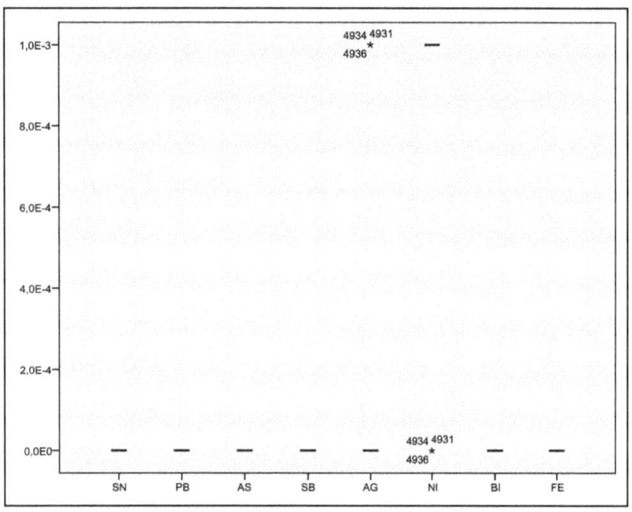

Fig. 10.5 Box plot of the metal group *pure copper* (numbering refers to ANo at the catalogue).

The two clusters of **pure copper** include objects that merely have traces, i.e. 0.001%, of nickel (Cluster 48) and of silver (Cluster 49); both contain finds chiefly dating to the 5th and 4th millennium BC. Pure copper corresponds to Cluster 21 (Fig. 10.5).

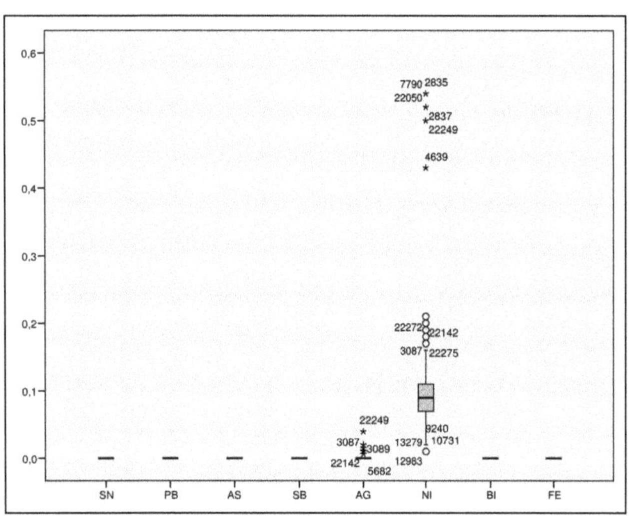

Fig. 10.6 Box plot of the metal group *pure copper with nickel* (numbering refers to ANo at the catalogue).

Both clusters that are labelled as **pure copper with nickel** (Cluster 22) and **pure copper with silver** (Cluster 23) contain slightly more of those elements. Cluster 22 contains a mean value of around 0.1% concentrations of nickel and traces of silver (Fig. 10.6), while Cluster 23 has a mean of only c. 0.01% silver and a few traces of nickel (Fig. 10.7). Despite the fact that Cluster 22 includes the many analysed beads from the Corded Ware site at Vinelz (CH), this cluster mainly comprises finds that date to the first half of the 3rd millennium BC. Cluster 23 contains finds from more or less all chronological and geographical areas, albeit with a bias towards early finds. We can discuss all four clusters together, because their properties are quite similar. Since the size of atomic radii and the chemical characteristics of both nickel and silver are analogous to copper, they hardly affect the mechanical properties of copper, and therefore it is also difficult to separate them from copper. Consequently, it is suggested that the absence of impurityes apart from nickel and/or silver indicates carefully refined or often re-melted metal, because every refining and re-melting procedure removes more and more trace elements. Procedures that involve the heating of copper allow volatile and easily oxidised impurities to disappear. This has been demonstrated, for instance, on a wide range of trace elements by Pernicka (1999, 165). The amounts of nickel as they are contained in Cluster 22 increase hardness only gradually. Because nickel is hard to remove form copper this may indicate that copper which only contains nickel was probably often re-heated under oxidising conditions, so that all the other, more volatile impurities have disappeared.

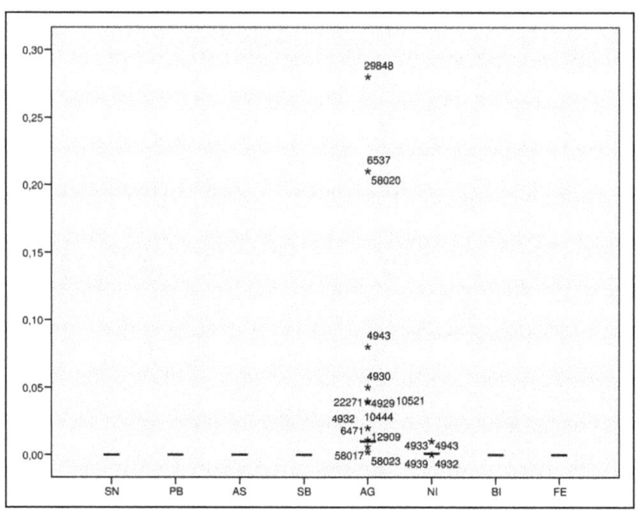

Fig. 10.7 Box plot of the metal group *pure copper with silver* (numbering refers to ANo at the catalogue).

Cluster 9 is defined as **almost pure copper** with low concentrations of various trace elements (generally less than 0.01% of each; Fig. 10.8). It was also used in every period and region, but with a slight peak of finds from the early 4th millennium BC. Since nearly all trace elements are present in relatively small amounts and the small peak reflects the nickel concentration, it can be suggested that this type of copper may represent an often re-melted or refined metal. From a metallurgical point of view, the mean content of trace elements is so low that the objects constituting this metal group do not have remarkably changed properties.

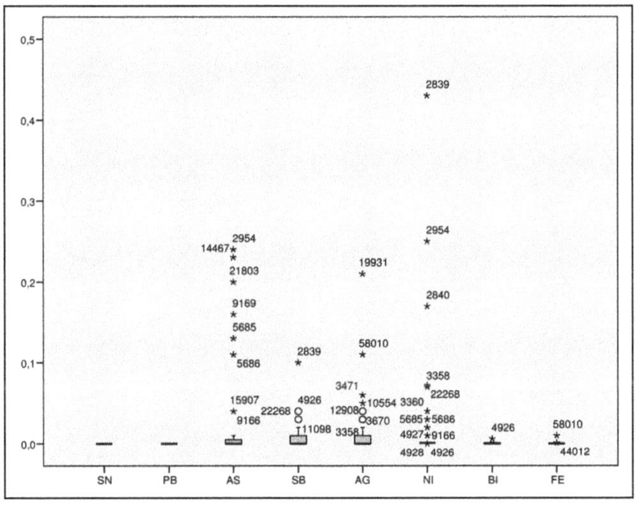

Fig. 10.8 Box plot of the metal group *almost pure copper* (numbering refers to ANo at the catalogue).

To summarise, pure and almost pure types of copper do not remarkably affect the properties of copper artefacts. Nickel and/or silver – two elements that are chemically and physically comparable to copper – are mostly contained in these coppers. This allows the conclusion that pure types of copper have been continually refined or often reused, as these trace elements are very similar to copper and therefore hardly removable by smelting. For this group, only the relatively high content of nickel in Cluster 22 may increase the hardness slightly.

From an archaeological point of view, the purer types of copper include objects from all areas of research (geographically and chronologically), but most finds date to the 5th and 4th millennium BC.

There are 15 finds from Cluster 13 which have to be excluded and discussed. These are made of an almost pure copper, but with significant contents of iron (termed **almost pure copper with iron**; Fig. 10.9). Nearly all of the data come from double-analyses of objects carried out by various different laboratories. If only one laboratory had carried out these analyses, we would assume the measurements are caused by a specific method that better measures iron concentrations, but that is not the case. The results of the double-analyses, which were carried out on artefacts of this metal group by different laboratories, are comparable. The referenced analyses nearly all indicate pure or almost pure copper, so that the dissimilarity is only caused by the iron concentration. That, finally, indicates that the difference is created by either a mistake in the measuring (which here seems unlikely) or by a small variation within the artefact: because iron precipitates easily and in small traces, it may accumulate at a certain point on the artefact. Nevertheless, these 15 analyses can be seen as genuine outliers, because in general the results of other double-analyses are indeed comparable.

10.5.2 SILVER-COPPER

Copper containing silver has not been sufficiently discussed in archaeometallurgy, particularly if silver appears in copper artefacts in combination with further impurities. However, because silver has similar chemical and physical properties, it does not affect copper very much. Although previous studies have not focused in any detail on the effects of silver on copper, our cluster analyses have produced some silver-dominated groups. The statistical study defined two types of silver-bearing coppers. These groups summarise finds that are defined by higher silver contents than antimony, while containing generally lower concentrations of impurities (cf. Clusters 24 and 27) than the other analyses of Clusters 39 and 40.

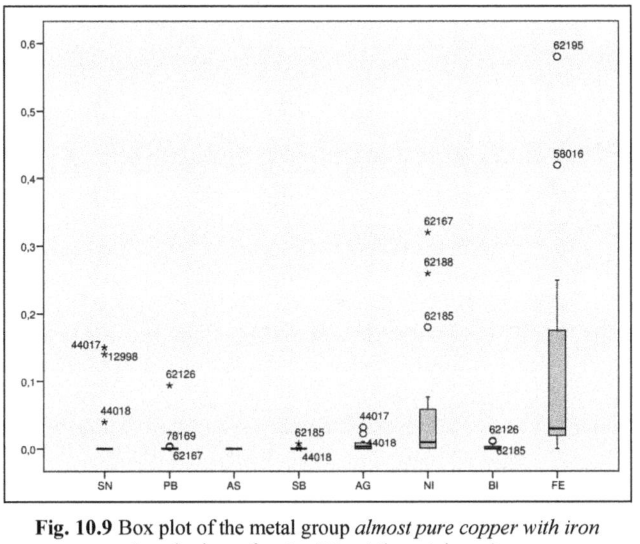

Fig. 10.9 Box plot of the metal group *almost pure copper with iron* (numbering refers to ANo at the catalogue).

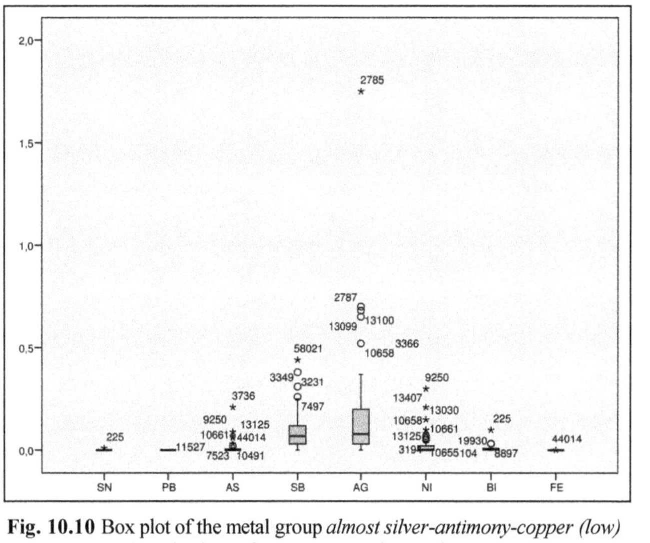

Fig. 10.10 Box plot of the metal group *almost silver-antimony-copper (low)* (numbering refers to ANo at the catalogue).

The copper finds containing silver and antimony (**silver-antimony-copper [low]**) are spatially and temporally distributed over the whole study area (Fig. 10.10). To this group a flat axe from the Corded Ware settlement at Nieder-Ramstadt (D; ANo 225) must be added. The analysis of this find fits better into the silver-antimony-group than to its previous grouping in Cluster 32.

In contrast, objects with higher concentrations of silver and antimony are characterised as **silver-antimony-copper (high)** (Fig. 10.11). The artefacts are mainly dated to the 3rd millennium BC and come from throughout the research area. The group contains finds from the Earliest Bronze Age cemetery of Mierzanowice (PL).

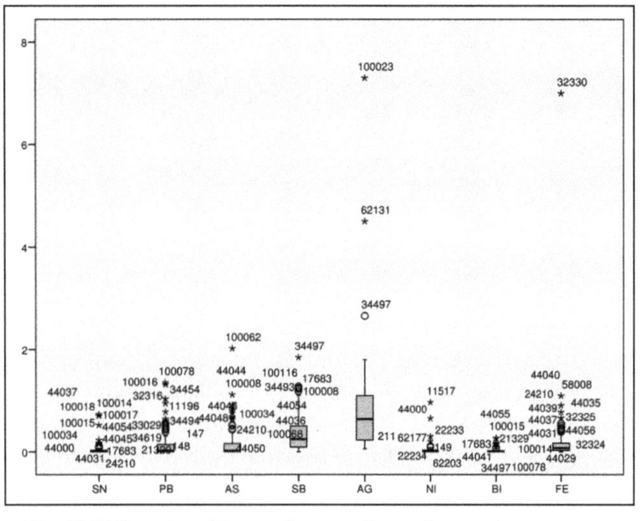

Fig. 10.11 Box plot of the metal group *almost silver-antimony-copper (high)* (excluding ANo 100032; numbering refers to ANo at the catalogue).

It can be assumed that the properties of copper containing mainly silver and antimony are not greatly altered, as silver does not influence copper, but it does counteract the negative effects of antimony. Arsenic, however, even in relatively low concentrations, may improve copper as it interacts with antimony. This should be taken into account especially in the case of silver-antimony-copper (high), which contains trace elements other than silver and antimony in significant amounts. It can be supposed that high levels of silver – in contrast to antimony and arsenic – result from repeated melting and recycling processes because, while the volatile impurities, such as antimony and arsenic, tend to disappear when copper is heated, silver shows a tendency to remain in copper. Consequently, silver-antimony-copper could be a recycled fahlore-copper.

10.5.3 ARSENIC-COPPER

It has been mentioned before that copper with significant arsenic concentrations has often been discussed in archaeometallurgy (e.g. Charles 1967; Selimchanow 1977; Budd 1991a; Lechtman 1996; Klassen et al. 1998). The present study has indeed identified various clusters of arsenical copper. These are pooled into several metal groups: low and high level arsenic-coppers that contain nearly no other impurities, arsenic-copper with impurities, arsenic-antimony-copper, arsenic-nickel-copper and, finally, arsenic-copper with lead.

The **arsenic-copper (low)** group combines 269 trace element analyses from Clusters 13, 28 – 31, and 38. This group is defined by an average arsenic content of around 1%, while other impurities are only present in trace amounts (Fig. 10.12)[169]. Matuschik defined typical arsenical copper (so-called Mondsee-copper) by the following thresholds: 0.27 – 2.55% As; 0.009 – 0.06% Sb; 0.005 – 0.038% Ag; and 0.004 – 0.025% Ni (Matuschik 1998, 239 – 244; Krause 2003, 152, note 213). Despite some outliers, the type with lower levels of arsenic copper fits generally within Matuschik's thresholds. According to, for example, Budd (1991a, 132 f.), typical arsenic copper with less than 2% arsenic changes the properties only slightly compared to purer copper. Malleability, according to Hanson and Marryat (1927, 126), increases especially up to 1% concentration of arsenic. The arsenic level, however, seems to be deliberate, because arsenic is a very volatile trace element and can simply be removed from copper as an oxide. This type of arsenic-copper must have been produced under reducing conditions, since experiments have shown that under oxidising circumstances arsenic disappears easily (McKerrell and Tylecote 1972, 210 ff.). The method of producing arsenic-copper can only be assumed, however, as archaeological evidence is still missing for this area of research. Because of the volatility of arsenic, special techniques are required to keep arsenic inside the copper.

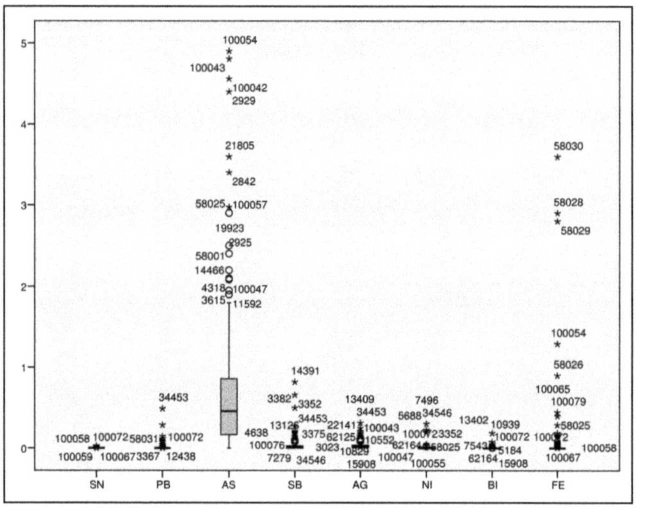

Fig. 10.12 Box plot of the metal group *arsenic-copper (low)* (numbering refers to ANo at the catalogue).

This type of low concentration arsenic-copper is found in artefacts from more or less all over the studied area. Around one third of all analyses are from 3rd millennium contexts (e.g. from the Nitra group), although most of the analysed artefacts date to the 4th millennium BC.

169 Nevertheless some outliers contain up to 5% of arsenic, whereas the other impurities are of very low concentrations.

In contrast to arsenic-copper (low) that is relatively uncontaminated by other trace elements, the three Clusters 33, 36 and 43 have to be discussed separately. Although these also contain on average around 1% arsenic, they cannot be included in the previous grouping and have to be addressed as **arsenic-copper with impurities**, because they have varying and higher contents of several other trace elements than the previously discussed metal group (Fig. 10.13).

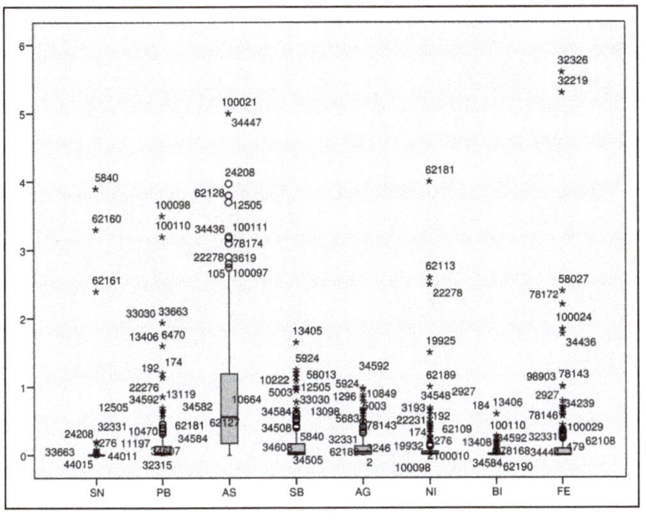

Fig. 10.13 Box plot of the metal group *arsenic-copper with impurities* (numbering refers to ANo at the catalogue).

While Cluster 36 spatially and chronologically covers the entire study area, Cluster 33 encompasses some analyses on the Drosa material and from other objects of the 3rd millennium BC. Cluster 43 also consists of objects mostly dated to the 3rd millennium BC, with a regional focus on Middle Germany and Slovakia. Whereas the arsenic, antimony and lead content of Cluster 33 points to an harder and more workable copper, the properties of the objects of Cluster 43 and especially Cluster 36, probably do not differ from those of the lower arsenical copper.

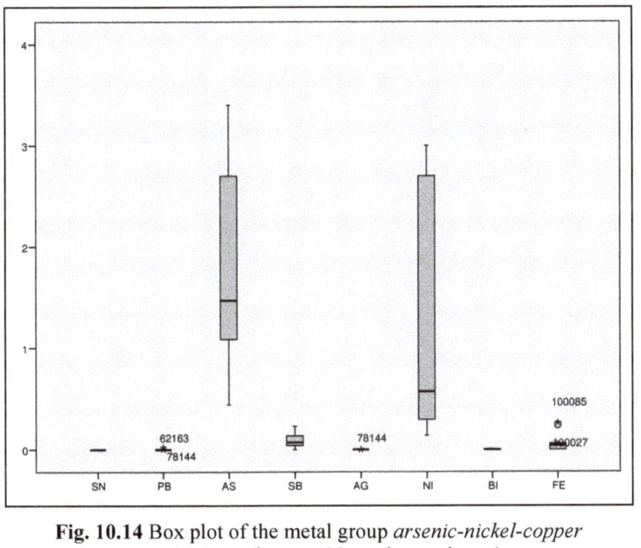

Fig. 10.14 Box plot of the metal group *arsenic-nickel-copper* (numbering refers to ANo at the catalogue).

The **arsenic-nickel-copper** (Cluster 37) – including 11 trace element analyses – also needs to be distinguished, because the objects of that cluster primarily contain impurities of arsenic and nickel – and little else – so that it cannot be compared properly with other clusters. As mentioned above, the cluster includes finds with higher and lower levels of impurities (Tab. 10.40a and 10.40b), whereby the finds with higher rates of impurities tend to be younger (Chapter 10.4). The box plot reflects this by the great dispersion of arsenic and nickel (Fig. 10.14). The higher content of arsenic and nickel should raise hardness, ductility and malleability, especially in the case of the artefacts that are made of higher concentrated arsenic-nickel-copper and that are generally dated to the 3rd millennium BC.

Clusters 50 – 53, although initially appearing to differ in terms of metallurgy have to be discussed together – not only because they are archaeologically comparable, but also because the PCA shows that these analyses are similar (Fig. 10.15). These clusters, however, also show parallels to the arsenic-copper with impurities.

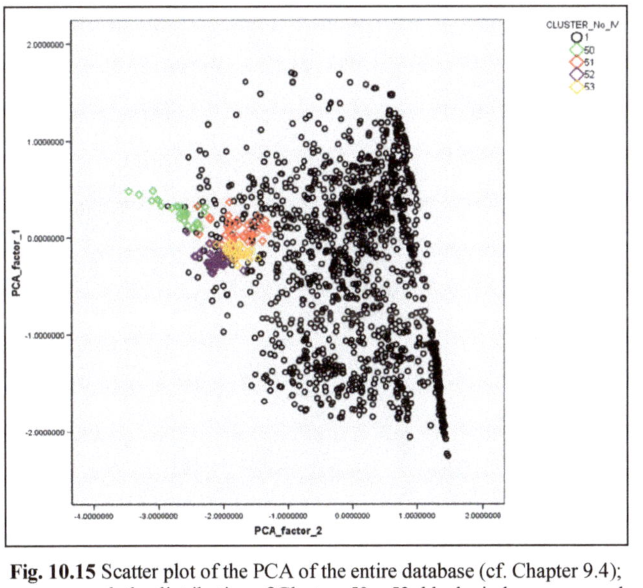

Fig. 10.15 Scatter plot of the PCA of the entire database (cf. Chapter 9.4); squares mark the distribution of Clusters 50 – 53; black circles correspond with the trace element analyses of the other clusters.

The clusters all have significant lead and bismuth values. Because the component *PCA_factor_2* resulting from the PCA chiefly reflects the lead and bismuth content, these objects are located at the far left of the scatter plot (Fig. 10.15). Moreover, with the exception of Cluster 51, all are characterised by considerably more than 1% of arsenic. In order to account for these clusters properly in the further discussion and in the catalogue, this group is named differently. The two Clusters 50 and 51 are labelled **arsenic-lead-copper** (Fig. 10.16), and Clusters 52, 53 and 55 are labelled as **arsenic-copper (high)** (Fig. 10.17).

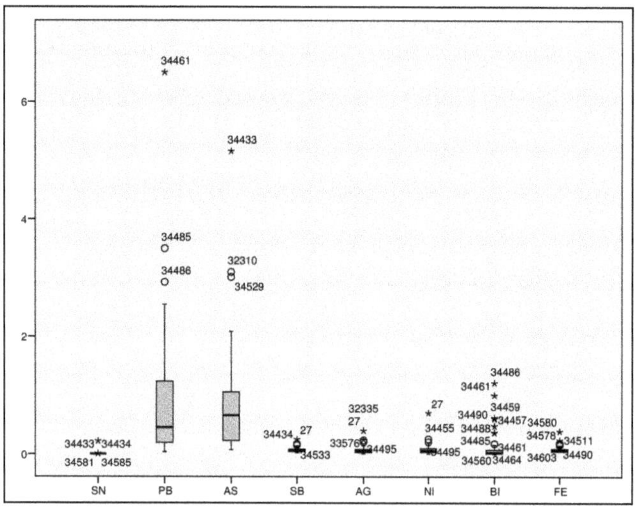

Fig. 10.16 Box plot of the metal group *arsenic-lead-copper* (numbering refers to ANo at the catalogue).

The arsenic and the lead contents of Cluster 50 and 51, but also the arsenic and nickel concentrations of most artefacts in Clusters 52 and 53, suggest a harder material and a reduced melting temperature which improves the casting properties. In Chapter 7.3.5 lead on its own was considered to be a harmful impurity; however the arsenic content, as present in these types of copper, neutralises this. Lead should also improve the cold-working properties of copper.

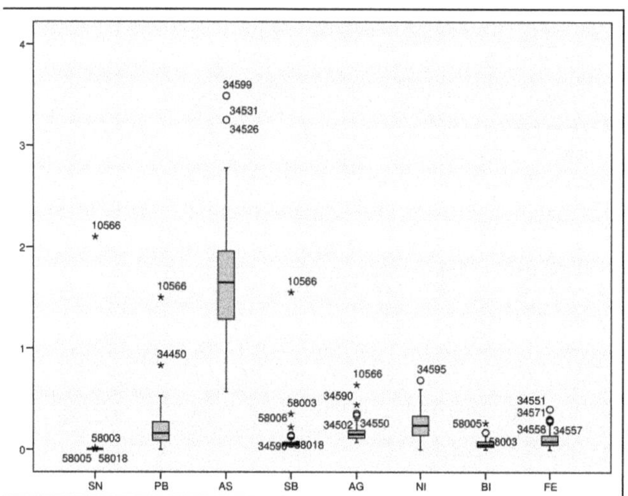

Fig. 10.17 Box plot of the metal group *arsenic- copper (high)* (numbering refers to ANo at the catalogue).

Nearly all analyses of Clusters 50 – 53 have been made on finds from the graves from Drosa (140 analyses) and the others (12 analyses) are either also related to Corded Ware or to Bell Beaker contexts, or are at least from the same period. The significant high bismuth content in Clusters 53 and 50 should actually make the metal very brittle, but overall the arsenic concentrations – which are far higher – counteract and neutralise the negative bismuth effect.

Finally, Cluster 54 has to be mentioned. Metallurgically, it is very similar to Cluster 68 (discussed below), although it is dominated by its arsenic content. Cluster 54 objects chiefly contain arsenic, antimony and some silver[170] (Fig. 10.18). We list it as **arsenic-antimony-copper**. Although Pernicka (1995) called it '*Fahlerzmetall ohne Nickel* (fahlore-metal without nickel)', it is labelled differently in the present work to avoid interpretations of copper types as specific ores. The metallurgical composition indicates that, in comparison with pure copper, this type has improved hardness and working properties. Cluster 54 encompasses 37 trace element analyses primarily from Earliest Bronze Age contexts, but also a few earlier finds, and almost all from the northalpine region.

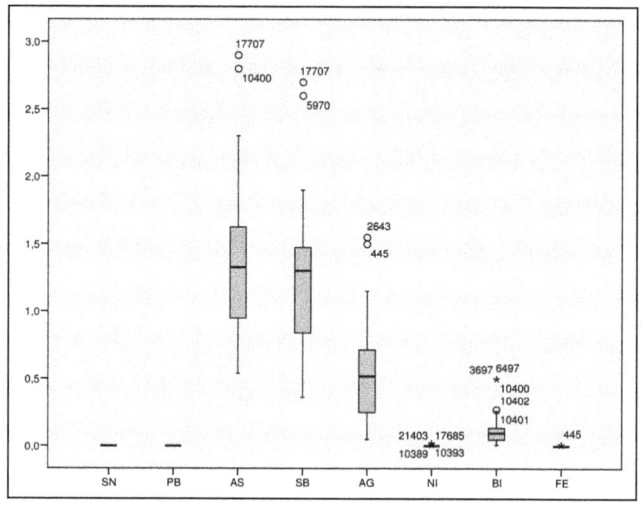

Fig. 10.18 Box plot of the metal group *arsenic-antimony-copper* (numbering refers to ANo at the catalogue).

Finally, looking at the entire database and checking which objects of which periods contain more arsenic, the following conclusions can be drawn. No artefact of the database in this study contains more than 5% arsenic – and only around 7% of all analyses contain more than 2% arsenic – whereas less than one quarter of these 7% is dated to the 4th millennium or older. As a result, it is obvious that artefacts that contain more arsenic are predominantly more recent. In contrast, however, many objects from contexts of the 3rd millennium BC also have low levels of arsenic generally – whereby arsenic is the main impurity. (Around 40% of all objects that contain less than 2% arsenic are dated to the 3rd millennium BC). Several previous studies postulated that copper which contains mainly arsenic is typical for central Europe during the 4th millennium BC (e.g. Sangmeister 1971; Ottaway 1978; Matuschik 1998). The results of the present study confirm this. Moreover, the type of copper, which contains predominantly arsenic and only traces of other impurities (so-called: arsenic-copper [low]), was still commonly used throughout the 3rd millennium BC. It is even known from finds classified as Earliest Bronze

[170] Cluster 54 includes one outlier: a flanged axe of type Salez from a hoard found at Feldkirch 'Tillis' (A) that contains a remarkable 6% of antimony; the rest of the concentrations are analogous to Cluster 68.

Age. That indicates either a recycling of older artefacts, or more likely, the ability of later Chalcolithic metallurgists to deal with similar techniques and/or to exploit the same resources. As arsenic is very volatile, it needs a special technique in order to keep arsenic inside the copper.

10.5.4 ANTIMONY-COPPER

The clusters that are summarised by the label antimony-copper contain antimony as the predominant trace element. Here as well, the groups can be divided into various types of antimony-copper based on their specific trace element concentrations. Whereas the group with less impurities has to be differentiated between an antimony-copper with silver, and an antimony-arsenic-copper, the types of antimony-copper with higher concentrations of trace elements are sub-grouped as various types of fahlore-copper – both with and without nickel. Therefore, the results of the TSCs are comparable to the SAM-group's *Stammbaum* (Chapter 8.1.1)[171]. Although Krause combined several types of metal that are dominated by arsenic, antimony, nickel and silver under the umbrella term 'fahlore-metal' (Krause 2003, 90, Fig. 40), the statistical analysis has separated out clusters with very high concentrations of impurities that are dominated by antimony. Consequently, in comparison with the other clusters, the copper types that contain the greatest amounts of trace elements noted in the present study contain antimony as the main impurity.

As has been identified in Chapter 7.3.8, antimony on its own is one of the harmful trace elements. It is not surprising, therefore, that none of the artefacts in the entire database contains only antimony without even traces of other elements. Antimony appears in copper artefacts only in combination with other impurities, especially with nickel and arsenic. Therefore, the main group is divided into a group of antimony copper with both arsenic and silver, and two groups of antimony copper with silver – one with higher, and one with lower, concentrations.

The metal group **antimony-arsenic-copper with silver** is defined by some analyses in Clusters 25 and 68 (Fig. 10.19). Cluster 25 includes 30 analyses, mostly from earlier periods. The impurities slightly improve the properties of copper, but not to a great extent as the overall content is not very high. Three quarters of the finds of Cluster 25 are dated to the 5th and 4th millennium BC, and most of these are single finds (flat and hammer axes) from the Czech Republic and Slovakia. Cluster 68 has to be compared with Cluster 25, because its structure is similar, although the antimony, arsenic and silver levels are about twice as high[172]. Therefore, the antimony-arsenic-copper of Cluster 68 has better hardness. Most of the finds of that type of copper are from Earliest Bronze Age contexts of the northalpine region.

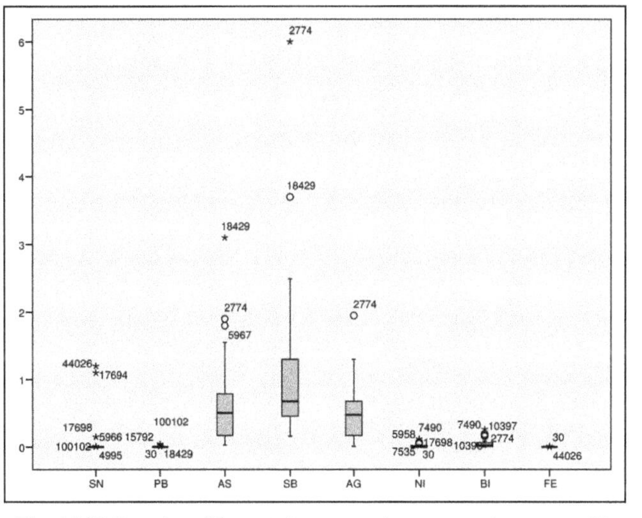

Fig. 10.19 Box plot of the metal group *antimony-arsenic-copper with silver* (numbering refers to ANo at the catalogue).

Comparing Cluster 68 with arsenic-antimony-copper (cf. Chapter 10.5.3), both are of different structure. Their total amount of impurities is similar, but one group is dominated by arsenic and the other by antimony. Nonetheless, it can be assumed that both metal groups have increased hardness and similar metallurgical characteristics. Additionally, their archaeological context is similar.

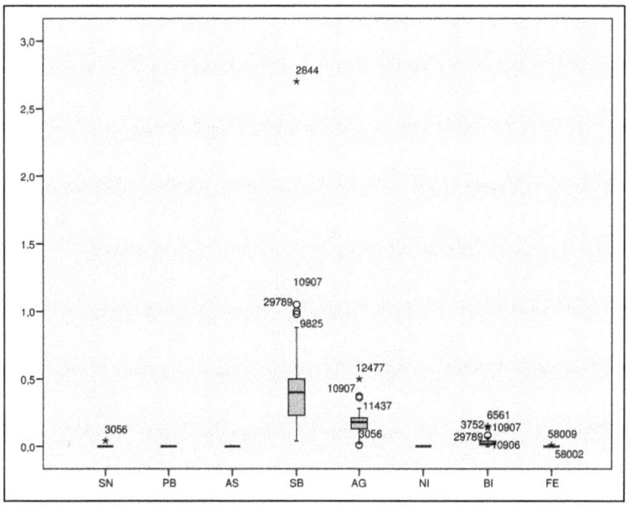

Fig. 10.20 Box plot of the metal group *antimony-silver-copper* (numbering refers to ANo at the catalogue).

The **antimony-silver-copper** based on Cluster 26 is another antimony-dominated copper that contains, apart from traces of nickel and small amounts of bismuth, only silver and antimony (Fig. 10.20). The archaeological

171 The comparison between names of types of copper such as fahlore- or arsenic-copper etc and the SAM-grouping is based on an article by Pernicka (1995). According to him, the various C_2- and C_6-groups of the *SAM-Stammbaum* correspond to *fahlore-metal without nickel* and, in contrast, the A-, B_2-, C_5-, E_{11}-, FB- and FG-groups of the *SAM-classification* correspond to *fahlore-metal with nickel* (Pernicka 1995, 97 ff.).

172 Four outliers (ANo 5840, 17694, 17698 and 44026) are significant, because of their remarkably high tin content (up to 3.9%).

configuration primarily encompasses axes dated to the 4th millennium BC, while more recent objects are less frequently included. The two remaining analyses from Cluster 32 fit into Cluster 26. Thus, the flat axes from Dolné Srnie and Oravská Polhora (both SK) are considered here[173]. Silver may counteract the negative effects of antimony in copper. However, it is unlikely that both trace element affects the properties of the metal very much if compared with pure copper, because silver has little effect, although it might slightly bleach the colour of the metal.

The picture changes with the remaining types of copper. These contain higher amounts of impurities. These major groups are characterised by dominant antimony and nickel contents, followed by smaller concentrations of trace elements. Besides antimony and nickel, arsenic and silver and sometimes lead and tin is present. This group can be split up into three types of antimony-nickel-copper according to the differing concentrations of impurities.

The **antimony-nickel-copper (low)** group includes Clusters 66, 67 and 69 – 71 (Fig. 10.21). Apart from some outliers, none of the trace elements exceeds a concentration of around 2%. This primarily includes Earliest Bronze Age artefacts (mainly rings) from southern Germany and western Slovakia. This group also includes parts of Cluster 37 that were defined as antimony-nickel-copper. The analysed objects are distributed over the entire area of research, although Cluster 37 contains only eleven analyses.

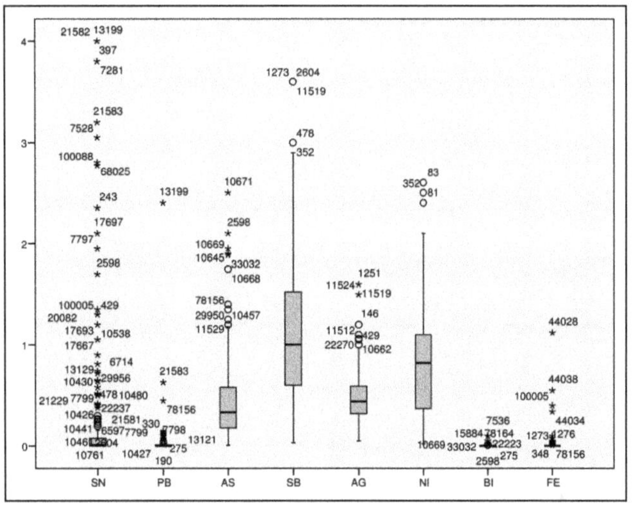

Fig. 10.21 Box plot of the metal group *antimony-nickel-copper (low)* (numbering refers to ANo at the catalogue).

The **antimony-nickel-copper (intermediate)** group, encompassing Clusters 58 – 62, includes more or less the same spectrum of artefacts, apart from a slight peak of pins. The chronological and the geographical distribution are comparable to antimony-nickel-copper (low). Within this metal group the general antimony and nickel concentrations range between about 2 and 4% (Fig. 10.22).

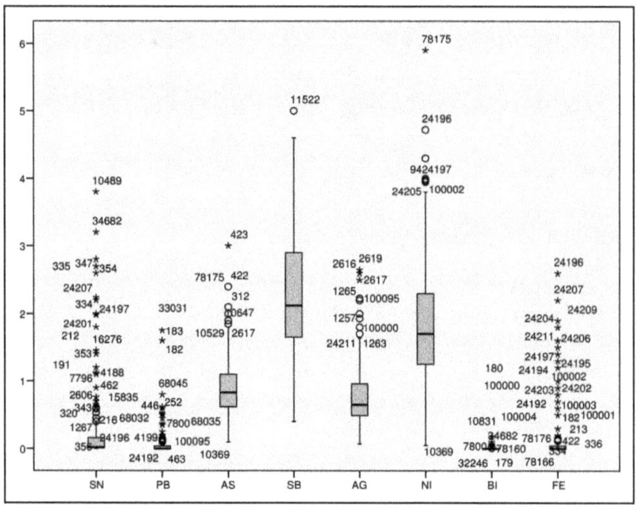

Fig. 10.22 Box plot of the metal group *antimony-nickel-copper (intermediate)* (numbering refers to ANo at the catalogue).

The **antimony-nickel-copper (high)** (Clusters 56, 57 and 59) contains nickel contents of usually more than 2%, while antimony concentrations are beyond 4% (Fig. 10.23). This type of copper is almost exclusively used in flanged axes – the so called Salez axes – from south-west Germany and Switzerland. According to Kienlin, these axes are made of very hard copper (Kienlin et al, 2006). Bill (1997, 251) reports that copper containing antimony and nickel has a silverish colour, which differs from the orange colour of pure copper, and the golden colour of tin-bronze. This may be caused by the high antimony and nickel concentration.

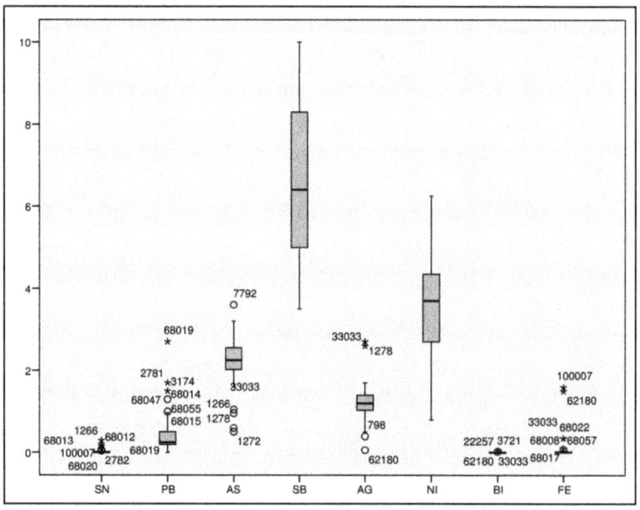

Fig. 10.23 Box plot of the metal group *antimony-nickel-copper (high)* (numbering refers to ANo at the catalogue).

In conclusion, the metal groups which contain antimony as the principal trace element are mainly provided by artefacts that generally date to the younger half of the studied periods. Whereas antimony-silver-copper and antimony-arsenic-copper with silver – both types contain mostly few impurities – were also used during PER 25, copper with higher concentrations of trace elements and metallurgically-improved types was only used during the 3rd millennium BC – and especially by Earliest Bronze

[173] ANo 10386 and ANo 10392.

Age people. Antimony-arsenic-copper with silver (mainly Cluster 68) is also dated to the 3rd millennium BC. Overall, then, finds of the three types of antimony-nickel-copper date almost exclusively to the 3rd millennium BC.

10.5.5 Additional Metal Groups

Nearly all of the 42 clusters could be grouped according to the scheme above. However, some clusters, and the trace elements bismuth and iron, have to be discussed separately as they do not fit into the system.

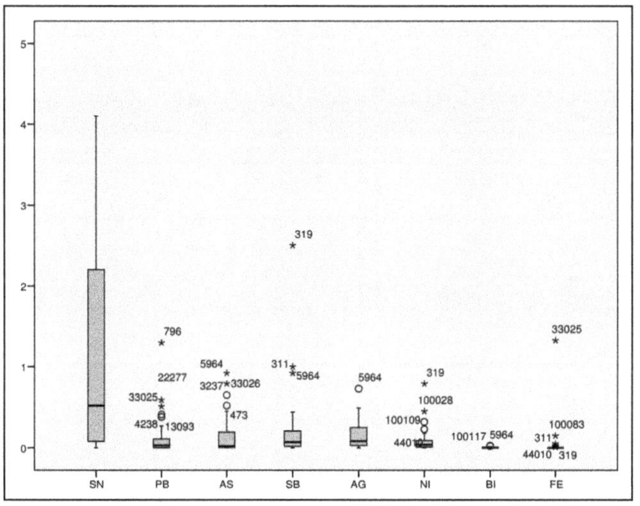

Fig. 10.24 Box plot of the metal group *tin-rich-copper (incl. As, Sb etc)* (numbering refers to ANo at the catalogue).

Cluster 41 is interesting, because of its remarkable tin concentrations – although the objects of that cluster have generally not more than 1% of tin (**tin-rich-copper [incl. As, Sb etc]**). Not only is it relatively tin-rich, but also antimony, arsenic and silver appear in various quantities (Fig. 10.24). This tin content, as well as the antimony-arsenic concentrations, probably meant that these objects had improved hardness. The vast majority of the finds of that cluster date to the 3rd millennium BC.

The small Cluster 55 is difficult to compare with others. Apart from the arm spiral from Beizkofen (D)[174] which can be classified as antimony-arsenic-copper with silver, the trace element analyses seem to be difficult to compare as they appear as outliers in the double-analyses in Chapter 8.3. Nevertheless, those four analyses are arsenic-copper (high). Three of the five analyses of Cluster 55 are double-analyses made by Novotná (1970). Those three are also outliers of the cluster analysis of the double-analyses[175].

It is also important to examine trace elements that do not specifically define a cluster by statistical analysis. The two impurities of bismuth and iron do not appear as cluster-defining elements, but since both are considered as trace elements that negatively affect the properties of copper, and because iron may be a marker for the smelting process, these elements need to be explored in more detail. In Chapter 7 it was highlighted that if bismuth is the only impurity, even in small traces, it causes brittleness and makes copper difficult to work. In combination, however, especially with arsenic, this negative effect is neutralised. None of the clusters is dominated by bismuth and not a single object contains bismuth as the only impurity. So, Chalcolithic copper was not affected by the harmful influence of bismuth. Moreover, all the clusters that show significant concentrations of bismuth also contain much more arsenic. As a consequence, bismuth does not negatively affect those types of copper.

The issues are similar for iron. All the clusters that show remarkable quantities of iron include nickel as well (cf. Clusters 33, 36 and 59). Nickel raises the solubility of iron in solid copper (Dies 1967, 806), and thus the extremely harmful effect of precipitating iron and the resulting brittleness is counteracted by the presence of nickel. The increased solubility caused by nickel explains why those clusters which incorporate more iron also contain more nickel, as soluble iron is harder to remove by smelting and refining. Apart from certain clusters, single finds with extreme iron contents are also conspicuous, as these were comparably purer in terms of other trace elements. While the three finds with the highest trace element content [176] are found in areas and periods where metallurgy appeared only sporadically in archaeological records, the others have been discovered in the context of smelting sites[177]. In both cases, the reason for the high iron content seem to be connected to the production; either because metallurgical techniques have not been used frequently and, therefore, they have probably been carried out less accurately, or else the analysed objects are remains from smelting processes, with the iron concentrations resulting from the reduction of copper ores.

[174] ANo 30; PER 40.

[175] This issue is similar to the problem previously discussed regarding the metal group 'almost pure with iron' in Chapter 10.5.1.

[176] ANo 32219, 32330 and 100032; ANo 32326 is also an iron-rich double-analyses of ANo 32219 – a reel from Gotha-Ostheim (D).

[177] Crucibles and so-called wall plaster fragments from Makotrasy (CZ) (ANo 58028 – 58030) (Pleslová-Štiková 1985, 175 ff.).

10.5.6 SUMMARY OF THE METAL GROUPS

The groups of Cluster Level IV and Clusters 9 and 13 of Level III have been combined to create certain metal groups reflecting both their archaeological and especially metallurgical characteristics, as the following table shows (Tab. 10.61). Table 10.62 displays the mean values of the trace elements that were consulted in order to characterise the 19 metal groups. The table can be used as orientation for comparing the results of the present evaluation with other analyses

By considering the chronological distribution of these metal groups, the following bar charts emphasise again (Fig. 10.25). Especially from PER 31 (c. 2800 BC) onwards, types of copper with higher concentrations of impurities had been used. Those types indicate improved metallurgical properties, such as hardness or workability. With the appearance of the Earliest Bronze Age contexts types of copper were commonly used that have generally improved properties. Just before the establishment of tin-alloyed copper – the classical bronze – copper types bearing substantial quantities of antimony, arsenic and other impurities were used. These metal groups are characterised by better mechanical and working properties and dealing with them can be seen as metallurgical progress (cf. Chapter 7.4).

Metal group	Mainly included clusters	Number of analyses
Pure copper	48, 49	243
Almost pure copper	9	62
Almost pure copper with Iron	13	15
Pure copper with Nickel	22	69
Pure copper with Silver	23	53
Antimony-arsenic-copper with Silver	25, 68	41
Antimony-nickel-copper (high)	56, 57, 59	55
Antimony-nickel-copper (intermediate)	58 – 62	221
Antimony-nickel-copper (low)	66, 67, 69 – 71	224
Antimony-silver-copper	26, 32	66
Arsenic-copper (high)	52, 53, 55	74
Arsenic-copper (low)	13, 28 – 31, 38	269
Arsenic-copper with Impurities	25, 33, 36, 43	214
Arsenic-antimony-copper	54	36
Arsenic-lead-copper	50, 51	80
Arsenic-nickel-copper	37	11
Silver-antimony-copper (high)	39, 40	99
Silver-antimony-copper (low)	24, 27, 32 (only ANo 225)	70
Tin-rich-copper (incl. Ag, Sb etc)	41	41

Tab. 10.61 Metal groups and related clusters.

		SN	PB	AS	SB	AG	NI	BI	FE
		Mean	Mean	Mean	Mean	Mean	Mean	Mean	Mean
Metal_groups	Almost pure	,0000	,0000	,0205	,0065	,0115	,0190	,0004	,0002
	Almost pure with Iron	,0220	,0069	,0000	,0006	,0067	,0634	,0024	,1209
	Antimony-arsenic-copper with Silver	,0604	,0028	,6476	1,0110	,4922	,0147	,0460	,0003
	Antimony-nickel-copper (high)	,0195	,4533	2,2083	6,6262	1,2008	3,5829	,0061	,0824
	Antimony-nickel-copper (intermediate)	,2327	,0666	,9368	2,2837	,7816	1,8639	,0060	,1536
	Antimony-nickel-copper (low)	,2510	,0284	,4315	1,1074	,4860	,8271	,0025	,0138
	Antimony-silver-copper	,0006	,0000	,0000	,4353	,1771	,0010	,0292	,0003
	Arsenic-antimony-copper	,0000	,0000	1,3686	1,2417	,5500	,0023	,1303	,0003
	Arsenic-copper (high)	,0317	,2081	1,6896	,0817	,1671	,2616	,0586	,1080
	Arsenic-copper (low)	,0008	,0072	,6932	,0283	,0320	,0095	,0037	,0543
	Arsenic-copper with Impurities	,0528	,3804	,8941	,1679	,1540	,1132	,0277	,1843
	Arsenic-lead-copper	,0062	,8461	,8096	,0692	,0690	,0804	,1023	,0844
	Arsenic-nickel-copper	,0000	,0025	1,7945	,0955	,0001	1,3027	,0017	,0668
	Pure	,0000	,0000	,0000	,0000	,0000	,0010	,0000	,0000
	Pure with Nickel	,0000	,0000	,0000	,0000	,0023	,1220	,0000	,0000
	Pure with Silver	,0000	,0000	,0000	,0000	,0268	,0011	,0000	,0000
	Silver-antimony-copper (high)	,0437	,1449	,1619	,3528	,7913	,0393	,0256	,3927
	Silver-antimony-copper (low)	,0001	,0000	,0107	,0893	,1601	,0206	,0072	,0000
	Tin-rich-copper (incl. Ag, Sb etc)	1,1125	,1194	,1458	,2080	,1540	,0836	,0052	,0415

Tab. 10.62 Mean values of the trace elements of the 19 metal groups.

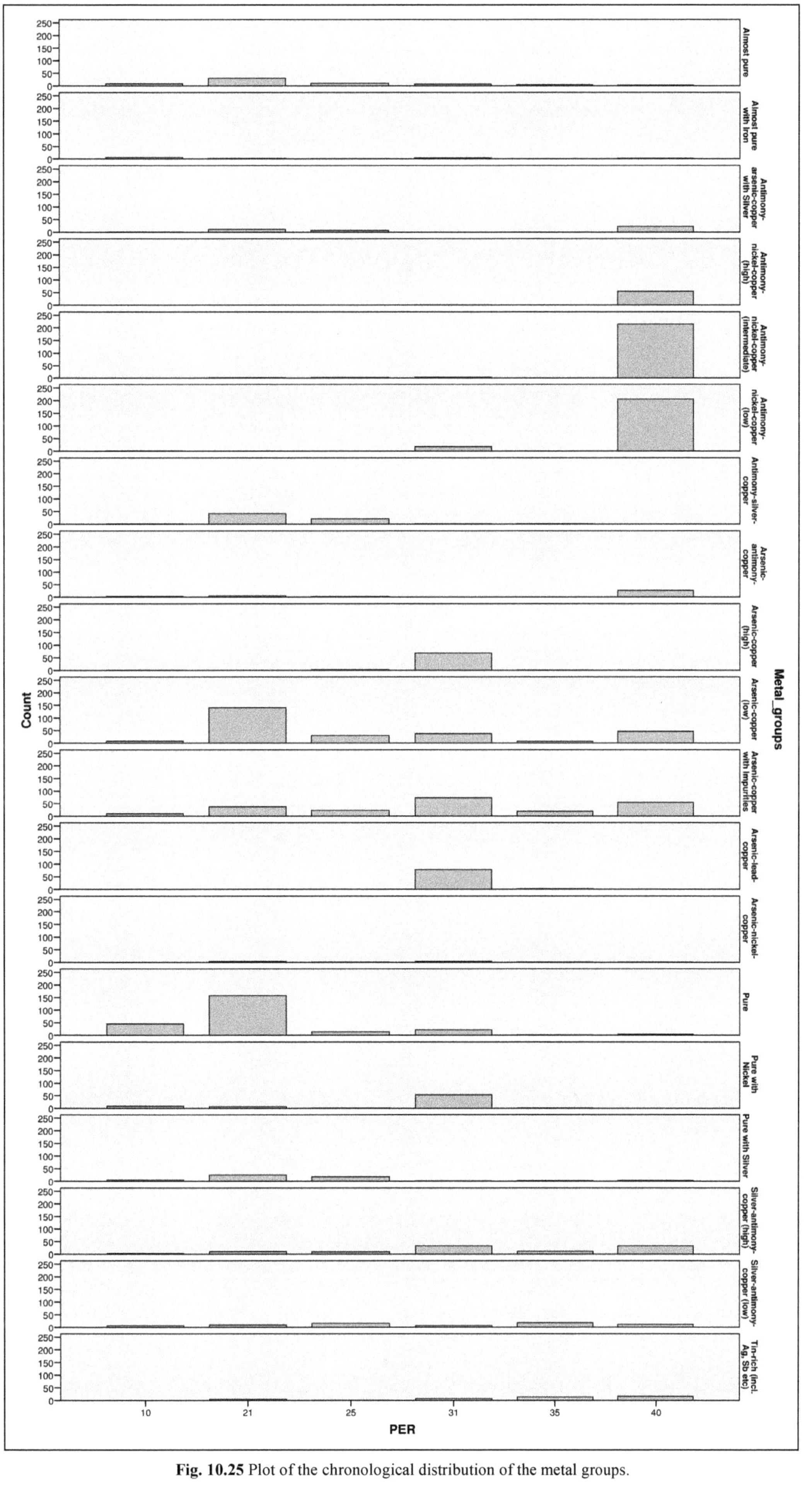

Fig. 10.25 Plot of the chronological distribution of the metal groups.

10.6 ARCHAEOLOGICAL REFLECTION ON THE CHALCOLITHIC TYPES OF COPPER

The clusters have been compared and rearranged in terms of their metallurgical compositions and properties. The following section will now summarise these results in terms of their archaeological interpretation. The chronological background of the metal groups is illustrated very clearly by the scatter plot of the PCA referring to the entire database (Fig. 10.26).

The comparison and re-combination of the clusters of Level IV shows a picture that is more or less similar to the previous Cluster Level III (Appendices E.30 and Fig. 10.27)[178]. Comparing the results of the TSC with the SAM classification indicates similar tendencies, although slight variations can be recognised which are caused by consideration of different impurities (Appendix E.31 – E.76). Looking at the scatter plot resulting from the PCA that displays both the chronological and metallurgical categorisations (Fig. 10.27), it can be observed that from Period 31 onwards artefacts incorporate more impurities than before – especially with the beginning of what we recognise as the Earliest Bronze Age. The black dashed line symbolises that transition (Fig. 10.27). Of course this is not a fixed break, but rather an area of transition and a few older outliers are below this line, but the vast majority of the older objects contain fewer impurities. On the other hand, the metallurgists of the 3rd millennium BC still used copper with low levels of impurities, as had been used in earlier periods.

In contrast to the result that 3rd millennium BC objects are characterised by higher levels of impurities, Cluster 22 – a type identified as pure copper with nickel[179] (Fig. 10.22) – is formed mainly by finds from Corded Ware sites in Switzerland; most of the analysed object were found in Vinelz (CH) (Fig. 10.28). The significant nickel content in this otherwise very pure copper does not affect the properties of the copper (cf. Chapter 7.3.6). It can be inferred that this cluster reflects a copper that has probably been recycled many times, as all the more volatile and easily removable impurities are missing. However, nickel, which is hard to remove, is present. Thus, it can be assumed that communities which used this metal had no access to fresh supplies of copper; consequently they were forced to reuse old metal. This assumption can be supported by three facts: firstly, the artefacts of Cluster 22 are dated to a time[180] when objects were generally produced with highly contaminated types of copper; and, secondly, most of the analyses were made of finds that were found at one site in Vinelz (CH), whereas at nearby sites other, 'better' types of copper were used. Moreover, a few tools from Vinelz also contain more impurities. Finally, some artefacts indicate that they have been made of recycled artefacts. An awl, for example, was made out of four other awls (Fig. 10.28m). Consequently, other types of copper must have been in contemporary use. This leads to the conclusion that this particular community is, in terms of their metallurgy, separated from the rest, although it is not clear why.

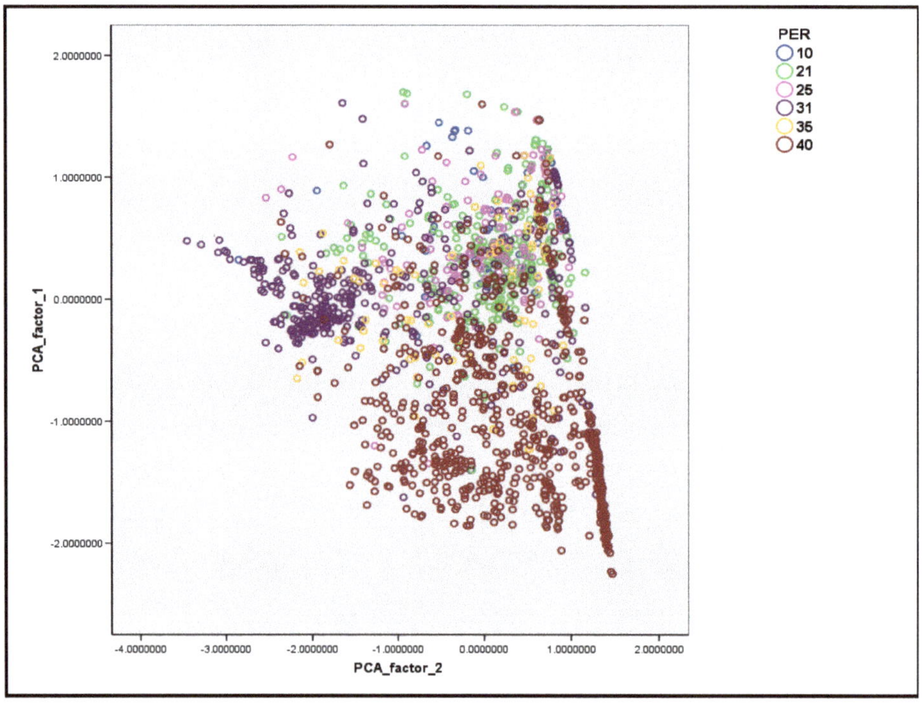

Fig. 10.26 Scatter plot resulting from the PCA of Chapter 9.5 displaying the chronological spread of the cases (PER).

178 On the scatter plot of Appendix E.102 the types of almost pure copper are marked in reddish colours, the metal groups with lower levels of impurities in green, the higher concentration fahlore-coppers in blue, and the three types of antimony-nickel-copper in orange/yellow.

179 The so-called FC-copper according to the SAM-classification.

180 The older objects of Cluster 22 are grouped in this cluster, because they also contain nickel as the only measured impurity, but with these the quantity of nickel is only around half that of the finds of PER 31.

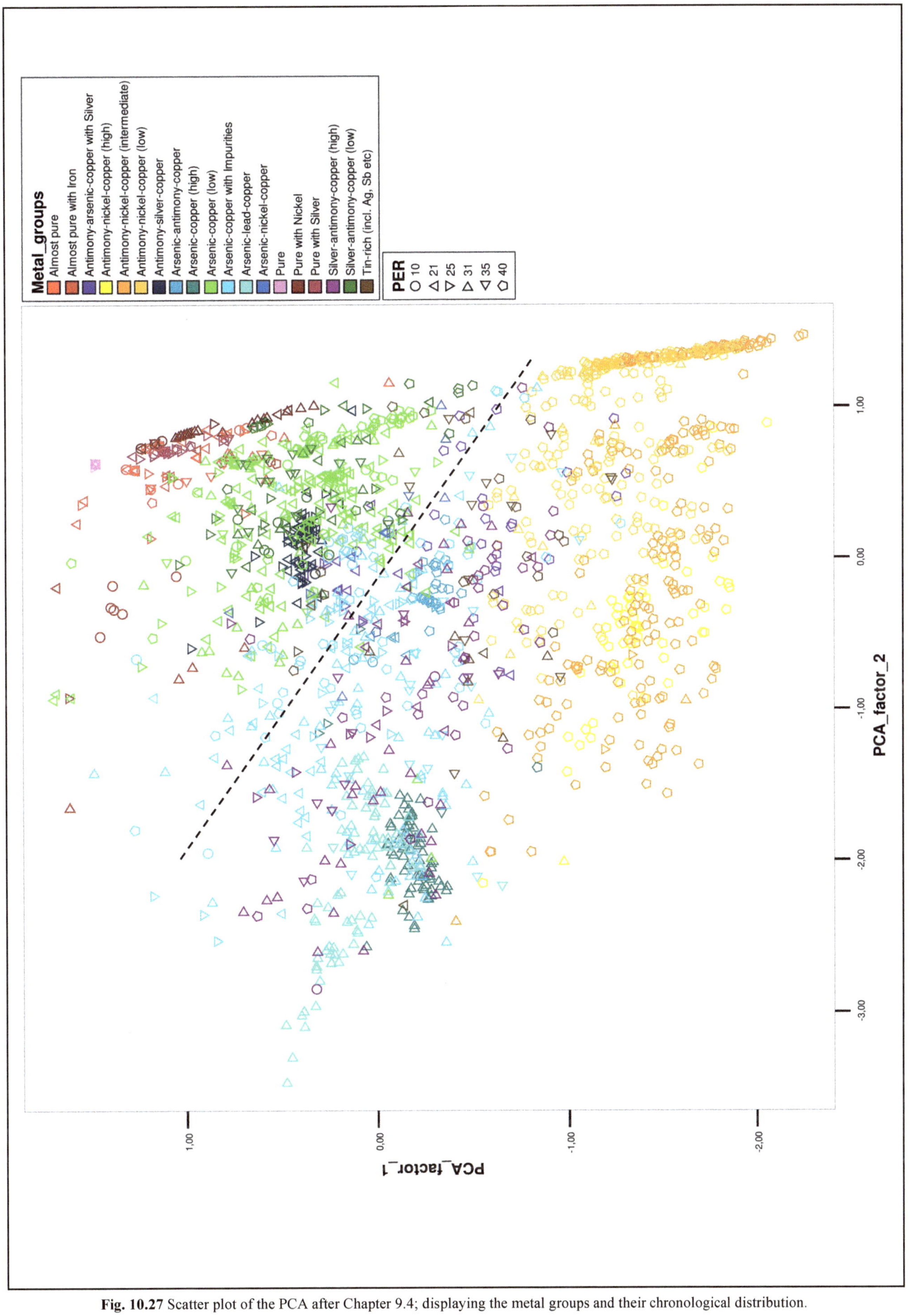

Fig. 10.27 Scatter plot of the PCA after Chapter 9.4; displaying the metal groups and their chronological distribution.

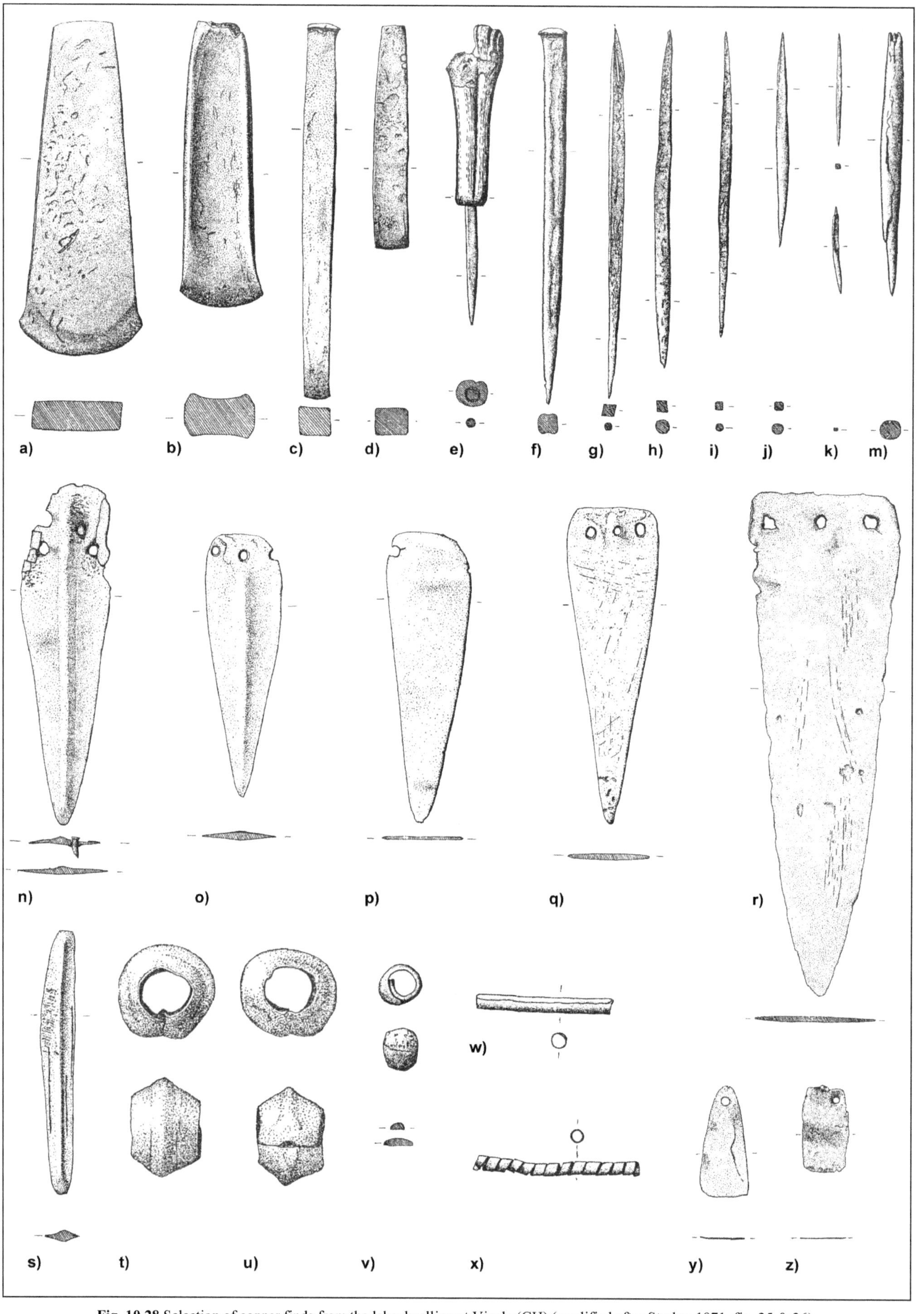

Fig. 10.28 Selection of copper finds from the lake dwelling at Vinelz (CH) (modified after Strahm 1971, fig. 25 & 26).

Further recently analysed objects from Switzerland are of similar composition. These finds are chiefly contemporary to the site of Vinelz[181]. These objects are also made of relatively pure copper that contains significant amounts of nickel. Therefore, it seems that during that first half of the 3rd millennium BC, access to copper supply ended for people in north-western Switzerland, and metal workers were forced to recycle more often than usual. In his analysis of finds from Britain and Ireland Bray (2009) could demonstrate that with increasing distance from the ore source at Ross Island (IRL), the proportional concentration of nickel and silver is either constant or increases, whereas the volatile trace elements arsenic and antimony decrease. Since Bray was able to locate the origin of the ores by lead isotope analyses, his interpretation of the enrichment of nickel and silver is that copper from Ross Island was re-used and re-melted several times before it reached Scotland[182] (Bray 2009).

If we look at the geographical distribution of the types of copper detailed in the previous section, we can see that the chronologically varying types (in particular the copper containing rather lower levels of impurities) are distributed across the study area as a whole and, with the exception of Cluster 22, no centre can be located. This picture changes slightly during the 3rd millennium BC. The scatter plot (Fig. 10.29) shows a centre in Middle Germany (AREA 103 – 105). Although this is chiefly defined by the finds from Drosa (D; AREA 103), objects from other sites cluster there as well. So, it seems that in that area during the first half of the 3rd millennium BC a local, lead-rich arsenical copper was used. For now, that seems to point to an area of local metallurgy.

Some clusters of Level IV indicate a regional type of copper as well. The Earliest Bronze Age Clusters 58 – 71, for instance, diverge regionally – mainly between southern Germany and Slovakia – but in a broader sense, these varying antimony-nickel-coppers appear all over the study area. Here it is likely that a general metallurgy was common, but with local diversity caused by slightly different resources or technology, although the metal working skills, and the intention to produce an antimony-nickel-copper, must have been the same. This is emphasised by the fact that the metal workers of those archaeological groups were able to deal with other copper types, e.g. arsenic and pure copper, so that it is obvious that they had not lost the older techniques, and/or still had

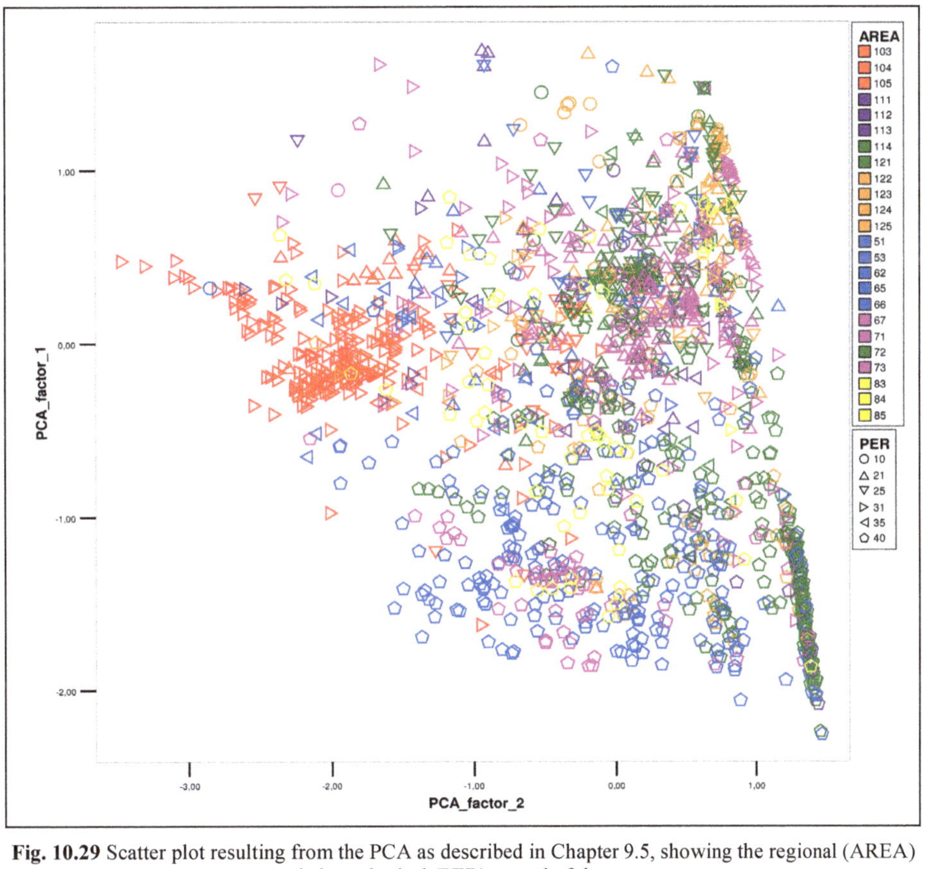

Fig. 10.29 Scatter plot resulting from the PCA as described in Chapter 9.5, showing the regional (AREA) and chronological (PER) spread of the cases.

[181] I would like to thank Dr. S. van Willigen and Prof. Dr. Chr. Strahm for offering me the opportunity to compare my results with their new data, although it was not possible to include their data in my database. Results of their project, which has been carried out in cooperation with other colleagues, have been recently published (Cevey et al, 2006).

[182] At this point, I would like to thank Dr. P. J. Bray for his helpful comments, the discussions on this research and for making some of the results of his unpublished PhD-thesis available to me.

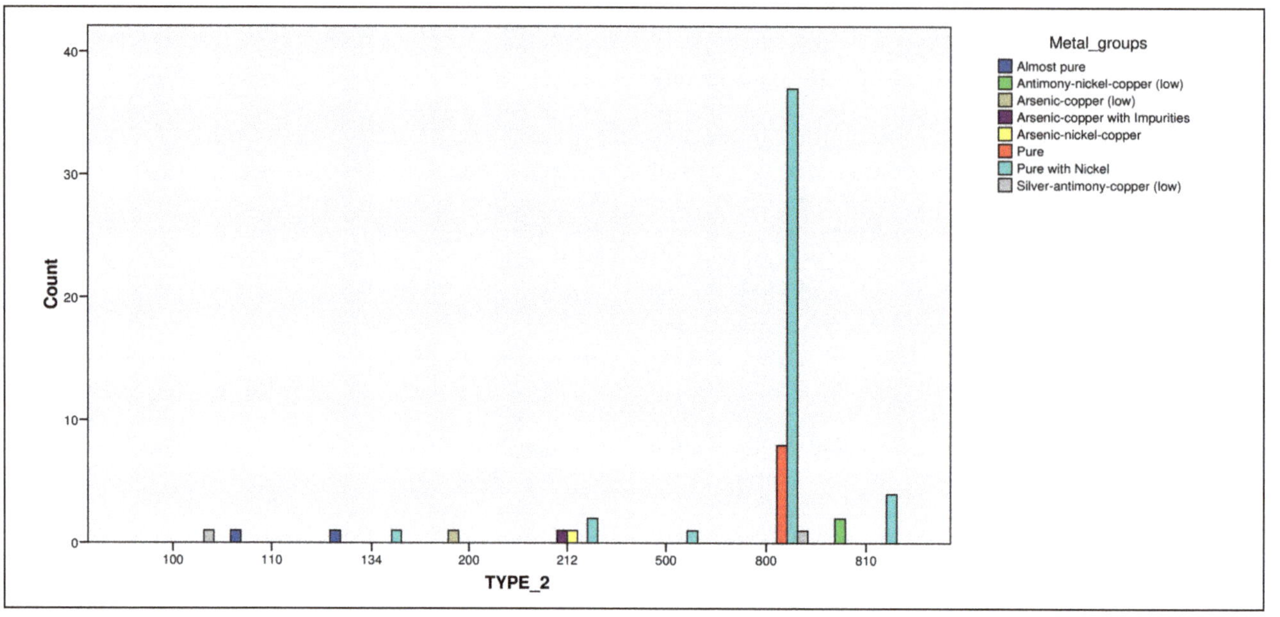

Fig. 10.30 Distribution of metal groups by artefact type in the Corded Ware site Vinelz (CH).

access to those resources. If late Chalcolithic metallurgists were still able to produce artefacts in the same way as in previous times, the question becomes why they used other types of copper, and did these copper types have advantages over the others. A higher concentration of nickel (> c. 2%) has been proven as being intentionally added to a sample of *Spangenbarren* (Moosleitner and Moesta 1988, 66 f.). This could have been possible in this research area as well. This concentration of nickel also brightens the copper, and for that reason, nickel-bearing types of copper are easier to recognise and select. Beside the enhanced hardness and tensile strength, the brighter colour of copper containing more than around 2% nickel was probably also a decisive factor in its use.

Proving whether or not Chalcolithic metallurgists applied their knowledge of the properties of copper in order to produce artefacts with specific characteristics is very difficult across the entire database. This is because, before the appearance of the Earliest Bronze Age groups, the variation of analysable artefacts within one period is fairly limited. Therefore, it is difficult to compare differing types of copper in terms of the function of objects. For example, the dominant objects of PER 10 are heavy shaft-hole axes, while PER 21 includes chiefly flat axes and PER 35 mainly daggers (cf. Chapter 5.2.2). Consequently, a comparative and diachronic analysis of the relationship between copper types and the function of artefacts reflects more the chronological configuration of the type of copper than the functional background – and this can give rise to circular arguments. Nevertheless, at this point this issue will briefly be discussed using two examples that provide various types of artefacts of one site or one period: firstly, the trace element analyses from the Corded Ware site of Vinelz (CH) and, secondly, a comparison of the trace element analyses of the Nitra group with the *Blechkreis*.

Because the trace element analyses of Vinelz are centred in Cluster 22, they seem to indicate a local copper metallurgy that was probably forced to recycle the metal due to its dwindling supply. The very few highly contaminated objects are all daggers and awls – weapons and tools respectively (Fig. 10.30). The two daggers with more than 3% of arsenic are especially significant (metal groups: arsenic-copper with impurities and arsenic-nickel-copper). However, as nearly all types of object are made of relatively pure copper, it points to a primary utilisation of pure copper with nickel, even when considering that all the beads reflect just one object type. It can be assumed that all the beads are made of the same type of copper, because they were produced from one piece of raw material.

In the case of the many objects from contexts of the Earliest Bronze Age *Blechkreis* and Nitra group (both PER 40), which are made from a greater diversity of copper, the picture is slightly different. Jewellery, pins, daggers and axes reflect more or less the regularly distributed type of antimony-nickel-copper that was used by these communities. The willow-leaf-shaped knives and rings of the Nitra group (TYPE 2: 300 and 420), however, seem to be less restricted to antimony-nickel copper (Fig. 10.31a). The *Blechkreis* finds are also largely made of antimony-nickel copper, but other types of copper have also been used. It is notable that the very highly concentrated antimony-nickel copper (Clusters 56 and 57) were only used for flanged axes (especially axes of Salez type). Despite this, flanged axes (TYPE 2: 110) are also made of other types of copper (Fig. 10.31b). The 'Salez copper' in particular indicates the intentional utilization of copper, as this antimony-nickel-copper (high) was almost exclusively used for these axes.

It has been speculated that these standard-shaped axes, chiefly found as hoards, were used and traded as ingots in order to process them into other finished products

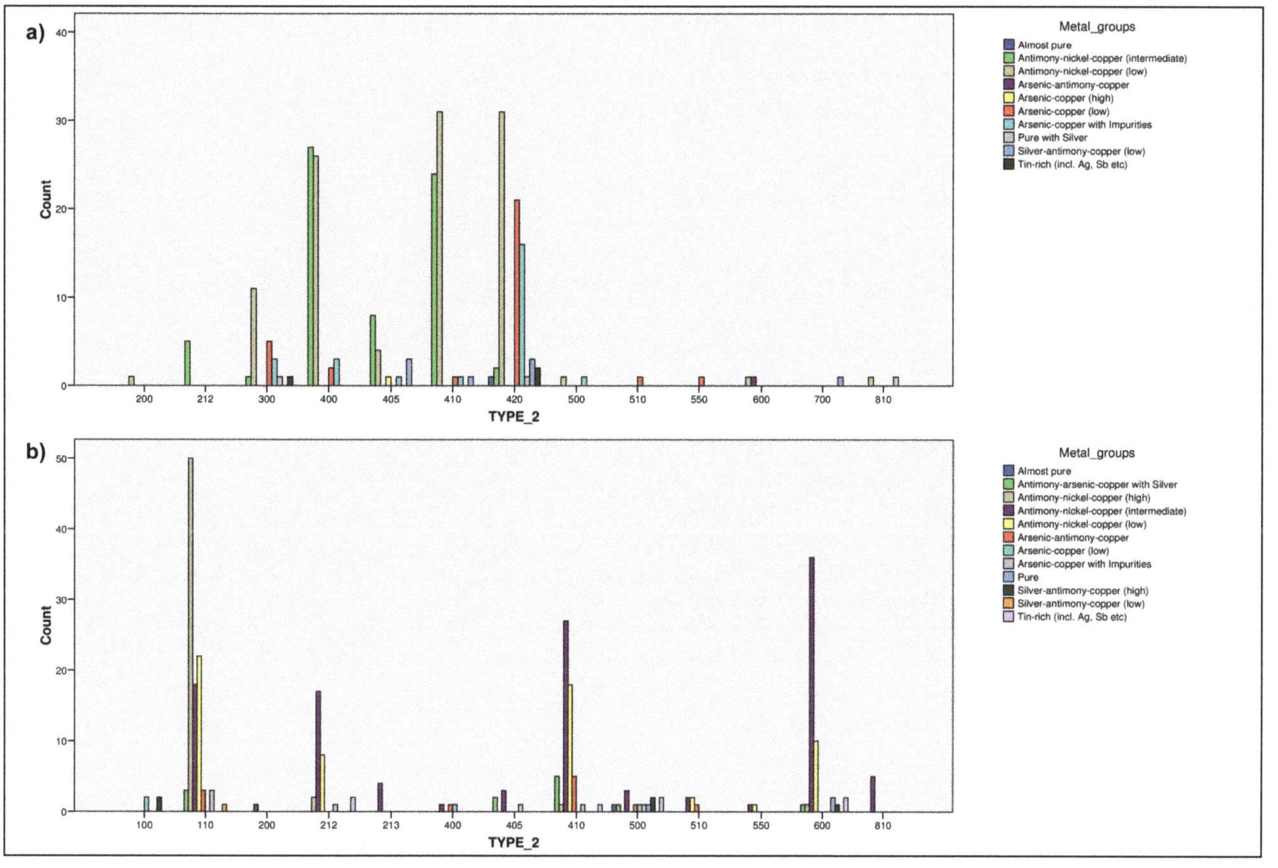

Fig. 10.31 Distribution of metal groups for artefact types of of **a)** the Nitra group and **b)** the *Blechkreis.*

(Krause 1988, 235 ff.; Bill 1997, 251). Since other copper objects of that time show similar ratios of impurities, but in smaller amounts (cf. Clusters 58 – 67 and 69 – 71), this suggests that flanged axes were probably processed further, during which the copper was cleared and impurities were removed by heating, hot working and casting. Hence, the provinces of that metal cannot yet be located.

As for the results of the 1943 Chalcolithic trace element analyses from roughly the northalpine region and the Carpathian Basin, we can after all draw the conclusion that during the 3rd millennium BC there was a transition, not only in terms of the archaeological evidence, but also concerning the types of copper used. Whereas the types of copper in use during earlier periods usually contain fewer trace elements which only slightly affect the properties of copper, the types of copper with more and higher concentrations of impurities are of remarkably improved quality. Despite this rather evolutionary development, it must be taken into account that copper types of earlier periods (PER 10 – 25), such as pure or low-concentration arsenical copper (in which the impurities have less influence on copper properties), were also being employed by metallurgists of later periods (PER 31 – 40). This reflects the ability of metallurgists of subsequent periods to access previous types of copper. Either they were able to deal with the old 'recipes', or could exploit different resources, and/or recycled scrap metal. In particular, as the relatively new copper types were used throughout the area, it seems that we are dealing with a deliberate decision to use improved types of copper, rather than the accidental exploitation of a particular local ore deposit.

Previous studies explained the emergence of fahlore-copper – copper that contains a variety of impurities – as the result of the accomplishment of mining and smelting of more complex, especially sulphide, ores (e.g. Strahm 1994). From the current perspective, however, this was probably not the motivation. Other scholars have demonstrated that smelting complex ores and cleaning them of their impurities may not have been all that complicated (Rostoker and Dvorak 1991; Ryndina et al. 1999). The metallurgists of the late 5th/early 4th millennium BC, for example the Gumelnista and Varna groups in the Balkans, were able to do this (Ryndina et al. 1999). If it is assumed that early metallurgy spread over Europe from the Balkans westwards, than there is no reason why this knowledge could not have been transferred as well. In order to clarify both issues ultimately, more analyses – lead isotope, microstructure and experimental analyses – need to be undertaken. Additionally, sulphide concentrations of copper finds and, of course, local ore deposits have to be studied. For now, it can be asserted that Chalcolithic metallurgists were able to choose particular types of copper intentionally, and that their selection of copper types was deliberate and based upon metallurgical properties. Therefore, we cannot exclude the possibility that they were also intentionally using particular ore types that provide those properties. The properties of copper had been improved over the centuries. Gradually, the artefacts were made of copper with changed and improved smelting, casting, working and optical properties, such as a lowered smelting point, increased hardness and tensile strength, and varying colours.

It appears that in central Europe Chalcolithic metal workers used metallurgically-improved copper – arsenic-containing types of copper – from the 4th millennium BC onwards. However, at least during the 3rd millennium BC and especially by the appearance of what is today defined as Earliest Bronze Age groups, copper types with significantly improved qualities were used (e.g. the various types of copper containing higher concentrations of impurities). This metallurgical development is significant throughout the study area. The actual 'recipe' to produce these types of copper cannot yet be identified, because the sources and the particular copper manufacturing techniques are still not known. We can, however, propose that these 'recipes' are a mixture of using certain ore types, processing them in a special way and producing the artefacts – by melting and casting as well as merging and recycling of ores and various types of copper. Now, however, we can be sure that newer 'recipes' were developed and used, as metallurgists could produce qualitatively improved copper, with compositions featuring higher concentrations of trace elements.

It is likely that various (trace) elements were not intentionally added, but that particular ores or copper types were mixed, with the benefit of experience and traditional expertise, in order to produce copper with special properties. This marks an extraordinary stage in the history of metallurgy in the region: it probably facilitated the introduction and uptake of new metallurgical knowledge – the intentional alloying of tin and copper, to make tin-bronze. It has to be supposed that prehistoric societies did not accept new developments easily as those communities were based on traditions (e.g. Kristiansen and Larsson 2005, 51 – 60). The metalworking tradition may have been passed on from generation to generation, but an interaction of new experiences and the impact from other communities may have improved the knowledge.

Perhaps one of these impacts came with the appearance of Bell Beaker-using metal workers. As we have seen that the knowledge and the technology of using fahlerz ores was introduced during roughly the first half of the 3rd millennium BC, the question is now, whether this is linked to the contemporaneous spread of the Bell Beaker phenomenon. In other words: did the Bell Beaker users play a key role in transmitting or establishing this know-how?

11. THE QUESTION OF BELL BEAKER METAL IN CENTRAL EUROPE

Ever since the first studies on early metallurgy in Europe, the question of typical Bell Beaker metal has been asked over and over again. This question, however, has not been answered satisfactorily, as most of these studies focused either on Bell Beaker related objects to the exclusion of all others, or on broadly contemporary finds that date very close to the Bell Beaker horizon within geographically restricted areas. With the vast number of easily accessible data published by Krause (2003), there are now many more trace element analyses available for comprehensive research. Including these and those from additional publications, a database covering 82 trace element analyses provided by 80 Bell Beaker objects was available for this study.

This section of the book looks at the configuration of the new set of Bell Beaker data and compares it with the results of previous studies. The Bell Beaker copper compositions are set within the context of the entire database of 1943 trace element analyses in order to explore characteristics of the development of copper use during the Chalcolithic in central Europe, and to explore whether Bell Beaker metal work played an exclusive role in this. These issues are perennially relevant to the question of whether there was a distinct use of copper by Bell Beaker people.

11.1 REVIEW OF PREVIOUS STUDIES ON THE COMPOSITION OF BELL BEAKER COPPER IN CENTRAL EUROPE

Before discussing the properties of the types of copper used for Bell Beaker metal objects and comparing them with other finds, it is necessary to evaluate the results of earlier studies that have examined trace element analyses of central European Bell Beaker metal. Four studies focused on the compositions of Bell Beaker related copper finds (Kuna and Matoušek 1978; Bertemes and Šebela 1998; Matuschik 2004; Metzinger-Schmitz 2004). All of these studies, however, differ from the present research in that they dealt with fewer data, used different analytical methods and/or followed different research questions. In contrast to the present study, they chiefly compared metallurgical data with artefact typology and aimed to locate the provenance of the copper used that had been used.

Kuna and Matoušek (1978) only took into account the trace element analyses of typical Bell Beaker objects, mainly tanged daggers. They compared the composition of the finds with both the typology and the regional distribution of the finds. Matuschik (2004), who recently reviewed their paper, argued that various aspects of their classification, listing and mapping are insufficient.

About 20 years later Bertemes and Šebela (1998) evaluated trace element analyses from Corded Ware, Bell Beaker and Earliest Bronze Age contexts in central Europe. By graphical comparisons they assumed that Bell Beaker metal contains more arsenic than Corded Ware metal, whereas in Corded Ware finds silver is more frequent. A generally larger difference is seen between both of these and Earliest Bronze Age copper. They mentioned that these results could only be regarded as preliminary, because their available database was too small for comprehensive statements (Bertemes and Šebela 1998, 236 f.; Bertemes and Heyd 2002, 215). In contrast to the present research they could only refer to 34 trace element analyses from Corded Ware artefacts, 56 from Bell Beaker and 9 analyses classified as Earliest Bronze Age (Bertemes and Šebela 1998, Tab. II – IV) – although they included some analyses from the Otto/Witter-project which are actually not comparable to those of other projects (cf. Chapter 8.3).

Metzinger-Schmitz's PhD thesis (2004) investigated the typo-chronology of the Lower Austria and Moravian Bell Beaker group and included a brief metallurgical excursus. Her research is based on 108 trace element analyses of objects exclusively found in contexts of the Eastern Bell Beaker group which have been carried out by the Otto/Witter- and the SAM-projects[183] (Metzinger-Schmitz 2004, 242). She grouped her dataset by Next-Neighbour cluster analysis into 13 clusters and postulated that arsenic, tin and lead characterise some clusters (Metzinger-Schmitz 2004, 247 ff.). In order to clarify whether these three elements had been deliberately added, three further cluster analyses were run, considering each element individually. According to the thresholds that were thereby obtained between clusters containing different amounts of trace elements it was concluded that copper containing over 1.7% arsenic and

[183] It is not clear how this large database had been compiled, because the analyses named here as "Suffix b" are published neither by SAM I nor by SAM II (cf. Metzinger-Schmitz 2004, 242). Most of the finds have been analysed at least twice by SAM and/or Otto and Witter.

2.5% tin are intentionally alloyed, and those with over 0.03% lead indicate, in her opinion, a special type of copper (Metzinger-Schmitz 2004, 254 ff.). A metallurgical explanation of this theory, however, is not given, and so one can only interpret these results as a statistical threshold which, in the case of the lead-rich Bavarian copper, may be regionally specific. The entire study only examines copper artefacts from Bell Beaker contexts and does not provide a wider geographical or chronological comparison of the Bell Beaker metallurgy. Metzinger-Schmitz's work can only be seen as a 'palaeometallurgical excursus', within a study mainly focusing on sequences of pottery typology.

The most important recently published investigation on Bell Beaker metal in the northalpine region was carried out by Matuschik (2004). He investigated the copper of 22 Bell Beaker finds from the Czech Republic and southern Germany concerning their trace element composition and possible provenance. Although Matuschik mainly focused on the composition of Bell Beaker finds, analyses of objects found in other contexts were also considered. Based on a cluster analysis of trace element analyses obtained by Pernicka, the author recognised that the composition of Bell Beaker finds from southern Germany is similar in terms of arsenic, antimony, silver and nickel content to Corded Ware finds from south-west Germany and Switzerland, but differs in its bismuth and lead contents (Matuschik 2004, 290 f.). This is comparable to Metzinger-Schmitz's results, as she also recognised that lead is significant for the Bavarian finds. As a next step, he excluded the Bell Beaker objects from southern Germany and concentrated on the significance of bismuth and lead – especially with the Bell Beaker artefacts from the Czech Republic (Matuschik 2004, 292 f.). In the end, Matuschik showed plausibly that the copper composition of southern German Bell Beaker objects differs from Bohemian and Moravian finds, because the former contain significantly more bismuth, lead and nickel. He argues that – mainly because it differs in composition from contemporaneous types of copper used in Switzerland – the 'southern German Bell Beaker copper' ('*A-Kupfer*' after Matuschik 2004, 293) was of north-east Alpine provenance, the only ore region which is supposed not to have been exploited at that time (Matuschik 2004, 296). This, however, can only be seen as an assumption, since it is not documented archaeologically.

Although Matuschik developed his study out of a broader database and the results of a cluster analysis of Chalcolithic and Bronze Age trace element analyses, he chiefly focuses on specific Bell Beaker related finds and their composition in order to locate the provenance of the copper. As demonstrated in detail (cf. Chapter 6), without consulting other methods, such as lead isotope analyses, trace element analyses of finished artefacts are inadequate for identifying the provenance of ores. Consequently, it is necessary to explore different questions in order to shed light on the problem of Bell Beaker copper use.

11.2 BELL BEAKER METAL AND ITS PROPERTIES

In this study the methodological process of searching for a specific Bell Beaker copper is different from other investigations. Two different statistical methods are applied and their outcome compared in order to define metal groups. Whereas the cluster analyses used in Matuschik's work on typical Bavarian Bell Beaker copper was based on 600 trace element analyses of Chalcolithic (*spätneolithisch*) copper objects from the northalpine region (Matuschik 2004, 290), the present research benefits from almost 2000 trace element analyses, spread over an area from the Rhine to the Carpathians, and from the Alps to the Northern German Plain. This section will describe the metal groups individually, with the clusters evaluated by TSC and PCA as set out in to Chapter 10[184].

11.2.1 PURE AND ALMOST PURE COPPER

Nine trace element analyses can be summarised as relatively pure or almost pure copper from Clusters 9, 22, 23 and 48. The two analyses of an awl from Ludeřov[185] (CZ) indicate that this find is made of almost pure copper with some traces of silver. Clusters of pure and almost pure copper consist of artefacts from all periods and regions of the study.

11.2.2 ARSENIC-COPPER

Arsenic-copper containing concentrations of arsenic both above and below the 1% level had been used by Bell Beaker metallurgists. In total, eight Bell Beaker finds are characterised by a dominant arsenic content, whereby other impurities are nearly absent (Clusters 28 – 31 and 52). Arsenic-copper has commonly been located at sites dating from the Mondsee-Altheim-Pfyn-horizon onwards, but even finds from the Nitra group were made of low level arsenic-copper. Copper with more than c. 1.5% arsenic, however, is rarely used before at least 3000 BC. The Bell Beaker objects of arsenic-copper were found throughout the research area. It is interesting that three of the five daggers from Predmostí[186] (CZ) contain relatively high amounts of arsenic. In contrast, the other two daggers from Predmostí are made of arsenic-nickel copper (Cluster 37) and silver-antimony copper (Cluster 24).

[184] The correlations of trace element analyses, archaeological information and metal groups of the Bell Beaker finds are displayed in the correlation tables at the end of this book. The correlation table resulted from the statistical evaluations in Chapter 9 and 10.
[185] ANo 19931 and 29848.
[186] ANo 19920 – 19924.

11.2.3 ARSENIC-COPPER WITH IMPURITIES

Apart from the typical arsenic-copper with nearly no other impurities, a copper characterised by overshadowing arsenic content, but also including higher amounts of other trace elements can be identified[187]: arsenic-copper with impurities. Objects of this group primarily contain antimony, silver and often lead (Clusters 33, 36 and 43 and outliers of Cluster 69 and 70). Most of the Bell Beaker artefacts that are grouped under this label were discovered either in Bavaria or in the Middle-Elbe-Saale-region. The comparable trace element analyses of these clusters come from objects chiefly found both in central Germany and in Slovakia. All clusters contain hardly any objects dated older than PER 31.

11.2.4 TIN-RICH-COPPER

Several artefacts found in Bell Beaker contexts contain significant amounts of tin. Therefore, 13 Bell Beaker objects of tin-rich copper can be clustered together; nine of them are included in Cluster 41, while the rest had to be re-classified[188]. Apart from two Bavarian finds, all were discovered in the Czech Republic. Objects with analogue compositions are spread throughout the area of research, but chiefly date to the periods PER 31 – 40.

11.2.5 ANTIMONY-SILVER-COPPER

The flat axe from Koštov[189] (CZ) is made of antimony-silver copper and is thus the only object with a dominant antimony concentration (Cluster 26). Antimony-silver-copper is commonly used for flat axes found in the Czech Republic, Slovakia and Hungary, although these finds are usually dated to the 4th millennium BC. It may be that this artefact is a re-used object that had been produced much earlier, and had circulated for a long time. Since Hájek (1968, 44) mentioned that he is unsure of its archaeological context, its archaeological classification needs to be questioned.

11.2.6 SILVER-ANTIMONY-COPPER

Copper characterised by higher concentrations of antimony and silver appears often among Bell Beaker objects (29 of the 82 trace element analyses). As the antimony-silver-copper (Cluster 26 and 32) and silver-antimony-copper (Cluster 24, 27, 39 and 40), defined in Chapter 10.5, are metallurgically similar, this may lead to some problems during the TSC. For example, two of the three Bell Beaker finds in Cluster 26 should actually be classified as silver-antimony-copper. The reason for this is a statistical misclassification during the grouping of Cluster 10 and 11.

In general, Clusters 24 and 27, which nearly exclusively contain silver and antimony, can be distinguished from Cluster 40 that contains objects with some arsenic, bismuth and lead besides higher concentrations of the main impurities of silver and antimony. The 'silver-antimony-copper (high)' group encompasses not only the Bavarian and Austrian Bell Beaker objects that Matuschik (2004) covers in his article but also one Hungarian and four Czech Bell Beaker finds. All of these artefacts also contain comparably high amounts of lead. Most of the other artefacts of Cluster 40 are classified as of Corded Ware or Earliest Bronze Age date and were excavated in central Germany and Poland (cemetery II at Mierzanowice). Furthermore, finds from Switzerland and the Czech Republic are also included.

The Bell Beaker objects made of 'silver-antimony-copper (low)' are – with the exception of a riveted dagger from Oggau 'Seestrasse'[190] (A) – all from graves in the Czech Republic, and apart from a tanged dagger from Prosimerice[191] (CZ) all cases are part of Clusters 24 and 27. The group 'silver-antimony copper (low)' is the largest one of the Bell Beaker groups including 17 analysed objects – though objects dated to all other periods under consideration, with a peak at the late 4th and 3rd millennium BC, and were found throughout the area of research.

How can the geographical difference between silver-antimony-copper of low and high concentrations be explained? It could be that the purer copper finds of Cluster 24 and 27 had been recycled and, thereby, 'cleaner' versions of the same raw material as the metal of Cluster 40, which contains more impurities. This can be assumed on the grounds that both groups have a similar ratio of impurities, while the concentration of the volatile elements antimony and arsenic are lower in Clusters 24 and 27, and bismuth and lead levels are also lower: these tend to precipitate and are thus removed during melting. On the other hand, the silver content tends not to vary between the groups. This is unsurprising, as silver is not significantly affected by re-melting. The nickel content, which varies remarkably between the groups, runs counter to that explanation, however. Furthermore, the near-complete absence of arsenic, lead and nickel in the metal containing fewer impurities indicates a different source and/or production technique.

[187] There are 19 trace element analyses of 18 objects included. The tanged dagger from Mühlhausen (D) provides two analyses which are quite similar – apart from the arsenic concentration which, however, is significant in both cases (ANo 32228 and 32328).

[188] This metal group contains the analysis of a tanged dagger from Praha-Kopylisy (CZ; ANo 100028). Although no tin is measured here, it is included in this cluster. Because its singular composition of only containing a small amount of nickel and silver, it cannot be classified otherwise; statistically, however, this dagger was misclassified in the Cluster 41 of tin-rich copper.

[189] ANo 7546.

[190] ANo 11527.

[191] ANo 19934.

11.3 THE QUESTION OF A SPECIFIC TYPE OF BELL BEAKER METAL

As seen above, the database of this research contains the data of Matuschik's study and additional analyses, especially from the Middle-Elbe-Saale region and from Moravia. A PCA run on the sample of all 82 Bell Beaker analyses (cf. PER 35 in the database) compares the trace element compositions of the Bell Beaker copper finds (Appendix D). The scatter plot of the results of this PCA confirms the variation between the Czech and Hungarian objects (AREA 111 – 125) and the German samples[192] (AREA 62 – 103) (Fig. 11.1). However, as Matuschik (2004, 293 ff.) also pointed out, the particular lead- and bismuth-rich copper is not exclusively used in these areas. In general, it can be summarised that there is an East-West discrepancy. It must be emphasised, however, that south-western German, Bavarian or Lower Austrian Bell Beaker finds (AREA 62, 66 and 72) are not alone in containing significant bismuth, lead and nickel concentrations:the samples from the Middle-Elbe-Saale

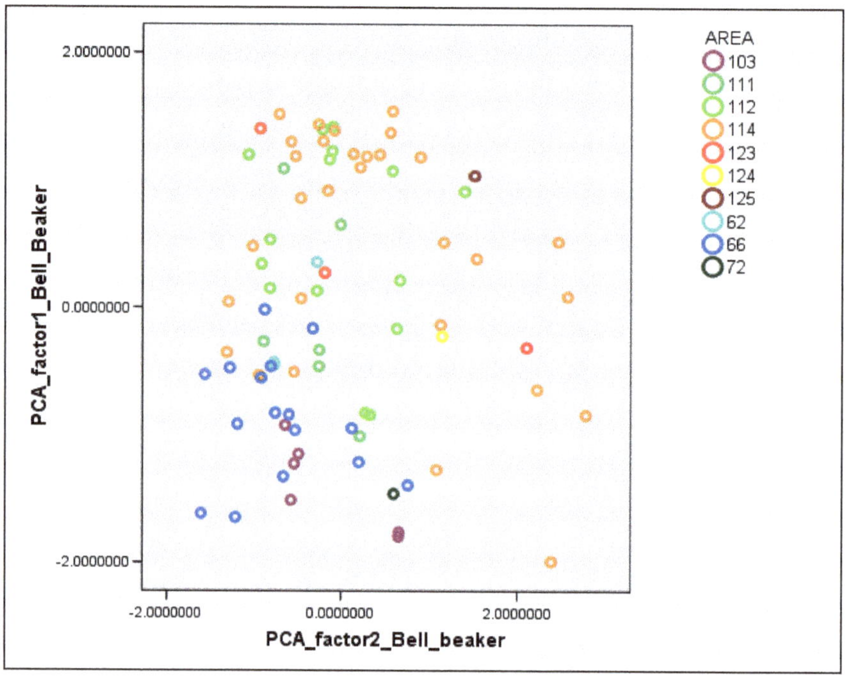

Fig. 11.1 Scatter plot of the PCA on trace element analyses carried out only on Bell Beaker finds; displaying the regional distribution (AREA).

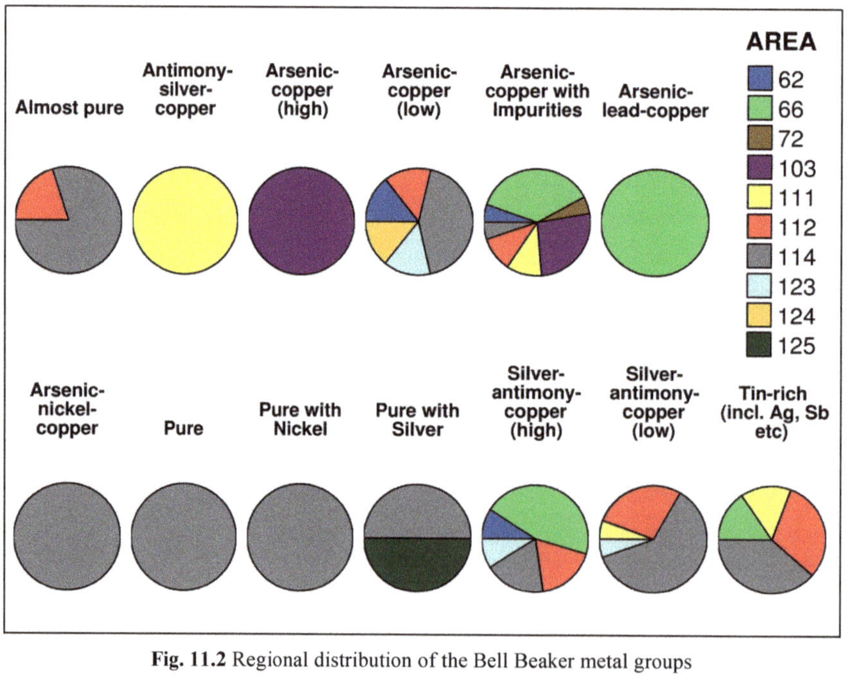

Fig. 11.2 Regional distribution of the Bell Beaker metal groups

[192] This difference between trace element analyses is not caused by inter-laboratory variability, because both the SAM- and the FMZM-data overlap.

region (AREA 103) and some analyses of Czech finds share similar characteristics (AREA 111 – 114) (Fig. 11.2). Indeed an artefact-specific copper usage cannot be identified, simply because by far most of the analysed objects have been daggers. Consequently, if one only focuses on the copper compositions of Bell Beaker related objects within the area of study, the southern and central German metal contrasts with the technology used in the Eastern regions of the Czech Republic, Austria and Hungary. Whether this reflects a difference that is specific to Bell Beaker objects, or else a general regional difference can only be ascertained by looking at the evaluation of the entire database of 1943 trace element analyses including Bell Beaker copper.

The first impression, of rather heterogeneous metal groups, has already been remarked upon in the previous paragraphs. Bell Beaker metal objects do not consist of one specific type of copper. This variation is also emphasised by displaying the relationship between the copper compositions of Bell Beaker and other Chalcolithic objects in the scatter plot resulting from the PCA (Fig. 11.3).

Although the types of antimony-nickel copper of the Earliest Bronze Age finds were not used by Bell Beaker metallurgists, a clearly defined cluster of Bell Beaker finds cannot be localised in any of the sectors of the scatter plot. However, finds with higher levels of impurities were discovered in southern and central Germany, while the artefacts generally found in the eastern part of the area of research consist of copper with lower levels of impurities[193] (Fig. 11.4). Nevertheless, it is not possible to identify a specific type of copper that had only used by metal workers or for artefacts of the Eastern Bell Beaker group, because 'Bell Beaker copper' was also used for other artefacts found in other regions, and also in non-Bell-Beaker contexts of various periods.

To summarise the comparison of Bell Beaker copper with types of copper used for artefacts of earlier and contemporary archaeological groups, both statistical methods, TSC and PCA, have clearly shown that there is no specific Bell Beaker type of copper that was exclusively used by Bell Beaker metal workers. Despite that, the highly concentrated antimony-nickel-rich copper of the Earliest Bronze Age seems to have been avoided,

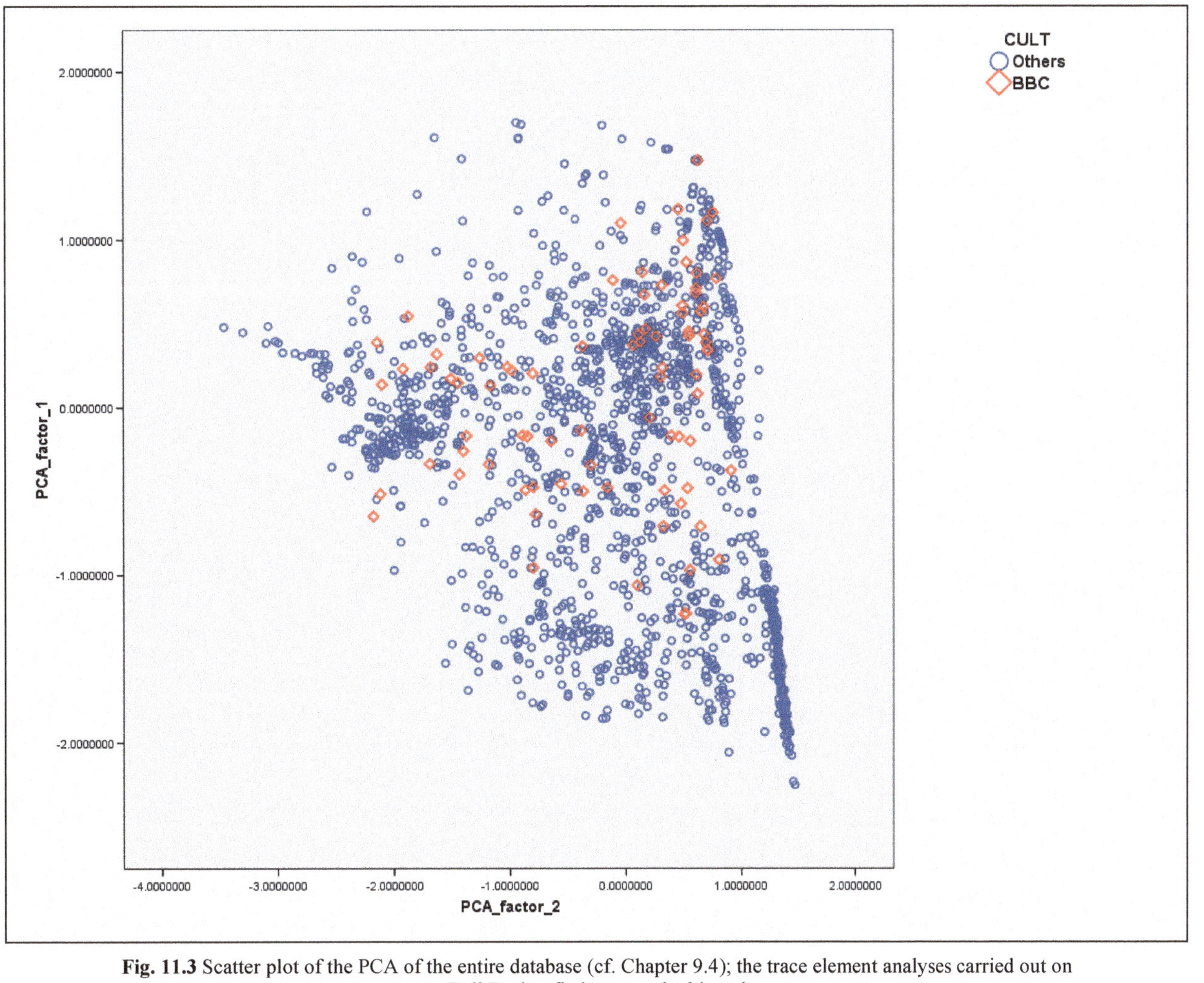

Fig. 11.3 Scatter plot of the PCA of the entire database (cf. Chapter 9.4); the trace element analyses carried out on Bell Beaker finds are marked in red.

[193] Concerning the results of Chapter 9.4 the scatter plot shows that the trace element analyses situated on the right side of the upper half contain lower amounts of impurities than the samples on left side and in the lower half (see also: Figure 10.27).

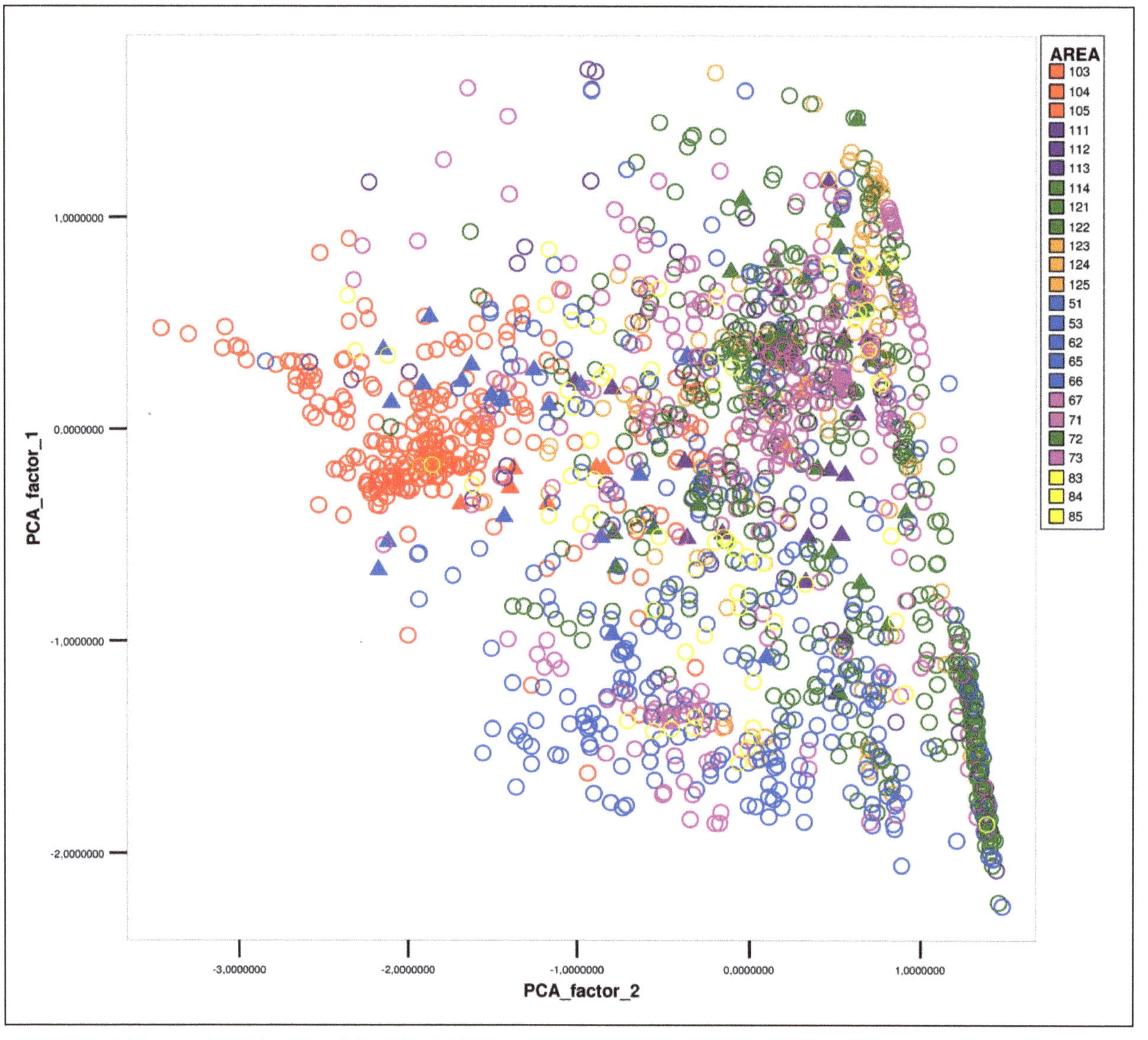

Fig. 11.4 Scatter plot of the PCA of the entire database (see Chapter 9.4), highlighting the trace element analyses carried out on Not Bell Beaker (**O**) and Bell Beaker finds (▲) and the regional distribution of the entire database (AREA).

or not yet known, by Bell Beaker metallurgists. They produced artefacts of copper which was generally used during the 3rd millennium BC and of copper that had also been used in previous times. Throughout central Europe the distribution of various types of copper used by Bell Beaker metallurgists coincides chiefly with the copper types of PER 31, so that we can speak of a metallurgy generally used during both PER 31 and 35 (and Bell Beaker copper is typical for that time), but not specifically used by an archaeological group.

As a result, metallurgy cannot be seen as a 'culture-defining' element or as the chief reason for the appearance and distribution of the Bell Beaker phenomenon in central Europe. In other areas this may be different. In Britain for instance, where the appearance of the Beaker phenomenon was roughly contemporaneous with the appearance of the first metal objects, Bell Beaker using communities may well have been more strongly connected to a specific Bell Beaker metallurgy (cf. Needham 2002).

For a case study on the Eastern Bell Beaker group, the question of whether Bell Beaker metallurgists supplied other communities with copper must be left unanswered, since no significant amount of archaeological evidence, such as castings or crucibles, differentiates the Bell Beaker group from other groups such as the Corded Ware complex. Given the current state of research, there is no indication that Bell Beaker people were prospectors, metalworkers or traders, since they did not have special metallurgical expertise or use specific types of copper. Bell Beaker metallurgists had no exclusive knowledge and exported their copper to other communities, because in that case Bell Beaker copper would be more uniform and different from copper of previous groups.

We can state that there was a remarkable change in metallurgy during the 3rd millennium BC and artefacts were produced of copper with improved melting, casting and mechanical properties. Bell Beaker metal fits perfectly into this system and is not distinguishable from Corded Ware metal – but it is distinguishable from

Earliest Bronze Age metal work. The results of the present research suggest that metallurgical knowledge made a first step towards the intentional use of types of copper due to their chemical and physical properties. That lead to a further improvement by the metallurgists of the early *Blechkreis* and the Nitra group, which cumulated in the emergence and establishment of typical tin-bronze with mainly around 10% tin.

The pan-European network of the Bell Beaker communities may have provided the basis for the acquisition and trade of copper and, especially, of the far rarer tin. However, the trace element analyses do not confirm a copper metallurgy that allows us to characterise central European Bell Beaker communities as being significantly different from previous communities. With the emergence of the Bell Beaker phenomenon in central Europe a distinct metallurgy – e.g. the use of fahlore-copper – does not appear. Fahlore-copper was already in use. As a result, we can conclude that there is no indication for a connection between the distribution of the Bell Beaker phenomenon and a distinct metallurgical tradition.

The comparison and interpretation of trace element analyses of Chalcolithic copper artefacts only implies that, because of the broad use of different types of copper, Bell Beaker metallurgists had both access to several resources and were able to deal with distinct types of copper. This reflects the Bell Beaker network, with its various overlaps, not only geographical but also cultural (cf. Vander Linden 2007b). In terms of their copper use, Bell Beaker communities in central Europe are not distinct from other archaeologically-evident communities. Certainly, the number of copper, gold and silver grave goods and the graves with metal workers' tools indicate that metal played an important role for Bell Beaker people, at least for their burial rites and their religious beliefs. In the case of the Eastern Bell Beaker group those objects do point to an outstanding position of the buried person, but Bell Beaker people in general cannot be seen as metal workers or traders with special knowledge that explains metallurgy as the motivation for the spread of the Bell Beaker phenomenon in our study area. Access to distant resources or technology is not particularly evident and the compositions of Bell Beaker metal finds show no indication that its users played a special role in distributing fahlore-copper. In contrast, Bell Beaker metallurgy is part of the general metallurgical traditions of the 3rd millennium BC. Especially, because of the comparable burial rites of the Eastern Bell Beaker group and the partly contemporaneous Corded Ware complex, it is suggested that the adoption of the Bell Beaker phenomenon in central Europe had been encouraged by religious motives and not by metallurgy.

12. Conclusions

12.1 Discussion

Earlier surveys chiefly studied the trace element composition of prehistoric copper finds in order to ascertain the provenance of the copper, to trace chronological developments, or to investigate the relationship between types of copper and artefact typology (e.g. Otto and Witter 1952; SAM I & II; Ottaway 1982; Schalk 1998; Krause 2003). Important results were obtained regarding the issues of chronological and chorological diversity of individual copper types. However, even if these studies succeeded in demonstrating that certain types of copper were applied preferably at different periods, they were unable to explain why those different types had been used.

Archaeological records and terminologies differ considerably within the area of this research, so that it was necessary to re-evaluate both and to develop a comprehensive, uniform definition for the studied area (Chapter 2 – 4). The term 'Chalcolithic' was chosen as a label to encompass all the copper–using (not tin-bronze-using) archaeological groups of central Europe (Chapter 3).

The present book has focused primarily on constructing a broad database, covering a wide chronological and geographical area, in order to ascertain whether phase-specific 'recipes' had been used for producing copper types with improved metallurgical properties. This research reviewed the archaeological interpretability of trace element analyses and aimed to clarify two crucial questions concerning the spread of metallurgy in central Europe from the 5th to the late 3rd millennium BC, namely: were Chalcolithic metallurgists able to use specific types of copper for specific purposes relating to the properties of the metal; and do Bell Beaker copper types differ from those used by other archaeological groups (e.g. Corded Ware communities) in central Europe?

If so, then the question becomes whether this could be one of the reasons for the extensive spread of the Bell Beaker phenomenon, and for its diversity when compared to other contemporary archaeological groups. Throughout the history of Bell Beaker research, it has often been assumed that metallurgy was the motivation for the extensive spread of the Bell Beaker phenomenon (e.g. Butler and van der Waals 1966; Sangmeister 1976; Needham 2002), but this has never been proved in the case of the central European Bell Beaker communities.

Several studies have been able to describe the occurrence of different types of copper, distinguished by their trace element compositions, during the Chalcolithic (e.g. SAM I & II; Ottaway 1978a; Krause 2003), but they were not able to show why these copper types had been used. These studies showed that, after the earliest use of almost pure copper during the 4th millennium BC, copper containing arsenic as the main impurity was used all over central Europe. Krause showed that types of so-called fahlore-copper (containing significant concentrations of antimony, arsenic, silver, bismuth and often nickel) were in use from about 2500 BC onwards. Since in central Europe the appearance of fahlore-copper is roughly contemporary with that of Bell Beaker finds, it seems likely that the two phenomena were closely connected. It was suggested that the Bell Beaker users brought the fahlore technology to central Europe.

This development of early metallurgy ended with the full establishment of classical bronze containing at least 8% tin. In central Europe, this is most unlikely to have occurred before 2200 BC (Krause 2003; Strahm and Hauptmann 2009). Several theories have been proposed to account for the variation in Chalcolithic copper types. Initially, attempts were made to determine the origin of the deposits by comparing the trace element compositions of copper artefacts with those of copper ores. The different types of copper were explained in terms of a change in the exploitation of ore deposits and in the changing abilities of the metal workers to deal with different types of copper oxide and sulphide ores (cf. Otto and Witter 1952; Pittioni 1957). However, it was subsequently demonstrated that various factors, in particular the smelting and casting processes, affect the trace element compositions to such an extent that it is impossible to determine the origin of the raw material from the composition of the artefacts (cf. Sperl 1975, 16; Tylecote 1980b; and see Chapter 6 for a detailed discussion). Accordingly, other ways to arrive at archaeological interpretations of trace element analyses of prehistoric finds had to be considered (Chapter 7).

One commonly-used explanation for the use of different types of copper at different times has been that metalworkers used a certain 'recipe' – although, given that the source of the raw material and the specific smelting techniques could not be identified, these explanations have been unable to clarify what the ingredients were or why they were used[194]. The present research has focused on identifying the reasons for the use of those 'recipes'.

Since nearly all available trace element analyses have been carried out on finished products and because it is unknown how the copper of those objects had been produced, it has only been possible to proceed by considering the composition of finished objects. Consequently, this study has addressed the question of whether a certain type of copper had been chosen because of the properties imparted by its impurities. It was assumed that copper from specific deposits, or a mixture of various types of copper, may have been processed intentionally in order to produce copper with special characteristics. In Chapter 7 it was argued that a variety of impurities, such as antimony, arsenic, bismuth, iron, and nickel, significantly affect copper properties – especially if copper contains over c. 2% of impurities. Accordingly, typical trace element compositions characterise the quality of copper in terms both of the manufacturing processes and of the finished product. Melting temperature, malleability, and hardness are affected by several combinations of impurities. Those properties are crucial for the colour, smelting, casting and working behaviour of copper.

In order to address the research questions set out in Chapter 1, a database of trace element analyses incorporating the general metallurgical evidence for the area of research was compiled and evaluated using two methods of statistical grouping. The results of both two-step cluster (TSC) and principle component analysis (PCA) on these 1943 Chalcolithic trace element analyses have been evaluated for the purpose of defining metal groups according to similar compositions. This made the large database more manageable and allowed comparison and interpretation without losing important information.
The statistical evaluation has confirmed, using distinct and new methods, what previous researchers had also postulated: namely that different types of copper were indeed used at different periods (e.g. SAM I & II; Ottaway 1982; Krause 2003). In addition, the present study showed that types of copper that were typical for a specific period were sometimes used in later periods as well.

Whereas copper types with no or low levels of impurities were employed during all parts of the Chalcolithic, those with higher impurity levels do not seem to have been used until after c. 2800 BC. Copper with higher quantities of impurities, including arsenic, antimony and nickel – particularly in combination – has better metallurgical qualities, e.g. enhanced hardness, tensile strength and malleability. Therefore, we can postulate that a change in metalworking practice, featuring the use of copper with improved working and use properties, occurred during the first half of the 3rd millennium BC. Apart from evincing a general interest in producing better types of metal, it is hard to say what impulse caused the introduction of new types of copper. In central Europe certain technological and cultural changes appeared in the archaeological record over this time: settlement evidence becomes rarer, while funerary practices that featured gender-differentiated individual interment, accompanied by grave goods that varied in amount and quality emerged. This corresponds to the disappearance of the Baden complex and the emergence, in its place, of the Corded Ware complex and the Hungarian Early Bronze Age groups, such as the Makó-Kosiny-Caka or Somogyvár-Vinkovci group, around 2800 BC. We cannot exclude the possibility that techniques were transferred along with these new cultural impulses. In non-literate societies knowledge is only passed on orally. Consequently, the transmission of knowledge relies on a very fragile basis. Cultural changes may interrupt, change or develop technical traditions. We could postulate that, along with the changes that are archaeologically visible, invisible changes were taking place, as a result of the influx of new ideas and practices to central Europe. These impulses and ideas from outside the area of research served to advance copper metallurgy.

It therefore seems likely that, in contrast to the situation in previous periods, 3rd millennium BC metalworkers were able to produce improved copper. Consequently, it can be suggested (although unfortunately not proved, since mines of this period are missing in central Europe) that at this time metallurgists had access to types of copper, and had the technological expertise to work with them. Earlier metalworkers had either lacked the skills to deal with copper bearing higher amounts of impurities, or else they had been unable to obtain that material; a combination of both factors may have been involved.

We can now argue that this change was a technological development, an advance in metallurgical expertise that exploited the different chemical, physical and mechanical properties of copper with different trace element compositions. After around 2800 BC and at least after 2500 BC, metallurgists were able to deal with types of copper that were – thanks to their compositions – easier to cast and shape, and which probably also had improved mechanical properties. It is clear that this metallurgical expertise was not restricted to a certain archaeological group. It can be assumed that metallurgical know-how was kept as a secret, to be passed on between metalworkers – a segment of the population that is archaeologically invisible. Their knowledge was essential for subsequent developments approximately half a millennium later, when 10% tin-bronze was commonly used all over central Europe. The groundwork for the metallurgical expertise required to produce tin-bronze, and for the development of networks which were necessary to organise the relatively scarce tin ores, was laid during the 3rd millennium BC and can be seen in the deliberate use of various types of copper. Therefore, tin-bronze use is a consequence of a development process

[194] Those 'recipes' may have featured a mixture of various types of raw material, and the choice of specific techniques of smelting in order to eliminate or retain individual trace elements within copper. We do not know how these processes took place, but it is clear that various types of copper have been produced.

that featured the successive improvement of copper types. From this we may postulate that late Chalcolithic/early Bronze Age metal workers accepted the new technology of tin-bronze metallurgy as if the metal were a new type of copper with much better properties.

Although previous scholars had argued that there had been an evolutionary development of the skills to extract copper, firstly from easier available oxide, then from sulphide ores (Strahm 1994; idem, 2007; Strahm and Hauptmann 2009), this was not, in fact, a linear progression involving the smelting of increasingly complex copper ores (cf. Ottaway and Roberts 2008, 206), because experiments have persuasively demonstrated that this developmental step was rather small (e.g. Rostoker et al. 1989; Rostoker and Dvorak 1991; Klein and Lechtman 1999). A new technological development that made it possible to smelt copper sulphide ores can be excluded, because several experiments have shown that smelting these ores is not difficult: it can be done using simple methods of crucible smelting (e.g. Rostoker et al. 1989; Lorscheider et al. 2003). Recent work has clearly demonstrated that during the early 4th millennium BC the ability already existed to exploit complex poly-metallic sulphide copper ores (Klein and Lechtman 1999; Ryndina et al. 1999; Moesta 2004). It is therefore essential to remember that 'older' types of copper continued to be used by metallurgists of subsequent periods, as shown for example in the fact that the Earliest Bronze Age artefacts of the Nitra group included a few objects of relatively pure copper[195]. The phases of using pure, arsenical or fahlore-copper were not separated and earlier types of copper could have been used alongside these. Consequently, the shortage of pure copper or oxide ores was not the reason to force prehistoric metallurgists to work with more complex copper ores. In contrast, co-smelting of oxide and sulphide ores has been shown to be easier, since it requires lower temperatures (Rostoker et al. 1989). The sequence of using firstly almost pure copper, then of arsenic-richer copper, and finally of copper containing higher concentrations of various impurities, especially antimony, nickel, arsenic and lead, indicates a progression in producing and using types of copper that were increasingly easier to work and mechanically better. A further reason why copper with specific levels of impurities was used may have been the effects of these impurities on the copper's colour. The colour of copper may have been important for aesthetic reasons on the one hand, and for indicating a specific type of copper with specific properties, on the other. Prehistoric metallurgists may have selected types of copper according to the characteristic colours of the material. The present research has allowed us to add that Chalcolithic metallurgists used various types of copper on account of their distinct metallurgical properties. Behind the apparent chronological development in copper use there was indeed a metallurgical development, featuring the ability to work with improved copper types.

[195] For example, artefacts found at the cemetery of the Nitra group in Výčapy-Opatovce (CZ) had been made from both low concentrated arsenic-copper and more complex antimony-nickel copper.

The development from pure copper to qualitatively-improved tin-bronze is indeed an evolutionary process, but this does not mean that the general spread of metallurgy was a linear development. On the contrary, we have seen in Chapter 4.1 that the amount of evidence, and the type-spectrum of copper, fluctuates throughout the Chalcolithic. The adoption and evolution of copper metallurgy depended on various factors including cultural change, and are not exclusively connected to metallurgical expertise. Roberts (2008a; 2008b) has clearly shown that in prehistoric times the transmission of metallurgical expertise between both individuals and communities was very fragile. Crucial information about what constituted an appropriate raw material, how to find it and how to smelt and cast it was probably transferred by word of mouth, from one person to another. It may well be that this knowledge was only accessible to a restricted group of experts. Thus, cultural changes may interfere with that fragile framework.

Compared with the Bronze Age, the Chalcolithic has little metallurgical evidence and the evidence that does exist is far less comprehensive than that for later periods. Therefore, it may be that the metallurgical impact on Chalcolithic societies was relatively small. There is precious little evidence from the central European Chalcolithic for the extensive extraction and production of copper, and it can therefore be assumed that the number of people who would need to be exempted from other tasks such as food production in order to extract and work metal would have been small – too small to have a major impact on the social structure (Bartelheim 2007).

Even if it is not yet possible to clarify beyond doubt from the trace element analyses whether or not metal was intentionally alloyed, the use of several, often improved, types of copper over a wide area indicates networking and a deliberate use of those types. For several reasons we can assume that these types of copper were deliberately used. Firstly, various types were in use in the same area and at the same time, so metallurgists must have had access to various raw materials. Moreover, metal workers deliberately retained impurities that are very volatile, easy to remove and fugitive (e.g. arsenic) in copper, because they improve its quality. Finally, the continuous substitution of types of purer copper by copper with improved properties indicates that better materials were intentionally chosen. In central Europe the development of metallurgical expertise took several centuries and was probably influenced by impulses from outside (perhaps the Balkan region). Metallurgical inventions may have occurred sporadically and accidentally, while metallurgists adopted new, improved technologies intentionally. There was surely a deliberate selection and use of certain types of copper due to their metallurgical properties. In the cases of the copper containing lower concentrations of trace elements, this represents a selection of ores that contained the desired impurities, and the working of these ores in such a way as to produce copper with the required properties, rather than the addition of other elements to a pure copper melt. It is suggested that the intentional addition of a compound is indicated only when more than 2% of a particular element is present (Rovira and Montero 2003, 15). The experience of merging different ore types probably resulted in better copper. Deciding to work with one or another type of copper would have been founded in both experience and tradition.

After at least a thousand years of metallurgical experience, it is most likely that the metal workers of the 3rd millennium BC were well aware of what they were doing and had the skills to deal expertly with specific types of copper. This can be seen most obviously in the case of copper containing rather volatile impurities, such as arsenic or antimony. High arsenic-copper needs to be treated particularly carefully and melted under reducing conditions, if one wants to keep those elements within the metal. Therefore, prehistoric metal workers must have been able to control the copper composition, succeeding in retaining those impurities that improved the material. They understood the characteristics of different kinds of copper and made informed decisions about their use. Although prehistoric metallurgists may not have known that it was the trace elements that directly affected the character of the material, they may have been aware of the fact that mixtures of certain types of copper, indicated by colour or provenance, improved smelting, casting, working or other properties. They likely produced those coppers by selecting specific ores or by mixing individual types of copper. That knowledge was built up by experience and transferred from generation to generation through tradition.

Was this development restricted to specific archaeological groups? It would appear not, since similar results were obtained by Ottaway (1982) in her study of metalworking in the northalpine region, albeit using a smaller database and only considering the earliest finds. That study and the present research, based on the 1943 trace element analyses and covering the whole of the Chalcolithic, indicate that the development of metallurgy in and around central Europe at this time can be characterised as a series of broadly contemporary advances shared among interacting and archaeologically-differing communities, rather than as advances made by individual communities who then played the role of the distributors of metal and of metallurgical knowledge. It is likely that the people who exploited copper mines, smelted copper and distributed the metal provided communities with material regardless of their cultural identity (as archaeologically-visible today).

Whether a certain group played a special role in that network is hard to prove, because even if one group supplied the others, the material did not change; and in the case of this research area it is unknown who produced and distributed the copper. The extensive similarity of particular copper types indicates a commonality in the mode of copper production. Presumably, a relatively small number of people, not restricted to a specific archaeological group, provided others with, at least, raw metal; otherwise much more regional diversity would be recognisable in the types of copper used. This may reflect a great mobility of metal exchange and networking. Metallurgists used all types of available copper.

What seemed like a 'culture-overlapping' phenomenon for the entire central European Chalcolithic was tested more precisely to examine the possibility of a distinct Bell Beaker metal use. Within the archaeological-cultural contexts that provide evidence for the spread and development of copper metallurgy, the Bell Beaker phenomenon has been awarded a significant role (cf. Chapter 1). This is primarily based on two facts: firstly, its wide distribution and networks; and secondly, the regular presence of metal objects and of metalworking-related objects, such as copper daggers, gold rings and metal workers' tools, which have been repeatedly interpreted as symbols of status. Thus, metallurgy has been seen as an important aspect of the Bell Beaker phenomenon (e.g. Burgess and Shennan 1976; Benz et al. 1998; Heyd 2007; Vander Linden 2007b).

The Bell Beaker phenomenon is dispersed over a huge area, from Sicily to Scotland and from Portugal to Poland, and displays a remarkable homogeneity in its archaeological remains. When trying to explain this uniform distribution, previous researchers have posited the existence of an underlying network of contacts, with metalworkers playing a key role. Not only graves with metal workers' tools, but also standardised depositions of copper and gold artefacts in graves, support the argument that the Bell Beaker phenomenon was distributed by metal prospectors, workers and traders (e.g. Childe 1929; Sangmeister 1976). Because Bell Beaker graves vary in terms of the quantity and quality of grave goods, the Bell Beaker communities are seen as hierarchically stratified. Because the 'richer' furnished often contain metal artefacts, metallurgy is interpreted as the decisive factor in that stratification (Dornheim et al. 2005, 52 f.; Heyd 2007, 358 ff). However, these previous studies have not definitively ascertained precisely what role metallurgy played within the Bell Beaker phenomenon.

In Chapter 11, the evaluation of compositions of copper finds which are seen as closely related to the Bell Beaker phenomenon (e.g. tanged daggers) has comprehensively demonstrated that those particular objects do not differ from other roughly contemporary artefacts in terms of the concentrations of their impurities, especially those in the Corded Ware complex. Subsequently, it was concluded that a distinct metallurgy of the Eastern Bell Beaker group could not be identified. The present study was for the first time able to prove, using a comprehensive database, that the spread of the Eastern Bell Beaker group was not based on a distinct metallurgy: in other words, there was no distinct tradition or special metallurgical expertise among Bell Beaker metal workers. All types of copper used for Bell Beaker objects were also used for finds belonging to other archaeological groups. In general, all of these types are spread throughout the area of research, although there is a negligible quantitative difference between the eastern and the western part of our area of research – but only if Bell Beaker metal is considered exclusively (Chapter 11). The overlapping distribution of Bell Beaker copper indicates an exchange of metal objects and metallurgical know-how both within the Eastern Bell Beaker group, and with other archaeological groups of the 3rd millennium BC. The area-wide and overlapping circulation of copper types reflects the great mobility and interaction of the Bell Beaker users. This mobility has also been indicated by the results of strontium isotope analyses of these people's tooth enamel (Price et al. 2004).

Whereas Bell Beaker metallurgists worked the same types of copper as both earlier and contemporary archaeological groups, antimony-nickel-copper, which is typical of the Earliest Bronze Age, seems to have been avoided by them. During the Earliest Bronze Age (the *Blechkreis* and the Nitra group), it is likely that new and further improved types of copper appeared, including this antimony-nickel-copper, indicating that new resources and technologies were being explored.

Concerning the spread and development of metallurgical techniques the conclusion has to be drawn that the Bell Beaker communities did not play an exceptional role in central Europe. This is shown by the fact that there are no types of copper specifically used for Bell Beaker finds. In contrast, all types of copper of which Bell Beaker finds consist have regularly been used by other groups. Rather, new impulses emerged both earlier and later. As other archaeological groups also used 'Bell Beaker metal', it cannot be justified to assume that Bell Beaker users separated themselves from other communities by metallurgy, or that they had special knowledge with would explain their proposed 'wealth' and 'status'. In contrast, it seems that the central European Bell Beaker group and other communities of the 3rd millennium BC shared the same metallurgical expertise. The graves with metal workers' tools might indicate that Bell Beaker using communities were in charge of the copper production of that period and 'delivered' copper products to other communities. However, because burial rites and arrangements of grave goods do not necessarily reflect everyday life (Thomas 1991, 129; Härke 1993), answering this question is problematic. There is no evidence to prove that Bell Beaker users were metal prospectors, workers or trades. On the contrary, it seems that the difference that we recognise between Corded Ware and Bell Beaker communities is rather based on burial rites or 'religion' than on copper technology. The archaeological difference reflects, at least in central Europe, an altered 'religion' or 'ideology' (Dornheim et al. 2005). While there is a lack of analyses comparing the remains of Corded Ware and Bell Beaker communities' everyday life in order to make comprehensive statements about how these groups interacted outside of the sphere of funerary practices, at least the metallurgy indicates that both groups seem only to differ in ritual behaviour. For the future, this has to be investigated in more detail, especially by analysing and comparing domestic sites of both groups. Finally, copper was not a 'culture-defining' aspect and its use was not the main reason for the establishment of the Bell Beaker ideology in central Europe.

Apart from the typology of its finished products, central European Bell Beaker copper metallurgy is not exceptional. It is now clear that in central Europe we cannot speak of a specific 'Bell Beaker metal', in contrast to the situation in the Netherlands and Britain (Butler and van der Waals 1965; Needham 2002). The same types of fahlore-copper which were used for Bell Beaker objects were used for other objects as well, so it is clear that the spread of the Bell Beaker phenomenon was not motivated by specific metallurgical technology, but rather by advanced access to resources and by the need to produce copper with better properties. The metallurgical record of the Eastern Bell Beaker group does not suggest that these people differed from those of other archaeological groups by having special skills or playing a distinct role as copper prospectors, copper miners or copper traders. Subsequently, how and why the Beaker using people of the Eastern Bell Beaker group distinguished themselves from their Corded Ware-using and other neighbours has to be sought in other areas of life.

We can draw the conclusion that metallurgy was indeed important and precious to Bell Beaker communities. This is reflected by the rich graves containing both prestigious copper and gold objects and metal workers' tools. Displaying the dead as metalworkers indicates the status of metallurgy within the Bell Beaker communities. Ethnographic examples demonstrate that crafting has often been presented in mythology as a way of explaining the creation of the world[196] (Helms 1993, 24 ff.). Metal working displayed in Bell Beaker graves may symbolise creation in a spiritual sense. Interestingly, it is only the smiths, and not the actual miners or smelters, who have been portrayed in rich graves. In other words, it is only the craftsmen who shape the finished product, and not the smelter who transform solid stone to fluid melted copper by "magic", who were thus honoured. Finally, I presume that in the case of the Eastern Bell Beaker group, metal objects and metalworking tools were used to express the religious or social role of the buried person and do not necessarily indicate any special metallurgical skills in his community.

Nevertheless, this interpretation does not exclude the possibility that metallurgical know-how was restricted to a distinct group of specialists, who passed on their techniques and knowledge from one generation to the next (cf. Roberts 2008b, 364 f.). This group, however, is not visible as a specific archaeological group in our study area. In particular, it cannot be confirmed that the Eastern Bell Beaker group provided specialised metallurgists, as we have no definite evidence for mining, smelting or metalworking sites relating to this group. The individuals buried with metal workers' tools indicate the importance of metalwork within their burial rites and religious beliefs. Direct extrapolation from this to all members of the Bell Beaker phenomenon is unconvincing. It may be that 'metal workers' carried out their craft on a part-time basis. However, even they had to have specialist

[196] Helms (1993, 24) mentioned, for instance, the Fang in West Africa who believe man was created by forging.

knowledge to identify the adequate ores, to mine and smelt them and to cast an artefact. This may have earned them a special status in society. In prehistoric communities metal working was probably connected with 'magic' and 'ritual' that additionally supported its social status. This expertise could have been passed on within a restricted group in order to keep it exclusive (cf. Budd and Taylor 1995). The archaeological and metallurgical record, however, does not show that metallurgical expertise was exclusively restricted to Bell Beaker people.

In terms of the broader picture of Chalcolithic metalwork in central Europe, it can be suggested tentatively that metal workers had been a kind of itinerant craftsman, supplying wider areas with copper and possibly objects according to the norms of local traditions and the 'wishes' of their 'customers'. As argued above, there is no indication that Bell Beaker people played that role of accomplished metallurgists. It is essential to keep in mind that, for central Europe during the 3rd millennium BC, most of the available finds have been discovered in grave contexts, and are therefore only a biased selection of the types and numbers of copper artefacts that had been circulating. Objects in everyday use, perhaps not differing between Corded Ware, Bell Beaker and the Earliest Bronze Age groups, may have been re-cycled and re-used and have thus not survived. Since copper can be re-used over and over again, it is not inconceivable that copper mined in prehistoric times is probably still in use today. Moreover, because many prehistoric mines were re-used in later periods, obliterating the early traces of extraction, we can only speculate as to how much copper circulated during the Chalcolithic. Using Chalcolithic Italy as an example, a recent paper has evaluated the amount of copper that may have been mined and distributed, based on an analysis of finds and mining sites (Pearce 2009). Pearce concluded far more copper – thousands of tons' worth – had been circulating than previously assumed (ibid., 2009, 282 f.).

Metallurgical developments throughout the central European Chalcolithic were long-lasting and appeared at small scales, probably without direct effects on the social structures for that period (Bartelheim 2007, 257). It is suggested that, in general, technological changes do not affect culture very much, but conversely the cultural environment may provide the basis for adopting technological modernisation. For instance, the emergence of tin-bronze needed more effective long-distance networks to obtain the required tin and the other resources that were required. Thus, metallurgy itself probably did not affect social structures, but the assumed stratified society and the networks of the 3rd millennium BC provided the basis for more extensive metallurgy with all its related aspects, such as specialised workers, mining, and the widespread trading which results from a continuous demand of metal. These required long-distance networks which may have been based on the much earlier established Bell Beaker networks – links stretching from Portugal to Hungary and from Italy to Britain.

12.2 FUTURE RESEARCH

This study has examined whether Chalcolithic metallurgists used special copper types, selected because of the properties of the copper. The results have shown that the chronological and chorological occurrence of various types of copper suggests that metal workers were developing chemically- and physically-improved copper types. It has been possible to trace this thanks to the existence of several metallurgical (e.g. Archbutt and Prytherch 1937) and archaeometallurgical studies (e.g. Budd 1991a; Junk 2003; Kienlin 2008). Although these studies undertook fundamental research on the issue of the effects of trace elements on copper, the archaeometallurgical studies only focused on either a small selection of impurities or on a specific type of artefact, and so the broader picture that the present study has been able to trace was not apparent to these previous researchers.

Now it is clear that in central Europe there is no evidence for distinct Bell Beaker copper use, and copper metallurgy was not a defining feature of the Eastern Bell Beaker group, although metallurgy played a certain role in the burial rites of this group. Other ideas that have been suggested to explain the spread of the Bell Beaker phenomenon can be examined in further studies. For instance, distinct social structure or religious beliefs as indicated by burial rites – especially in comparison with the Corded Ware complex – may provide the differences between the central European 'cultures' during the 3rd millennium BC.

Additionally, the present study has shown that antimony-nickel-copper was almost exclusively used during the Earliest Bronze Age and its composition indicates that this copper had good tensile strength, hardness and malleability. It would be useful to explore its use with regard to specific artefact types, and to study the properties of metal finds – including the effects of working – from a closed archaeological context, such as an Early Bronze Age cemetery in which various types of metal object are found. Such research should not only focus solely on one type of artefact or on a small number of trace elements; and it should also examine the metallurgical properties, as the brief excursus on the finds the *Blechkreis* and Nitra contexts has indicated.

Due to the lack of archaeological evidence for Chalcolithic mining, smelting and metal working, it is generally necessary to combine several methods for investigating prehistoric metallurgy. We have seen that various metallurgical treatments influence the characteristics of copper finds, hence examining prehistoric metallurgy using only one method would miss much valuable information; several aspects have to be taken into account. In this research an attempt has been made to minimise information loss by combining two statistical methods, but even this may not be enough to understand prehistoric metallurgy fully. Therefore, it is also essential to apply several different methods of analysing metal. Accordingly, it is important to produce and publish many more lead-isotope analyses and micro- and macroscopic analyses, and to evaluate these results comprehensively as regards their archaeological contexts, as has been done for the trace element analyses. As some scholars have demonstrated in various key studies (e.g. Staniaszek and Northover 1983; Wang and Ottaway 2004; Kienlin 2008), further metallographic and experimental work will be crucial in order to enhance our knowledge of prehistoric metallurgy. For now, however, the only systematic dataset for archaeometallurgical studies is based on trace element analyses, although Budd and his colleagues have suggested the compilation of a broad database of lead-isotope results (Budd et al. 1996).

Based on the database of the present research and its interpretation, several regional and chronological studies could follow, as is illustrated in detail by the present work on the copper metallurgy of the Eastern Bell Beaker group; the scope for this further study has been indicated by the discussion of the finds from Vinelz (CH) and the short comparison between the *Blechkreis* and the Nitra group (cf. Chapter 10.6). Indeed, such studies should be extended by adding other archaeological and metallurgical methods, for example more detailed absolute dating, lead-isotope and macro- and microscopic analyses, which should be a research priority.

BIBIOGRAPHY

ABBREVIATIONS

Arch. USTAV Nitra:	Institute of Archaeology of the Academy of Science of Slovakia, Nitra.
BVBl.:	Bayerische Vorgeschichtsblätter.
Fundber. Baden-Württemberg:	Fundberichte aus Baden-Württemberg.
Fundber. Schwaben NF:	Fundberichte aus Schwaben, Neue Folge.
MNM:	Collection of the National Museum of Hungary, Budapest.
Nat. Museum Prague:	Collection of the National Museum, Prague.
Niederösterr. Landesmuseum:	Collection of the Niederösterreichisches Landesmuseum, Asparn/Zaya.
PBF:	Prähistorische Bronzefunde.
RGK Frankfurt:	Pers. comm. K. Rassmann (Römisch-Germanische Kommission, Frankfurt/Main).
SAM:	Studie zu den Anfängen der Metallurgie.
Stslg. München:	Bavarian State Archaeological Collection, Munich.

LITERATURE

Abels, B.-U. (1972): *Die Randleistenbeile in Baden-Württemberg, dem Elsaß, der Franche Comté und der Schweiz. PBF IX, Vol. 4.* Beck, München.

Aitchison, T., Ottaway, B. S. and Al-Ruzaiza, A. S. (1991): Summarizing a group of ^{14}C dates on the historical time scale: with a worked example from the Late Neolithic of Bavaria. *Antiquity 65*, 108 – 116.

Anderson, C. A. (1955): Oxidation of Copper Sulphides and Secondary Sulphide Enrichment. *Economic Geology 50*, 324 – 340.

Anderson, G. M. (2005): *Thermodynamics of Natural Systems*, 2nd edition. Cambridge University Press, Cambridge.

Archbutt, S. L. and Prytherch, W. E. (1931): Investigation of the Effects of Impurities on Copper. Part VII. – The Effect of Antimony on Copper. Part VIII. – The Combined Effect of Antimony and Arsenic on Copper. *Journal of the Institute of Metals XLV/1*, 265 – 305.

Archbutt, S. L. and Prytherch, W. E. (1937): *Effect of Impurities in Copper.* British Non-Ferrous Metals Research Association, London.

Armbruster, B. (2006): Steingeräte des bronzezeitlichen Metallhandwerks. *Ethnographisch-archaologische Zeitschrift 47/2*, 163 – 191.

Aspes, A. and Fasani, L. (eds.) (1982): *Il Pasaggio dal Neolitjico all'Età del Bronzo nell'Europa Centrale e nella Regione Alpina. Atti del X Simposio Internazionale sulla fine del Neolithico e gli inizi dell'etá del Bronzo in Europa. Lazise-Verona 8-12 aprile 1980.* Museo Civico di Storia Naturale, Verona.

Bach, A., Bach, H., Gall, W., Feustel, R. and Teichert, M. (1975): Beiträge zur Kultur und Anthropologie der mitteldeutschen Schnurkeramiker II. *Alt-Thüringen 13*, 43 – 107.

Bacher, J., Wenzig, K. and Vogler, M. (2004): *SPSS TwoStep Cluster – A First Evaluation. Arbeits- und Diskussionspapiere 2004-2. Lehrstuhl für Soziologie der Friedrich-Alexander-Universität Erlangen-Nürnberg.* http://www.opus.ub.uni-erlangen.de/opus/volltexte/2004/81/pdf/a_04-02.pdf [04/04/2009]

Bachmann, H.-G. (1993): Vom Erz zum Metall (Kupfer, Silber, Eisen). Die chemischen Prozesse im Schaubild, in: Steuer, H. and Zimmermann, U. (eds.), *Alter Bergbau in Deutschland. Sonderheft der Archäologie in Deutschland 1993.* Theiss, Stuttgart, 35 – 40.

Bamberger, M. (1985): The Working Conditions of the Ancient Copper Smelting Process, in: Craddock, P. T. and Hughes, M. J., *Furnaces and Smelting Technology in Antiquity.* British Museum, London, 151 – 157.

Banner, J. (1956): *Die Péceler Kultur.* Akadémiai Kiadó, Budapest.

Banner, J. (1960): The Settlement on the Kremenyák Hill at Csóka (Čoka). The excavations of F. Móra in the years 1907 to 1913. *Acta Archaeologica Academiae Scientiarum Hungaricae 12*, 1960, 1 – 56.

Bartelheim, M. (1998): *Studien zur böhmischen Aunjetitzer Kultur – Chronologische und chorologische Untersuchungen. Universitätsforschungen zur prähistorischen Archäologie 46.* Habelt, Bonn.

Bartelheim, M. (2002): Metallurgie und Gesellschaft in der Frühbronzezeit Mitteleuropas, in: Müller, J. (ed.) (2002), *Vom Endneolithikum zur Frühbronzezeit: Muster sozialen Wandels? (Tagung Bamberg 14.-16. Juni 2001). Universitätsforschungen zur prähistorischen Archäologie 90.* Habelt, 29-43.

Bartelheim, M. (2007): *Die Rolle der Metallurgie in vorgeschichtlichen Gesellschaften. Sozioökonomische und Kulturhistorische Aspekte der Ressourcennutzung.* Leidorf, Rahden.

Bartelheim, M., Eckstein, K., Huijsmans, M., Krauß, R. and Pernicka, E. (2002): Kupferzeitliche Metallgewinnung in Brixlegg, Österreich, in: Bartelheim, M., Krause, R. and Pernicka, E. (eds.), *Die Anfänge der Metallurgie in der Alten Welt. The Beginnings of Metallurgy in the Old World, Forschungen zur Archäometrie und Altertumswissenschaft. Vol. 1.* Leidorf, Rahden, 33 – 82.

Bátora, J. (1991): The Reflection of Economy and Social Structure in the Cemeteries of the Chłopice-Veselé and Nitra Cultures. *Slovenská Archeológia XXXIX*, 91 – 142.

Bátora, J. (2002): Contribution to the Problem of 'Craftsmen' Graves at the End of the Aeneolithic and in the Early Bronze Age in Central, Western and Eastern Europe. *Slovenská Archeológia L*, 179 – 228.

Bátora, J. (2003): Kupferne Schaftlochäxte in Mittel-, Ost- und Südeuropa (Zu Kulturkontakten und Datierung – Äneolithikum/Frühbronzezeit). *Slovenská Archeológia LI*, 1 – 38.

Baxter, M. J. (1994): *Exploratory Multivariante Analysis in Archaeology.* Edinburgh University Press, Edinburgh.

Baxter, M. J. (1995): Standardization and Transformation in Principal Component Analysis, with Applications to Archaeometry. *Applied Statistics 44/4*, 513 – 527.

Baxter, M. J. (2003): *Statistics in Archaeology*. Arnold, London.

Becker, B., Krause, R. and Kromer, B. (1989): Zur absoluten Chronologie der Frühen Bronzezeit. *Germania 67*, 1989, 421 – 442.

Begemann, F., Pernicka, E. and Schmitt-Strecker, S. (1995): Searching for the Ore Source of Eneolithic and EBA copper artefacts from Serbia, in: Petrović, P. and Đurđekanović, S. (eds.), *Ancient Mining and Metallurgy in Southeast Europe*. Archaeological Institute/Museum of Mining and Metallurgy, Belgrade/Bor, 143 – 149.

Behrens, G. (1916): *Bronzezeit Süddeutschlands. Kataloge des Römisch-Germanischen Zentralmuseums No. 6.* Wilkens, Mainz.

Behrens, H. (1973): *Die Jungsteinzeit im Mittelelbe-Saale-Gebiet*. Deutscher Verlag der Wissenschaft, Berlin.
Beier, H.-J. (1988): *Die Kugelamphorenkultur im Mittelelbe-Saale-Gebiet und in der Altmark*. Deutscher Verlag der Wissenschaft, Berlin.

Benz, M., Gramsch, A. and Wiermann, R. R. (1998): Rethinking Bell Beakers, in: Benz, M. and van Willigen, S. (eds.), *Some New Approaches to the Bell Beaker 'Phenomenon'. Lost Paradise...? Proceedings of the 2nd Meeting of the „Association Archéologie et Gobelets". Feldberg (Germany), 18th – 20th April 1997. BAR 690*, Oxford, 181 – 185.

Bernhart, R. (1968): *Die Pfahlbausammlung des Heimathauses Vöcklabruck. Ihre Geschichte und ihr Bestand. Beilage zu den Mitteilungen der Prähistorischen Kommission der Österreichischen Akademie der Wissenschaft, Bd. XI-XII, 1963 – 1968*, Wien.

Bertemes, F. (1989a): *Das frühbronzezeitliche Gräberfeld von Gemeinlebarn. Kulturhistorische und paläometallurgische Studien. Text. Saarbrücker Beiträge zur Altertumskunde 45*. Habelt, Bonn.

Bertemes, F. (1989b): *Das frühbronzezeitliche Gräberfeld von Gemeinlebarn. Kulturhistorische und paläometallurgische Studien. Katalog. Saarbrücker Beiträge zur Altertumskunde 45*. Habelt, Bonn.

Bertemes, F. (2000): Zur Entstehung der Danubischen Frühbronzezeit in Mitteleuropa, in: *In Memoriam Jan Rulf, Památky Archeologické – Supplementum 13.* Archeologický Ústav AVČR, Praha, 25 – 37.

Bertemes, F. and Heyd, V. (1996): Définition et origine de L'Age du Bronze ancien en Europe central, in: Mordant, C. and Gaiffe, O. (eds.), *Cultures et Sociétés du Bronze Ancien en Europe, Actes de 117e Congrès National des Sociétés savantes, Clermont-Ferrand 27 – 29 octobre 1992*. Éd. du C.T.H.S., Paris, 13 – 36.

Bertemes, F. and Heyd, V. (2002): Der Übergang Kupferzeit/Frühbronzezeit am Nordwestrand des Karpatenbeckens – Kulturgeschichtliche und paläometallurgische Betrachtung, in: Bartelheim, M., Krause, R. and Pernicka, E. (eds.), *Die Anfänge der Metallurgie in der Alten Welt. The Beginnings of Metallurgy in the Old World. Forschungen zur Archäometrie und Altertumswissenschaft. Vol. 1.* Leidorf, Rahden, 185 – 228.

Bertemes, F. and Šebela, L. (1998): Quelques aspects de la métellurgie du Chalcolithique Récent et le début de l'Age du Bronze Ancie en Autriche, Bohème et Moravie, in: Mordant, C., Pernot, M. and Rychner, V. (eds.), *L'Atelier du bronzier en Europe du XXe au VIIIe siècle avant notre ère. Actes du colloque international Bronze 96. Neuchâtel et Dijon II: Du mineria au métal, du à l'objet.* Comité des Travaux Historiques et Scientifiques, Paris, 227 – 239.

Bertemes, F., Schmotz, K. and Thiele, W.-R. (2000): Das Metallurgengrab 9 des Gräberfeldes der Glockenbecherkultur von Künzing, Lkr. Deggendorf, in: Chytráąek, M., Michálek, J. and Schmotz, K. (eds.), *Archäologische Arbeitsgemeinschaft Ostbayern/West- und Südböhmen – Archeologická pracovní skupina východni Bavorsko/západní a jižní Čechy. 9. Treffen 23. bis 26. Juni 1999 in Neukirchen b. Hl. Blut*, 53 – 60.

Besse, M. (2004): Bell Beaker common ware during the third Millennium BC in Europe, in: Czebreszuk, J. (ed.), *Similar but Different. Bell Beakers in Europe*. Adam Mickiewicz University, Poznań, 127 – 148.

Besse, M. and Strahm, Chr. (2001): The components of the Bell Beaker complex, in: Nicolis, F. (ed.), *Bell Beakers today. Pottery, people, culture, symbols in prehistoric Europe. Proceedings of the International Colloquium Riva del Garda (Trento, Italy) 11 – 16 May 1998, Vol. 1.* Provincia Autonoma di Trento, Trient,103 – 110.

Bill, J. (1973): *Die Glockenbecherkultur und die frühe Bronzezeit im französischen Rhonebecken und ihre Beziehungen zur Südwestschweiz*. Schweizerische Gesellschaft für Ur- u. Frühgeschichte, Basel.

Bill, J. (1997): Die Bronzebeile von Salez. Das 1883 gefundene Depot aus der Frühbronzezeit. *Werdenberger Jahrbuch 10*, 247 – 261.

Billamboz, A. (1998): Die jungneolithischen Dendrodaten der Pfahlbausiedlungen Südwestdeutschlands als Zeitrahmen für die Einflüsse der Michelsberger Kultur in ihrem südlichen Randgebiet, in: Biel, J., Schlichterle, H., Strobel, M. and Zeeb, A. (eds.), *Die Michelsberger Kultur und ihre Randgebiete – Probleme der Entstehung, Chronologie und des Siedlungswesens. Materialhefte zurArchäologie in Baden-Württemberg 43.* Theiss, Stuttgart, 159 – 168.

Binford, L. R. (1965): Archaeological Systematic and the Study of Culture Prozess. *American Antiquity 31/2*, 203 – 210.

Bognár-Kutzián, I. (1963): *The Copper Age Cemetery of Tiszapolgár-Basatanya.* Akadémiai Kiadó, Budapest.

Bognár-Kutzián, I. (1972): *The Early Copper Age Tiszapolgár Culture in the Carpathian Basin.* Akadémiai Kiadó, Budapest.

Bóna, I. (1965): The People of Southern Origin of the Early Bronze Age in Hungary I-II, *Alba Regia IV-V*, 1965, 17 – 63.

Bóna, I. (1992): Bronzezeitliche Tell-Kulturen in Ungarn, in: Meier-Arendt, W. (ed.) (1992), *Bronzezeit in Ungarn. Forschungen in Tell-Siedlungen an der Donau und Thiess.* Pytheas, Frankfurt/M., 9 – 39.

Boomert, A. (1975): A Contribution to the Classification of Prehistoric Metal Objects. *Helinium 15*, 134 – 161.

Boroffka, N. (2009): Simple Technology: Casting Moulds for Axe-adzes, in: Kienlin, T. L. and Roberts, B., *Metals and Societies. Studies on honour of Barbara S. Ottaway.Unversitätsforschungen zur prähistorischen Archäologie 169*, Habelt, Bonn, 246 – 257.

Bourgarit, D. (2007): Chalcolithic Copper Smelting, in: La Niece, S., Hook, D. and Craddock, P. (eds.), *Metals and Mines. Studies in Archaeometallurgy.* Archetype, London, 3 – 14.

Bradley, R. (1990): *The Passage of Arms. An archaeological analysis of prehistoric hoards and votive deposits.* Cambridge University Press, Cambridge.

Brather, S. (2001): Kulturgruppe und Kulturbegriff, in: *Reallexikon der Germanischen Altertumskunde 17, Kleinere Götter – Landschaftsarchäologie*, 442 – 452. de Gruyter, Berlin.

Bray, P. J. (2009): *Exploring the social basis of technology: Reanalysing regional archaeometric studies of the first copper and tin-bronze use in Great Britain and Ireland. Unpublished PhD-Thesis at the University of Oxford*, Oxford.

Brodie, N. (1994): *The Neolithic – Bronze Age Transition in Britain. A critical review of some archaeological and craniological concept. BAR 238*, Oxford.

Buchvaldek, M. (1986): Zum gemeineuropäischen Horizont der Schnurkeramik. *Praehistorische Zeitschrift 61*, 142 – 179.

Buchvaldek, M. and Strahm, Chr. (eds.) (1992): *Die kontinentaleuropäischen Gruppen der Kultur mit Schnurkeramik. Schnurkeramik – Symposium 1990, Praehistorica 19.* Univerzita Karlova, Praha.

Bücke, S., Barthel, H.-J. and Gall, W. (1989): Beiträge zur Kultur der mitteldeutschen Schnurkeramiker III, *Alt-Thüringen 24*, 33 – 116.

Budd, P. (1991a): *A Metallographic Investigation of Eneolithic Arsenical Copper. Determination of the Manufacturing Processes of Early Prehistoric Arsenical Copper Artefacts by Microstructural Analysis based on a Re-evaluation of the Mechanical Properties of Copper-arsenic Alloys. Unpublished PhD-thesis at the University of Bradford*, Bradford.

Budd, P. (1991b): Endneolithic arsenical Copper: Heat Treatment and the metallographic Interpretation of manufacturing Processes, in: Pernicka, E. and Wagner, G. A. (eds.), *Archaeometry'90. International Symposium on Archaeometry Heidelberg 2. – 6. April 1990.* Birkhäuser, Basel, 35 – 44.

Budd, P. and Ottaway, B. S. (1991): The Properties of Arsenical Copper Alloys: Implications for the Development of Eneolithic Metallurgy, in: Budd, P., Chapmann, B. and Jackson, C. (eds.), *Archaeological Sciences 1989.* Oxbow, Oxford, 132 – 142.

Budd, P. and Taylor, T. (1995): The faerie smith meets the bronze industry: magic versus science in the interpretation of prehistoric metal-making. *World Archaeology 27/1*, 133 – 143.

Budd, P., Haggerty, R., Pollard, A. M., Scaife, B. and Thomas, R. G. (1996): Rethinking the quest for provenance. *Antiquity 70*, 168 – 174.

Budinsky-Krička, V. (1965): Gräberfeld der späten schnurkeramischen Kultur in Veselé. *Slovenská Archeológia XIII*, 51 – 106.

Burger, I. (1988): *Die Siedlung der Chamer Gruppe von Dobl, Gemeinde Prutting, Landkreis Rosenheim und ihre Stellung im Endneolithikum Mitteleuropas.* Graf, Fürth.

Burgees, C. and Shennan, A. (1976): The Beaker Phenomenon: Some Suggestions, in: Burgess, C. and Miket, R. (eds.), *Settlement and Economy in the Third and Second millennium BC.* BAR 33, Oxford, 309 – 327.

Butler, J. J. and Van der Waals, J. D. (1966): Bell Beakers and Early Metalworking in the Netherlands. *Paleohistoria 12*, 42 – 139.

Cartailhac, E. (1886): *Les âge préhistoriques de l'Espagne et du Portugal.* Reinwald, Paris.

Case, H. (2004): Beakers and Beaker Culture, in: Czebreszuk, J. (ed.), *Similar but Different. Bell Beakers in Europe.* Adam Mickiewicz University, Poznań, 11 – 34.

Cattin, F. (2008): *Modalités d'approvisionnement et modalités de consommation du cuivre dans les Alpes au 3e millénaire avant notre ère: apport des analyses métalliques à la connaissance des peuplements du Néolithique final, du Campaniforme et du Bronze ancien. Unpublished PhD-thesis at the Université de Genève*, Geneve.

Černych, E. N. (1978): Aibunar – a Balkan Copper Mine of the fourth millennium BC (Investigations of the years 1971, 1972 and 1974). *Proceedings of the Prehistoric Society 44*, 203 – 217.

Černych, E. N. (1992): *Ancient Metallurgy in the USSR*. Cambridge University Press, Cambridge.

Cevey, Chr., Gunther, D., Hubert, V., Hunger, K., Hildbrand, E., Käser, M.-A., Lehmann, E., Müller-Scheesel, N., Wörle-Soares, M., Strahm, Chr. and Van Willigen, S. (2006): Neue archäometallurgische Untersuchungen zum Beginn der Kupferverarbeitung in der Schweiz. *Archäologie in der Schweiz 29*, 24 – 33.

Charles, J. A. (1967): Early arsenical Bronzes – A metallurgical view. *American Journal of Archaeology 71/1*, 21 – 26.

Charles, J. A. (1980): The Coming of Copper and Copper-Base Alloys and Iron: A Metallurgical Sequence, in: Wertime, T. A. and Muhly, J. D. (eds.), *The Coming of the Age of Iron*. Yale University Press, New Haven/London, 151 – 181.

Childe, V. G. (1925): *The Dawn of European Civilisation*. Paul, Trench and Trubner, London.

Childe, V. G. (1929): *The Danube in Prehistory*. Clarendon Press, Oxford.

Chiu, T., Fang, D., Chen, J., Wang, Y. and Jeris, Chr. (2001): A Robust and Scalable Clustering Algorithm for Mixed Type Attributes in Large Database Enviroment, in: Provost, F. and Srikant R. (eds.): *Proceedings of the seventh ACM SIGKDD International Conference on Knowledge Discovery and Data Mining*. ACM, New York, 263 – 269.

Christoforidis, A. and Pernicka, E. (1988): Gruppierung von Metallanalysen mit Hilfe der Clusteranalyse, in: Krause, R., *Die endneolithischen und frühbronzezeitlichen Grabfunde auf der Hochterrasse von Singen am Hohentwiel*. Theiss, Stuttgart, 252 – 262.

Chropovský, B. (ed.) (1973): *Symposium über die Entstehung und Chronologie der Badener Kultur*. Slovenská akadémia Archeologický ústav, Nitra.

Clark, G. (1957): *Archaeology and Society*. 3rd edition. Methuen, London.

Clarke, D. L. (1968): *Analytical Archaeology*, Methuen, London.

Clarke, D. L. (1970): *Beaker Pottery of Great Britain and Ireland*. Cambridge University Press, Cambridge.

Clarke, D. L. (1976): The Beaker network – social and economic models, in: Lanting, J. N. and van der Waals, J. D. (eds.), *Glockenbecher Symposium. Oberried 1974*. Fibula-Van Dishoeck, Bussum/Haarlem, 459 – 477.

Coghlan, H. H. and Willows, R. (1962): A note upon native copper: Its occurrence and properties. *Proceedings of the Prehistoric Society 28*, 58 – 67.

Corsi, M., Cristoforetti, G., Giuffrida, M., Hidalgo, M., Legnaioli, S., Masotti, L., Palleschi, V., Salvetti, A., Tognoni, E., Vallebona, Chr. and Zanini, A. (2005): Archaeometric Analysis of Ancient Copper Artefacts by Laser-Induced Breakdown Spectroscopy Technique. *Microchim Acta 152*, 105 – 111.

Craddock, P. T. (1995): *Early Metal Mining and Production*. Edinburgh University Press, Edinburgh.

Craddock, P. T. and Meeks, N. (1987): Iron in Ancient Copper. *Archaeometry 29*, 187 – 204.

Craig, O.E., Chapman, J., Heron, C.P., Willis, L.H., Taylor, G., Whittle, A. and Collins, M.J. (2005): Did the first farmers of central and eastern Europe produce dairy foods? *Antiquity 79*, 882 – 894.

Czebreszuk, J. (2004a): Corded Ware from East to West, in: Bogucki, P. and Crabtree, P.J. (eds.), *Ancient Europe 8000 B.C. – A.D. 1000. Encyclopedia of the Barbarian World. Vol. I. The Mesolithic to Copper Age (c. 8000 – 2000 B.C.)*. Thompson and Gale, New York, 467 – 475.

Czebreszuk, J. (2004b): Bell Beakers from West to East, in: Bogucki, P. and Crabtree, P.J. (eds.), *Ancient Europe 8000 B.C. – A.D. 1000. Encyclopedia of the Barbarian World. Vol. I. The Mesolithic to Copper Age (c. 8000 – 2000 B.C.)*. Thompson and Gale, New York, 476 – 485.

Dąbrowski, J. and Hensel, Z. (1983): Epineolityczne materiały ze Strzyzowa, woj. Zamość. *Sprawozdania Archeologiczne 35*, 69 – 85.

Dalidowski, A. (2006): Die Gräber der Glockenbecherkultur, in: Meller, H. (ed.), *Archäologie XXL. Archäologie an der B6n im Landkreis Quedlinburg*. Landesamt für Denkmalpflege und Archäologie Sachsen-Anhalt, Halle, 116 – 120.

Dannatt, C. W. and Ellingham, H. J. T. (1948): Roasting and Reduction Processes – A General Survey. *Discussions of the Faraday Society 1948/4*, 126 – 139.

David-Elbiali, M. (2000): *La Suisse occidentale au IIe millénaire av. J.-C. Chronologie, culture, intégration européenne*. Cahiers d'archéologie romande, Lausanne.

Dehn, W. (1952): Eine frühbronzezeitliche Siedlungsgrube bei Gaimersheim, Ldkr. Ingolstadt. Zur Siedlungskeramik des Straubinger Kreises. *BVbl. 18/19, 1951/1952*, 1- 25.

Dies, K. (1967): *Kupfer und Kupferlegierungen in der Technik*. Springer, Berlin.

Dobeš, M. (1989): Zu den äneolithischen Kupferflachbeilen in Mähren, Polen und in der DDR – K eneolitickým plochým měděným sekerám na Moravě, v Čechách, Polsku a NDR, in: Richter, M. and Buchvaldek, M. (eds.), *Das Äneolithikum und die früheste Bronzezeit (C^{14} 3000 – 2000 b.c.) in Mitteleuropa: kulturelle und chronologische Beziehungen. Acta des XIV. Internationalen Symposiums Prag-Liblice 20.-24.10.1986. Praehistorica 15.* Univerzita Karlova, Praha, 39 – 48.

Dobeš, M. (1990): Nálezy měděných seker a sekeromlatů z oblasti a muzeí severozápadních Čech. *Litoměřicko 25*, 131 – 153.

Dobeš, M. (1992): Die Beziehungen der Kupferindustrie Böhmens und Mährens zu dem Balkan zur Zeit des älteren Kupferhorizontes. *Studia Praehistorica 11-12*, 334 – 338.

Dombay, J. (1960): *Die Siedlung und das Gräberfeld in Zengővárkony. Beiträge zur Kultur des Aeneolithikums in Ungarn.* Akadémiai Kiadó, Budapest.

Dornheim, S., Lißner, B., Metzler, S., Müller, A., Ortolf, S., Sprenger, S., Stadelbacher, A., Strahm, Ch., Wolters, K. and Wiemann, R. (2005): *Sex* und *gender*, Alter und Kompetenz, Status und Prestige: Soziale Differenzierung im 3. vorchristlichen Jahrtausend, in: Müller, J. (ed.), *Alter und Geschlecht in ur- und frühgeschichtlichen Gesellschaften.* Habelt, Bonn, 27 – 71.

Dresely, V. (2004): *Schnurkeramik und Schnurkeramiker im Taubertal.* Theiss, Stuttgart.

Driehaus, J. (1960): *Die Altheimer Gruppe und das Jungneolithium in Mitteleuropa.* Römisch-Germanisches Zentralmuseum, Bonn.

Dvořák, P. (ed.) (1992): *Gräberfelder der Glockenbecherkultur in Mähren I (Bez. Blansko, Brno-město, Brno-venkov).* Dvořák, Brno.

Dvořák, P. (ed.) (2006): Siedlungen der Glockenbecherkultur in Mähren. Ústav Archeologické Památkové Péče, Brno.

Dvořák, P. and Hájek, L. (1990): *Die Gräberfelder der Glockenbecherkultur bei Šlapanice (Bez. Brno-venkov). Katalog der Funde.* Dvořák, Brno.

Eibner, C. (1982): Kupferbergbau in Österreichs Alpen, in: Hänsel, B. (ed.): *Südosteuropa zwischen 1600 und 1000 v. Chr. Prähististorische Archäologie in Südosteuropa 1*, Moreland Eds., Berlin, 399 – 408.

Endrődi, A. (1998): Results of settlement archaeology in Bell Beaker Culture research in Hungary, in: Benz, M. and van Willigen, S. (eds.), *Some new Approaches to the Bell Beaker 'Phenomenon' Lost Paradise...?* Proceedings of the 2nd Meeting of the "Association Archéologie et Gobelets" Feldberg (Germany), 18th – 20th April 1997, *BAR 690,* Oxford, 141 – 160.

Endrődi, A. (ed.) (2004): *Everyday life and spirituality at the end of the Copper Age. 5000 years old remains of the Baden culture in Budapest.* BTM, Budapest.

Engelhardt, B. (1994): Die Altheimer Feuchtbodensiedlung Ergolding-Fischergasse bei Landshut und ihr Hinterland. Die Entwicklung eines Kleinraumes von der Lienearbandkeramik bis zum Mittleren Spätneolithikum. *Vorträge des 12. Niederbayerischen Archäologentages*, 41 – 87.

Engelhardt, B. (1998): Gräber der Schnurkeramik und der Glockenbecherkultur in Südbayern, in: Chytráąek, M., Michálek, J. and Schmotz, K. (eds.), *Archäologische Arbeitsgemeinschaft Ostbayer/West- und Südböhmen – Archeologická pracovní skupina východni Bavorsko/západní a jižní Čechy. 7. Treffen 11. bis 14. Juni 1997 in Lindau a. d. Isar*, 71 – 87.

Erickson, P. A. (1998): *A History of Anthropological Theory.* Broadview Press, Ontario.

Evans, A. M. (1980): *An Introduction to Ore Geology.* Blackwell, Oxford.

Fischer, U. (1953): Die Orientierung der Toten in den neolithischen Kulturen des Saalegebietes, *Jahresschrift der mitteldeutsche Vorgeschichte 37*, 49 – 66.

Fischer, U. (1976): Die Dialektik der Becherkulturen. *Jahresschrift der mitteldeutsche Vorgeschichte 60*, 235 – 245.

Fokkens, H., Achterkamp, Y. and Kuijpers, M. (2008): Bracers or Bracelets? About the Functionality and Meaning of Bell Beaker Wrist-guards. *Proceedings of the Prehistoric Society 74*, 109 – 140.

Foltiny, I. (1941): Koraréz- és Bronzkori Temetö Deszken (Torontál M.). Frühkupferzeitliches und Bronzezeitliches Gräberfeld in Deszk (Kom. Torontál). *Folia Archaeologica III-IV*, 69 – 98.

Forenbaher, S. (1993): Radiocarbon dates and absolute chronology of the central European Early Bronze Age. *Antiquity 67*, 218–256.

Forrer, R. (1923): Nouvelles Découvertes et Acquisitions du Musée Préhistorique et Gallo-Romain de Strasbourg, *Cahiers d'Archéologie et d'Historie d'Alsace 53 – 56*, 88 – 124.

Frána, J. , Jiráň, L., Maštalka, A. and Moucha, V. (1995): Artefacts of copper and copper alloys in prehistoric Bohemia from the viewpoint of analyses of element composition. *Památky archeologické Supplementum 3*, Praha, 125 – 205.

Freudenberg, M. (2009): Steingeräte zur Metallbearbeitung – Einige neue Aspekte zum spätneolithischen und frühbronzezeitlichen Metallhandwerk vor dem Hintergrund des schleswig-holsteinischen Fundmaterials. *Archäologisches Korrespondenzblatt 39/3*, 341 – 359.

Friedman, A. M., Conway, M., Kastner, M., Milsted, J., Metta, D., Fields, P. R. and Olsen, E. (1966): Copper Artifacts: Correlation with Source Types of Copper Ores. *Science 152*, 1504 – 1506.

Furholt, M. (2003): *Die absolutchronologische Datierung der Schnurkeramik in Mitteleuropa und Südskandinavien. Universitätsforschungen zur prähistorischen Archäologie 101.* Habelt, Bonn.

Furmánek, V. (1980): *Die Anhänger in der Slowakei, PBF XI, Vol. 3.* Beck, München.

Furmánek, V., Veliačik, L. and Vladár, J. (1999): *Die Bronzezeit im slowakischen Raum*, Leidorf, Rahden.

Gebers, W. (1978): *Endneolithikum und Frühbronzezeit im Mittelrheingebiet. Katalog. Saarbrücker Beiträge zur Altertumskunde 28.* Habelt, Bonn.

Gebers, W. (1984): *Das Endneolithikum im Mittelrheingebiet. Typologische und chronologische Studien. Text. Saarbrücker Beiträge zur Altertumskunde 28. Habelt,* Bonn.

Gerloff, S. (1975): *The Early Bronze Age Daggers in Great Britain and a Reconsideration of the Wessex Culture, PBF VI, Vol. 2.* Beck, München.

Gleser, R. and Schmitz, A. (2001): Schernau und Varna. Überlegungen zur Herkunft des ältesten Metalls in Süddeutschland, *Archäologisches Korrespondenzblatt 31*, 365 – 376.

Götze, A., Höfer, P. and Zschiesche, P. (1909): *Die vor- und frühgeschichtlichen Altertümer Thüringens.* Kabitzsch, Würzburg.

Griffith, A. (2007): *SPSS for Dummies.* Wiley & Sons, Hoboken.

Grupe, G., Price, T. D. and Schröter, P. (1998): Migration in the Bell Beaker period of central Europe. Recent strontium isotope analysis of Bell Beaker burials from Bavaria raises important new questions about prehistoric migrations in Europe. *Antiquity 72*, 405 – 411.

Guerra-Doce, E. (2006) : Exploring the Significance of Beaker Pottery through Residue Analyses. *Oxford Journal Archaeology 25/3*, 247 – 259.

Hafner, A. (2001) : Archäologische Belege der älteren Frühbronzezeit aus der Westschweiz, in: Eberschweiler, B. and Sangmeister, E. (eds.), *Aktuelles zur Frühbronzezeit und frühen Mittelbronzezeit im nördlichen Alpenvorland. Rundgespräch Hemmenhofen 6. Mai 2000.* Janus, Freiburg, 155 – 164.

Hafner, A. and Suter, P. J. (2003): *Das Neolithikum in der Schweiz* (27/11/2003).
www.jungsteinsite.de
[10/07/2006]

Hafner A. and Suter, P. J. (2004a): Vom Spät- zum Endneolithikum: Wandel und Kontinuität um 2700 v. Chr., in: Beier, H.-J. and Einicke, R. (eds.), *Varia Neolithica III – Gedenkschrift für Annemarie Häußer und Helmut Spatz.* Beier und Beran, Langenweissbach, 213 – 232.

Hafner, A. and Suter, P. J. (2004b): Vom Endneolithikum zur Frühbronzezeit: Wandel und Kontinuität zwischen 2400 und 1500 v. Chr., in: Beier, H.-J. and Einicke, R. (eds.), *Varia Neolithica III – Gedenkschrift für Annemarie Häußer und Helmut Spatz.* Beier und Beran, Langenweissbach, 233 – 249.

Hájek, L. (1957): Knoflíky středoevropské skupiny kultury zvoncovitých pohárů – Die Knöpfe der mitteleuropäischen Glockenbecherkultur, *Památky archeologické 48*, 1957, 389 – 424.

Hájek, L. (1966): Die älteste Phase der Glockenbecherkultur in Böhmen und Mähren, *Památky archeologické 57*, 210 – 241.

Hájek, L. (1968): *Kultura Zvoncovitých Pohárů v Čechách*, Archeologický Ústav ČSAV, Praha.

Hanson, D. and Ford, G. W. (1924): Investigation of the Effects of Impurities on Copper. Part II. – The Effect of Iron on Copper. *Journal of the Institute of Metals XXXII/2*, 335 – 365.

Hanson, D. and Ford, G. W. (1927): Investigation of the Effects of Impurities on Copper. Part V. – The Effect of Bismuth on Copper. *Journal of the Institute of Metals XXXVII/1*, 169 – 181.

Hanson, D. and Marryat, C. B. (1927): Investigation of the Effects of Impurities on Copper. Part III. – The Effect of Arsenic on Copper. Part IV. – The Effect of Arsenic plus Oxygen on Copper. *Journal of the Institute of Metals XXXVII/1*, 121 – 181.

Härke, H. (1978): Probleme der optischen Emissionsspektralanalyse in der Urgeschichtsforschung. Technische Möglichkeiten und methodische Fragestellungen. *Praehistorische Zeitschrift 53*, 165 – 276.

Härke, H. (1993): Intentionale und funktionale Daten. Ein Beitrag zur Theorie und Methodik der Gräberarchäologie. *Archäologisches Korrespondenzblatt 23*, 141 – 146.

Härke, H. (1997): The Nature of Burial Data, in: Jensen, C. K. and Nielsen, K. H. (eds.), *Burial and Society. The Chronological and Social Analysis of Archaeological Burial Data*. Aarhus University Press, Aarhus, 19 – 27.

Harrison, R. and Heyd, V. (2007): The Transformation of Europe in the Third Millennium BC. The example of 'Le Petit-Chasseur I + III'. *Praehistorische Zeitschrift 82*, 130 – 214.

Hauptmann, A. (2007): *The Archaeometallurgy of Copper. Evidence from Faynan, Jordan*. Spronger, Berlin.

Häusler, A. (1991): Geschlechtsdifferenzierte Bestattungssitten im Neolithikum und in der frühen Bronzezeit Mitteleuropas. *Saeculum 41/3, 1990*, 332 – 348.

Havel, J. (1978): Pohřební ritus kultury zvoncovitých pohárů v Čechách a na Moravě – The Burial Rite of the Bell Beaker Culture in Bohemia and Moravia. *Praehistorica 7,* 91 – 117.

Helms, M. W. (1993): *Craft and the Kingly Ideal. Art, Trade and Power*. University of Texas Press, Austin.

Henderson, J. (2000): *The Science and Archaeology of Materials. An Investigation of Inorganic Materials*. Routledge, London/New York.

Heyd, V. (2000a): *Die Spätkupferzeit in Süddeutschland. Untersuchungen zur Chronologie von der ausgehenden Mittelkupferzeit bis zum Beginn der Frühbronzezeit im süddeutschen Donaueinzugsgebiet und den benachbarten Regionen bei besonderer Berücksichtigung der keramischen Funde. Textband. Saarbrücker Beiträge zur Altertumskunde 73*. Habelt, Bonn.

Heyd, V. (2000b): *Die Spätkupferzeit in Süddeutschland. Untersuchungen zur Chronologie von der ausgehenden Mittelkupferzeit bis zum Beginn der Frühbronzezeit im süddeutschen Donaueinzugsgebiet und den benachbarten Regionen bei besonderer Berücksichtigung der keramischen Funde. Tafelband. Saarbrücker Beiträge zur Altertumskunde 73*. Habelt, Bonn.

Heyd, V. (2004): Soziale Organisation im 3. Jahrtausend v. Chr. entlang der oberen Donau. Der Fall Schnurkeramik und Glockenbecher. *Das Altertum 49/3*, 183 – 214.

Heyd, V. (2007): Families, Prestige Goods, Warriors & Complex Societies: Beaker Groups of the 3rd Millennium cal BC along the Upper & Middle Danube. *Proceedings of the Prehistoric Society 73*, 327 – 379.

Heyd, V., Husty, L. and Kreiner, L. (2004): *Siedlungen der Glockenbecherkultur in Süddeutschland und Mitteleuropa*. Faustus, Büchenbach.

Hicke, W. (1987): Hügel- und Flachgräber der Frühbronzezeit aus Jois und Oggau. Frühbronzezeitliche Bevölkerungsgruppen im archäologischen Wandel. *Wissenschaftliche Arbeiten aus dem Burgenland 75*, 5 – 229.

Higham, T., Chapman, J., Slavchev, V., Gaydarska, B., Honch, N., Yordanov, Y. and Dimitrova, B. (2007): New perspectives on the Varna cemetery (Bulgaria) – AMS dates and social implications. *Antiquity 81*, 640 – 654.

Hillebrand, J. (1929): *Das frühkupferzeitliche Gräberfeld von Pusztaistvánháza*, Budapest.

Hochuli, St. (2001): Archäologische Belege der älteren Frühbronzezeit aus der Zentral- und Ostschweiz, in: Eberschweiler, B. (ed.), *Aktuelles zur Frühbronzezeit und frühen Mittelbronzezeit im nördlichen Alpenvorland. Rundgespräch Hemmenhofen 6. Mai 2000*. Janus, Freiburg, 137 – 146.

Hodson, F. R. (1969a): Searching for Structure in Multivariate archaeological Data. *World Archaeology 1/1*, 90 – 105.

Hodson, F. R. (1969b): Cluster Analysis and Archaeology: some new Developments and Applications. *World Archaeology 1/3*, 299 – 320.

Hoffmann, W. and Schmidt, B. (1961): Wichtige Fundmitteilungen und Neuerwerbungen des Jahres 1958. *Jahresschrift der Mitteldeutschen. Vorgeschichte 45*, 278 – 298.

Hofman, H. O. (1924): *Metallurgy of Copper*. McGraw-Hill, New York/London.

Höppner, B., Bartelheim, M., Huijsmans, M., Krause, R., Martinek, K.-P., Pernicka, E. and Schwab, R. (2005): Prehistoric Copper Production in the Inn Valley (Austria), and the Earliest Copper in Central Europe. *Archaeometry 47/2*, 293 – 315.

Horváth, L. A. and Simon, K. H. (2003): *Das Neolithikum und die Kupferzeit in Südwesttransdanubien*. Magyar Nemzeti Múzeum, Budapest.

Hughes M. J., Cowell, M. R. and Craddock, P. T. (1976): Atomic absorbtion techniques in archaeology. *Archeometry 18/1*, 19 – 37.

Hundt, H.-J. (1958): *Katalog Straubing I. Die Funde der Glockenbecherkultur und der Straubinger Kultur.* Materialhefte zur bay. Vorgeschichte 11. Lassleben, Kallmünz.

Hundt, H.-J. (1975): Steinerne und kupferne Hämmer der frühen Bronzezeit. *Archäologisches Korrespondenzblatt 5,* 115 – 120.

Husty, L. (2004): Glockenbecherzeitliche Funde aus Landau a. d. Isar, in: Heyd, V., Husty, L. and Kreiner, L., *Siedlungen der Glockenbecherkultur in Süddeutschland und Mitteleuropa.* Faustus, Büchenbach, 15 – 102.

Itten, M. (1970): *Die Horgener Kultur.* Birkhäuser, Basel.

Jacobs, J. (1989): Jungsteinzeitliche Metallfunde auf dem Gebiet der DDR. *Zeitschrift für Archäologie 23,* 1 – 17.

Johnson, M. (1999): *Archaeological Theory. An Introduction.* Blackwell, Oxford.

Jones, A. (2007): *Memory and Material Culture.* Cambridge University Press, Cambridge.

Jovanovoć, B. (1982): *Rudna Glava, najstarije rudarstvo bakra na Centralnom Balkanu.* Muzej rudarstva i metalurgije, Beograd.

Jovanović, B. and Ottaway, B. S. (1976): Copper mining and metallurgy in the Vinča group. *Antiquity 50,* 104 – 113.

Junk, M. (2003): *Material properties of copper alloys containing arsenic, antimony and bismuth. The material of Early Bronze Age ingot torques.* PhD-thesis at the Technische Universität Bergakademie Freiberg, Freiberg. http://www.cez-archaeometrie.de/dipl-diss/diss-junk-2003.pdf
[11.05.2007]

Junghans, S., Klein, H. and Scheufele, E. (1954): Untersuchungen zur Kupfer- und Frühbronzezeit Süddeutschlands, *Berichte der Römisch-Germanischen Kommission 34,* 1951 – 1953, 77 – 114.

Kadrow, S. (1991): *Iwanowice. Stanowisko Babia Góra, czesc I. Rozwój przestrzenny osady z wczesnego okresu epoki brązu.* Polskiej Akademii Nauk, Kraków.

Kadrow, S. (1997): Bemerkungen zur Chronologie der Frühbronzezeit in Kleinpolen und in der Slowakei, in: Becker, C., Dunkelmann, M.-L., Metzner-Nebelsick, C., Peter-Röcher, H.,. Roeder, M. and Teržan, B. (eds.), *Χρόνος. Beiträge zur Prähistorischen Archäologie zwischen Nord- und Südosteuropa. Festschrift für Bernhard Hänsel.* Leidorf, Espelkamp, 229 – 235.

Kadrow, S., Machnik, J. and Jan, A. (1992): *Iwanowice. Stanowisko Babia Góra, Cz. II. Rozwój przestrzenny osady z wczesnego okresu epoki brązu.* Polskiej Akademii Nauk, Kraków.

Kadrow, S. and Machnik, J. (1992): *Kultura Mierzanowicka. Chronologia, taksonomia i rozwoj przestrzenny.* Polskiej Akademii Nauk, Kraków.

Kadrow, S. and Peška, J. (1999): Älteste Frühbronzezeitliche Siedlungen in Mähren und der Anfang des Epischnurkeramischen Karpatenländischen Kulturkreises, in: Bátora, I. and Peška, J. (eds.), *Aktuelle Probleme der Erforschung der Frühbronzezeit in Böhmen und Mähren und in der Slowakei.* Instituti Archaeologici Nitriensis Academiae Scientiarum Slovacae, Nitra, 33 – 40.

Kaiser, G. (1962): Das frühbronzezeitliche Gräberfeld von Unterwölbing, p. B. St. Pölten, N.Ö. *Archaeologia Austriaca 32,* 35 – 61.

Kalicz, N. (1959): Die Frühbronzezeitlichen Brandbestattungen in der Umgebung der Gemeinde Alsónémedi. *Acta Archaeologica Academiae Scientiarum Hungaricae 9,* 195 – 209.

Kalicz, N. (1982): Die terminologischen und chronologischen Probleme der Kupfer- und Bronzezeit in Ungarn, in: Aspes, A. and Fasani, L. (eds.), *Il Pasaggio dal Neolitjico all'Età del Bronzo nell'Europa Centrale e nella Regione Alpina. Atti del X Simposio Internazionale sulla fine del Neolithico e gli inizi dell'etá del Bronzo in Europa. Lazise-Verona 8-12 aprile 1980.* Museo Civico di Storia Naturale Verona, 117 – 137.

Kalicz, N. (1992): The Oldest Metal Finds in Southeastern Europe and the Carpathian Basin from the 6th to 5th Millennium BC, *Archaeologiai Értesitö 119,* 3 – 14.

Kalicz, N. (1995): Die Balaton-Lasinja-Kultur in der Kupferzeit Süd- und Mitteleuropas, in: Kovás, T. (ed.), *Neuere Daten zur Siedlungsgeschichte und Chronologie der Kupferzeit des Karpatenbeckens.* Magyar Nemzeti Múzeum, Budapest, 37 – 49.

Kalicz, N. and Raczky, P. (1987): The Late Neolithic of the Tisza Region: A Survey of Recent Archaeological Research, in: Raczky, P. (ed.), *The Late Neolithic of the Tisza Region.* Raczky *Budapest-Szolnok,* 11 – 30.

Kalicz-Schreiber, R. (1976): Die Probleme der Glockenbecherkultur in Ungarn, in: Lanting, J. N. and van der Waals, J. D. (eds.), *Glockenbecher Symposium. Oberried 1974.* Fibula-Van Dishoeck, Bussum/Haarlem, 183 – 215.

Kalicz-Schreiber, R. and Kalicz, N. (2001): Were the Bell Beakers as Social Indicators of the Early Bronze Age in Budapest, in: Nicolis, F. (ed.), *Bell Beakers today. Pottery, people, culture, symbols in prehistoric Europe. Proceedings of the International Colloquium Riva del Garda (Trento, Italy) 11 – 16 May 1998, Vol. 2.* Provincia Autonoma di Trento, Trento, 439 – 458.

Kalicz-Schreiber, R. and Kalicz, N. (2002): The Role of Bell Beakers in Reflecting social Relations in the Early Bronze Age of Budapest, *Archaeologiai Értesitö 125*, 45 – 78.

Kibbert, K. (1980): *Die Äxte und Beile im mittleren Westdeutschland I. PBF IX, Vol. 10*. Beck, München.

Kienlin, T. L. (2008): *Frühes Metall im nordalpinen Raum. Eine Untersuchung zu technischen und kognitiven Aspekten früher Metallurgie anhand der Gefüge frühbronzezeitlicher Beile. Universitätsforschungen zur prähistorischen Archäologie 162*. Habelt, Bonn.

Kienlin, T. L. and Pernicka, E. (2009): Aspects of the Production of Copper Age Jászladány Type Axes, in: Kienlin, T. L. and Roberts, B. W. (eds.), *Metals and Societies. Studies in honour of Barbara S. Ottaway. Universitätsforschungen zur prähistorischen Archäologie 169*. Habelt, Bonn, 258 – 276.

Kienlin, T. L., Bischoff, E. and Opielka, H. (2006): Copper and Bronze during the Eneolithic and Early Bronze Age: A Metallographic Examination of Axe from the Northalpine Region. *Archaeometry 48/3*, 453 – 468.

Kim, J.-I. (2005): *Formation and Change in Individual Identity between the Bell Beaker Culture and the Early Bronze Age in Bavaria, South Germany. BAR 1450*, Oxford.

Kinnear, P. R. and Gray, C. D. (2006): *SPSS 14 Made Simple*. Psychology Press, New York.

Klassen, L. (2000): *Frühes Kupfer im Norden*. Aarhus University Press, Aarhus.

Klassen, L. and Stürup, S. (2001): Decoding the Riesebusch-copper: Lead-Isotope Analysis applied to Early Neolithic Copper Finds from South Scandinavia. *Praehistorische Zeitschrift 76*, 55 – 73.

Klassen, L., Matuschik, I., Pernicka, E. and Rassmann, K. (1998): Die frühesten Metallfunde in Mecklenburg-Vorpommern im Lichte neuer Metallanalysen. Vom Endneolithikum bis zur Bronzezeit. *Jahrbuch der Bodendenkmalpflege in Mecklenburg-Vorpommern 1997-45*, 41 – 67.

Klein, S. and Lechtman, H. (1999): The Production of Copper-Arsenic Alloys (Arsenic Bronze) by Cosmelting: Modern Experiment, Ancient Practice. *Journal of Archaeological Science 26*, 497 – 526.

Kossinna, G. (1926): *Ursprung und Verbreitung der Germanen in vor- und frühgeschichtlicher Zeit. Irminsul 1*. Germanen-Verlag, Berlin-Licherfelde.

Köster, C. (1966): Beiträge zum Endneolithikum und zur Frühen Bronzezeit am nördlichen Oberrhein. *Praehistorische Zeitschrift 43/44*, 2 – 95.

Kovács, T. (1977): *The Bronze Age in Hungary*. Corvina, Budapest.

Kraume, E. (1964): *Kupfer. Die metallischen Rohstoffe ihre Lagerungsverhältnisse und ihre wirtschaftliche Bedeutung*. Enke, Stuttgart.

Krause, R. (1988): *Die endneolithischen und frühbronzezeitlichen Grabfunde auf der Hochterrasse von Singen am Hohentwiel*. Theiss, Stuttgart.

Krause, R. (2003): *Studien zur kupfer- und frühbronzezeitlichen Metallurgie zwischen Karpatenbecken und Ostsee (incl. CD-ROM "Stuttgarter Metallanalysen Datenbank")*. Leidorf, Rahden.

Krause, R. and Pernicka, E (1996): Das neue Stuttgarter Metallanalyseprojekt „SMAP". *Archäologisches Nachrichtenblatt 3*, 274 – 291.

Kreiner, L. (2003): Die Gräber der Glockenbecherkultur im Landkreis Dingolfing-Landau, Niederbayern. *Preistoria Alpina 37. 2001*, 337 – 363.

Kristiansen, K. and Larsson, T. B. (2005): *The Rise of Bronze Age Society. Travels, Transmission and Transformations*, Cambrigde Univerisity Press, Cambridge.

Krut'ová, M. (2003): Bell Beaker and Únětice burial rites. Continuity and change in funerary practices at the beginning of Bronze Age, in: Czebreszuk, J. and Szmyt, M., *The Northern Frontier of Bell Beakers. Proceedings of the symposium held at the Adam Mickiewicz University Poznań (Poland) 26 – 29 May 2002, BAR 1155*, Oxford, 209 – 214.

Kubach, W. (1977): *Die Nadeln in Hessen und Rheinhessen. PBF XIII, Vol. 3*. Beck, München.

Kulcsár, G. (1998): The Distribution of the Makó Culture, in: Ciugudean, H. and Gogáltan, F. (eds.), *The Early and Middle Bronze Age in the Carpathian Basin. Proceedings of the International Symposium in Alba Iulia 24.-28. September 1997*. Muzeul Nat˜io nal al Unirii Alba Iulia, Alba Iulia, 31 – 53.

Kuna, M. (1989): Soziale und ökonomische Faktoren der Entwicklung der frühen Kupfermetallurgie in Südost- und Mitteleuropa, in: Richter, M. and Buchvaldek, M. (eds.), *Das Äneolithikum und die früheste Bronzezeit (C^{14} 3000 – 2000 b.c.) in Mitteleuropa: kulturelle und chronologische Beziehungen. Acta des XIV. Internationalen Symposiums Prag-Liblice 20.-24.10.1986. Praehistorica 15*. Univerzita Karlova, Praha, 33 – 38.

Kuna, M. and Matoušek, V. (1978): Měděná industrie kultury zvoncovitých pohárů ve střední Evropě – Das Kupferinventar der Glockenbecherkultur in Mitteleuropa. *Praehistoria 7.*, 65 – 89.

Kytlicová, O. (1960): Eneolithické pohřebiště v Brandýsku, *Památky archeologické 51*, 442 – 474.

Lanting, J. N. and van der Waals, J. D. (1976): Beaker Culture relationships in the Lower Rhine Basin, in: Lanting, J. N. and van der Waals, J. D. (eds.), *Glockenbecher Symposium. Oberried 1974*. Fibula-Van Dishoeck, Bussum/Haarlem, 1 – 80.

Lauermann, E. (2003): *Studien zur Aujetitz-Kultur im nördlichen Niederösterreich. Universitätsforschungen zur prähistorischen Archäologie 99*, Bonn.

Lechtman, H. (1996): Arsenic Bronze. Dirty Copper or Chosen Alloy? A View from the Americas. *Journal of Field Archaeology 23/4*, 477 – 514.

Lehrberger, G., Fridrich, J., Gebhard, R. and Hrala, J. (eds.) (1997): *Das Prähistorische Gold in Bayern, Böhmen und Mähren. Herkunft - Technologie - Funde, Památky archeologické, Supplementum 7*. Praha.

Lenneis, E., Neugebauer-Maresch, Ch. and Ruttkay, E. (1995): *Jungsteinzeit im Osten Österreichs*. Niederösterreichisches Pressehaus, St. Pölten.

Levy, T. E., Adams, R.B., Hauptmann, A., Prange, M., Schmitt-Strecker, S. and Najjar, M. (2002): Early Bronze Age Metallurgy: a newly discovered copper manufactory in southern Jordan. *Antiquity 76*, 425 – 437.

Lichardus, J. (1991a): Die Kupferzeit als historische Epoche. Ein forschungsgeschichtlicher Überblick, in: Lichardus, J. (ed.), *Die Kupferzeit als historische Epoche. Saarbrücker Beiträge zur Altertumskunde 55*. Habelt, Bonn, 13 – 32.

Lichardus, J. (1991b): Die Kupferzeit als historische Epoche. Versuch einer Deutung, in: Lichardus, J. (ed.), *Die Kupferzeit als historische Epoche. Saarbrücker Beiträge zur Altertumskunde 55*. Habelt, Bonn, 763 – 800.

Lichter, C. (2001): *Untersuchungen zu den Bestattungssitten des südosteuropäischen Neolithikums und Chalkolithikums*. von Zabern, Mainz.

Lißner, B. (2004): *Analyse der frühbronzezeitlichen Gruppen in Süddeutschland. Archäologie Digital 2(2)*, Freiburg.

Liversage, D. (1994): Interpreting composition Patterns in Ancient Bronze: The Carpathian Basin, *Acta Archaeologica 65*, 57 – 134.

Lock, G. (1991): An Introduction to Statistics for Archaeologists, in: Ross, S., Moffet, J. and Henderson, J., *Computing for Archaeologists*. Oxford University, Oxford, 57 – 95.

Lorscheider, F., Maass, A. and Steiniger, D. (2003): Frühe Kupferproduktion – archäologischer Befund und Experiment: Versuche zur Fahlerzverhüttung in einem einzigen Ofengang, in: Stöllner, T., Körlin, G., Steffens, G. and Cierny, J. (eds.), *Man and Mining – Mensch und Bergbau. Studies in honour of Gerd Weisgerber on occasion of the 65th birthday. Der Anschnitt. Beiheft 16*. Deutsches Bergbaumuseum, Bochum, 301 – 307.

Lucas, H. (1965): *Kataloge zur mitteldeutschen Schnurkeramik. Teil II: Saalemündungsgebiet*. Deutscher Verlag der Wissenschaft, Berlin.

Lüning, J. (1996): Neue Gedanken zur Benennung der neolithischen Perioden. *Germania 74*, 233 – 237.

Lutz, J. and Pernicka, E. (1996): Energy Dispersive X-ray Fluorescence Analysis of Ancient Copper Alloys: Empirical Values for Precision and Accuracy. *Archaeometry 38/2*, 313 – 232.

Mainberger, M. (1998): *Das Moordorf von Reute. Steinzeit in Oberschwaben*. Teraqua CAP, Staufen i. Br.

Maran, J. (1998a): *Kulturwandel auf dem griechischen Festland und den Kykladen im späten 3. Jahrtausend v. Chr. Universitätsforschungen zur prähistorischen Archäologie 53*. Habelt, Bonn.

Maran, J. (1998b): Die Badener Kultur und der ägäisch-anatolische Bereich. Eine Neubewertung eines alten Forschungsproblems. *Germania 76/2*, 497 – 525.

Mason, B. (1966): *Principles of Geochemistry*. 3rd edition. Wiley and Sons, New York.

Matthias, W. (1968): *Kataloge zur mitteldeutschen Schnurkeramik. Teil III: Nordharzgebiet*. Deutscher Verlag der Wissenschaft, Berlin.
Matthias, W. (1974): *Kataloge zur mitteldeutschen Schnurkeramik. Teil IV: Südharz-Unstrut-Gebiet*. Deutscher Verlag der Wissenschaft, Berlin.

Matthias, W. (1982): *Kataloge zur mitteldeutschen Schnurkeramik. Teil V: Mittleres Saalegebiet*. Deutscher Verlag der Wissenschaft, Berlin.

Matthias, W. (1987): *Kataloge zur mitteldeutschen Schnurkeramik. Teil VI: Restgebiete und Nachträge*. Deutscher Verlag der Wissenschaft, Berlin.

Matuschik, I. (1990): *Die neolithische Besiedlung in Riekofen-„Kellerfeld". Beiträge zur Kenntnis des Spätneolithikums im südlichen Bayern. Unpublished PhD-thesis at the Albert-Ludwigs Universität Freiburg*, Freiburg.

Matuschik, I. (1991): Grabenwerke des Spätneolithikums in Süddeutschland. *Fundberichte aus Baden-Württemberg 16*, 27 – 55.

Matuschik, I. (1992): Die Chamer Kultur Bayerns und ihre Synchronisation mit den östlich und südöstlich benachbarten Kulturen. *Studia Praehistoria 11-12*, 200 – 220.

Matuschik, I. (1996): Brillen- und Hakenspiralen der frühen Metallzeit Europas, *Germania 74/1*, 1 – 43.

Matuschik, I. (1997a): Eine donauländische Axt vom Typ Şiria aus Überlingen am Bodensee. Ein Beitrag zur Kenntnis des frühesten kupferführenden Horizontes im zentralen Nordalpengebiet. Mit einem Beitrag von J. Matschullat. *Praehistorische Zeitschrift 72*, 81 – 105.

Matuschik (1997b): Der neue Werkstoff – Metal, in: Archäologisches Landesmuseum Baden-Württemberg (ed.), *Goldene Jahrhunderte. Die Bronzezeit in Südwestdeutschland. ALManach 2.* Archäologisches Landesmuseum Baden-Württemberg, Stuttgart, 16 – 25.

Matuschik, I. (1998): Kupferfunde und Metallurgie-Belege, zugleich ein Beitrag zur Geschichte der kupferzeitlichen Dolche Mittel-, Ost- und Südeuropas, in: Mainberger, M., *Das Moordorf von Reute. Steinzeit in Oberschwaben.* Teraqua CAP, Staufen i. Br., 207 – 261.

Matuschik, I. (2004): Kupfer der Glockenbecherkultur im Nordalpengebiet. Zur endneolithischen Kupfergewinnung in den nordöstlichen Alpen, in: Weisgerber, G. and Goldenberg, G. (eds.) *Alpenkupfer – Rame delle Alpi. Der Anschnitt, Beiheft 17.* Deutsches Bergbaumuseum, Bochum, 285 – 302.

Mayer, E. F. (1977): *Die Äxte und Beile in Österreich, PBF IX, Vol. 9.* Beck, München.

Maynard, J. B. (1983): *Geochemistry of sedimentary ore deposits.* Springer, New York.

McKerell, H. and Tylecote, R. F. (1972): The Working of Copper-Arsenic Alloys in the Early Bronze Age and the Effects on the Determination of Provenance. *Proceedings of the Prehistoric Society 38*, 209 – 218.

McNairn, B. (1980): *The Method and Theory of V. Gordon Child.* Edinburgh University Press, Edinburgh.

Menghin, O. (1937): *Die vorgeschichtlichen Funde Vorarlbergs.* Rohrer, Baden bei Wien.

Merkl, M. (2005): *Die Metallurgie der Glockenbecher-kultur in Bayern. Unpublished master's dissertion at the Institut für Ur- und Frühgeschichte und Archäologie des Mittelalters at the Albert-Ludwigs-Universität Freiburg*, Freiburg.

Metzinger-Schmitz, B. (2004): *Die Glockenbecherkultur in Mähren und Niederösterreich. Typologische und chronologische Studien auf dem Hintergrund der kulturhistorischen Abläufe während der späten Kupferzeit im Untersuchungsgebiet. Mit einem paläo-metallurgischen Exkurs*, Saarbrücken. http://scidok.sulb.uni-saarland.de/volltexte/2004/320/ [22.07.2006]

Midgley, M. S. (1992): *TRB Culture. The First Farmers of the North European Plain*, Edinburgh University Press, Edinburgh.

Moesta, H. (2004): Bermerkungen zu bronzezeitlichen Metallen mit hohem Gehalt an Arsen und/oder Antimon, den sog. Fahlerzen, in: Weisgerber, G. and Goldenberg, G. (eds.) *Alpenkupfer – Rame delle Alpi. Der Anschnitt, Beiheft 17.* Deutsches Bergbaumuseum, Bochum, 269 – 272.

Montero-Ruiz, I. and Rodríguez de la Esperanza, M. J. (2004): Der prähistorische Kupferbergbau in Spanien. Ein Überblick über den Forschungsstand. *Der Anschnitt 56/2-3*, 54 – 63.

Moosleitner, F. and Moesta, H. (1988): Vier Spangenbarrendepots aus Obereching, Land Salzburg. *Germania 66/1*, 29 – 67.

Mortimer, Ch. E. (1983): *Chemie. Basiswissen der Chemie in Schwerpunkten. Mit Übungsaufgaben.* Translation: Schilling, G. and Schweizer, H. J, 4th edition. Thieme, Stuttgart.

Moucha, V. (1963): Die Periodisierung der Úněticer Kultur in Böhmen. *Sbornik čs. Společnosti archeologické 3*, 9 – 60.

Moucha, V. (1989): Böhmen am Ausklang des Äneolithikums und am Anfang der Frühbronzezeit, in: Richter, M. and Buchvaldek, M. (eds.), *Das Äneolithikum und die früheste Bronzezeit (C^{14} 3000 – 2000 b.c.) in Mitteleuropa: kulturelle und chronologische Beziehungen. Acta des XIV. Internationalen Symposiums Prag-Liblice 20.-24.10.1986. Praehistorica 15.* Univerzita Karlova, Praha, 213 – 218.

Moucha, V. (2005): *Hortfunde der frühen Bronzezeit in Böhmen.* Archaeologicky Ústav, Praha.

Mozsolics, A. (1967): *Bronzezeit des Karpatenbeckens. Depotfundhorizonte von Hajdúsámson und Kosziderpadlás.* Akadémiai Kiadó, Budapest.

Müller, A. (1993): *Die Glockenbecherkultur in Bayern unter besonderer Berücksichtigung der Begleitkeramik. Unpublished master's dissertion at the Institut für Ur- und Frühgeschichte und Archäologie des Mittelalters Freiburg*, Freiburg.

Müller, A. (2001): Gender Differentiation in burial rites and grave-goods in the Eastern or Bohemian-Morovian Group of the Bell Beaker Culture, in: Nicolis, F. (ed.), *Bell Beakers today. Pottery, people, culture, symbols in prehistoric Europe. Proceedings of the International Colloquium Riva del Garda (Trento, Italy) 11 – 16 May 1998, Vol. 2.* Provincia Autonoma di Trento, Trento, 589 – 599.

Müller, J. (1999): Zur Radiocarbondatierung des Jung- bis Endneolithikums und der Frühbronzezeit im Mittelelbe-Saale-Gebiet (4100 – 1500 v. Chr.), *Berichte der Römisch-Germanischen Kommission 80*, 31 –90.

Müller, J. (2001): *Soziochronologische Studien zum Jung- und Spätneolithikum im Mittelelbe-Saale-Gebiet (4100 – 2700 v. Chr.). Eine sozialhistorische Interpretation prähistorischer Quellen.* Leidorf, Rahden.

Müller, J. (2002): Modelle zur Einführung der Zinnbronzetechnologie und zur sozialen Differenzierung der mitteleuropäischen Frühbronzezeit, in: Müller, J. (ed.) *Vom Endneolithikum zur Frühbronzezeit: Muster sozialen Wandels? (Tagung Bamberg 14.-16. Juni 2001). Universitätsforschungen zur prähistorischen Archäologie 90.* Habelt, Bonn, 267 – 289.

Müller, J. (ed.) (2002): *Vom Endneolithikum zur Frühbronzezeit: Muster sozialen Wandels? (Tagung Bamberg 14.-16. Juni 2001). Universitätsforschungen zur prähistorischen Archäologie 90.* Habelt, Bonn.

Müller, J. and van Willigen, S. (2001): New radiocarbon evidence for European Bell Beakers and the consequences for the diffusion of the Bell Beaker Phenomenon, in: Nicolis, F. (ed.), *Bell Beakers today. Pottery, people, culture, symbols in prehistoric Europe. Proceedings of the International Colloquium Riva del Garda (Trento, Italy) 11 – 16 May 1998, Vol. 2.* Provincia Autonoma di Trento, Trento, 59 – 80.

Needham, S. (2002): Analytical Implications for Beaker Metallurgy in North-West Europe, in: Bartelheim, M., Pernicka, E. and Krause, R. (eds.), *Die Anfänge der Metallurgie in der Alten Welt. The Beginnings of Metallurgy in the Old World. Forschungen zur Archäometrie und Altertumswissenschaft. Vol. 1.* Leidorf, Rahden, 99 – 133.

Niederschlag, E. and Pernicka, E. (2002): The provenance of Early Bronze Age copper from Bohemia and central Germany: first results from physic-chemical investigations, in: Ottaway, B. S. and Wager, E. C. (eds.): *Metals and Society. Papers from a session held at the European Association of Archaeologists Sixth Annual Meeting in Lisbon 2000.* BAR Series 1061, 51 – 59.

Neugebauer, J.-W. (1994): *Bronzezeit in Ostösterreich.* Niederösterreichisches Pressehaus, St. Pölten.

Neugebauer, J.-W. and Neugebauer-Maresch, Chr. (1998): Zum Forschungsstand der Glockenbecherkultur in Ostösterreich, in: Fritsch, B., Maute, M., Matuschik, I., Müller, J. and Wolf, C. (eds.), *Tradition und Innovation.* Prähistorische Archäologie als historische Wissenschaft. Festschrift für Chr. Strahm, Studia Honoraria Bd. 3. Leidorf, Rahden, 307 – 324.

Nocete, F., Queipo, G., Sáez,, R., Nieto, J.M., Inácio, N., Bayona, M.R., Peramo, A., Vargas, J.M., Cruz-Auñón, Gil-Ibarguchi, J.I. and Santos, J.F. (2008): The Smelting Quarter of Valencina de la Concepción (Sevilla, Spain). The Specialised Copper Industry in a Political Centre of the Guadalquivir Valley during the Third millennium BC (2750 – 2550 BC). *Journal of Archaeological Science 35*, 717 – 732.

Northover, J. P. (1989): Properties and Use of Arsenic-Copper Alloys, in: Hauptmann, A., Pernicka, E. and Wagner, G.A. (eds.), *Old World Archaeometallurgy. Proceedings of the International Symposium "Old World Archaeometallurgy" Heidelberg 1987.* Deutsches Bergbaumuseum, Bochum, 111 – 118.

Northover, J. P., O´Brien W. and Stos, S., (2001): Lead Isotopes and Metal Circulation in Beaker/Early Bronze Age Ireland. *Journal of Irish Archaeology X*, 25 – 47.

Norušis, M. J. (2008): *SPSS 16.0 Guide to Data Analysis.* Prentice Hall, Upper Saddle River.

Novotná, M. (1955): Medené nástroje a problem najstaršej Ťažby medi. *Slovenská Archeológia III*, 70 – 100.

Novotná, M. (1970): *Äxte und Beile in der Slowakei. PBF IX, Vol. 3.* Beck, München.

Novotná, M. and Novotný, B. (1996): Karparten- und Donauregion in der Frühen und Mittleren Bronzezeit. *Acta Archaeologica 67*, 87 – 98.

Obereder, J. Pernicka, E. and Ruttkay, E. (1993): Die Metallfunde und die Metallurgie der kupferzeitlichen Mondseegruppe. Ein Vorbericht. *Archäologie Österreichs 4/2*, 5 – 9.

O'Brien, W. (2004): *Ross Island. Mining, Metal and Society in Early Ireland. Bronze Age Studies 6.* National University of Ireland, Galway.

Ottaway, B. S. (1974): Cluster analysis of Impurity Patterns in Armorico-British Daggers. *Archaeometry 16*, 221 – 231.

Ottaway, B. S. (1978a): *Aspects of the Earliest Copper Metallurgy in the Northern sub-Alpine Area in its Cultural Setting, Vol. I & II. Unpublished PhD-Thesis at the University of Edinburgh,* Edinburgh.

Ottaway, B. S. (1978b): Frühes Arsenkupfer im nordalpinen Raum. *Thracia Praehistorica Supplementum 3*, 177 – 188.

Ottaway, B. S. (1982): *Earliest Copper Artefacts of the Northalpine Region: Their Analysis and Evolution.* Seminar für Urgeschichte, Bern.

Ottaway, B. S. (1992): Copper artefacts oft he Corded Ware complex. In: Buchvaldek, M. and Strahm, Chr. (eds.), *Die kontinentaleuropäischen Gruppen der Kultur mit Schnurkeramik. Schnurkeramik – Symposium 1990, Praehistorica 19.* Univerzita Karlova, Praha, 283 – 289.

Ottaway, B. S. (1999): *A Changing Place. The Galgenberg in Lower Bavaria from the fifth to the first millennium BC.* BAR 752, Oxford.

Ottaway, B. S. (2001): Innovation, production and specialization in early prehistoric copper metallurgy. *European Journal of Archaeology 4/1*, 87 – 112.

Ottaway, B. S. and Roberts, R. (2008): The Emergence of Metalworking, in: Jones, A. (ed.), *Prehistoric Europe: theory and practice.* Blackwell, Edinburgh, 193 – 225.

Otto, H. and Witter, W. (1952): *Handbuch der ältesten vorgeschichtlichen Metallurgie in Mitteleuropa.* Barth, Leipzig.

Parkinson, W. A. (2006): *The Social Organization of Early Copper Age Tribes on the Great Hungarian Plain.* BAR 1573, Oxford.

Parzinger, H. (1992): Hornstaad – Hlinsko – Stollhof. Zur absoluten Datierung eines vor-Baden-zeitlichen Horizontes. *Germania 70/2*, 241 – 257.

Pare, Chr. (2000): Bronze and the Bronze Age, in: Pare, Chr. (ed.), *Metals Make the World Go Round. The Supply and Circulation of Metals in Bronze Age Europe. Proceedings of a conference held at the University of Birmingham in June 1997.* Oxbow, Oxford, 1 – 38.

Pástor, J. (1965): Frühbronzezeitliches Gräberfeld in Všechsvätych. *Folia Archaeologica 17*, 37 – 50.

Pástor, J. (1969): *Košické Pohrebisko.* Východoslovenské Múzeum Košice, Košické.

Pászthory, K. and Meyer, E. F. (1998): *Die Beile und Äxte in Bayern. PBF IX, Vol. 20.* Steiner, Stuttgart

Patay, P. (1938): *Korai Bronzkori Kultúrák Magyarországon. Frühbronzezeitliche Kulturen in Ungarn, Dissationes Pannonicae II 13*, Budapest.

Patay, P. (1961): *A Bodrogkereszturi Kultura Temetöi.* Magyar Nemzeti múzeum, Budapest.

Patay, P. (1975): Die hochkupferzeitliche Bodrogkeresztúr-Kultur. *Berichte der Römisch-Germanischen Kommission 55, 1974*, 1 – 72.

Patay, P. (1984): *Kupferzeitliche Meißel, Beile und Äxte in Ungarn. PBF IX, Vol. 15.* Beck, München.

Patay, P., Zimmer, K., Szabó, Z. and Sinay, G. (1963): Spektrographische und Metallographische Untersuchung Kupfer- und Frühbronzezeitlicher Funde. *Acta Archaeologica Academiae Scientiarum Hungaricae 15*, 37 – 64.

Pearce, M. (2009): How much Metal was there in Circulation in Copper Age Italy? in: Kienlin, T. L. and Roberts, B. W. (eds.), *Metals and Societies. Studies in honour of Barbara S. Ottaway.* Habelt, Bonn, 277 – 284.

Pernicka, E. (1984): Instrumentelle Multi-Elementanalyse archäologischer Kupfer- und Bronzeartefakte: Ein Methodenvergleich. *Jahrbuch des Römisch Germanischen Zentralmuseums 31,1984*, 517 – 531.

Pernicka, E. (1989): Erzlagerstätten in der Ägäis und ihre Ausbeutung im Altertum. Geochemische Untersuchungen zur Herkunftsbestimmung archäologischer Metallobjekte. *Jahrbuch des Römisch Germanischen Zentralmuseums 34/2, 1987*, 607 – 714.

Pernicka, E. (1995): Gewinnung und Verbreitung der Metalle in prähistorischer Zeit, *Jahrbuch des Römisch Germanischen Zentralmuseums 37/1, 1990*, 21 – 129.

Pernicka, E. (1999): Trace Element Fingerprinting of Ancient Copper. A Guide to Technology or Provenance? In: Young, S. M. M., Pollard, A. M., Budd, P. and Ixer, R. A.: *Metals in Antiquity. BAR S792*, Oxford, 22 – 32

Pernicka, E. (2000): Isotope archchaeology, in: Rammelmair, D., Mederer, J., Oberthür, Th. Heimann, D. B. and Pentinghaus, H. (eds.), *Applied Mineralogy in Research, Economy, Technology, Ecology and Culture. Vol. 2.* A. A. Balkema, Rotterdam/Brookfield, 1025 – 1028.

Pernicka, E., Begemann, F., Schmitt-Strecker, S. and Wagner, G.A. (1993): Eneolithic and Early Bronze Age copper and their relation to Serbian copper ores. *Praehistorische Zeitschrift 68*, 1 – 54.

Peška, J. and Šebela, L. (1992): Funde der Chłopice-Vesele-Gruppe in Mähren und ihre Stellung in der Endphase des Endneolithikums, in: Buchvaldek, M. and Strahm, Chr. (eds.), *Die kontinentaleuropäischen Gruppen der Kultur mit Schnurkeramik. Schnurkeramik – Symposium 1990, Preahistorica 19.* Univerzita Karlova, Praha, 131 – 141.

Petrascheck, W. and Pohl, W. (1982): *Lagerstättenlehre. Eine Einführung in die Wissenschaft von den mineralischen Bodenschätzen.* 3rd edition. Schweizerbart, Stuttgart.

Pike, A. W. G. and Richards, M. P. (2002): Diagenetic Arsenic Uptake in Archaeological Bone. Can we really Identify Copper Smelters? *Journal of Archaeological Science 29*, 607 – 611.

Pittioni, R. (1954): *Urgeschichte des österreichischen Raumes.* Deuticke, Wien.

Pittioni, R. (1957): *Urgeschichtlicher Bergbau auf Kupfererz und Spurenanalyse. Beiträge zum Problem der Relation Lagerstätte – Fertigobjekt. Archaeologia Austriaca. Beiheft 1*, Wien.

Pleslová-Štiková, E. (1985): *Makotřasy. A TRB Site in Bohemia*, Museum Nationale Pragae, Praha.

Pollard, A. M. and Heron, C. (2008): *Archaeological Chemistry*. Royal Society of Chemistry, Cambridge.

Pollard, A. M., Thomas, R.G. and Williams, P.A. (1991): Some Experiments Concerning the Smelting of Arsenical Copper, in: Budd, P., Chapmann, B. and Jackson, C., (eds.): *Archaeological Sciences 1989*. Oxbow, Oxford, 169 – 174.

Prange, M. (2001): *Vergleichende Untersuchungen zur Charakterisierung des omanischen Kupfers mittels chemischer und isotopischer Analysemethoden. 5000 Jahre Kupfer im Oman, Vol. II, Metalla 8 1/2*, Bochum.

Price, T. D., Knipper, C., Grupe, G. and Smrcka, V. (2004): Strontium Isotopes and perhistoric human migration: the Bell Beaker period in central Europe. *European Journal of Archaeology 7*, 9 – 40.

Raetzel-Fabian, D. (2001): Anmerkungen zur Interpretation von ^{14}C-Daten. In: Czebreszuk, J. and Müller, J. (eds.) *Die absolute Chronologie in Mitteleuropa 3000 – 2000 v. Chr.*. Leidorf, Rahden, 11 – 23.

Rapp Jr., G. (1988): On the Origins of Copper and Bronze Alloying, in: Maddin, R. (ed.), *The Beginning of the Use of Metals and Alloys*. MIT Press, Massachusetts, 21 – 27.

Rassmann, K. (1996): Zum Forschungsstand der absoluten Chronologie der frühen Bronzezeit in Mitteleuropa auf der Grundlage von Radiokarbondaten. *Acta Archaeologica 67*, 199 – 210.

Rassmann, K. (2005): Zur Chronologie der Hortfunde der Klassischen Aunjetitzer Kultur. Eine Auswertung von Metallanalysen aus dem Forschungsvorhaben „Frühe Metallurige im zentralen Mitteleuropa", in: Horejs, B., Jung, R., Kaiser, E. and Teržan, B. (eds.), *Interpretationsraum Bronzezeit. Bernhard Hänsel von seinen Schülern gewidmet. Universitätsforschungen zur prähistorischen Archäologie 121*. Habelt, Bonn, 463 – 480.

Rau, P. and Willing, B. (1991): Statistische und archäologische Bearbeitung von prähistorischen Kupferanalysen. Ein neues Stuttgarter Metallanalysen-projekt, *Archäologisches Korrespondenzblatt 21*, 357 – 367.

Regenye, J. (ed.) (2001): *Sites and stones. Lengyel culture in Western Hungary and beyond. A review of the current research*, Veszprém Megyei Múzeum Igazgotóság, Veszprém.

Reinecke, P. (1930): Die Stufe der spätneolithischen Glockenbecherkultur in Bayern. *Bayerischer Vorgeschichtsfreund 9*, 16 – 29.

Renfrew, C. (1986): Varna and the emergence of wealth in prehistoric Europe, in: Appadurai, A. (ed.), *The social life of things. Commodities in Cultural Perspective*. Cambridge University Press, Cambridge, 141 – 146.

Richter, M. and Buchvaldek, M. (eds.) (1989): *Das Äneolithikum und die früheste Bronzezeit (C^{14} 3000 – 2000 b.c.) in Mitteleuropa: kulturelle und chronologische Beziehungen. Acta des XIV. Internationalen Symposiums Prag-Liblice 20.-24.10.1986. Praehistorica 15*. Univerzita Karlova, Praha.

Říhovský, J. (1992): *Die Äxte, Beile, Meißel und Hämmer in Mähren. PBF. IX, Vol. 17*. Steiner, Stuttgart.

Robb, L. (2005): *Introduction to Ore-Forming Processes*. Blackwell, Oxford.

Roberts, B. (2008a): Migration, Craft Expertise and Metallurgy: Analysing the 'Spread' of Metal in Western Europe. *Archaeological Reviews from Cambridge 23/2*, 27 – 45.

Roberts, B. (2008b): Creating traditions and shaping technologies: understanding the earliest metal objects and metal production in Western Europe. *World Archaeology 40/3*, 354 – 372.

Rosenqvist, T. (1983): *Principles of Extractive Metallurgy*. 2nd edition. McGraw-Hill, New York.

Rothenberg, B. (1990): Copper Smelting Furnaces, Tuyeres, Slags, Ingot-Moulds and Ingots in the Arabah: Archaeological Data, in: Rothenberg, B. (ed.) (1990): *The Ancient Metallurgy of Copper. Archaeology – Experiment – Theory*. Institute for Archaeo-Metallurgical Studies, London, 1 – 77.

Rothenberg, B. (ed.) (1990): *The Ancient Metallurgy of Copper. Archaeology – Experiment – Theory*. Institute for Archaeo-Metallurgical Studies, London.

Rostoker, W. and Dvorak, J. R. (1991): Some Experiments with Co-smelting to Copper Alloys. *Archeomaterials 5*, 10 – 37.

Rostoker, W., Pigott, V.C. and Dvorak, J.R. (1989): Direct reduction to copper metal by oxide-sulfide mineral interaction, *Archaeomaterials 3/1*, 69 – 87.

Rovira Llorens, S. and Delibes de Castro, G. (2005): Beaker Metallurgical Technology in the Iberian Peninsula, in: García Martínez de Lagrán, Í., Garrido Pena, R. and Rojo Guerra, M. A. (eds.), *El Campaniforme en la Península Ibérica y su Contexto Europeo. Bell Beakers in the Iberian Peninsula and their European Context*. Universidad de Valladolid, Valladolid, 513 – 521.

Rovira, S. and Montero, I. (2003): Natural Tin-Bronze Alloy in Iberian Peninsula Metallurgy: Potentiality and Reality, in: Guilmia-Mair, A. and Lo Schiavo, F. (eds.): *The Problem of Tin.* BAR 1199, Oxford, 15 – 22.

Ruckdeschel, W. (1978a): *Die frühbronzezeitlichen Gräber Südbayerns. Ein Beitrag zur Kenntnis der Straubinger Kultur. Text. Antiquitas II.* Habelt, Bonn.

Ruckdeschel, W. (1978b): *Die frühbronzezeitlichen Gräber Südbayerns. Ein Beitrag zur Kenntnis der Straubinger Kultur. Katalog, Antiquitas II.* Habelt, Bonn.

Ruhrmann, J. (1925): Über Arsen und Nickel sowie deren Sauerstoffverbindungen im Kupfer und ihren Einfluß in geringen Mengen auf seine mechanischen Eigenschaften, *Metall und Erz 22/14*, 339 – 348.

Ruttkay, E. (1981): Typologie und Chronologie der Mondsee-Gruppe. In: Straub, D., *Das Mondsee-Land. Geschichte und Kultur.* Oberösterreichischer Landesverlag, Linz, 269 – 294.

Ryndina, N., Indenbaum, G. and Kolosova, V. (1999): Copper Production from Polymetallic Sulphide Ores in the Northeastern Balkan Eneolithic Culture. *Journal of Archaeological Science 26*, 1059 – 1068.

SAM I: Junghans, S., Sangmeister, E. and Schröder, M. (1960): *Metallanalysen kupferzeitlicher und frühbronzezeitlicher Bodenfunde aus Europa.* Mann, Berlin.

SAM II: Junghans, S., Sangmeister, E. and Schröder, M. (1968 & 1974): *Kupfer und Bronze in der frühen Metallzeit Europas.* Vol. 1 – 4. Mann, Berlin.

Sangmeister, E. (1951): *Die Jungsteinzeit im nordmainischen Hessen. Die Glockenbecherkultur und die Becherkulturen.* Bernecker, Melsungen.

Sangmeister, E. (1957): Ein geschlossener Becherfund im Museum Cordova (Ein Beitrag zur Untersuchung der ausserspanischen Beziehungen der Glockenbecherkultur). *Zephyrus 8*, 257 – 267.

Sangmeister, E. (1966a): Die Glockenbecher im Oberrheintal. *Jahrbuch des Römisch-Germanischen Zentralmuseums 11, 1964*, 81 – 114.

Sangmeister, E. (1966b): Die Datierung des Rückstroms der Glockenbecher und ihre Auswirkung auf die Chronologie der Kupferzeit in Portugal. *Palaeohistoria 12*, 397 – 407.

Sangmeister, E. (1971): Aufkommen der Arsenbronze in SO-Europa. *Actes du VIIIe Congres International des Sciences Prehistoriques et Protohistoriques. Beograd 9 – 15 Septembre 1971*, Beograd, 109 – 129.

Sangmeister, E. (1972): Sozial-ökonomische Aspekte der Glockenbecherkultur. *Homo 23*, 188 – 203.

Sangmeister, E. (1999): Bemerkungen zur Vergleichbarkeit von Analysegruppen mit archäologischen Daten, in: Hauptmann, A., Pernicka, E., Rehren, Th. and Yalcin, Ü. (eds.), *The Beginning of Metallurgy. Proceedings of the International Conference „The Beginning of Metallurgy", Bochum 1995.* Deutsches Bergbaumuseum, Bochum, 333 – 351.

Sangmeister, E. and Strahm, Chr. (1973): Die Funde aus Kupfer in Seeberg, Burgäschisee-Süd, *Acta Bernensia II/6*, 189 – 260.

Schalk, E. (1998): *Die Entwicklung der prähistorischen Metallurgie im nördlichen Karpatenbecken. Eine typologische und metallanalytische Untersuchung.* Leidorf, Rahden.

Scharbert, H. G. (1984): *Einführung in die Petrologie und Geochemie der Magmatite. Allgemeine Probleme der magmatischen Petrologie und Geochemie, Vol. 1.* Deuticke, Wien.

Schibler, J., Jacomet, S. and Choyke, A. (2004): Neolithic Lake Dwellings in the Alpine Region, in: Bogucki, P. and Crabtree, P.J. (eds.), *Ancient Europe 8000 B.C. – A.D. 1000. Encyclopedia of the Barbarian World. Vol. I. The Mesolithic to Copper Age (c. 8000 – 2000 B.C.).* Thompson and Gale, New York, 385 – 392.

Schickler, H. (1981): ‚Neolithische' Zinnbronze, in: Lorenz, H. (ed.), *Studien zur Bronzezeit. Festschrift für Wilhelm Albert v. Brunn.* von Zabern, Mainz, 419 – 445.

Schlichterle, H. (1990): Aspekte der siedlungsarchäologischen Erforschung von Neolithikum und Bronzezeit im südwestdeutschen Alpenvorland. *Berichte der Römisch-Germanischen Kommission 71/1*, 208 – 244.

Schlichterle, H. (1999): Die Goldberg III Gruppe in Oberschwaben, in: Schlichterle, H. and Strobel, M., *Horgen – Cham – Goldberg III – Schnurkeramik in Süddeutschland. Rundgespräch Hemmenhofen 26. Juni 1998.* Janus, Freiburg, 35 – 48.

Schlichtherle, H. and Rottländer, R. (1982): Gusstiegel der Pfyner Kultur in Südwestdeutschland. *Fundberichte aus Baden-Württemberg 7*, 1982, 59 – 71.

Schliz, A. (1906): Der schnurkeramische Kulturkreis und seine Stellung zu anderen neolithischen Kulturformen in Südwestdeutschland. *Zeitschrift für Ethnologie 38*, 312 – 345.

Schmidt, H. (1909): Der Bronzefund von Canena (Bez. Halle). *Praehistorische Zeitschrift 1*, 113 – 139.

Schmotz, K. (1992): Eine Gräbergruppe der Glockenbecherkultur von Künzing, Lkr. Deggendorf. *Vorträge des 10. Niederbayerischen Archäologentages*, 41 – 68.

Schreiber, R. (1973): *Die Glockenbecherkultur in Budapest*. Budapesti Történeti Múzeum, Budapest.

Schreiner, M. (2007): *Erzlagerstätten im Hrontal, Slowakei. Genese und prähistorische Nutzung*. Leidorf, Rahden.

Schröder, M. (1991): Die Stuttgarter Gruppe. 40 Jahre Metallanalysen vor- und frühgeschichtlicher Bodenfunde. *Fundber. Baden-Württemberg 16*, 1 – 26.

Schubert, E. (1974): Studie zur frühen Bronzezeit an der mittleren Donau. *Berichte der Römisch-Germanischen Kommission 54*, 1 – 105.

Schubert, E. (1981): Zur Frage der Arsenlegierungen in der Kupfer- und Frühbronzezeit Südosteuropas, in: Lorenz, H. (ed.), *Studien zur Bronzezeit. Festschrift für W. A. v. Brunn*. von Zabern, Mainz, 447 – 459.

Selimchanow, I. R. (1977): Zur Frage einer Kupfer-Arsen-Zeit, *Germania 55*, 1 – 6.

Seracsin, A. (1929): Bronzezeitliche Funde aus dem Leithagebirge. *Wiener Praehistorische Zeitschrift XVI*, 95-102.

Shennan, S. (1975a): The social organization at Branč. *Antiquity 49*, 279 – 288.

Shennan, S. (1975b): Die soziale Bedeutung der Glockenbecherkultur in Mitteleuropa. *Acta Archaeologica Carpatica 15*, 173 – 179.

Shennan, S. (1986): Central Europe in the Third Millennium B.C.: An Evolutionary Trajectory for the Beginning of the European Bronze Age. *Journal of Anthropological Archaeology 5*, 115 – 146.

Shennan, S. (1995): *Bronze Age copper production of the Eastern Alps. Excavation at St. Veit-Klinglberg. Universitätsforschungen zur prähistorischen Archäologie 27*. Habelt, Bonn.

Shennan, S. (1997): *Quantifying Archaeology*. 2nd edition. Edinburgh Univerisity Press, Edinburgh.

Sherratt, A. (1981): Plough and pastoralism. Aspects of the secondary products revolution, in: Hammond, N., Hodder, I. and Isaac, G. (eds.): *Pattern of the Past. Studies in honour of D. Clarke*. Cambridge University Press, Cambridge, 261 – 305.

Sherratt, A. (1982): The Development of Neolithic and Copper Age Settlement in the Great Hungarian Plain. Part I: The Regional Setting. *Oxford Journal of Archaeology 1/3, 1982*, 287 – 316.

Sherratt, A. (1983a): The secondary exploitation of animals in the Old World. *World Archaeology 15*, 90 – 104.

Sherratt, A. (1983b): The Development of Neolithic and Copper Age Settlement in the Great Hungarian Plain. Part II: The Survey and Settlement Dynamics. *Oxford Journal of Archaeology 2/1*, 13 – 41.

Sherratt, A. (1987): Cubs that cheered, in: Waldren, W. H. and Kennard, R. C. (eds.), *Bell Beakers of the West Mediterranean. Definition, Interpretation, Theory and New Site Data. Part i*. BAR 331, Oxford, 81 – 114.

Sherratt, A. (1996): „Das sehen wir auch den Rädern ab": some thoughts on M. Vosteen's „Unter die Räder gekommen". *Archäologische Informationen 19/1&2*, 155 – 172.

Sládek, V., Berner, M. and Sailer, R. (2006): Mobility in Central European Late Eneolithic and Early Bronze Age: Femoral Cross-Section Geometry. *American Journal of Physical Anthropology 130*, 320 – 332.

Slater, E. A. and Charles, J. A. (1970): Archaeological Classification by Metal Analysis. *Antiquity 44*, 207 – 213.

Smart, J. S. (1970a): The Physical Properties of Copper, in: Butts, A. (ed.), *Copper. The Science and Technology of the Metal, its Alloys and Compounds*. Hafner, New York, **360 – 378.**

Smart, J. S. (1970b): The Effect of Impurities in Copper, in: Butts, A. (ed.), *Copper. The Science and Technology of the Metal, its Alloys and Compounds*. Hafner, New York, 410 – 416.

Sperl, G. (1975): Metallurgische Untersuchungskriterien metallischer Funde der Kupfer- und Bronzezeit Europas, in: Jeglitsch, F. and Petzow, G., *Fortschritte der Metallographie, Praktische Metallographie 4*. Riederer, Stuttgart, 3 – 26.

Spindler, K. (1971): Zur Herstellung der Zinnbronze in der frühen Metallurgie Europas. *Acta Praehistorica et Archaeologica 2*, 199 – 253.

Stahl, W. (1918): Erörterung und Ergänzung der Angaben über den Einfluß des Antimons auf die mechanischen Eigenschaften des Kupfers, *Metall und Erz 15/21*, 395 – 396.

Staniaszek, B. E. P. and Northover, J. P. (1982): The Properties of Leaded Bronze alloys, in: *Proceedings of the 22nd Symposium on Archaeometry, University of Bradford March 30th – April 3rd 1982*. University of Bradford, Bradford, 262 – 272.

Stöckli, W. (eds.) (1995): *Die Schweiz vom Paläolithikum bis zum Mittelalter. Vom Neandertaler bis zu Karl dem Großen. SPM II. Neolithikum*. Schweizerische Gesellschaft für Ur- und Frühgeschichte, Basel.

Strahm, Chr. (1971): *Die Gliederung der Schnurkeramischen Kultur in der Schweiz.* Acta Bernensia IV. Staempfli, Bern.

Strahm, Chr. (1979): Kalibration und Herkunft der Glockenbecher. *Archive suisse d'anthropologie général 43/2*, 285 – 293.

Strahm, Chr. (1981): Die Bedeutung der Begriffe Kupferzeit und Bronzezeit. *Slovenská Archeológia XXIX*, 191 – 202.

Strahm, Chr. (1982): Zu den Begriffen Chalkolithikum und Metallikum, in: Aspes, A. and Fasani, L. (eds.), *Il Pasaggio dal Neolitico all'Età del Bronzo nell'Europa Centrale e nella Regione Alpina. Atti del X Simposio Internazionale sulla fine del Neolithico e gli inizi dell'etá del Bronzo in Europa. Lazise-Verona 8-12 aprile 1980.* Museo Civico di Storia Naturale, Verona, 13 – 26.

Strahm, Chr. (1988): Chalkolitikum und Metallikum: Kupferzeit und frühe Bronzezeit in Südwestdeutschland und der Schweiz. *Età del Rame (Rassegna Arch.) 7*, 1988, 175 – 192.

Strahm, Chr. (1994): Die Anfänge der Metallurgie in Mitteleuropa, *Helvetia Arch. 25*, 2 – 39.

Strahm, Chr. (2002): Tradition und Wandel der sozialen Strukturen vom 3. zum 2. vorchristlichen Jahrtausend, in: Müller, J. (ed.), *Vom Endneolithikum zur Frühbronzezeit: Muster sozialen Wandels? Tagung Bamberg 14. – 16. Juni 2001.* Habelt, Bonn, 175 – 194.

Strahm, Chr. (2004): Das Glockenbecher-Phänomen aus Sicht der Komplementär-Keramik, in: Czebreszuk, J. (ed.), *Similar but Different. Bell Beakers in Europe.* Adam Mickiewicz University, Poznań, 101 – 126.

Strahm, Chr. and Hauptmann, A. (2009): The Metallurgical Developmental Phases in the Old World, in: Kienlin, T. L. and Roberts, B. W. (eds.), *Metals and Societies. Studies in honour of Barbara S. Ottaway. Universitätsforschungen zur prähistorischen Archäologie 169.* Habelt, Bonn, 116 – 128.

Strahm, Chr. (ed.) (1995): *Das Glockenbecher-Phänomen. Ein Seminar.* Institut für Ur- und Frühgeschichte der Universität Freiburg, Freiburg.

Stroh, F. (1941): Funde der Glockenbecherkultur in Oberdonau. Skelettgräber in Scharlinz. *Wiener Prähistorische Zeitschrift 28*, 74 – 82.

Subramanian, P. R., Chakrabarti, D.J. and Laughlin, D.E. (eds.) (1994): *Phase Diagrams of Binary Copper Alloys.* ASM International, Materials Park, OH.

Sulimirski, T. (1961): Copper Hoard from Horodnica on the Dniester. *Mitteilungen der Anthropologischen Gesellschaft Wien 41*, 91 – 97.

Šumberová, R. and Hora, K. (1992): Typologie des Kupferschmucks und der Kupfergeräte in der schnurkeramischen Kultur Böhmens und Mährens, in: Buchvaldek, M. and Strahm, Chr. (eds.), *Die kontinentaleuropäischen Gruppen der Kultur mit Schnurkeramik. Schnurkeramik – Symposium 1990, Praehistorica 19.* Univerzita Karlova, Praha, 117 – 125.

Taylor, T. (1999): Envaluing metal: theorizing the Eneolithic 'hiatus'. In: Young, S. M. M., Pollard, A. M., Budd, P. and Ixer, R. A.: *Metals in Antiquity.* BAR S792, Oxford, 22 – 32.

Tasić, N. (ed.) (1984): *Kulturen der Frühbronzezeit des Karpatenbeckens und Nordbalkans. Balkanološki Institut Sanu, Sonderausgabe 22*, Beograd.

Thomas, J. (1991): *Rethinking the Neolithic.* Cambridge University Press, Cambridge.

Thomsen, C. J. (1836): *Ledetraat til Nordisk Oldkyndighet.* S.L. Møllers bogtr., Kopenhagen.

Tillmann, A. (1996): Schnurkeramische Bestattungen aus Kösching, Lkr. Eichstätt, und Bergheim, Lkr. Neuburg-Schrobenhausen, Oberbayern. Versuch einer Gliederung der Schnurkeramik in Südbayern, in: Campen, I., Hahn, J. and Uerpmann, M. (eds.), Spuren *der Jagd – Die Jagd nach Spuren. Festschrift für Hansjürgen Müller-Beck.* Mo-Vince, Tübingen, 363 – 380.

Timberlake, S. (2007): The Use of Experimental Archaeology/Archaeometallurgy for the Understanding and Reconstruction of Early Bronze Age Minning and Smelting Technologies, in: La Niece, S., Hook, D. and Craddock, P., *Metals and Mines. Studies in Archaeometallurgy.* Archetype, London, 27 – 36.

Točik, A. (1963): Die Nitra-Gruppe. *Archeologicke Rozhledy XV*, 716 – 774.

Točik, A. (1964): Záchranny výskum v Bajč-Vlkanove 1959-1960. *ŠZ AÚSAV 12*, 5 – 185.

Točik, A. (1979): *Výčapy-Opatovce a dal'šie pohrebiská zo staršej doby bronzovej na juhozápadnom Slovensku/und weitere altbronzezeitliche Gräberfelder in der Südwestslowakei.* Archeologický Ústav Slovenskej Akadémie, Nitra.

Torgmayer, O. (1957): Ásatás Tápé-Lebő. Ausgrabung auf Tápé-Lebő. *SZME 1957*, 19 – 60.

Turek, J. and Černý, V. (2001): Society, gender and sexual dimorphism of the Corded Ware and the Bell Beaker populations, in: Nicolis, F. (ed.), *Bell Beakers today. Pottery, people, culture, symbols in prehistoric Europe. Proceedings of the International Colloquium Riva del Garda (Trento, Italy) 11 – 16 May 1998, Vol. 2.* Provincia Autonoma di Trento, Trento, 601 – 612.

Turek, J. and Peška, J. (2001): Bell Beaker settlement pattern in Bohemia and Moravia, in: Nicolis, F. (ed.), *Bell Beakers today. Pottery, people, culture, symbols in prehistoric Europe. Proceedings of the International Colloquium Riva del Garda (Trento, Italy) 11 – 16 May 1998, Vol. 2*. Provincia Autonoma di Trento, Trento, 411 – 428.

Tylecote, R. F. (1980a): Furnaces, Crucibles, and Slags, in: Wertime, T. A. and Muhly, J. D. (eds.), *The Coming of the Age of Iron*. Yale University Press, New Haven/London, 183 – 228.

Tylecote, R. F. (1980b): Summary of Results of Experimental Work on Early Copper Smelting, in: Oddy, W. A. (ed.), *Aspects of Early Metallurgy*. British Museum, London, 5 – 62.

Tylecote, R. F. (1987): *The Early History of Metallurgy in Europe*. Longman, London/New York.

Tylecote, R. F., Ghaznavi, H.A. and Boydell, P.J. (1977): Partioning of Trace Elements between the Ores, Fluxes and Metal during the Smelting of Copper. *Journal of Archaeological Science 4*, 305 – 333.

Vander Linden, M. (2003): Competing Cosmos. On the Relationships between Corded Ware and Bell Beaker Mortuary Practice, in: Czebreszuk, J. and Szmyt, M. (eds.), *The Northern Frontier of Bell Beakers. Proceedings of the symposium held at the Adam Mickiewicz University Poznań (Poland) 26 – 29 May 2002*. BAR 1155, Oxford, 11 – 19.

Vander Linden, M. (2004): Polythetic Networks, Coherent People: A new historical hypothesis for the Bell Beaker phenomenon, in: Czebreszuk, J. (ed.), *Similar but Different. Bell Beakers in Europe*. Adam Mickiewicz University, Poznań, 35 – 62.

Vander Linden, M. (2007a): For equalities are plural: reassessing the social in Europe during the third millennium BC. *World Archaeology 39/2*, 177 – 193.

Vander Linden, M. (2007b): What linked the Bell Beakers in the third millennium BC Europe? *Antiquity 81*, 343 – 352.

Visy, Z. (2003): *Hungarian Archaeology at the Turn of the Millennium*. Ministry of National Cultural Heritage, Budapest.

Vladár, J. (1971): *Pohrebiská zo staršej doby bronzovej v Branči. МОГИЛЬНИКИ РАНЕБРОНЗОВОГО ВЕКА В БРАНЧЕ. Gräberfelder aus der älteren Bronzezeit in Branč*, Bratislava.

Vladár, J. (1974): *Die Dolche in der Slowakei. PBF VI, Vol. 3*. Beck, München.

Vogt, E. (1948): Die Gliederung der schweizerischen Frühbronzezeit. In: *Festschrift für O. Tschumi zum 22. November 1948*. Huber, Frauenfeld, 53 – 69.

Vollmann, D. (2005): *Studien zum Übergang von der späten Kupferzeit zur frühen Bronzezeit im östlichen Mitteleuropa. Saarbrücker Beiträge zur Altertumskunde 77*. Habelt, Bonn.

Vosteen, M. (1996a): *Unter die Räder gekommen. Untersuchungen zu Sherratts 'Secondary Product Revolution'. Archäologische Berichte 7*. Holos, Bonn.

Vosteen, M. (1996b): Taken the Wrong Way: einige Bemerkungen zu A. Sherratts „Das sehen wir auch den Rädern ab". *Archäologische Informationen 19/1&2*, 173 – 186.

Vulpe, A. (1975): *Die Äxte und Beile in Rumänien II*. PBF IX, Vol. 5. Beck, München.

Wang, Q. and Ottaway, B. S. (2004): *Casting Experiments and Microstructure of archaeologically relevant Bronzes*. BAR 1331, Oxford.

Waterbolk, H. T. and Butler, J. J. (1965): Comments on the Use of Metallurgical Analyses in Prehistoric Studies. *Helinium 5*, 227 – 251.

Wiermann, R. R. (1998): An anthropological approach to burial customs of the Corded Ware in Bohemia, in: Benz, M. and van Willigen, S. (eds.), *Some new Approaches to the Bell Beaker 'Phenomenon' Lost Paradise...? Proceedings of the 2nd Meeting of the "Association Archéologie et Gobelets" Feldberg (Germany), 18th – 20th April 1997*, BAR 690, Oxford, 129 – 140.

Wiermann, R. R. (2004): *Die Becherkulturen in Hessen. Glockenbecher, Schnurkeramik, Riesenbecher. Freiburger Studien 4*. Leidorf, Rahden.

Willvonseder, K. (1937): Zwei Grabfunde der Badener Kultur mit Metallbeigaben aus Niederösterreich. Wiener *Praehistorische Zeitschrift 26*, 15 – 28.

Willvonseder, K. (1968): *Die jungsteinzeitlichen und bronzezeitlichen Pfahlbauten des Attersees in Oberösterreich. Mitteilungen der Prähistorischen Kommission der Österreichischen Akadademie der Wissenschaft, Vol. XI-XII*. Böhlau, Wien.

Wüstemann, H. (1995): *Die Dolche und Stabdolche in Ostdeutschland, PBF VI, Vol. 8*. Steiner, Stuttgart.

Yazawa, A. (1980): Distribution of various elements between copper, matte and slag. *Erzmetall 33*, 377 – 382.

Zalai-Gaál, I. (1996): Die Kupferfunde der Lengyel-Kultur im südlichen Transdanubien. *Acta Archaeologica Academiae Scientiarum Hungaricae 48*, 1 – 34.

Zich, B. (1996): *Studine zur regionalen und chronologischen Gliederung der nördlichen Aunjetitzer Kultur. Vorgeschichtliche Forschungen 20*, de Gruyter, Berlin.

Zhang, T., Ramakrishnan, R. and Livny, M. (1996): BIRCH. An Efficient Data Clustering Method for Very Large Databases, in: Jagadish, H. V. (ed.): *Proceedings of the 1996 ACM SIGMOD International Conference on Management of Data (Montreal)*. ACM Press, New York, 103 – 114.

Zhang, T., Ramakrishnan, R. and Livny, M . (1997): BIRCH. A new Data Clustering Algorithm and its Applications. *Data Mining and Knowledge Discovery 1*, 141 – 182.

Zumstein, H. (1965): L'Age du bronze dans le département du Haut-Rhin. *Revue Archéologique de l'Est et Centre-Est 16*, 7 – 56.

GLOSSARY OF VARIOUS METAL PROPERTIES AND TESTS

ANNEALING is a process of heat treatment of metal in order to change its properties. The work piece is heated above its typical re-crystallisation temperature. In this process, metal softens and cracks are neutralised by re-crystallisation. Cold-working metal causes dislocations within the metal lattice. Annealing is primarily used to improve the working properties and to counteract fracturing.

BEND TEST (or reverse bend test) is used to examine brittleness, fatigue, ductility and cold-working properties. A test sample is stressed by bending until it breaks. Its resistance indicates strength.

BRITTLENESS defines the scale of fracture when objects are stressed. Brittleness can be measured in various ways including by notched-bar impact tests or bend tests.

COLD- AND HOT-SHORTNESS: Cold-shortness is the tendency to crack when metal is hammered or rolled under cold conditions. Hot-shortness is brittleness caused by working under hot conditions.

COLD- AND HOT-WORKING: Working, especially hammering, at low temperatures (below the metal specific re-crystallisation temperature) is called cold-working. Cold-working is used to shape and harden an artefact. If the work piece is heated to above its softening temperature and then processed, this is named hot-working.

DUCTILITY describes the ability of a material to be deformed without cracking. This characteristic is essential for shaping objects, especially drawing wires. Metals with high ductility can be hammered, drawn or bent without breaking.

ELASTICITY describes the ability of metal to return to its original shape after being deformed by stress. This is especially important for tools that undergo force effect during their use life, such as hammers or axes. Elastic objects can withstand harder treatment without deformation or cracking until their elastic limit is reached.

ELONGATION is used to measure the strength and ductility of metals. Materials with high elongation and elasticity are more resistant and can be stressed to a greater degree before breaking.

FATIGUE is a type of damage caused by constant loading. It is tested by fatigue tests which investigate the stability of materials by endurance strain.

HARDNESS reflects the resistance of materials to loading, stress and cutting action. It depends on properties such as ductility, elasticity, toughness and strength. Hardness can be increased by working (e.g. hammering) and by certain impurities, such as arsenic or tin.

MALLEABILITY is the property that enables metals to be formed into various shapes – even into thin sheet metals – without brittleness or cracking. Malleability is similar to ductility, but specifically emcompasses methods of shaping such as hammering (from the Latin *malleus*, hammer).

MELTING POINT (or freezing point) is a material-specific temperature at which materials change their phase from the solid to the liquid stage. The freezing point of copper is 1084 °C.

NOTCHED-BAR IMPACT TEST is used to determine the brittleness, ductility, hardness, strength and toughness of materials. Various types of impact test are applied to measure the energy needed to deform a material. Archbutt and Prytherch (1937, 9) provide an in-depth description of this test.

OVERPOLING means an elimination of all the oxygen by over-lengthy poling. This makes copper more brittle. Poling is a technique to reduce oxygen from copper by stirring the molten metal with a green-wood pole.

RE-CRYSTALLISATION is defined as the process of heat treatment in order to replace grains which have been deformed especially by cold working. Re-crystallisation, as with all kinds of heat treatment, reduces strength and hardness, but increases ductility. The critical temperature when re-crystallisation takes place is about 50% of the respective softening temperature.

SOFTENING TEMPERATURE (also freezing temperature) is the characteristic temperature from which a substance (in our case a copper alloy) starts to soften, but is not yet fully melted. In phase diagrams this temperature is marked by the solidus line. Below the solidus, a metal is solid and stabile.

SOLIDIFICATION TEMPERATURE is the temperature above which a substance is completely melted and below which it begins to solidify. In phase diagrams this temperature is marked by the liquidus line. Between the liquidus and solidus lines substances are generally soft and cross over several partly soluble and metastabile phases.

STRAIN describes the force that has to be applied in order to deform materials.

STRENGTH expresses the ability of materials to withstand stress without breaking.

STRESS is defined as a force operating on an object and – possibly – causing its deformation. Stress and strain have similar meanings.

TEMPERING is a type of heat treatment used to counteract brittleness that can occur after a piece of metal has been worked (for example by hammering). Tempering softens and re-crystallises the metal, allowing cracks to be melted out. Tempering and annealing are comparable processes.

TENSILE STRENGTH is a measurement of the ultimate stress a material can resist before it perishes. It is measured by determining the energy required to break a test piece by elongation.

TOUGHNESS is a measurement to indicate a lasting resistance of materials against stress or other forces without cracking. The terms toughness and strength can be used synonymously.

Comments on the Catalogue

The following listing explains the information of the catalogue and the abbreviations that have been used in the text, graphs, appendices and catalogues.

ANo
Unique ID-number that references the analysis-number in the *Stuttgarter Datenbank* (cf. *Handbuch zur Stuttgarter Datenbank*: Krause 2003, 293 – 296; ibid. CD-Rom)[1].

ANo 100000 – 100131 refer to trace element analyses which were not published by Krause (2003).
The ANo is used in both the text and figures in order to identify the trace element analyses. The catalogue is arranged according to the ANo.

ANo	Publications of Trace Element Analyses
1 - 98903	Krause 2003, CD-Rom.
100000 - 100010	Krause 1988.
100012	Tillmann 1996, 378.
100013 - 100020	Ottaway 1992, 289.
100021 - 100022	Matuschik 1998.
100023 - 100027	Frána et al. 1995.
100028 - 100086	Bartelheim et al. 2002.
100087	Novotná 1982, 315.
100088 - 100131	Pers. comm. R. Wiermann

LABNo
Laboratory-numbers given by the laboratory, the author or the project that generated the relevant analysis (LABNo 00017 – 78176 correspond to SMAP coding; cf. Krause 2003, 293 – 296).

LABNo	Laboratory/Project that carried out the Trace Element Analyses
00017 - 22362	SAM-project,
29777 - 29996	SAM-project (Double-analyses).
49900 - 49901	Gleser and Schmitz 2001.
58001 - 58023	Novotná 1955.
58024 - 58031	published in Pleslová-Štiková 1985, 176.
62102 – 62204	Ottaway 1989.
68004 – 68057	Schweizer Landesmuseum Zürich
78142 – 78176	Selimchanov (various publications)
WA	Laboratory in Warsaw (cf. Krause 2003, 294):
FMZM	Project: *Frühe Metallurgie in Zentralen Mitteleuropa.*
HDM	Heidelberg (Max-Planck-Institute of Nuclear Physics, undertaken by E. Pernicka).
FRA	Analyses published by Frána et al. (1995).
FG	Institute of Archaeometry, TU Bergakademie Freiberg (Bartelheim et al. 2002).
Halle	Pers. comm. R. R. Wiermann; Landesmuseum Sachsen-Anhalt in Halle.

[1] The use of 'ANo' and 'LABNo' is borrowed from Krause's publications of trace element analyses in order to avoid causing confusion and to ease comparability (Krause used the abbreviations "ANR" and "LABNR").

FIND PLACE
Place (town, village etc) where the analysed object was found. The find place have been confirmed by literature review and/or the records in museum collections.

SITE
This specifies the location of the find, for example a grave, or an area within a settlement; it also includes comments on the site. The site description is taken from the relevant publication, and is presented in the original language of the publication (e.g. Czech, English, German etc).

FEATURE
This specifies the context within which the object had been found (i.e *Grave*, *Settlement* or *Hoard)*. If no clear context could be certified by literature review or in museum collections, the term *Single Find* is used.

NAT
Names the country in which the analysed object was found.

A:	Austria	F:	France	SK:	Slovakia
CH:	Switzerland	H:	Hungary	SRB:	Serbia
CZ:	Czech Republic	PL:	Poland	UA:	Ukraine
D:	Germany	RO:	Romania		

DIST
Names the district of the find place (e.g. the German federal state, the Swiss canton or the French department).

Austrian federal state (Bundesland):
- BL: Burgenland
- NÖ: Niederösterreich
- OÖ: Oberösterreich
- SL: Salzburg-Land
- TY: Tirol
- VA: Vorarlberg
- VI: Vienna

Czech Republic district (Okres):
- BCV: Břeclav
- BRN: Brno-město
- BRV: Brno-venkov
- CB: České Budejovice
- CHO: Chomutov
- JI: Jičín
- HK: Hrradec Králové
- HO: Hodonín
- KL: Kolín
- KLD: Kladno
- KR: Kroměříž
- KV: Karlovy Vary
- LIT: Litoměřic
- LOU: Louny
- MB: Mladá Boleslav
- MEL: Mělník
- MO: Most
- NY: Nymburk
- OLO: Olomouc
- PO: Prostějov
- PA: Pardubice
- PRA: Praha
- PRV: Praha-východ
- PRZ: Praha-západ
- TB: Třebíč
- TEP: Teplice
- UH: Uherské Hradíště
- UnL: Ústí nad Labem
- VN: Vsetín
- VY: Vyškov
- Zl: Zlonice
- ZN: Znojmo

French department (Département):
- BR: Bas-Rhin
- HR: Haut-Rhin

German federal state (Bundesland):
- BW: Baden-Württemberg
- BA: Bayern
- HE: Hessen
- NI: Niedersachsen
- RP: Rheinland-Pfalz
- SaAn: Sachsen-Anhalt
- SA: Sachsen
- TH: Thüringen

Hungarian county (Megye):
- Bar: Baranya
- BAZ: Borsod-Abaúj-Zemplén
- BK: Bács-Kiskun
- BKS: Békés
- CS: Csongrád
- FJ: Fejér
- GS: Györ-Sopron
- HB: Hajdú-Bihar
- HV: Heves
- ND: Nógrád
- PS: Pest
- SZ: Szolnok
- SY: Somogy
- SSB: Szabolcs-Szatmár-Bereg
- TOL: Tolna
- VES: Veszprém

Poland (Gmina):
- Kra: Kraków
- POD: podkarpackie
- WO: Wojciechowice

Romanian district (Judeţ):
- AD: Arad
- BAN: Banat
- BI: Bihor
- CaS: Caraş-Severin
- CEI: Ceica
- CT: Cetariu
- ME: Mehedinţi
- TA: Tarcea
- TM: Timiş

Serbian district (Okruzi):
- NB: North Banat

Slovakian district (Okres):
- BB: Banská Bystrica
- BV: Bratislava
- DK: Dolný Kubín
- GA: Galanta
- KS: Košice
- KO: Komárom
- LE: Levice
- LC: Lučenec
- LM: Liptovsky Mikulás
- MA: Martin
- MI: Michalovce
- ML: Malacky

- NA: Námestovo
- NI: Nitra
- NM: Nové Mesto
- NZ: Nové Zámky
- PB: Považská Bystrica
- PP: Poprad
- PR: Prievidza
- RS: Rimavská Sobota
- SB: Sobrance
- SE: Senica
- SNV: Spišská Nová Ves
- TC: Trenčin
- TO: Topol'čany
- TR: Trnava
- UH: Uherské Hradiště
- ZI: Žilina
- ZM: Zlaté Moravce
- ZV: Zvolen

Swiss canton (Kanton):
- AG: Aargau
- BE: Bern
- FR: Fribourg
- GE: Geneva
- LU: Luzern
- NE: Neuchâtel
- SG: St. Gallen
- SH: Schaffhausen
- SO: Solothurn
- TG: Thurgau
- VD: Vaud
- VS: Valais
- ZG: Zug
- ZH: Zürich

Ukraine (Rajon):
- HUS: Husiatyński

AREA
Denotes the so-called *SMAP-areas*. The *SAM-project* separated Europe into different areas on the basis of major landscape divisions and on archaeological criteria, and allocated a number code to every area (Junghans et al. 1960, 59 ff.). Since the 1960s the division has been changed and extended several times. The coding and map used in this study is based on the *SMAP-project* published by Sangmeister (1999, Fig. 2) and by Krause (2003, 290 – 292).
Note that AREA 67 (Switzerland) partly falls outside our study area, and so has had to be split. Only the part north of the river Rhone and the Lake Geneva is used here.

The codes for AREA correspond to the regions shown in the following map:

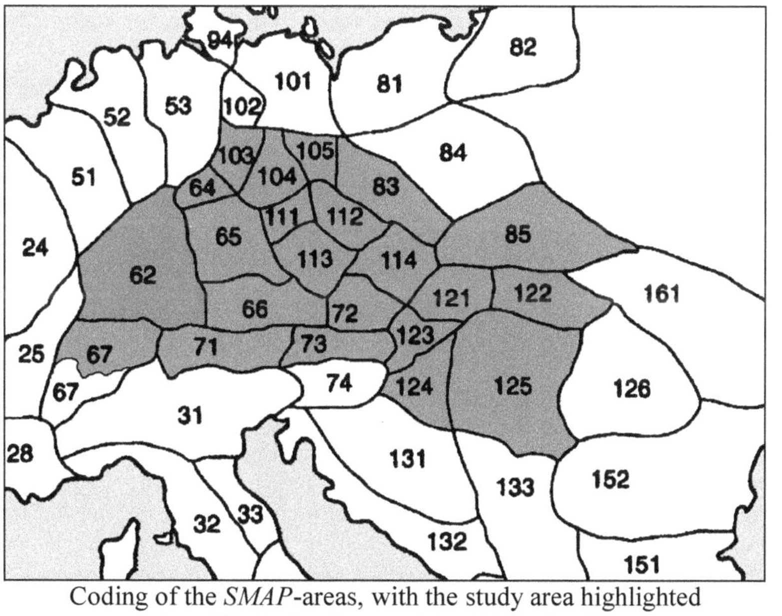

Coding of the *SMAP*-areas, with the study area highlighted
(modified after Sangmeister 1999, Fig. 2).

CULT
Names the cultural attribution of the analysed find. Objects must have been found in an 'archaeocultural' context, or else must be typical for an archaeological group. Not every archaeological group within the area of research is listed; only those that provided analysed objects to the database are included.
The classification relates to the literature referenced in the catalogue (category: LITERATURE) and in the text (particularly Chapter 4.1).

The archaeological groups included in the database are:

ABC: Adlerberg group
ALT: Altheim group
BAA: Baalberge group (late phase of the Middle German TRB complex)
BAD: Baden complex
BBC: Bell Beaker culture (Eastern Bell Beaker group)
BER: Bernburg group (early phase of the Middle German TRB complex)
BK: *Blechkreis* (Early Bronze Age of the northalpine region)
BOD: Bodrogkeresztúr group
CHA: Cham group
COR: Cortailliod group
CWC: Corded Ware complex

GIII: Goldberg-III-group
KST: Košt'any group
LEN: Lengyel complex
LLG: Loretto-Leithaprodersdorf group
Mier: Mierzanowice group
MKC: Makó-Kosiny-Čaka group
MON: Mondsee group
MUE: Münchshöfen group
NGY: proto-Nagyrév group
NIT: Nitra group
PFY: Pfyn group
PIT: Pitvaros group
RIV: Řivnáč group
SAL: Salzmünde group (middle phase of the Middle German TRB complex)
SOP: Stroke Ornamented Pottery
TC: Tiszapolgár group
TZC: Tisza group

PER

Names the period to which the finds are dated.Many artefacts that in the database can only be roughly dated. Furthermore, the date ranges for the periods given below can only be regarded as approximate; the periods should not be seen as phases defined by fixed thresholds, but rather as providing a general classification to facilitate chronological comparisons. A more precise dating is given by the cases that can be set into an archaeological group. Archaeological groups, however, are not necessarily limited by absolute chronological dates; the date ranges used are those within which most of the dates for that particular group fit.

While in general the database uses general phases of c. 500 years, the 3rd millennium BC is split into the earlier phase (PER 31), the Bell Beaker related objects (PER 35) and the Earliest Bronze Age (PER 40).

PER 10 = before c. 4000 BC	LEN; MUE; TC; TZC
PER 21 = c. 4000 – 3500 BC	ALT; BAA; BOD; COR; MON; PFY; SOP
PER 25 = c. 3500 – 2800 BC	BAD; BERN; CHA; GIII; RIV; SAL
PER 31 = c. 2800 – 2300 BC	CWC; MKC
PER 35 = c. 2500 – 2000 BC	BBC
PER 40 = c. 2200 – 2000 BC	Earliest Bronze Age groups: ABC; BK; KST; LLG; Mier; NGY; NIT; PIT

TYPE

The figures provide the numerical code, as used in the *Handbuch zur Stuttgarter Datenbank* (Krause 2003, 292 f.), for artefact types, as follows:

Axes
101 = Axe (indeterminate type)
104 = Axe, flat axe
106 = Axe, adze
110 = Axe, flanged
112 = Axe, slightly flanged
122 = Axe, shaft hole, single-edged
127 = Axe, double axe
128 = Axe, shaft-hole axe-adzes
131 = Axe, axe-hammer
134 = Chisel

Daggers and swords
201 = Dagger or blade (indeterminate type)
212 = Dagger, riveted
213 = Rivet
215 = Dagger, tanged

Knives
310 = Knife
317 = Sickle
320 = Knife, willow-leaf-shaped (no special type in "Stuttgarter Handbuch")

Rings
401 = Ring (indeterminate type)
402 = Ring, neck ring
403 = Ring, torc
404 = Arm-ring, fragment
405 = Arm-ring
408 = Bracelet
407 = Arm-ring, massive
409 = Arm spiral
413 = Ring, spiral finger ring, large
414 = Ring, spiral finger ring, small
416 = Ring, spiral
417 = Bead, tubular sheet bronze bead
418 = Ring, opened, small
420 = Ring, earring, willow-leaf-shaped ("Stuttgarter Handbuch": No. 521)

Sheets, plate and wire metal objects
501 = Sheet metal (indeterminate type)
503 = Reel
504 = Sheet metal, rolled
505 = Sheet metal, two-sided rolled
506 = Sheet metal, rolled in several directions
507 = Pendant
512 = Pendant, lunate
513 = Disc
515 = Disc-pendant
517 = Spiral
518 = Tutulus
519 = Tutulus, spiral
522 = Diadem
523 = Sheet band

527 = Wire
528 = Wire, spiral
529 = Spectacle spiral

Pins
602 = Pin, shaft
603 = Pin, disc-/oar-headed
604 = Pin, disc-/oar-headed, decorated
607 = Pin, knot-headed, Cypriot Type
610 = Pin, roll-headed
615 = Pin, lugged-headed

Castings, crucibles etc.
710 = Casting
711 = Sprule
712 = Bifurcate header (casting), Slag, Smelting crucible
714 = Coarse copper
716 = Copper point

Diverse objects
801 = Bead
804 = Fish hook
806 = Awl
807 = Awl with lug

fragm. = fragment

TYPE 2

The category TYPE 2 summarises objects according to typological and in particular functional criteria. Thus, several TYPE-categories can be combined as TYPE 2.

Axes
100 = Flat axes (TYPE 101, 104, 106)
110 = Flanged axe (TYPE 110, 112)
122 = Shaft-hole axes (TYPE 122)
130 = axe-adzes, axe-hammers (TYPE 127, 128, 131)
134 = Chisels (TYPE 134)

Daggers and knives
200 = Dagger (TYPE 201)
212 = Riveted dagger (TYPE 212)
213 = Rivet (TYPE 213)
215 = Tanged dagger (TYPE 215)
300 = Knives (TYPE 311, 317, 320)

Rings
400 = Rings (TYPE 401, 402, 403, 418)
405 = Massive rings (e.g. arm rings) (TYPE 404, 405, 407, 408)
410 = Spiral rings (TYPE 409, 413, 414, 416, 417)
420 = Willow-leaf-shaped rings (TYPE 420)

Sheet metal and wire objects
500 = Sheet metal (TYPE 501 – 515, 518, 522, 523)
510 = Wires and small spirals (TYPE 517, 519, 527, 528, 529)
550 = Reels (TYPE 503)

Pins
600 = Pins (TYPE 602 – 615)

Evidence of copper production
700 = Castings, crucibles etc. (TYPE 707 – 716)

Other finds
800 = Beads (TYPE 801)
810 = Awls/Fish hook (TYPE 804, 806, 807)

OBJECT

Description of the analysed object.

PLACENo

Each site, e.g. cemetery or settlement, is labelled with an individual number (PLACENo). The figures refer mostly to Krause (2003). Sites added by the author hold numbers from no. 7201 onwards.

AM

Method by which trace element analyses have been carried out (insofar as known).

OES: Optical emission spectroscopy
XRF: X-ray fluorescence spectroscopy
NAA: Neutron activation analysis

MG
Material group, according to the *Stammbaum* that was developed by the SAM-group (see Chapter 8.1.1).

00: A	24: C_2 C	41: E_{01}	52: FB_1
01: A_1	25: C_2 D	42: E_{10}	53: FB_2
02: A_2	26: C_3	43: E_{11} A	54: FC
10: B_2	27: C_4	44: E_{11} B	55: FD
20: C_1 A	28: C_5	45: E_{00} FC	56: FG
21: C_1 B	29: C_6 A	46: E_{01}A	60: G
22: C_2	30: C_6 B	50: FA_1	61: G_1
23: C_2 A/B	40: E_{00}	51: FA_2	62: G_2

MG: 49 (not defined by SAM)

As	Sb	Ag	Ni	Bi
0 – 0.02	0 – 0.02	0 – 0.29	0.001 – 0.002	0 – 0.002

Material group N is not present, because nickel is always at least in small traces (0.001%) present (= E_{00}).

CU, SN, PB, AS, SB, AG, NI, BI, AU, ZN, CO, FE
SI-Unit of copper and the other analysed trace elements. The concentration of the particular element is listed in percent by weight.

0%: Element not included or not measured/unknown
0.001%: only 0.001% or less of the element has been measured.
If in the literature the trace element concentration has been presented with the algebraic sign "<", then the total value is used here, because this ensures that the concentration of the impurity does not exceed that maximum content.
The concentration of copper is often not displayed, but in the case of all the objects that have been collected and analysed here, copper is always the main element.

ng Log (X)
Negative decade logarithms of the value of the relevant element's concentration X.
If an element concentration is listed as 0%, than the value is replaced by 3.5, because the logarithm of 0 does not exist (in detail: Chapter 8.4.1).

LITERATURE
Names the principle reference where the type of find, its find place or the analyses are published or the collections where the objects could be checked by the author. The chronological and archaeological-cultural classification is defined by these references.

DOUBLEANALYSES
If more than one analysis has been taken from an object, this variable labels it (modified after Krause 2003, 289).

* = Several laboratories carried out analyses of one object.
# = Several analyses of one object were carried out by the same laboratory/project.
+ = Different parts of an object were analysed e.g. blade and rivet of a dagger.

INVENTARYNo
This names the inventory number as listed in the museum or archive where the object is kept.

COMMENT
Important comments on the object, the analysis, the find place or absolute-chronological date of the site in question.

Randomised
Displays the decimal number that has been developed by randomising the arrangements of every case in the database. In particular, this is explained in Chapter 8.2.1.

PCA_factor_1, PCA_factor_2
The regression factor scores of the first and the second principle component resulting from the principle component analysis run on the entire database (cf. Chapter 8.4.3).

CLUSTER_No_I, CLUSTER_No_II, CLUSTER_No_III, CLUSTER_No_IV
The cluster number of the respective Cluster Level I to IV resulting from the two-step cluster analyses (cf. Chapter 10.1 – 10.4; Fig. 10.1).

Metal_groups
Every trace element analysis is classified as a particular metal group defined by its compositional configuration (cf. Chapter 10.5).

PCA_factor1_Bell_Beaker, PCA_factor2_Bell_Beaker
This category is only displayed in the catalogue of the Bell Beaker finds. It shows the regression factor scores of the first and the second principle component as they resulted from the principle component analysis run on Bell Beaker analyses (cf. Chapter 11.2).

Correlation tables of the Trace Element Analyses, Find places, Clusters and Metal Groups of the entire Database

ANo	FIND_PLACE	AREA	CULT	PER	TYPE_2	CL_No_I	CL_No_II	CL_No_III	CL_No_IV	Metal_groups
2839	Vinelz, 'Amt Erlau'	67	CWC	31	134	1	3	9	9	Almost pure
2840	Vinelz, 'Amt Erlau'	67	CWC	31	110	1	3	9	9	Almost pure
2845	Risch, 'Schwarzenbach'	67	PFY	21	100	1	3	9	9	Almost pure
2954	Chevroux	67		31	134	1	3	9	9	Almost pure
3239	Praha-Prosek	112	BBC	35	215	1	3	9	9	Almost pure
3341	Brno	114		21	100	1	3	9	9	Almost pure
3358	Mistrín I	114	BBC	35	810	1	3	9	9	Almost pure
3360	Brno-Holásky II	114	BBC	35	200	1	3	9	9	Almost pure
3372	Strážnice	114		21	130	1	3	9	9	Almost pure
3376	Rosice	114		21	130	1	3	9	9	Almost pure
3393	Ždánice	114		10	130	1	3	9	9	Almost pure
3471	Vyskov	114	BBC	35	600	1	3	9	9	Almost pure
3670	Holic	114		10	130	1	3	9	9	Almost pure
3698	Bratislava	121		10	130	1	3	9	9	Almost pure
4242	Ayent, 'Les Place'	67	BK	40	500	1	3	9	9	Almost pure
4926	Stollhof, 'Lange Wand', p. B. Wiener Neustadt	72	BAD	25	100	1	4	12	30	Almost pure
4927	Stollhof, 'Lange Wand', p. B. Wiener Neustadt	72	BAD	25	510	1	3	9	9	Almost pure
4928	Stollhof, 'Lange Wand', p. B. Wiener Neustadt	72	BAD	25	510	1	3	9	9	Almost pure
5685	Horodnica	85		31	500	1	3	9	9	Almost pure
5686	Horodnica	85		31	500	1	3	9	9	Almost pure
6466	Visegrad	124		21	100	1	3	9	9	Almost pure
6477	Kiskörös	125	BOD	21	130	1	3	9	9	Almost pure
6492	Oravita Montana	125		21	130	1	3	9	9	Almost pure
6521	Tiszapolgár-Basatanya	125	TC	10	400	1	3	9	9	Almost pure
6525	Tiszapolgár-Basatanya	125	BOD	21	215	1	3	9	9	Almost pure
6527	Tiszapolgár-Basatanya	125	BOD	21	405	1	3	9	9	Almost pure
6528	Tiszapolgár-Basatanya	125	BOD	21	405	1	3	9	9	Almost pure
6529	Tiszapolgár-Basatanya	125	BOD	21	810	1	3	9	9	Almost pure
6544	Kerepes	125		21	130	1	3	9	9	Almost pure
6564	Dabas	125		21	130	1	3	9	9	Almost pure
8718	Oradea	125		10	130	1	3	9	9	Almost pure
9166	Chisoda	125		21	130	1	3	9	9	Almost pure
9167	Ciacova	125		21	130	1	3	9	9	Almost pure
9169	Orsova	125		21	130	1	3	9	9	Almost pure
9233	Hencida	125		21	130	1	3	9	9	Almost pure
10388	Lisov	121		25	100	1	3	9	9	Almost pure
10394	Kláštor pod Znievom	121		21	130	1	3	9	9	Almost pure
10554	Výcapy-Opatovce	121	NIT	40	420	1	3	9	9	Almost pure
10815	Gánovce	122		21	130	1	3	9	9	Almost pure
10852	Rimavská Sec	122		25	100	1	3	9	9	Almost pure
11098	Nymburk	112		10	130	1	3	9	9	Almost pure
11506	Zwerndorf a. d. March, 'Pfarrwiese', p. B. Gänserndorf	72		21	100	1	3	9	9	Almost pure
12905	Budakalász	124	BAD	25	550	1	3	9	9	Almost pure
12908	Budakalász	124	BAD	25	550	1	3	9	9	Almost pure
12963	Edelény	122		21	130	1	3	9	9	Almost pure
12965	Emöd	122	BOD	21	510	1	3	9	9	Almost pure
12967	Emöd	122	BOD	21	510	1	3	9	9	Almost pure
12986	Székely, 'Szolohegy'	122	BOD	21	130	1	3	9	9	Almost pure
12987	Szabolcs, 'Fecsképart'	122		21	130	1	3	9	9	Almost pure
13031	Tarcal, 'Nyavalyateto'	122		10	130	1	3	9	9	Almost pure
13041	Kisvárda, 'Umgebung von Kisvárda'	122		21	130	1	3	9	9	Almost pure
13049	Hajdúhadház	125	BOD	21	130	1	3	9	9	Almost pure
13200	Szeged	125		21	130	1	3	9	9	Almost pure

ANo	FIND_PLACE	AREA	CULT	PER	TYPE_2	CL_No_I	CL_No_II	CL_No_III	CL_No_IV	Metal_groups
13284	Mezoberény	125		21	122	1	3	9	9	Almost pure
14382	Jászladány	125	TC	10	600	1	3	9	9	Almost pure
14467	Attersee, p. B. Vöcklabruck	73	MON	21	100	1	3	9	9	Almost pure
15907	Wildenroth, Lkr. Fürstenfeldbruck	66		25	100	1	3	9	9	Almost pure
19931	Luderov	114	BBC	35	810	1	3	9	9	Almost pure
21803	Goldburghausen, 'Goldberg', Ostalbkreis	62	GIII	25	810	1	3	9	9	Almost pure
22268	St. Blaise	67	CWC	31	100	1	3	9	9	Almost pure
44012	Strzyzów, 'Horodlo'	84	Mier	40	420	1	3	9	9	Almost pure
58010	Durciná	121		25	100	1	5	13	13	Almost pure
3353	Tibava, 'Hrun za cintirom'	122	TC	10	130	1	5	13	13	Almost pure with Iron
6545	Budapest-III-Obuda	124		21	130	1	5	13	13	Almost pure with Iron
12998	Székely, 'Szolohegy'	122	BOD	21	130	1	5	13	13	Almost pure with Iron
44017	Strzyzów, 'Horodlo'	84	Mier	40	550	1	5	13	13	Almost pure with Iron
44018	Strzyzów, 'Horodlo'	84	Mier	40	500	1	5	13	13	Almost pure with Iron
58016	Trhovište	121		10	130	1	5	13	13	Almost pure with Iron
62126	Morges	67		31	300	1	5	13	13	Almost pure with Iron
62167	Bevaix	67	CWC	31	212	1	5	13	13	Almost pure with Iron
62185	Delley, 'Portalban'	67		31	100	1	5	13	13	Almost pure with Iron
62188	Yvonand, 'La Peupleraie; Grabung 1973'	67	CWC	31	212	1	5	13	13	Almost pure with Iron
62195	Linz-St. Peter, p. B. Linz	66	BAD	25	100	1	5	13	13	Almost pure with Iron
78169	Tibava, 'Hrun za cintirom'	122	TC	10	130	1	5	13	13	Almost pure with Iron
78170	Tibava, 'Hrun za cintirom'	122	TC	10	130	1	5	13	13	Almost pure with Iron
78171	Tibava, 'Hrun za cintirom'	122	TC	10	130	1	5	13	13	Almost pure with Iron
78173	Tibava, 'Hrun za cintirom'	122	TC	10	130	1	5	13	13	Almost pure with Iron
30	Beizkofen, Kr. Sigmaringen	62	BK	40	410	2	6	18	55	Antimony-arsenic-copper with Silver
181	Manching, 'Hundsrückenacker i. d. Au', Lkr. Ingolstadt	66	BK	40	600	1	4	10	25	Antimony-arsenic-copper with Silver
251	Ravensburg, Lkr. Ravensburg	62	BK	40	410	1	4	10	25	Antimony-arsenic-copper with Silver
394	Straubing, 'Jungmeier', Lkr. Straubing-Bogen	66	BK	40	410	1	4	10	25	Antimony-arsenic-copper with Silver
2641	Osterbuch, 'Altes Mahd', Lkr. Augsburg	62	BK	40	500	1	4	10	25	Antimony-arsenic-copper with Silver
2774	Feldkirch, 'Tillis', p. B. Feldkirch	71	BK	40	110	2	6	18	54	Antimony-arsenic-copper with Silver
4995	Hippersdorf, 'Plexenthal', p. B. Tulln	72	LLG	40	212	2	7	20	68	Antimony-arsenic-copper with Silver
5004	Hippersdorf, 'Plexenthal', p. B. Tulln	72	LLG	40	410	2	7	20	68	Antimony-arsenic-copper with Silver
5904	Gemeinlebarn, p. B. St. Pölten	72	BK	40	410	1	4	10	25	Antimony-arsenic-copper with Silver
5958	Unterwölbling, p. B. St. Pölten	72		40	410	2	7	20	68	Antimony-arsenic-copper with Silver
5966	Unterwölbling, p. B. St. Pölten	72		40	410	2	7	20	68	Antimony-arsenic-copper with Silver
5967	Unterwölbling, p. B. St. Pölten	72		40	410	2	7	20	68	Antimony-arsenic-copper with Silver
5968	Unterwölbling, p. B. St. Pölten	72		40	500	2	7	20	68	Antimony-arsenic-copper with Silver
6483	Nitrianske Pravno	121		21	130	1	4	10	25	Antimony-arsenic-copper with Silver
6498	Kežmarok	122		21	130	1	4	10	25	Antimony-arsenic-copper with Silver
7490	Praha-Sárka	112		21	100	2	7	20	68	Antimony-arsenic-copper with Silver
7495	Praha-Sárka	112		21	100	1	4	10	25	Antimony-arsenic-copper with Silver
7525	Kunetice	112		10	130	1	4	10	25	Antimony-arsenic-copper with Silver
7535	Cisteves	112		40	110	2	7	20	68	Antimony-arsenic-copper with Silver
10396	Handlová, 'Na Pstruhároch'	121		25	100	1	4	10	25	Antimony-arsenic-copper with Silver
10397	Handlová, 'Na Pstruhároch'	121		25	100	1	4	10	25	Antimony-arsenic-copper with Silver
10398	Handlová, 'Na Pstruhároch'	121		21	130	1	4	10	25	Antimony-arsenic-copper with Silver
10820	Cigel	121		25	100	1	4	10	25	Antimony-arsenic-copper with Silver
10854	Bajc-Vlkanovo, 'Sonda III'	121	SOP	21	510	1	4	10	25	Antimony-arsenic-copper with Silver
12166	Horné Lefantovce	121		25	100	1	4	10	25	Antimony-arsenic-copper with Silver
14385	Nógrádkövesd	121		21	100	1	4	10	25	Antimony-arsenic-copper with Silver
15792	Nattenhausen, Lkr. Günzburg	62		40	410	2	7	20	68	Antimony-arsenic-copper with Silver
15836	Manching, 'Hundsrückenacker i. d. Au', Lkr. Ingolstadt	66	BK	40	410	1	4	10	25	Antimony-arsenic-copper with Silver
15844	München-Sendling, 'Lindwurmstr. 76', Stkr. München	66	BK	40	405	2	7	20	68	Antimony-arsenic-copper with Silver

ANo	FIND_PLACE	AREA	CULT	PER	TYPE_2	CL_No_I	CL_No_II	CL_No_III	CL_No_IV	Metal_groups
15846	München-Sendling, 'Lindwurmstr. 76', Stkr. München	66	BK	40	405	2	7	20	68	Antimony-arsenic-copper with Silver
17694	Vsech Svätych	122	KST	40	212	2	7	20	68	Antimony-arsenic-copper with Silver
17698	Vsech Svätych	122	KST	40	410	2	7	20	68	Antimony-arsenic-copper with Silver
18429	Erpfingen, Lkr. Reutlingen	62	BK	40	110	2	7	20	68	Antimony-arsenic-copper with Silver
20085	Neureut, 'Herdenschlag', Kr. Karlsruhe	62	BK	40	110	2	7	20	68	Antimony-arsenic-copper with Silver
21567	Tlumacov	114		21	100	1	4	10	25	Antimony-arsenic-copper with Silver
21568	Tlumacov	114		21	100	1	4	10	25	Antimony-arsenic-copper with Silver
29787	Horné Lefantovce	121		25	110	1	4	10	25	Antimony-arsenic-copper with Silver
29788	Horné Lefantovce	121		25	110	1	4	10	25	Antimony-arsenic-copper with Silver
44026	Mierzanowice II	85	Mier	40	510	2	7	20	68	Antimony-arsenic-copper with Silver
100087	Nowa Huta-Wyciaze	83	BOD	21	200	1	4	10	25	Antimony-arsenic-copper with Silver
100102	Halle-Giebichenstein, Stkr. Halle	103		21	100	2	7	20	68	Antimony-arsenic-copper with Silver
798	Salez, 'Sennwald'	67	BK	40	110	2	7	19	57	Antimony-nickel-copper (high)
1266	Singen a. Hohentwiel, Kr. Konstanz	62	BK	40	410	2	7	19	59	Antimony-nickel-copper (high)
1272	Singen a. Hohentwiel, Kr. Konstanz	62	BK	40	212	2	7	19	59	Antimony-nickel-copper (high)
1278	Singen a. Hohentwiel, Kr. Konstanz	62	BK	40	212	2	7	19	58	Antimony-nickel-copper (high)
2768	Salez, 'Sennwald'	67	BK	40	110	2	7	19	57	Antimony-nickel-copper (high)
2779	Salez, 'Sennwald'	67	BK	40	110	2	7	19	57	Antimony-nickel-copper (high)
2780	Salez, 'Sennwald'	67	BK	40	110	2	7	19	57	Antimony-nickel-copper (high)
2781	Salez, 'Sennwald'	67	BK	40	110	2	7	19	57	Antimony-nickel-copper (high)
2782	Salez, 'Sennwald'	67	BK	40	110	2	7	19	57	Antimony-nickel-copper (high)
2783	Salez, 'Sennwald'	67	BK	40	110	2	7	19	56	Antimony-nickel-copper (high)
2784	Salez, 'Sennwald'	67	BK	40	110	2	7	19	57	Antimony-nickel-copper (high)
3174	Meckenbeuren-Liebenau, Bodenseekreis	62	BK	40	110	2	7	19	57	Antimony-nickel-copper (high)
3721	Salez, 'Sennwald'	67	BK	40	110	2	7	19	57	Antimony-nickel-copper (high)
3761	Salez, 'Sennwald'	67	BK	40	110	2	7	19	57	Antimony-nickel-copper (high)
3762	Salez, 'Sennwald'	67	BK	40	110	2	7	19	57	Antimony-nickel-copper (high)
7791	Hindelwangen, Lkr. Konstanz	62	BK	40	110	2	7	19	57	Antimony-nickel-copper (high)
7792	Hindelwangen, Lkr. Konstanz	62	BK	40	110	2	7	19	57	Antimony-nickel-copper (high)
7793	Hindelwangen, Lkr. Konstanz	62	BK	40	110	2	7	19	57	Antimony-nickel-copper (high)
7794	Hindelwangen, Lkr. Konstanz	62	BK	40	110	2	7	19	57	Antimony-nickel-copper (high)
7795	Hindelwangen, Lkr. Konstanz	62	BK	40	110	2	7	19	57	Antimony-nickel-copper (high)
9280	Niedertraubling, 'Lohfeld', Lkr. Regensburg	66	BK	40	600	2	7	19	57	Antimony-nickel-copper (high)
18407	Riedlingen, Kr. Biberach	62	BK	40	110	2	7	19	57	Antimony-nickel-copper (high)
20077	Osterburken, Neckar-Odenwald-Kreis	62	BK	40	110	2	7	19	57	Antimony-nickel-copper (high)
22256	Salez, 'Sennwald'	67	BK	40	110	2	7	19	57	Antimony-nickel-copper (high)
22257	Salez, 'Sennwald'	67	BK	40	110	2	7	19	57	Antimony-nickel-copper (high)
29993	Salez, 'Sennwald'	67	BK	40	110	2	7	19	57	Antimony-nickel-copper (high)
29994	Salez, 'Sennwald'	67	BK	40	110	2	7	19	57	Antimony-nickel-copper (high)
29995	Salez, 'Sennwald'	67	BK	40	110	2	7	19	57	Antimony-nickel-copper (high)
29996	Salez, 'Sennwald'	67	BK	40	110	2	7	19	57	Antimony-nickel-copper (high)
33033	Jeßnitz, Lkr. Anhalt-Bitterfeld	105	CWC	31	400	2	7	19	58	Antimony-nickel-copper (high)
62180	Salez, 'Sennwald'	67	BK	40	110	2	7	19	58	Antimony-nickel-copper (high)
68004	Salez, 'Sennwald'	67	BK	40	110	2	7	19	56	Antimony-nickel-copper (high)
68005	Salez, 'Sennwald'	67	BK	40	110	2	7	19	56	Antimony-nickel-copper (high)
68006	Salez, 'Sennwald'	67	BK	40	110	2	7	19	56	Antimony-nickel-copper (high)
68007	Salez, 'Sennwald'	67	BK	40	110	2	7	19	56	Antimony-nickel-copper (high)
68008	Salez, 'Sennwald'	67	BK	40	110	2	7	19	56	Antimony-nickel-copper (high)
68009	Salez, 'Sennwald'	67	BK	40	110	2	7	19	57	Antimony-nickel-copper (high)
68010	Salez, 'Sennwald'	67	BK	40	110	2	7	19	56	Antimony-nickel-copper (high)
68011	Salez, 'Sennwald'	67	BK	40	110	2	7	19	56	Antimony-nickel-copper (high)
68012	Salez, 'Sennwald'	67	BK	40	110	2	7	19	56	Antimony-nickel-copper (high)
68013	Salez, 'Sennwald'	67	BK	40	110	2	7	19	56	Antimony-nickel-copper (high)

ANo	FIND_PLACE	AREA	CULT	PER	TYPE_2	CL_No_I	CL_No_II	CL_No_III	CL_No_IV	Metal_groups
68014	Salez, 'Sennwald'	67	BK	40	110	2	7	19	56	Antimony-nickel-copper (high)
68015	Salez, 'Sennwald'	67	BK	40	110	2	7	19	56	Antimony-nickel-copper (high)
68016	Salez, 'Sennwald'	67	BK	40	110	2	7	19	56	Antimony-nickel-copper (high)
68017	Salez, 'Sennwald'	67	BK	40	110	2	7	19	56	Antimony-nickel-copper (high)
68018	Salez, 'Sennwald'	67	BK	40	110	2	7	19	56	Antimony-nickel-copper (high)
68019	Auvernier	67	BK	40	110	2	7	19	56	Antimony-nickel-copper (high)
68020	Gams, 'Gasenzen'	67	BK	40	110	2	7	19	56	Antimony-nickel-copper (high)
68021	Gams, 'Gasenzen'	67	BK	40	110	2	7	19	56	Antimony-nickel-copper (high)
68022	Gams, 'Gasenzen'	67	BK	40	110	2	7	19	56	Antimony-nickel-copper (high)
68047	Meersburg, 'Haltau', Bodenseekreis	62	BK	40	110	2	7	19	57	Antimony-nickel-copper (high)
68055	Salez, 'Sennwald'	67	BK	40	110	2	7	19	56	Antimony-nickel-copper (high)
68056	Salez, 'Sennwald'	67	BK	40	110	2	7	19	56	Antimony-nickel-copper (high)
68057	Salez, 'Sennwald'	67	BK	40	110	2	7	19	56	Antimony-nickel-copper (high)
100007	Mengen, Umgebung, Kr. Sigmaringen	62	BK	40	110	2	7	19	59	Antimony-nickel-copper (high)
3	Adlerberg b. Worms, Kr. Alzey-Worms	62	ABC	40	212	2	7	20	60	Antimony-nickel-copper (intermediate)
4	Adlerberg b. Worms, Kr. Alzey-Worms	62	ABC	40	212	2	7	19	58	Antimony-nickel-copper (intermediate)
28	Beizkofen, Kr. Sigmaringen	62	BK	40	600	2	7	19	58	Antimony-nickel-copper (intermediate)
32	Berghausen, 'Roßweide', Lkr. Karlsruhe	62		40	600	2	7	19	58	Antimony-nickel-copper (intermediate)
71	Stuttgart-Bad Cannstatt, Stkr. Stuttgart	62	BK	40	600	2	7	19	58	Antimony-nickel-copper (intermediate)
82	Eriskirch, Bodenseekreis	62	BK	40	110	2	7	19	58	Antimony-nickel-copper (intermediate)
87	Galgweis-Gergweis, Lkr. Passau	66	BK	40	600	2	7	20	60	Antimony-nickel-copper (intermediate)
88	Galgweis-Gergweis, Lkr. Passau	66	BK	40	600	2	7	19	58	Antimony-nickel-copper (intermediate)
94	Grünwald-Geiselgasteig, Stkr. München	66	BK	40	600	2	7	20	60	Antimony-nickel-copper (intermediate)
132	Heroldingen, 'Kalbläcker', Donau-Ries Kreis	62	BK	40	600	2	7	19	59	Antimony-nickel-copper (intermediate)
151	Kirch-Brombach, Brombachtal, 'Im Eichels', Odenwaldkreis	62	BK	40	110	2	7	19	58	Antimony-nickel-copper (intermediate)
152	Kirchheim u. Teck, Kr. Esslingen	62	BK	40	410	2	7	20	61	Antimony-nickel-copper (intermediate)
179	Manching, 'Hundsrückenacker i. d. Au', Lkr. Ingolstadt	66	BK	40	600	2	7	19	58	Antimony-nickel-copper (intermediate)
180	Manching, 'Hundsrückenacker i. d. Au', Lkr. Ingolstadt	66	BK	40	600	2	7	19	58	Antimony-nickel-copper (intermediate)
182	Mengen, Umgebung, Kr. Sigmaringen	62	BK	40	110	2	7	19	58	Antimony-nickel-copper (intermediate)
183	Mengen, Umgebung, Kr. Sigmaringen	62	BK	40	110	2	7	19	58	Antimony-nickel-copper (intermediate)
191	Monsheim, 'Kahleberg', Kr. Alzey-Worms	62	ABC	40	600	2	7	19	58	Antimony-nickel-copper (intermediate)
209	München-Sendling, 'Implerplatz', Stkr. München	66	BK	40	212	2	7	19	58	Antimony-nickel-copper (intermediate)
212	Nähermemmingen, 'Feldwiesäcker', Donau-Ries Kreis	62	BK	40	550	2	7	19	58	Antimony-nickel-copper (intermediate)
213	Nähermemmingen, 'Feldwiesäcker', Donau-Ries Kreis	62	BK	40	410	2	7	19	58	Antimony-nickel-copper (intermediate)
214	Nähermemmingen, 'Feldwiesäcker', Donau-Ries Kreis	62	BK	40	410	2	7	19	58	Antimony-nickel-copper (intermediate)
215	Nähermemmingen, 'Feldwiesäcker', Donau-Ries Kreis	62	BK	40	405	2	7	19	58	Antimony-nickel-copper (intermediate)
216	Nähermemmingen, 'Feldwiesäcker', Donau-Ries Kreis	62	BK	40	212	2	7	19	58	Antimony-nickel-copper (intermediate)
252	Ravensburg, Lkr. Ravensburg	62	BK	40	110	2	7	19	58	Antimony-nickel-copper (intermediate)
253	Ravensburg, Lkr. Ravensburg	62	BK	40	110	2	7	19	58	Antimony-nickel-copper (intermediate)
254	Ravensburg, Lkr. Ravensburg	62	BK	40	110	2	7	19	58	Antimony-nickel-copper (intermediate)
312	Singen a. Hohentwiel, Kr. Konstanz	62	BK	40	600	2	7	19	58	Antimony-nickel-copper (intermediate)
320	Singen a. Hohentwiel, Kr. Konstanz	62	BK	40	213	2	7	19	58	Antimony-nickel-copper (intermediate)
331	Singen a. Hohentwiel, Kr. Konstanz	62	BK	40	600	2	7	19	58	Antimony-nickel-copper (intermediate)
332	Singen a. Hohentwiel, Kr. Konstanz	62	BK	40	212	2	7	19	58	Antimony-nickel-copper (intermediate)
333	Singen a. Hohentwiel, Kr. Konstanz	62	BK	40	410	2	7	19	58	Antimony-nickel-copper (intermediate)
334	Singen a. Hohentwiel, Kr. Konstanz	62	BK	40	410	2	7	19	59	Antimony-nickel-copper (intermediate)
335	Singen a. Hohentwiel, Kr. Konstanz	62	BK	40	810	2	7	19	59	Antimony-nickel-copper (intermediate)
336	Singen a. Hohentwiel, Kr. Konstanz	62	BK	40	410	2	7	19	58	Antimony-nickel-copper (intermediate)
343	Singen a. Hohentwiel, Kr. Konstanz	62	BK	40	212	2	7	19	58	Antimony-nickel-copper (intermediate)
345	Singen a. Hohentwiel, Kr. Konstanz	62	BK	40	405	2	7	19	59	Antimony-nickel-copper (intermediate)
346	Singen a. Hohentwiel, Kr. Konstanz	62	BK	40	212	2	7	19	59	Antimony-nickel-copper (intermediate)
347	Singen a. Hohentwiel, Kr. Konstanz	62	BK	40	212	2	7	19	59	Antimony-nickel-copper (intermediate)
353	Singen a. Hohentwiel, Kr. Konstanz	62	BK	40	810	2	7	19	59	Antimony-nickel-copper (intermediate)

ANo	FIND_PLACE	AREA	CULT	PER	TYPE_2	CL_No_I	CL_No_II	CL_No_III	CL_No_IV	Metal_groups
354	Singen a. Hohentwiel, Kr. Konstanz	62	BK	40	212	2	7	19	59	Antimony-nickel-copper (intermediate)
355	Singen a. Hohentwiel, Kr. Konstanz	62	BK	40	410	2	7	19	59	Antimony-nickel-copper (intermediate)
356	Singen a. Hohentwiel, Kr. Konstanz	62	BK	40	600	2	7	19	58	Antimony-nickel-copper (intermediate)
385	Straubing, 'Jungmeier', Lkr. Straubing-Bogen	66	BK	40	600	2	7	19	58	Antimony-nickel-copper (intermediate)
386	Straubing, 'Jungmeier', Lkr. Straubing-Bogen	66	BK	40	212	2	7	19	58	Antimony-nickel-copper (intermediate)
422	Straubing, 'Ortler', Lkr. Straubing-Bogen	66	BK	40	212	2	7	19	58	Antimony-nickel-copper (intermediate)
423	Straubing, 'Ortler', Lkr. Straubing-Bogen	66	BK	40	810	2	7	19	58	Antimony-nickel-copper (intermediate)
446	Veringenstadt, Kr. Sigmaringen	62	BK	40	410	2	7	19	58	Antimony-nickel-copper (intermediate)
462	Worms, 'Weinsheim', Kr. Alzey-Worms	62	ABC	40	212	2	7	19	58	Antimony-nickel-copper (intermediate)
463	Worms, 'Weinsheim', Kr. Alzey-Worms	62	ABC	40	600	2	7	19	58	Antimony-nickel-copper (intermediate)
472	Westhofen, Kr. Alzey-Worms	62	ABC	40	600	2	7	19	58	Antimony-nickel-copper (intermediate)
793	Conthey, 'Sensine T.2'	67	BK	40	600	2	7	19	58	Antimony-nickel-copper (intermediate)
1255	Singen a. Hohentwiel, Kr. Konstanz	62	BK	40	410	2	7	20	61	Antimony-nickel-copper (intermediate)
1257	Singen a. Hohentwiel, Kr. Konstanz	62	BK	40	410	2	7	19	58	Antimony-nickel-copper (intermediate)
1263	Singen a. Hohentwiel, Kr. Konstanz	62	BK	40	500	2	7	20	61	Antimony-nickel-copper (intermediate)
1264	Singen a. Hohentwiel, Kr. Konstanz	62	BK	40	410	2	7	20	61	Antimony-nickel-copper (intermediate)
1265	Singen a. Hohentwiel, Kr. Konstanz	62	BK	40	212	2	7	19	59	Antimony-nickel-copper (intermediate)
1267	Singen a. Hohentwiel, Kr. Konstanz	62	BK	40	600	2	7	20	61	Antimony-nickel-copper (intermediate)
1268	Singen a. Hohentwiel, Kr. Konstanz	62	BK	40	600	2	7	20	62	Antimony-nickel-copper (intermediate)
1271	Singen a. Hohentwiel, Kr. Konstanz	62	BK	40	600	2	7	19	58	Antimony-nickel-copper (intermediate)
1275	Singen a. Hohentwiel, Kr. Konstanz	62	BK	40	600	2	7	19	59	Antimony-nickel-copper (intermediate)
1277	Singen a. Hohentwiel, Kr. Konstanz	62	BK	40	600	2	7	20	61	Antimony-nickel-copper (intermediate)
1279	Singen a. Hohentwiel, Kr. Konstanz	62	BK	40	200	2	7	19	59	Antimony-nickel-copper (intermediate)
2601	Manching, 'Hundsrückenacker i. d. Au', Lkr. Ingolstadt	66	BK	40	410	2	7	19	58	Antimony-nickel-copper (intermediate)
2606	Kronwinkl, Kr. Landshut	66	BK	40	410	2	7	19	59	Antimony-nickel-copper (intermediate)
2616	Ortlfingen, Lkr. Augsburg	62	BK	40	212	2	7	20	61	Antimony-nickel-copper (intermediate)
2617	Ortlfingen, Lkr. Augsburg	62	BK	40	410	2	7	19	58	Antimony-nickel-copper (intermediate)
2619	Königsbrunn, 'Oberes Feld', Lkr. Augsburg	62	BK	40	410	2	7	19	58	Antimony-nickel-copper (intermediate)
2621	Kronwinkl, Kr. Landshut	66	BK	40	410	2	7	19	58	Antimony-nickel-copper (intermediate)
2963	Grandson-Corcelettes	67		31	212	2	7	20	60	Antimony-nickel-copper (intermediate)
4186	Ayent, 'Les Place'	67	BK	40	600	2	7	20	61	Antimony-nickel-copper (intermediate)
4188	Conthey, 'Erde'	67	BK	40	600	2	7	20	60	Antimony-nickel-copper (intermediate)
4189	Conthey, 'Erde'	67	BK	40	600	2	7	20	61	Antimony-nickel-copper (intermediate)
4192	Sion, 'Maladaires'	67	BK	40	600	2	7	20	61	Antimony-nickel-copper (intermediate)
4193	Sion, 'Maladaires'	67	BK	40	600	2	7	20	61	Antimony-nickel-copper (intermediate)
4198	Ayent, 'Les Place'	67	BK	40	212	2	7	20	60	Antimony-nickel-copper (intermediate)
4199	Ayent, 'Les Place'	67	BK	40	212	2	7	19	58	Antimony-nickel-copper (intermediate)
4445	Gollnsdorf, Gde. St. Valentin, p. B. Amstetten	72	BK	40	410	2	7	20	62	Antimony-nickel-copper (intermediate)
5000	Hippersdorf, 'Plexenthal', p. B. Tulln	72	LLG	40	600	2	7	20	61	Antimony-nickel-copper (intermediate)
5005	Hippersdorf, 'Plexenthal', p. B. Tulln	72	LLG	40	410	2	7	20	61	Antimony-nickel-copper (intermediate)
5006	Hippersdorf, 'Plexenthal', p. B. Tulln	72	LLG	40	410	2	7	20	61	Antimony-nickel-copper (intermediate)
5902	Gemeinlebarn, p. B. St. Pölten	72	BK	40	410	2	7	20	61	Antimony-nickel-copper (intermediate)
5916	Gemeinlebarn, p. B. St. Pölten	72	BK	40	410	2	7	20	61	Antimony-nickel-copper (intermediate)
5969	Unterwölbling, p. B. St. Pölten	72		40	550	2	7	20	61	Antimony-nickel-copper (intermediate)
6241	Roggendorf, p. B. Horn	72		40	110	2	7	20	61	Antimony-nickel-copper (intermediate)
7796	Böhringen-Rickelshausen, Kr. Konstanz	62	BK	40	110	2	7	19	58	Antimony-nickel-copper (intermediate)
7800	Böhringen-Rickelshausen, Kr. Konstanz	62	BK	40	110	2	7	19	58	Antimony-nickel-copper (intermediate)
10368	Výcapy-Opatovce	121	NIT	40	405	2	7	20	61	Antimony-nickel-copper (intermediate)
10369	Výcapy-Opatovce	121	NIT	40	212	2	7	19	58	Antimony-nickel-copper (intermediate)
10374	Výcapy-Opatovce	121	NIT	40	410	2	7	20	61	Antimony-nickel-copper (intermediate)
10379	Výcapy-Opatovce	121	NIT	40	410	2	7	19	58	Antimony-nickel-copper (intermediate)
10382	Výcapy-Opatovce	121	NIT	40	410	2	7	20	61	Antimony-nickel-copper (intermediate)
10384	Výcapy-Opatovce	121	NIT	40	410	2	7	20	61	Antimony-nickel-copper (intermediate)

ANo	FIND_PLACE	AREA	CULT	PER	TYPE_2	CL_No_I	CL_No_II	CL_No_III	CL_No_IV	Metal_groups
10385	Výcapy-Opatovce	121	NIT	40	410	2	7	20	60	Antimony-nickel-copper (intermediate)
10421	Výcapy-Opatovce	121	NIT	40	400	2	7	20	61	Antimony-nickel-copper (intermediate)
10423	Výcapy-Opatovce	121	NIT	40	405	2	7	20	61	Antimony-nickel-copper (intermediate)
10424	Výcapy-Opatovce	121	NIT	40	405	2	7	20	61	Antimony-nickel-copper (intermediate)
10437	Výcapy-Opatovce	121	NIT	40	400	2	7	20	61	Antimony-nickel-copper (intermediate)
10440	Výcapy-Opatovce	121	NIT	40	410	2	7	20	61	Antimony-nickel-copper (intermediate)
10442	Výcapy-Opatovce	121	NIT	40	400	2	7	20	61	Antimony-nickel-copper (intermediate)
10452	Výcapy-Opatovce	121	NIT	40	400	2	7	20	60	Antimony-nickel-copper (intermediate)
10453	Výcapy-Opatovce	121	NIT	40	405	2	7	20	60	Antimony-nickel-copper (intermediate)
10454	Výcapy-Opatovce	121	NIT	40	400	2	7	20	61	Antimony-nickel-copper (intermediate)
10455	Výcapy-Opatovce	121	NIT	40	400	2	7	20	61	Antimony-nickel-copper (intermediate)
10458	Výcapy-Opatovce	121	NIT	40	400	2	7	20	61	Antimony-nickel-copper (intermediate)
10460	Výcapy-Opatovce	121	NIT	40	400	2	7	20	61	Antimony-nickel-copper (intermediate)
10471	Výcapy-Opatovce	121	NIT	40	405	2	7	20	60	Antimony-nickel-copper (intermediate)
10472	Výcapy-Opatovce	121	NIT	40	410	2	7	20	60	Antimony-nickel-copper (intermediate)
10473	Výcapy-Opatovce	121	NIT	40	410	2	7	20	60	Antimony-nickel-copper (intermediate)
10475	Výcapy-Opatovce	121	NIT	40	405	2	7	20	60	Antimony-nickel-copper (intermediate)
10478	Výcapy-Opatovce	121	NIT	40	400	2	7	20	61	Antimony-nickel-copper (intermediate)
10482	Výcapy-Opatovce	121	NIT	40	400	2	7	20	61	Antimony-nickel-copper (intermediate)
10485	Výcapy-Opatovce	121	NIT	40	410	2	7	20	61	Antimony-nickel-copper (intermediate)
10487	Výcapy-Opatovce	121	NIT	40	410	2	7	20	62	Antimony-nickel-copper (intermediate)
10489	Výcapy-Opatovce	121	NIT	40	405	2	7	20	61	Antimony-nickel-copper (intermediate)
10503	Výcapy-Opatovce	121	NIT	40	400	2	7	20	61	Antimony-nickel-copper (intermediate)
10505	Výcapy-Opatovce	121	NIT	40	400	2	7	20	61	Antimony-nickel-copper (intermediate)
10507	Výcapy-Opatovce	121	NIT	40	400	2	7	20	61	Antimony-nickel-copper (intermediate)
10508	Výcapy-Opatovce	121	NIT	40	400	2	7	20	61	Antimony-nickel-copper (intermediate)
10512	Výcapy-Opatovce	121	NIT	40	400	2	7	20	61	Antimony-nickel-copper (intermediate)
10513	Výcapy-Opatovce	121	NIT	40	400	2	7	20	61	Antimony-nickel-copper (intermediate)
10514	Výcapy-Opatovce	121	NIT	40	400	2	7	20	61	Antimony-nickel-copper (intermediate)
10515	Výcapy-Opatovce	121	NIT	40	400	2	7	20	61	Antimony-nickel-copper (intermediate)
10517	Výcapy-Opatovce	121	NIT	40	410	2	7	20	61	Antimony-nickel-copper (intermediate)
10519	Výcapy-Opatovce	121	NIT	40	400	2	7	20	61	Antimony-nickel-copper (intermediate)
10529	Výcapy-Opatovce	121	NIT	40	410	2	7	20	61	Antimony-nickel-copper (intermediate)
10530	Výcapy-Opatovce	121	NIT	40	410	2	7	20	61	Antimony-nickel-copper (intermediate)
10533	Výcapy-Opatovce	121	NIT	40	410	2	7	20	61	Antimony-nickel-copper (intermediate)
10539	Výcapy-Opatovce	121	NIT	40	410	2	7	20	61	Antimony-nickel-copper (intermediate)
10540	Výcapy-Opatovce	121	NIT	40	410	2	7	20	61	Antimony-nickel-copper (intermediate)
10541	Výcapy-Opatovce	121	NIT	40	400	2	7	20	61	Antimony-nickel-copper (intermediate)
10544	Výcapy-Opatovce	121	NIT	40	410	2	7	20	61	Antimony-nickel-copper (intermediate)
10547	Výcapy-Opatovce	121	NIT	40	400	2	7	20	61	Antimony-nickel-copper (intermediate)
10555	Výcapy-Opatovce	121	NIT	40	410	2	7	20	61	Antimony-nickel-copper (intermediate)
10560	Výcapy-Opatovce	121	NIT	40	410	2	7	20	61	Antimony-nickel-copper (intermediate)
10561	Výcapy-Opatovce	121	NIT	40	400	2	7	20	61	Antimony-nickel-copper (intermediate)
10562	Výcapy-Opatovce	121	NIT	40	400	2	7	20	61	Antimony-nickel-copper (intermediate)
10564	Výcapy-Opatovce	121	NIT	40	400	2	7	20	61	Antimony-nickel-copper (intermediate)
10574	Komjatice, 'Homoky'	121	NIT	40	400	2	7	20	61	Antimony-nickel-copper (intermediate)
10647	Tvrdošovce, 'Körektó'	121	NIT	40	212	2	7	20	60	Antimony-nickel-copper (intermediate)
10652	Tvrdošovce, 'Körektó'	121	NIT	40	410	2	7	20	62	Antimony-nickel-copper (intermediate)
10653	Tvrdošovce, 'Körektó'	121	NIT	40	410	2	7	20	61	Antimony-nickel-copper (intermediate)
10667	Tvrdošovce, 'Körektó'	121	NIT	40	410	2	7	20	61	Antimony-nickel-copper (intermediate)
10831	Výcapy-Opatovce	121	NIT	40	212	2	7	20	60	Antimony-nickel-copper (intermediate)
10834	Výcapy-Opatovce	121	NIT	40	405	2	7	20	60	Antimony-nickel-copper (intermediate)
10836	Výcapy-Opatovce	121	NIT	40	410	2	7	20	60	Antimony-nickel-copper (intermediate)

ANo	FIND_PLACE	AREA	CULT	PER	TYPE_2	CL_No_I	CL_No_II	CL_No_III	CL_No_IV	Metal_groups
11511	Jois, p. B. Neusiedl a. See	123	LLG	40	600	2	7	20	60	Antimony-nickel-copper (intermediate)
11513	Loretto, p. B. Eisenstadt-Umgebung	123	LLG	40	405	2	7	19	58	Antimony-nickel-copper (intermediate)
11515	Loretto, p. B. Eisenstadt-Umgebung	123	LLG	40	405	2	7	19	58	Antimony-nickel-copper (intermediate)
11516	Loretto, p. B. Eisenstadt-Umgebung	123	LLG	40	405	2	7	19	58	Antimony-nickel-copper (intermediate)
11518	Loretto, p. B. Eisenstadt-Umgebung	123	LLG	40	600	2	7	20	60	Antimony-nickel-copper (intermediate)
11521	Loretto, p. B. Eisenstadt-Umgebung	123	LLG	40	400	2	7	19	58	Antimony-nickel-copper (intermediate)
11522	Loretto, p. B. Eisenstadt-Umgebung	123	LLG	40	405	2	7	20	60	Antimony-nickel-copper (intermediate)
11523	Loretto, p. B. Eisenstadt-Umgebung	123	LLG	40	405	2	7	19	58	Antimony-nickel-copper (intermediate)
11525	Loretto, p. B. Eisenstadt-Umgebung	123	LLG	40	550	2	7	20	60	Antimony-nickel-copper (intermediate)
11528	Loretto, p. B. Eisenstadt-Umgebung	123	LLG	40	410	2	7	19	58	Antimony-nickel-copper (intermediate)
15823	München-Langwied, 'Lochhausen', Stkr. München	66	BK	40	212	2	7	20	60	Antimony-nickel-copper (intermediate)
15827	München-Zamdorf, 'Emin-Pascha-Strasse', Stkr. München	66	BK	40	510	2	7	20	60	Antimony-nickel-copper (intermediate)
15835	Manching, 'Hundsrückenacker i. d. Au', Lkr. Ingolstadt	66	BK	40	410	2	7	20	61	Antimony-nickel-copper (intermediate)
15841	München-Sendling, 'Lindwurmstr. 76', Stkr. München	66	BK	40	600	2	7	19	58	Antimony-nickel-copper (intermediate)
15842	München-Sendling, 'Lindwurmstr. 76', Stkr. München	66	BK	40	500	2	7	20	60	Antimony-nickel-copper (intermediate)
15849	München-Sendling, 'Lindwurmstr. 76', Stkr. München	66	BK	40	600	2	7	20	61	Antimony-nickel-copper (intermediate)
15850	München-Sendling, 'Lindwurmstr. 76', Stkr. München	66	BK	40	410	2	7	20	60	Antimony-nickel-copper (intermediate)
15851	München-Sendling, 'Lindwurmstr. 76', Stkr. München	66	BK	40	410	2	7	20	61	Antimony-nickel-copper (intermediate)
15852	München-Sendling, 'Lindwurmstr. 76', Stkr. München	66	BK	40	410	2	7	20	60	Antimony-nickel-copper (intermediate)
15881	Kronwinkl, Kr. Landshut	66	BK	40	510	2	7	20	61	Antimony-nickel-copper (intermediate)
15913	Gilching, 'Steinbergäcker', Lkr. Starnberg	66	BK	40	410	2	7	20	60	Antimony-nickel-copper (intermediate)
16276	Horelice	111		40	110	2	7	20	61	Antimony-nickel-copper (intermediate)
21579	Griesheim, Ortenaukreis	62	BK	40	110	2	7	20	60	Antimony-nickel-copper (intermediate)
21580	Griesheim, Ortenaukreis	62	BK	40	110	2	7	19	58	Antimony-nickel-copper (intermediate)
24189	Singen a. Hohentwiel, Kr. Konstanz	62	BK	40	600	2	7	19	59	Antimony-nickel-copper (intermediate)
24190	Singen a. Hohentwiel, Kr. Konstanz	62	BK	40	410	2	7	19	58	Antimony-nickel-copper (intermediate)
24191	Singen a. Hohentwiel, Kr. Konstanz	62	BK	40	500	2	7	19	59	Antimony-nickel-copper (intermediate)
24192	Singen a. Hohentwiel, Kr. Konstanz	62	BK	40	600	2	7	19	58	Antimony-nickel-copper (intermediate)
24193	Singen a. Hohentwiel, Kr. Konstanz	62	BK	40	410	2	7	19	59	Antimony-nickel-copper (intermediate)
24194	Singen a. Hohentwiel, Kr. Konstanz	62	BK	40	810	2	7	19	59	Antimony-nickel-copper (intermediate)
24195	Singen a. Hohentwiel, Kr. Konstanz	62	BK	40	600	2	7	19	59	Antimony-nickel-copper (intermediate)
24196	Singen a. Hohentwiel, Kr. Konstanz	62	BK	40	600	2	7	19	59	Antimony-nickel-copper (intermediate)
24197	Singen a. Hohentwiel, Kr. Konstanz	62	BK	40	600	2	7	19	59	Antimony-nickel-copper (intermediate)
24199	Singen a. Hohentwiel, Kr. Konstanz	62	BK	40	410	2	7	19	59	Antimony-nickel-copper (intermediate)
24201	Singen a. Hohentwiel, Kr. Konstanz	62	BK	40	600	2	7	19	59	Antimony-nickel-copper (intermediate)
24202	Singen a. Hohentwiel, Kr. Konstanz	62	BK	40	600	2	7	19	59	Antimony-nickel-copper (intermediate)
24203	Singen a. Hohentwiel, Kr. Konstanz	62	BK	40	400	2	7	19	58	Antimony-nickel-copper (intermediate)
24204	Singen a. Hohentwiel, Kr. Konstanz	62	BK	40	600	2	7	19	59	Antimony-nickel-copper (intermediate)
24205	Singen a. Hohentwiel, Kr. Konstanz	62	BK	40	600	2	7	19	59	Antimony-nickel-copper (intermediate)
24206	Singen a. Hohentwiel, Kr. Konstanz	62	BK	40	810	2	7	19	59	Antimony-nickel-copper (intermediate)
24207	Singen a. Hohentwiel, Kr. Konstanz	62	BK	40	212	2	7	19	59	Antimony-nickel-copper (intermediate)
24209	Singen a. Hohentwiel, Kr. Konstanz	62	BK	40	213	2	7	19	59	Antimony-nickel-copper (intermediate)
24211	Singen a. Hohentwiel, Kr. Konstanz	62	BK	40	600	2	7	19	59	Antimony-nickel-copper (intermediate)
29777	München-Sendling, 'Lindwurmstr. 76', Stkr. München	66	BK	40	600	2	7	20	60	Antimony-nickel-copper (intermediate)
29955	Výcapy-Opatovce	121	NIT	40	400	2	7	20	61	Antimony-nickel-copper (intermediate)
32246	Issersheiligen, Unstrut-Hainich-Kreis	103		25	100	2	7	19	59	Antimony-nickel-copper (intermediate)
33031	Bautzen-Niederkaina, 'Schafberg, Qu III', Lkr. Bautzen	105	CWC	31	510	2	7	19	58	Antimony-nickel-copper (intermediate)
34134	Issersheiligen, Unstrut-Hainich-Kreis	103		25	100	2	7	19	59	Antimony-nickel-copper (intermediate)
34682	Halberstadt-Emersleben, Lkr. Harz	103		31	130	2	7	19	59	Antimony-nickel-copper (intermediate)
44001	Iwanowice, 'Babia Góra II	85	Mier	40	420	2	7	19	58	Antimony-nickel-copper (intermediate)
44002	Iwanowice, 'Babia Góra II	85	Mier	40	400	2	7	19	58	Antimony-nickel-copper (intermediate)
44004	Iwanowice, 'Babia Góra II	85	Mier	40	800	2	7	19	58	Antimony-nickel-copper (intermediate)
44005	Iwanowice, 'Babia Góra II	85	Mier	40	500	2	7	19	58	Antimony-nickel-copper (intermediate)

ANo	FIND_PLACE	AREA	CULT	PER	TYPE_2	CL_No_I	CL_No_II	CL_No_III	CL_No_IV	Metal_groups
44006	Iwanowice, 'Babia Góra II	85	Mier	40	510	2	7	19	58	Antimony-nickel-copper (intermediate)
44007	Iwanowice, 'Babia Góra II	85	Mier	40	500	2	7	19	59	Antimony-nickel-copper (intermediate)
44008	Iwanowice, 'Babia Góra II	85	Mier	40	500	2	7	19	58	Antimony-nickel-copper (intermediate)
68026	Regensdorf, 'Chatzensee'	67	BK	40	110	2	7	19	58	Antimony-nickel-copper (intermediate)
68032	Bronschhofen	67	BK	40	110	2	7	19	58	Antimony-nickel-copper (intermediate)
68035	Winterthur, 'Wülflingen'	67	BK	40	110	2	7	19	58	Antimony-nickel-copper (intermediate)
68045	Mels, 'Rossheld-Schlucht'	67	BK	40	110	2	7	19	58	Antimony-nickel-copper (intermediate)
68046	Meersburg, 'Haltau', Bodenseekreis	62	BK	40	110	2	7	20	62	Antimony-nickel-copper (intermediate)
68048	Salez, 'Sennwald'	67	BK	40	110	2	7	20	61	Antimony-nickel-copper (intermediate)
78160	Branc	121	NIT	40	420	2	7	19	58	Antimony-nickel-copper (intermediate)
78161	Branc	121	NIT	40	420	2	7	19	58	Antimony-nickel-copper (intermediate)
78165	Branc	121	NIT	40	400	2	7	19	58	Antimony-nickel-copper (intermediate)
78166	Branc	121	NIT	40	410	2	7	19	58	Antimony-nickel-copper (intermediate)
78167	Branc	121	NIT	40	212	2	7	19	58	Antimony-nickel-copper (intermediate)
78175	Tvrdošovce, 'Körektó'	121	NIT	40	212	2	7	19	59	Antimony-nickel-copper (intermediate)
78176	Tvrdošovce, 'Körektó'	121	NIT	40	300	2	7	19	58	Antimony-nickel-copper (intermediate)
100000	Gerlingen, Lkr. Ludwigsburg	62	BK	40	212	2	7	19	59	Antimony-nickel-copper (intermediate)
100001	Gerlingen, Lkr. Ludwigsburg	62	BK	40	213	2	7	19	59	Antimony-nickel-copper (intermediate)
100002	Gerlingen, Lkr. Ludwigsburg	62	BK	40	212	2	7	19	59	Antimony-nickel-copper (intermediate)
100003	Gerlingen, Lkr. Ludwigsburg	62	BK	40	213	2	7	19	59	Antimony-nickel-copper (intermediate)
100004	Gemmerigheim, Lkr. Ludwigsburg	62	BK	40	405	2	7	19	59	Antimony-nickel-copper (intermediate)
100006	Mengen, Umgebung, Kr. Sigmaringen	62	BK	40	110	2	7	19	59	Antimony-nickel-copper (intermediate)
100095	Burgwenden, 'Am Mühlberg', Lkr. Sömmerda	103		21	100	2	7	19	58	Antimony-nickel-copper (intermediate)
1	Adlerberg b. Worms, Kr. Alzey-Worms	62	ABC	40	810	2	7	20	70	Antimony-nickel-copper (low)
31	Berghausen, 'Roßweide', Lkr. Karlsruhe	62		40	600	2	7	20	70	Antimony-nickel-copper (low)
67	Stuttgart-Bad Cannstatt, Stkr. Stuttgart	62	BK	40	600	2	7	20	70	Antimony-nickel-copper (low)
68	Stuttgart-Bad Cannstatt, Stkr. Stuttgart	62	BK	40	550	2	7	20	70	Antimony-nickel-copper (low)
69	Stuttgart-Bad Cannstatt, Stkr. Stuttgart	62	BK	40	410	2	7	20	70	Antimony-nickel-copper (low)
70	Stuttgart-Bad Cannstatt, Stkr. Stuttgart	62	BK	40	410	2	7	20	70	Antimony-nickel-copper (low)
81	Eriskirch, Bodenseekreis	62	BK	40	110	2	7	20	71	Antimony-nickel-copper (low)
83	Eriskirch, Bodenseekreis	62	BK	40	110	2	7	20	71	Antimony-nickel-copper (low)
146	Kelsterbach, 'Siedl. Kesselschmiede', Lkr. Groß-Gerau	62	CWC	31	510	2	7	20	71	Antimony-nickel-copper (low)
150	Kelsterbach, 'Siedl. Kesselschmiede', Lkr. Groß-Gerau	62	CWC	31	800	2	7	20	69	Antimony-nickel-copper (low)
153	Kirchheim u. Teck, Kr. Esslingen	62	BK	40	212	2	7	20	71	Antimony-nickel-copper (low)
154	Kirchheim u. Teck, Kr. Esslingen	62	BK	40	212	2	7	20	66	Antimony-nickel-copper (low)
190	Monsheim, 'Kahleberg', Kr. Alzey-Worms	62	ABC	40	212	2	7	20	71	Antimony-nickel-copper (low)
243	Ostrach, Kr. Sigmaringen	62	BK	40	600	2	7	20	66	Antimony-nickel-copper (low)
273	Rottweil, Lkr. Rottweil	62	BK	40	110	2	7	20	71	Antimony-nickel-copper (low)
274	Rottweil, Lkr. Rottweil	62	BK	40	110	2	7	20	71	Antimony-nickel-copper (low)
275	Rottweil, Lkr. Rottweil	62	BK	40	110	2	7	20	71	Antimony-nickel-copper (low)
330	Singen a. Hohentwiel, Kr. Konstanz	62	BK	40	410	2	7	20	71	Antimony-nickel-copper (low)
348	Singen a. Hohentwiel, Kr. Konstanz	62	BK	40	600	2	7	20	71	Antimony-nickel-copper (low)
351	Singen a. Hohentwiel, Kr. Konstanz	62	BK	40	600	2	7	20	67	Antimony-nickel-copper (low)
352	Singen a. Hohentwiel, Kr. Konstanz	62	BK	40	212	2	7	20	70	Antimony-nickel-copper (low)
397	Straubing, 'Jungmeier', Lkr. Straubing-Bogen	66	BK	40	212	2	7	20	69	Antimony-nickel-copper (low)
429	Straubing, 'Ortler', Lkr. Straubing-Bogen	66	BK	40	212	2	7	20	71	Antimony-nickel-copper (low)
478	Wiesloch, Rhein-Neckar-Kreis	62	ABC	40	212	2	7	20	71	Antimony-nickel-copper (low)
1251	Singen a. Hohentwiel, Kr. Konstanz	62	BK	40	600	2	7	20	70	Antimony-nickel-copper (low)
1262	Singen a. Hohentwiel, Kr. Konstanz	62	BK	40	410	2	7	20	70	Antimony-nickel-copper (low)
1273	Singen a. Hohentwiel, Kr. Konstanz	62	BK	40	410	2	7	20	70	Antimony-nickel-copper (low)
1276	Singen a. Hohentwiel, Kr. Konstanz	62	BK	40	410	2	7	20	70	Antimony-nickel-copper (low)
2598	Manching, 'Hundsrückenacker i. d. Au', Lkr. Ingolstadt	66	BK	40	410	2	7	20	69	Antimony-nickel-copper (low)
2602	Manching, 'Hundsrückenacker i. d. Au', Lkr. Ingolstadt	66	BK	40	410	2	7	20	66	Antimony-nickel-copper (low)

ANo	FIND_PLACE	AREA	CULT	PER	TYPE_2	CL_No_I	CL_No_II	CL_No_III	CL_No_IV	Metal_groups
2603	Kronwinkl, Kr. Landshut	66	BK	40	600	2	7	20	70	Antimony-nickel-copper (low)
2604	Kronwinkl, Kr. Landshut	66	BK	40	212	2	7	20	71	Antimony-nickel-copper (low)
2605	Kronwinkl, Kr. Landshut	66	BK	40	410	2	7	20	71	Antimony-nickel-copper (low)
2769	Feldkirch, 'Tillis', p. B. Feldkirch	71	BK	40	110	2	7	20	67	Antimony-nickel-copper (low)
2843	Vinelz, 'Amt Erlau'	67	CWC	31	810	2	7	20	69	Antimony-nickel-copper (low)
2931	Lüscherz	67	CWC	31	212	2	7	20	69	Antimony-nickel-copper (low)
3088	Vinelz, 'Amt Erlau'	67	CWC	31	810	2	7	20	67	Antimony-nickel-copper (low)
4184	Ayent, 'Les Place'	67	BK	40	600	2	7	20	67	Antimony-nickel-copper (low)
4197	Ayent, 'Les Place'	67	BK	40	212	2	7	20	67	Antimony-nickel-copper (low)
4463	Gollnsdorf, Gde. St. Valentin, p. B. Amstetten	72	BK	40	600	2	7	20	69	Antimony-nickel-copper (low)
5002	Hippersdorf, 'Plexenthal', p. B. Tulln	72	LLG	40	410	2	7	20	69	Antimony-nickel-copper (low)
5007	Hippersdorf, 'Plexenthal', p. B. Tulln	72	LLG	40	405	2	7	20	69	Antimony-nickel-copper (low)
5008	Hippersdorf, 'Plexenthal', p. B. Tulln	72	LLG	40	400	2	7	20	69	Antimony-nickel-copper (low)
5009	Hippersdorf, 'Plexenthal', p. B. Tulln	72	LLG	40	410	2	7	20	67	Antimony-nickel-copper (low)
5931	Gemeinlebarn, p. B. St. Pölten	72	BK	40	410	2	7	20	67	Antimony-nickel-copper (low)
5932	Gemeinlebarn, p. B. St. Pölten	72	BK	40	410	2	7	20	67	Antimony-nickel-copper (low)
5933	Gemeinlebarn, p. B. St. Pölten	72	BK	40	410	2	7	20	67	Antimony-nickel-copper (low)
5934	Gemeinlebarn, p. B. St. Pölten	72	BK	40	410	2	7	20	67	Antimony-nickel-copper (low)
6596	Unterwölbling, p. B. St. Pölten	72	BK	40	410	2	7	20	66	Antimony-nickel-copper (low)
6597	Unterwölbling, p. B. St. Pölten	72	BK	40	410	2	7	20	66	Antimony-nickel-copper (low)
6714	Singen a. Hohentwiel, Kr. Konstanz	62	BK	40	410	2	7	20	67	Antimony-nickel-copper (low)
6715	Singen a. Hohentwiel, Kr. Konstanz	62	BK	40	212	2	7	20	71	Antimony-nickel-copper (low)
7281	Bex, 'Les Mûriers'	67	BK	40	110	2	7	20	69	Antimony-nickel-copper (low)
7512	Triskolupy	111		40	110	2	7	20	69	Antimony-nickel-copper (low)
7528	Predmerice nad Labem	112		40	110	2	7	20	69	Antimony-nickel-copper (low)
7534	Cisteves	112		40	110	2	7	20	69	Antimony-nickel-copper (low)
7536	Predmerice nad Labem	112		40	100	2	7	20	71	Antimony-nickel-copper (low)
7797	Böhringen-Rickelshausen, Kr. Konstanz	62	BK	40	110	2	7	20	66	Antimony-nickel-copper (low)
7798	Böhringen-Rickelshausen, Kr. Konstanz	62	BK	40	110	2	7	20	71	Antimony-nickel-copper (low)
7799	Böhringen-Rickelshausen, Kr. Konstanz	62	BK	40	110	2	7	20	71	Antimony-nickel-copper (low)
10371	Výcapy-Opatovce	121	NIT	40	420	2	7	20	66	Antimony-nickel-copper (low)
10372	Výcapy-Opatovce	121	NIT	40	420	2	7	20	66	Antimony-nickel-copper (low)
10373	Výcapy-Opatovce	121	NIT	40	420	2	7	20	66	Antimony-nickel-copper (low)
10376	Výcapy-Opatovce	121	NIT	40	410	2	7	20	66	Antimony-nickel-copper (low)
10377	Výcapy-Opatovce	121	NIT	40	410	2	7	20	66	Antimony-nickel-copper (low)
10383	Výcapy-Opatovce	121	NIT	40	300	2	7	20	71	Antimony-nickel-copper (low)
10407	Výcapy-Opatovce	121	NIT	40	410	2	7	20	66	Antimony-nickel-copper (low)
10408	Výcapy-Opatovce	121	NIT	40	410	2	7	20	66	Antimony-nickel-copper (low)
10409	Výcapy-Opatovce	121	NIT	40	400	2	7	20	66	Antimony-nickel-copper (low)
10410	Výcapy-Opatovce	121	NIT	40	400	2	7	20	66	Antimony-nickel-copper (low)
10412	Výcapy-Opatovce	121	NIT	40	400	2	7	20	67	Antimony-nickel-copper (low)
10414	Výcapy-Opatovce	121	NIT	40	420	2	7	20	69	Antimony-nickel-copper (low)
10416	Výcapy-Opatovce	121	NIT	40	420	2	7	20	69	Antimony-nickel-copper (low)
10419	Výcapy-Opatovce	121	NIT	40	420	2	7	20	71	Antimony-nickel-copper (low)
10420	Výcapy-Opatovce	121	NIT	40	400	2	7	20	71	Antimony-nickel-copper (low)
10422	Výcapy-Opatovce	121	NIT	40	420	2	7	20	66	Antimony-nickel-copper (low)
10425	Výcapy-Opatovce	121	NIT	40	410	2	7	20	66	Antimony-nickel-copper (low)
10426	Výcapy-Opatovce	121	NIT	40	420	2	7	20	66	Antimony-nickel-copper (low)
10427	Výcapy-Opatovce	121	NIT	40	420	2	7	20	69	Antimony-nickel-copper (low)
10428	Výcapy-Opatovce	121	NIT	40	420	2	7	20	69	Antimony-nickel-copper (low)
10429	Výcapy-Opatovce	121	NIT	40	410	2	7	20	66	Antimony-nickel-copper (low)
10430	Výcapy-Opatovce	121	NIT	40	405	2	7	20	66	Antimony-nickel-copper (low)
10434	Výcapy-Opatovce	121	NIT	40	420	2	7	20	69	Antimony-nickel-copper (low)

ANo	FIND_PLACE	AREA	CULT	PER	TYPE_2	CL_No_I	CL_No_II	CL_No_III	CL_No_IV	Metal_groups
10435	Výcapy-Opatovce	121	NIT	40	400	2	7	20	69	Antimony-nickel-copper (low)
10436	Výcapy-Opatovce	121	NIT	40	405	2	7	20	66	Antimony-nickel-copper (low)
10438	Výcapy-Opatovce	121	NIT	40	410	2	7	20	66	Antimony-nickel-copper (low)
10439	Výcapy-Opatovce	121	NIT	40	400	2	7	20	66	Antimony-nickel-copper (low)
10441	Výcapy-Opatovce	121	NIT	40	400	2	7	20	66	Antimony-nickel-copper (low)
10443	Výcapy-Opatovce	121	NIT	40	400	2	7	20	66	Antimony-nickel-copper (low)
10449	Výcapy-Opatovce	121	NIT	40	410	2	7	20	66	Antimony-nickel-copper (low)
10457	Výcapy-Opatovce	121	NIT	40	400	2	7	20	67	Antimony-nickel-copper (low)
10459	Výcapy-Opatovce	121	NIT	40	410	2	7	20	66	Antimony-nickel-copper (low)
10461	Výcapy-Opatovce	121	NIT	40	400	2	7	20	66	Antimony-nickel-copper (low)
10465	Výcapy-Opatovce	121	NIT	40	410	2	7	20	66	Antimony-nickel-copper (low)
10466	Výcapy-Opatovce	121	NIT	40	410	2	7	20	66	Antimony-nickel-copper (low)
10467	Výcapy-Opatovce	121	NIT	40	410	2	7	20	66	Antimony-nickel-copper (low)
10476	Výcapy-Opatovce	121	NIT	40	410	2	7	20	66	Antimony-nickel-copper (low)
10477	Výcapy-Opatovce	121	NIT	40	400	2	7	20	66	Antimony-nickel-copper (low)
10479	Výcapy-Opatovce	121	NIT	40	420	2	7	20	66	Antimony-nickel-copper (low)
10480	Výcapy-Opatovce	121	NIT	40	405	2	7	20	69	Antimony-nickel-copper (low)
10481	Výcapy-Opatovce	121	NIT	40	410	2	7	20	66	Antimony-nickel-copper (low)
10483	Výcapy-Opatovce	121	NIT	40	410	2	7	20	67	Antimony-nickel-copper (low)
10484	Výcapy-Opatovce	121	NIT	40	410	2	7	20	69	Antimony-nickel-copper (low)
10486	Výcapy-Opatovce	121	NIT	40	410	2	7	20	66	Antimony-nickel-copper (low)
10492	Výcapy-Opatovce	121	NIT	40	420	2	7	20	66	Antimony-nickel-copper (low)
10493	Výcapy-Opatovce	121	NIT	40	410	2	7	20	67	Antimony-nickel-copper (low)
10494	Výcapy-Opatovce	121	NIT	40	405	2	7	20	66	Antimony-nickel-copper (low)
10497	Výcapy-Opatovce	121	NIT	40	410	2	7	20	66	Antimony-nickel-copper (low)
10498	Výcapy-Opatovce	121	NIT	40	410	2	7	20	66	Antimony-nickel-copper (low)
10499	Výcapy-Opatovce	121	NIT	40	420	2	7	20	71	Antimony-nickel-copper (low)
10501	Výcapy-Opatovce	121	NIT	40	420	2	7	20	69	Antimony-nickel-copper (low)
10504	Výcapy-Opatovce	121	NIT	40	420	2	7	20	71	Antimony-nickel-copper (low)
10506	Výcapy-Opatovce	121	NIT	40	400	2	7	20	71	Antimony-nickel-copper (low)
10509	Výcapy-Opatovce	121	NIT	40	410	2	7	20	66	Antimony-nickel-copper (low)
10516	Výcapy-Opatovce	121	NIT	40	400	2	7	20	66	Antimony-nickel-copper (low)
10518	Výcapy-Opatovce	121	NIT	40	400	2	7	20	66	Antimony-nickel-copper (low)
10520	Výcapy-Opatovce	121	NIT	40	400	2	7	20	66	Antimony-nickel-copper (low)
10522	Výcapy-Opatovce	121	NIT	40	410	2	7	20	66	Antimony-nickel-copper (low)
10523	Výcapy-Opatovce	121	NIT	40	410	2	7	20	66	Antimony-nickel-copper (low)
10525	Výcapy-Opatovce	121	NIT	40	300	2	7	20	66	Antimony-nickel-copper (low)
10527	Výcapy-Opatovce	121	NIT	40	400	2	7	20	66	Antimony-nickel-copper (low)
10528	Výcapy-Opatovce	121	NIT	40	410	2	7	20	66	Antimony-nickel-copper (low)
10534	Výcapy-Opatovce	121	NIT	40	500	2	7	20	71	Antimony-nickel-copper (low)
10535	Výcapy-Opatovce	121	NIT	40	420	2	7	20	66	Antimony-nickel-copper (low)
10538	Výcapy-Opatovce	121	NIT	40	420	2	7	20	69	Antimony-nickel-copper (low)
10542	Výcapy-Opatovce	121	NIT	40	400	2	7	20	66	Antimony-nickel-copper (low)
10543	Výcapy-Opatovce	121	NIT	40	410	2	7	20	66	Antimony-nickel-copper (low)
10545	Výcapy-Opatovce	121	NIT	40	410	2	7	20	66	Antimony-nickel-copper (low)
10546	Výcapy-Opatovce	121	NIT	40	400	2	7	20	66	Antimony-nickel-copper (low)
10548	Výcapy-Opatovce	121	NIT	40	400	2	7	20	67	Antimony-nickel-copper (low)
10549	Výcapy-Opatovce	121	NIT	40	400	2	7	20	66	Antimony-nickel-copper (low)
10550	Výcapy-Opatovce	121	NIT	40	400	2	7	20	66	Antimony-nickel-copper (low)
10551	Výcapy-Opatovce	121	NIT	40	420	2	7	20	66	Antimony-nickel-copper (low)
10553	Výcapy-Opatovce	121	NIT	40	420	2	7	20	67	Antimony-nickel-copper (low)
10556	Výcapy-Opatovce	121	NIT	40	300	2	7	20	69	Antimony-nickel-copper (low)
10557	Výcapy-Opatovce	121	NIT	40	400	2	7	20	71	Antimony-nickel-copper (low)

ANo	FIND_PLACE	AREA	CULT	PER	TYPE_2	CL_No_I	CL_No_II	CL_No_III	CL_No_IV	Metal_groups
10558	Výcapy-Opatovce	121	NIT	40	400	2	7	20	66	Antimony-nickel-copper (low)
10559	Výcapy-Opatovce	121	NIT	40	400	2	7	20	71	Antimony-nickel-copper (low)
10563	Výcapy-Opatovce	121	NIT	40	400	2	7	20	66	Antimony-nickel-copper (low)
10568	Komjatice	121	NIT	40	420	2	7	20	66	Antimony-nickel-copper (low)
10569	Komjatice, 'Homoky'	121	NIT	40	420	2	7	20	69	Antimony-nickel-copper (low)
10570	Komjatice, 'Homoky'	121	NIT	40	420	2	7	20	69	Antimony-nickel-copper (low)
10571	Komjatice, 'Homoky'	121	NIT	40	410	2	7	20	69	Antimony-nickel-copper (low)
10575	Komjatice, 'Homoky'	121	NIT	40	410	2	7	20	69	Antimony-nickel-copper (low)
10643	Tvrdošovce, 'Körektó'	121	NIT	40	810	2	7	20	69	Antimony-nickel-copper (low)
10644	Tvrdošovce, 'Körektó'	121	NIT	40	300	2	7	20	71	Antimony-nickel-copper (low)
10645	Tvrdošovce, 'Körektó'	121	NIT	40	300	2	7	20	69	Antimony-nickel-copper (low)
10648	Tvrdošovce, 'Körektó'	121	NIT	40	200	2	7	20	71	Antimony-nickel-copper (low)
10654	Tvrdošovce, 'Körektó'	121	NIT	40	300	2	7	20	71	Antimony-nickel-copper (low)
10659	Tvrdošovce, 'Körektó'	121	NIT	40	300	2	7	20	66	Antimony-nickel-copper (low)
10662	Tvrdošovce, 'Körektó'	121	NIT	40	300	2	7	20	69	Antimony-nickel-copper (low)
10663	Tvrdošovce, 'Körektó'	121	NIT	40	420	2	7	20	71	Antimony-nickel-copper (low)
10668	Tvrdošovce, 'Körektó'	121	NIT	40	420	2	7	20	69	Antimony-nickel-copper (low)
10669	Tvrdošovce, 'Körektó'	121	NIT	40	420	2	7	20	69	Antimony-nickel-copper (low)
10671	Tvrdošovce, 'Körektó'	121	NIT	40	420	2	7	20	69	Antimony-nickel-copper (low)
10761	Komjatice, 'Argentina'	121	NIT	40	410	2	7	20	66	Antimony-nickel-copper (low)
10832	Výcapy-Opatovce	121	NIT	40	420	2	7	20	69	Antimony-nickel-copper (low)
10902	Abrahám	121	NIT	40	600	2	7	20	69	Antimony-nickel-copper (low)
10921	Abrahám	121	NIT	40	420	2	7	20	66	Antimony-nickel-copper (low)
10922	Abrahám	121	NIT	40	420	2	7	20	66	Antimony-nickel-copper (low)
11389	Breitenwaida, p. B. Hollabrunn	72		40	110	2	7	20	69	Antimony-nickel-copper (low)
11512	Loretto, p. B. Eisenstadt-Umgebung	123	LLG	40	410	2	7	20	71	Antimony-nickel-copper (low)
11519	Loretto, p. B. Eisenstadt-Umgebung	123	LLG	40	500	2	7	20	70	Antimony-nickel-copper (low)
11520	Loretto, p. B. Eisenstadt-Umgebung	123	LLG	40	600	2	7	20	67	Antimony-nickel-copper (low)
11524	Loretto, p. B. Eisenstadt-Umgebung	123	LLG	40	500	2	7	20	69	Antimony-nickel-copper (low)
11529	Loretto, p. B. Eisenstadt-Umgebung	123	LLG	40	600	2	7	20	69	Antimony-nickel-copper (low)
13117	Pitvaros	125	PIT	40	410	2	7	20	71	Antimony-nickel-copper (low)
13118	Pitvaros	125	PIT	40	410	2	7	20	71	Antimony-nickel-copper (low)
13121	Pitvaros	125	PIT	40	500	2	7	20	71	Antimony-nickel-copper (low)
13129	Pitvaros	125	PIT	40	410	2	7	20	69	Antimony-nickel-copper (low)
13199	Deszk, 'A'	125	TC	10	400	2	7	20	71	Antimony-nickel-copper (low)
15806	Obermeitingen, Kr. Landsberg a. Lech	66	BK	40	600	2	7	20	66	Antimony-nickel-copper (low)
15828	München-Zamdorf, 'Emin-Pascha-Strasse', Stkr. München	66	BK	40	510	2	7	20	67	Antimony-nickel-copper (low)
15829	Manching, 'Hundsrückenacker i. d. Au', Lkr. Ingolstadt	66	BK	40	410	2	7	20	66	Antimony-nickel-copper (low)
15847	München-Sendling, 'Lindwurmstr. 76', Stkr. München	66	BK	40	510	2	7	20	66	Antimony-nickel-copper (low)
15880	Königsbrunn, 'Oberes Feld', Lkr. Augsburg	62	BK	40	410	2	7	20	71	Antimony-nickel-copper (low)
15884	Obernau, Stkr.. Aschaffenburg	62		40	110	2	7	20	71	Antimony-nickel-copper (low)
16281	Libodrice	112		40	110	2	7	20	66	Antimony-nickel-copper (low)
17667	Kostany	122	KST	40	212	2	7	20	69	Antimony-nickel-copper (low)
17668	Kostany	122	KST	40	212	2	7	20	69	Antimony-nickel-copper (low)
17669	Kostany	122	KST	40	200	2	7	20	69	Antimony-nickel-copper (low)
17670	Kostany	122	KST	40	110	2	7	20	69	Antimony-nickel-copper (low)
17686	Veselé	121	CWC	31	400	2	7	20	69	Antimony-nickel-copper (low)
17687	Veselé	121	CWC	31	420	2	7	20	71	Antimony-nickel-copper (low)
17688	Veselé	121	CWC	31	400	2	7	20	67	Antimony-nickel-copper (low)
17693	Vsech Svätych	122	KST	40	212	2	7	20	69	Antimony-nickel-copper (low)
17697	Vsech Svätych	122	KST	40	410	2	7	20	69	Antimony-nickel-copper (low)
17699	Vsech Svätych	122	KST	40	410	2	7	20	69	Antimony-nickel-copper (low)
17701	Vsech Svätych	122	KST	40	410	2	7	20	69	Antimony-nickel-copper (low)

ANo	FIND_PLACE	AREA	CULT	PER	TYPE_2	CL_No_I	CL_No_II	CL_No_III	CL_No_IV	Metal_groups
18405	Bodnegg-Wollmarshofen, Lkr. Ravensburg	62	BK	40	110	2	7	20	69	Antimony-nickel-copper (low)
18428	Lippach, Ostalbkreis	62	BK	40	110	2	7	20	71	Antimony-nickel-copper (low)
20082	Neureut, 'Herdenschlag', Kr. Karlsruhe	62	BK	40	110	2	7	20	71	Antimony-nickel-copper (low)
20083	Neureut, 'Herdenschlag', Kr. Karlsruhe	62	BK	40	110	2	7	20	69	Antimony-nickel-copper (low)
21229	Pawlowiczki, 'Gnadenfeld II'	83		40	212	2	7	20	66	Antimony-nickel-copper (low)
21574	Kirchheim a. Neckar, Lkr. Ludwigsburg	62	BK	40	110	2	7	20	69	Antimony-nickel-copper (low)
21577	Griesheim, Ortenaukreis	62	BK	40	110	2	7	20	71	Antimony-nickel-copper (low)
21578	Griesheim, Ortenaukreis	62	BK	40	110	2	7	20	71	Antimony-nickel-copper (low)
21581	Griesheim, Ortenaukreis	62	BK	40	110	2	7	20	71	Antimony-nickel-copper (low)
21582	Griesheim, Ortenaukreis	62	BK	40	110	2	7	20	71	Antimony-nickel-copper (low)
21583	Griesheim, Ortenaukreis	62	BK	40	110	2	7	20	69	Antimony-nickel-copper (low)
21623	Strasbourg, 'Hafen vor dem Metzgertor	62	BK	40	110	2	7	20	69	Antimony-nickel-copper (low)
22223	Auvernier, 'La Saunerie; Grabung Vouga 1922'	67	CWC	31	810	2	7	20	69	Antimony-nickel-copper (low)
22224	Auvernier, 'La Saunerie; Grabung Strahm 1965'	67	CWC	31	810	2	7	20	69	Antimony-nickel-copper (low)
22237	Estavayer	67		31	212	2	7	20	69	Antimony-nickel-copper (low)
22270	Chevroux	67		31	212	2	7	20	69	Antimony-nickel-copper (low)
22274	Chevroux	67		31	212	2	7	20	67	Antimony-nickel-copper (low)
29778	München-Sendling, 'Lindwurmstr. 76', Stkr. München	66	BK	40	600	2	7	20	71	Antimony-nickel-copper (low)
29950	Tvrdošovce, 'Körekto'	121	NIT	40	410	2	7	20	71	Antimony-nickel-copper (low)
29956	Výcapy-Opatovce	121	NIT	40	400	2	7	20	66	Antimony-nickel-copper (low)
33032	Jeßnitz, Lkr. Anhalt-Bitterfeld	105	CWC	31	810	2	7	20	70	Antimony-nickel-copper (low)
44003	Iwanowice, 'Babia Góra II	85	Mier	40	500	2	7	20	70	Antimony-nickel-copper (low)
44028	Mierzanowice II	85	Mier	40	420	2	7	20	70	Antimony-nickel-copper (low)
44034	Mierzanowice II	85	Mier	40	420	2	7	20	70	Antimony-nickel-copper (low)
44038	Mierzanowice II	85	Mier	40	510	2	7	20	70	Antimony-nickel-copper (low)
62107	Chevroux	67		31	212	2	7	20	71	Antimony-nickel-copper (low)
62170	Estavayer	67		31	212	2	7	20	69	Antimony-nickel-copper (low)
68025	Schaffhausen	67	BK	40	110	2	7	20	70	Antimony-nickel-copper (low)
78156	Branc	121	NIT	40	300	2	7	20	71	Antimony-nickel-copper (low)
78157	Branc	121	NIT	40	300	2	7	20	70	Antimony-nickel-copper (low)
78158	Branc	121	NIT	40	300	2	7	20	71	Antimony-nickel-copper (low)
78162	Branc	121	NIT	40	420	2	7	20	70	Antimony-nickel-copper (low)
78164	Branc	121	NIT	40	410	2	7	20	71	Antimony-nickel-copper (low)
100005	Vilsingen-Thiergarten, Lkr. Sigmaringen	62		40	110	2	7	20	70	Antimony-nickel-copper (low)
100088	Allstedt, 'Staatsforst', Lkr. Mansfeld-Südharz	103	CWC	31	410	2	7	20	69	Antimony-nickel-copper (low)
2844	Cham, 'St. Andreas'	67	PFY	21	100	1	4	10	26	Antimony-silver-copper
3056	Estavayer	67		21	100	1	5	13	13	Antimony-silver-copper
3233	Teplicko	111		21	130	1	4	10	26	Antimony-silver-copper
3373	Malhostovice	114		25	100	1	4	10	26	Antimony-silver-copper
3377	Dzbánice	114		25	100	1	4	10	26	Antimony-silver-copper
3378	Ondratice	114		21	100	1	4	10	26	Antimony-silver-copper
3379	Krepice	114		25	100	1	4	10	26	Antimony-silver-copper
3385	Ostrožská Lhota	114		21	100	1	4	10	26	Antimony-silver-copper
3654	Szolnok	125		21	130	1	4	10	26	Antimony-silver-copper
3752	Nedakonice	114		25	100	1	4	10	26	Antimony-silver-copper
4920	Nedakonice	114		25	100	1	4	10	26	Antimony-silver-copper
4921	Nedakonice	114		21	700	1	4	10	26	Antimony-silver-copper
6373	Retz, 'Spittelmais', p. B. Hollabrunn	72		21	100	1	4	10	26	Antimony-silver-copper
6462	Dabas	125		21	100	1	4	10	26	Antimony-silver-copper
6499	Szilvásvárad	125		21	130	1	4	10	26	Antimony-silver-copper
6500	Komárom-Ujszöny	123		21	130	1	4	10	26	Antimony-silver-copper
6561	Alsóság	123		21	130	1	4	10	26	Antimony-silver-copper
7489	Praha-Sárka	112		21	100	1	4	10	26	Antimony-silver-copper

ANo	FIND_PLACE	AREA	CULT	PER	TYPE_2	CL_No_I	CL_No_II	CL_No_III	CL_No_IV	Metal_groups
7491	Bohosudov	111		21	100	1	4	10	26	Antimony-silver-copper
7492	Komorany	111		21	100	1	4	10	26	Antimony-silver-copper
7524	Prisimasy	112		21	100	1	4	10	26	Antimony-silver-copper
7526	Rohovládova Belá	112		21	100	1	4	10	26	Antimony-silver-copper
7529	Vysoká nad Labem	112		21	100	1	4	10	26	Antimony-silver-copper
7537	Blesno	112		21	100	1	4	10	26	Antimony-silver-copper
7545	Nechranice	111		21	100	1	4	10	26	Antimony-silver-copper
7546	Kostov	111	BBC	35	100	1	4	10	26	Antimony-silver-copper
9825	Holic	114		10	100	1	4	10	26	Antimony-silver-copper
10320	Domanovice	112		21	100	1	4	10	26	Antimony-silver-copper
10337	Mikula	121		21	130	1	4	10	26	Antimony-silver-copper
10386	Dolné Srnie	121		25	100	1	4	10	26	Antimony-silver-copper
10390	Oravsky Podzámok	121		25	100	1	4	10	26	Antimony-silver-copper
10392	Oravská Polhora	121		25	100	1	4	10	26	Antimony-silver-copper
10404	Dolmé Semerovce	121		21	215	1	4	10	26	Antimony-silver-copper
10405	Moravské Lieskové	121		25	510	1	4	10	26	Antimony-silver-copper
10842	Brestovany	121		25	100	1	4	10	26	Antimony-silver-copper
10851	Vysná Pokoradz	122		25	100	1	4	10	26	Antimony-silver-copper
10855	Bajc-Vlkanovo, 'Sonda III'	121	SOP	21	550	1	4	10	26	Antimony-silver-copper
10903	Malé Leváre, 'Topoly'	121		25	100	1	4	10	26	Antimony-silver-copper
10904	Malé Leváre, 'Topoly'	121		21	215	1	4	10	26	Antimony-silver-copper
10905	Malé Leváre, 'Topoly'	121		25	130	1	4	10	26	Antimony-silver-copper
10906	Malé Leváre, 'Topoly'	121		25	510	1	4	10	26	Antimony-silver-copper
10907	Kúty	121		21	215	1	4	10	26	Antimony-silver-copper
10988	Uherský Brod	114		25	100	1	4	10	26	Antimony-silver-copper
11437	Retz, p. B. Hollabrunn	72		21	100	1	4	10	26	Antimony-silver-copper
11471	Zöbing bei Koglberg, 'Koglberg' Niederösterr	72	BAD	25	510	1	4	10	26	Antimony-silver-copper
12073	Velehrad-Rákos	114		21	212	1	4	10	26	Antimony-silver-copper
12075	Tasov	114		21	100	1	4	10	26	Antimony-silver-copper
12466	Neverice	121		25	100	1	4	10	26	Antimony-silver-copper
12477	Veliki Gaj	125		31	122	1	4	10	26	Antimony-silver-copper
12521	Bratislava	122		21	212	1	4	10	26	Antimony-silver-copper
14386	Nógrádmarcal	121	BOD	21	130	1	4	10	26	Antimony-silver-copper
14388	Csitár	121		21	130	1	4	10	26	Antimony-silver-copper
14390	Nógrádmarcal	121		21	130	1	4	10	26	Antimony-silver-copper
14392	Nógrádmarcal	121	BOD	21	130	1	4	10	26	Antimony-silver-copper
14393	Nógrádmarcal	121	BOD	21	130	1	4	10	26	Antimony-silver-copper
14394	Nógrádmarcal	121	BOD	21	130	1	4	10	26	Antimony-silver-copper
17673	Malé Leváre, 'Topoly'	121		21	215	1	4	10	26	Antimony-silver-copper
17684	Skalica	121		21	212	1	4	10	26	Antimony-silver-copper
19926	Tršice	114		21	100	1	4	10	26	Antimony-silver-copper
19927	Slatinice	114		21	100	1	4	10	26	Antimony-silver-copper
19928	Droždín	114		21	100	1	4	10	26	Antimony-silver-copper
19929	Zlín-Príluky	114		21	100	1	4	10	26	Antimony-silver-copper
29789	Malé Leváre, 'Topoly'	121		25	100	1	4	10	26	Antimony-silver-copper
29790	Malé Leváre, 'Topoly'	121		25	100	1	4	10	26	Antimony-silver-copper
58002	Dolné Srnie	121		25	100	1	5	14	32	Antimony-silver-copper
58009	Oravská Polhora	121		25	100	1	5	14	32	Antimony-silver-copper
29	Beizkofen, Kr. Sigmaringen	62	BK	40	410	2	6	18	54	Arsenic-antimony-copper
111	Güglingen, Kr. Heilbronn	62	BK	40	110	2	6	18	54	Arsenic-antimony-copper
112	Güglingen, Kr. Heilbronn	62	BK	40	110	2	6	18	54	Arsenic-antimony-copper
113	Güglingen, Kr. Heilbronn	62	BK	40	110	2	6	18	54	Arsenic-antimony-copper
445	Veringenstadt, Kr. Sigmaringen	62	BK	40	500	2	6	18	54	Arsenic-antimony-copper

ANo	FIND_PLACE	AREA	CULT	PER	TYPE_2	CL_No_I	CL_No_II	CL_No_III	CL_No_IV	Metal_groups
2643	Nattenhausen, Lkr. Günzburg	62		40	410	2	6	18	54	Arsenic-antimony-copper
2648	Nattenhausen, Lkr. Günzburg	62		40	410	2	6	18	54	Arsenic-antimony-copper
3697	Lucenec	121		10	130	2	6	18	54	Arsenic-antimony-copper
4436	Gollnsdorf, Gde. St. Valentin, p. B. Amstetten	72	BK	40	410	2	6	18	54	Arsenic-antimony-copper
5841	Gemeinlebarn, p. B. St. Pölten	72	BK	40	400	2	6	18	54	Arsenic-antimony-copper
5842	Gemeinlebarn, p. B. St. Pölten	72	BK	40	410	2	6	18	54	Arsenic-antimony-copper
5955	Unterwölbling, p. B. St. Pölten	72		40	410	2	6	18	54	Arsenic-antimony-copper
5956	Unterwölbling, p. B. St. Pölten	72		40	410	2	6	18	54	Arsenic-antimony-copper
5962	Unterwölbling, p. B. St. Pölten	72	BK	40	410	2	6	18	54	Arsenic-antimony-copper
5963	Unterwölbling, p. B. St. Pölten	72	BK	40	410	2	6	18	54	Arsenic-antimony-copper
5970	Unterwölbling, p. B. St. Pölten	72		40	410	2	6	18	54	Arsenic-antimony-copper
6497	Budapest-III-Békásmegyer	124		10	130	2	6	18	54	Arsenic-antimony-copper
10389	Kláštor pod Znievom	121		21	130	2	6	18	54	Arsenic-antimony-copper
10393	Hontianske Nemce	121		25	100	2	6	18	54	Arsenic-antimony-copper
10399	Handlová, 'Na Pstruhároch'	121		21	130	2	6	18	54	Arsenic-antimony-copper
10400	Handlová	121		21	700	2	6	18	54	Arsenic-antimony-copper
10401	Handlová	121		21	700	2	6	18	54	Arsenic-antimony-copper
10402	Handlová	121		21	700	2	6	18	54	Arsenic-antimony-copper
10760	Komjatice, 'Argentina'	121	NIT	40	600	2	6	18	54	Arsenic-antimony-copper
15794	Nattenhausen, Lkr. Günzburg	62		40	410	2	6	18	54	Arsenic-antimony-copper
15848	München-Sendling, 'Lindwurmstr. 76', Stkr. München	66	BK	40	510	2	6	18	54	Arsenic-antimony-copper
17685	Sliac	121		25	100	2	6	18	54	Arsenic-antimony-copper
17696	Vsech Svätych	122	KST	40	410	2	6	18	54	Arsenic-antimony-copper
17700	Vsech Svätych	122	KST	40	410	2	6	18	54	Arsenic-antimony-copper
17702	Vsech Svätych	122	KST	40	400	2	6	18	54	Arsenic-antimony-copper
17703	Vsech Svätych	122	KST	40	400	2	6	18	54	Arsenic-antimony-copper
17704	Vsech Svätych	122	KST	40	400	2	6	18	54	Arsenic-antimony-copper
17705	Vsech Svätych	122	KST	40	410	2	6	18	54	Arsenic-antimony-copper
17706	Vsech Svätych	122	KST	40	410	2	6	18	54	Arsenic-antimony-copper
17707	Vsech Svätych	122	KST	40	410	2	6	18	54	Arsenic-antimony-copper
21403	Chlum	112		10	130	2	6	18	54	Arsenic-antimony-copper
32319	Buttelstedt, 'Ziegeleigrube Florstedt',Kr. Weimar	103	BBC	35	215	2	6	15	52	Arsenic-copper (high)
34450	Drosa 'Küsterberg', Lkr. Anhalt-Bitterfeld	103	CWC	31	800	2	6	15	52	Arsenic-copper (high)
34499	Drosa 'Küsterberg', Lkr. Anhalt-Bitterfeld	103	CWC	31	550	2	6	15	53	Arsenic-copper (high)
34500	Drosa 'Küsterberg', Lkr. Anhalt-Bitterfeld	103	CWC	31	550	2	6	15	52	Arsenic-copper (high)
34502	Drosa 'Küsterberg', Lkr. Anhalt-Bitterfeld	103	CWC	31	550	2	6	15	52	Arsenic-copper (high)
34504	Drosa 'Küsterberg', Lkr. Anhalt-Bitterfeld	103	CWC	31	550	2	6	15	52	Arsenic-copper (high)
34506	Drosa 'Küsterberg', Lkr. Anhalt-Bitterfeld	103	CWC	31	550	2	6	15	53	Arsenic-copper (high)
34507	Drosa 'Küsterberg', Lkr. Anhalt-Bitterfeld	103	CWC	31	550	2	6	15	52	Arsenic-copper (high)
34509	Drosa 'Küsterberg', Lkr. Anhalt-Bitterfeld	103	CWC	31	550	2	6	15	52	Arsenic-copper (high)
34510	Drosa 'Küsterberg', Lkr. Anhalt-Bitterfeld	103	CWC	31	550	2	6	15	52	Arsenic-copper (high)
34512	Drosa 'Küsterberg', Lkr. Anhalt-Bitterfeld	103	CWC	31	550	2	6	15	53	Arsenic-copper (high)
34513	Drosa 'Küsterberg', Lkr. Anhalt-Bitterfeld	103	CWC	31	550	2	6	15	52	Arsenic-copper (high)
34514	Drosa 'Küsterberg', Lkr. Anhalt-Bitterfeld	103	CWC	31	550	2	6	15	53	Arsenic-copper (high)
34515	Drosa 'Küsterberg', Lkr. Anhalt-Bitterfeld	103	CWC	31	550	2	6	15	53	Arsenic-copper (high)
34516	Drosa 'Küsterberg', Lkr. Anhalt-Bitterfeld	103	CWC	31	550	2	6	15	53	Arsenic-copper (high)
34517	Drosa 'Küsterberg', Lkr. Anhalt-Bitterfeld	103	CWC	31	550	2	6	15	53	Arsenic-copper (high)
34518	Drosa 'Küsterberg', Lkr. Anhalt-Bitterfeld	103	CWC	31	550	2	6	15	53	Arsenic-copper (high)
34519	Drosa 'Küsterberg', Lkr. Anhalt-Bitterfeld	103	CWC	31	550	2	6	15	53	Arsenic-copper (high)
34520	Drosa 'Küsterberg', Lkr. Anhalt-Bitterfeld	103	CWC	31	550	2	6	15	53	Arsenic-copper (high)
34521	Drosa 'Küsterberg', Lkr. Anhalt-Bitterfeld	103	CWC	31	550	2	6	15	53	Arsenic-copper (high)
34522	Drosa 'Küsterberg', Lkr. Anhalt-Bitterfeld	103	CWC	31	550	2	6	15	53	Arsenic-copper (high)
34523	Drosa 'Küsterberg', Lkr. Anhalt-Bitterfeld	103	CWC	31	550	2	6	15	53	Arsenic-copper (high)

ANo	FIND_PLACE	AREA	CULT	PER	TYPE_2	CL_No_I	CL_No_II	CL_No_III	CL_No_IV	Metal_groups
34526	Drosa 'Küsterberg', Lkr. Anhalt-Bitterfeld	103	CWC	31	550	2	6	15	53	Arsenic-copper (high)
34528	Drosa 'Küsterberg', Lkr. Anhalt-Bitterfeld	103	CWC	31	550	2	6	15	53	Arsenic-copper (high)
34530	Drosa 'Küsterberg', Lkr. Anhalt-Bitterfeld	103	CWC	31	550	2	6	15	52	Arsenic-copper (high)
34531	Drosa 'Küsterberg', Lkr. Anhalt-Bitterfeld	103	CWC	31	550	2	6	15	52	Arsenic-copper (high)
34539	Drosa 'Küsterberg', Lkr. Anhalt-Bitterfeld	103	CWC	31	550	2	6	15	53	Arsenic-copper (high)
34541	Drosa 'Küsterberg', Lkr. Anhalt-Bitterfeld	103	CWC	31	550	2	6	15	53	Arsenic-copper (high)
34542	Drosa 'Küsterberg', Lkr. Anhalt-Bitterfeld	103	CWC	31	550	2	6	15	53	Arsenic-copper (high)
34547	Drosa 'Küsterberg', Lkr. Anhalt-Bitterfeld	103	CWC	31	550	2	6	15	52	Arsenic-copper (high)
34549	Drosa 'Küsterberg', Lkr. Anhalt-Bitterfeld	103	CWC	31	550	2	6	15	52	Arsenic-copper (high)
34550	Drosa 'Küsterberg', Lkr. Anhalt-Bitterfeld	103	CWC	31	550	2	6	15	52	Arsenic-copper (high)
34551	Drosa 'Küsterberg', Lkr. Anhalt-Bitterfeld	103	CWC	31	550	2	6	15	53	Arsenic-copper (high)
34553	Drosa 'Küsterberg', Lkr. Anhalt-Bitterfeld	103	CWC	31	550	2	6	15	53	Arsenic-copper (high)
34555	Drosa 'Küsterberg', Lkr. Anhalt-Bitterfeld	103	CWC	31	550	2	6	15	53	Arsenic-copper (high)
34557	Drosa 'Küsterberg', Lkr. Anhalt-Bitterfeld	103	CWC	31	550	2	6	15	53	Arsenic-copper (high)
34558	Drosa 'Küsterberg', Lkr. Anhalt-Bitterfeld	103	CWC	31	550	2	6	15	53	Arsenic-copper (high)
34559	Drosa 'Küsterberg', Lkr. Anhalt-Bitterfeld	103	CWC	31	550	2	6	15	52	Arsenic-copper (high)
34561	Drosa 'Küsterberg', Lkr. Anhalt-Bitterfeld	103	CWC	31	550	2	6	15	52	Arsenic-copper (high)
34562	Drosa 'Küsterberg', Lkr. Anhalt-Bitterfeld	103	CWC	31	550	2	6	15	52	Arsenic-copper (high)
34563	Drosa 'Küsterberg', Lkr. Anhalt-Bitterfeld	103	CWC	31	550	2	6	15	52	Arsenic-copper (high)
34564	Drosa 'Küsterberg', Lkr. Anhalt-Bitterfeld	103	CWC	31	550	2	6	15	52	Arsenic-copper (high)
34565	Drosa 'Küsterberg', Lkr. Anhalt-Bitterfeld	103	CWC	31	550	2	6	15	52	Arsenic-copper (high)
34567	Drosa 'Küsterberg', Lkr. Anhalt-Bitterfeld	103	CWC	31	550	2	6	15	52	Arsenic-copper (high)
34569	Drosa 'Küsterberg', Lkr. Anhalt-Bitterfeld	103	CWC	31	550	2	6	15	52	Arsenic-copper (high)
34571	Drosa 'Küsterberg', Lkr. Anhalt-Bitterfeld	103	CWC	31	550	2	6	15	53	Arsenic-copper (high)
34573	Drosa 'Küsterberg', Lkr. Anhalt-Bitterfeld	103	CWC	31	550	2	6	15	52	Arsenic-copper (high)
34574	Drosa 'Küsterberg', Lkr. Anhalt-Bitterfeld	103	CWC	31	550	2	6	15	52	Arsenic-copper (high)
34575	Drosa 'Küsterberg', Lkr. Anhalt-Bitterfeld	103	CWC	31	550	2	6	15	52	Arsenic-copper (high)
34577	Drosa 'Küsterberg', Lkr. Anhalt-Bitterfeld	103	CWC	31	550	2	6	15	52	Arsenic-copper (high)
34579	Drosa 'Küsterberg', Lkr. Anhalt-Bitterfeld	103	CWC	31	550	2	6	15	52	Arsenic-copper (high)
34583	Drosa 'Küsterberg', Lkr. Anhalt-Bitterfeld	103	CWC	31	550	2	6	15	52	Arsenic-copper (high)
34586	Drosa 'Küsterberg', Lkr. Anhalt-Bitterfeld	103	CWC	31	550	2	6	15	52	Arsenic-copper (high)
34588	Drosa 'Küsterberg', Lkr. Anhalt-Bitterfeld	103	CWC	31	550	2	6	15	53	Arsenic-copper (high)
34589	Drosa 'Küsterberg', Lkr. Anhalt-Bitterfeld	103	CWC	31	550	2	6	15	53	Arsenic-copper (high)
34590	Drosa 'Küsterberg', Lkr. Anhalt-Bitterfeld	103	CWC	31	550	2	6	15	52	Arsenic-copper (high)
34591	Drosa 'Küsterberg', Lkr. Anhalt-Bitterfeld	103	CWC	31	550	2	6	15	52	Arsenic-copper (high)
34593	Drosa 'Küsterberg', Lkr. Anhalt-Bitterfeld	103	CWC	31	550	2	6	15	52	Arsenic-copper (high)
34594	Drosa 'Küsterberg', Lkr. Anhalt-Bitterfeld	103	CWC	31	550	2	6	15	53	Arsenic-copper (high)
34595	Drosa 'Küsterberg', Lkr. Anhalt-Bitterfeld	103	CWC	31	550	2	6	15	52	Arsenic-copper (high)
34596	Drosa 'Küsterberg', Lkr. Anhalt-Bitterfeld	103	CWC	31	550	2	6	15	53	Arsenic-copper (high)
34598	Drosa 'Küsterberg', Lkr. Anhalt-Bitterfeld	103	CWC	31	550	2	6	15	52	Arsenic-copper (high)
34599	Drosa 'Küsterberg', Lkr. Anhalt-Bitterfeld	103	CWC	31	550	2	6	15	52	Arsenic-copper (high)
34600	Drosa 'Küsterberg', Lkr. Anhalt-Bitterfeld	103	CWC	31	550	2	6	15	53	Arsenic-copper (high)
34601	Drosa 'Küsterberg', Lkr. Anhalt-Bitterfeld	103	CWC	31	550	2	6	15	53	Arsenic-copper (high)
34602	Drosa 'Küsterberg', Lkr. Anhalt-Bitterfeld	103	CWC	31	550	2	6	15	53	Arsenic-copper (high)
34604	Drosa 'Küsterberg', Lkr. Anhalt-Bitterfeld	103	CWC	31	550	2	6	15	53	Arsenic-copper (high)
34606	Drosa 'Küsterberg', Lkr. Anhalt-Bitterfeld	103	CWC	31	550	2	6	15	53	Arsenic-copper (high)
34610	Drosa 'Küsterberg', Lkr. Anhalt-Bitterfeld	103	CWC	31	550	2	6	15	52	Arsenic-copper (high)
58003	Handlová, 'Na Pstruhároch'	121		25	100	2	6	18	55	Arsenic-copper (high)
58005	Hontianske Nemce	121		25	100	2	6	18	55	Arsenic-copper (high)
58006	Lisov	121		25	100	2	6	18	55	Arsenic-copper (high)
58018	Jablonovce	121		10	130	2	6	18	55	Arsenic-copper (high)
10566	Nitra, 'Partisanenstraße'	121	NIT	40	405	2	6	17	40	Arsenic-copper (high)
8	Altheim, Lkr. Landshut	66	ALT	21	100	1	5	14	31	Arsenic-copper (low)

ANo	FIND_PLACE	AREA	CULT	PER	TYPE_2	CL_No_I	CL_No_II	CL_No_III	CL_No_IV	Metal_groups
106	Goldburghausen, 'Goldberg', Ostalbkreis	62	GIII	25	134	1	4	12	30	Arsenic-copper (low)
156	Kornwestheim Kr. Ludwigsburg	62		31	100	2	6	16	38	Arsenic-copper (low)
1944	Kassel-Waldau, Stkr. Kassel	53		21	100	1	4	12	29	Arsenic-copper (low)
2627	Ainring, 'Au', Lkr. Berchtesgadener Land	66	ALT	21	100	1	4	12	29	Arsenic-copper (low)
2637	Augsburg, 'Pfannenstiel', Lkr. Augsburg	62		25	100	1	4	12	30	Arsenic-copper (low)
2842	Vinelz, 'Amt Erlau'	67	CWC	31	200	1	4	11	28	Arsenic-copper (low)
2846	Zug-Otterswil	67		21	100	1	4	12	29	Arsenic-copper (low)
2860	Greng	67		21	100	1	4	12	30	Arsenic-copper (low)
2923	Sutz-Lattrigen	67	COR	21	212	1	4	12	29	Arsenic-copper (low)
2925	Mörigen	67	CWC	31	134	1	4	12	29	Arsenic-copper (low)
2929	Lüscherz	67	CWC	31	200	1	4	12	29	Arsenic-copper (low)
2953	Chevroux	67		31	134	1	4	12	30	Arsenic-copper (low)
2987	Seeberg 'Burgäschisee-Süd; Grabung 1957'	67	COR	21	800	1	4	12	29	Arsenic-copper (low)
2988	Seeberg 'Burgäschisee-Süd; Grabung 1957'	67	COR	21	800	1	4	12	29	Arsenic-copper (low)
2989	Seeberg 'Burgäschisee-Süd; Grabung 1957'	67	COR	21	800	1	4	12	29	Arsenic-copper (low)
2990	Seeberg 'Burgäschisee-Süd; Grabung 1957'	67	COR	21	800	1	4	12	29	Arsenic-copper (low)
2991	Seeberg 'Burgäschisee-Süd; Grabung 1957'	67	COR	21	800	1	4	12	29	Arsenic-copper (low)
2992	Seeberg 'Burgäschisee-Süd; Grabung 1957'	67	COR	21	800	1	5	14	31	Arsenic-copper (low)
2993	Seeberg 'Burgäschisee-Süd; Grabung 1957'	67	COR	21	800	1	4	12	29	Arsenic-copper (low)
2994	Seeberg 'Burgäschisee-Süd; Grabung 1957'	67	COR	21	800	1	4	12	29	Arsenic-copper (low)
2995	Seeberg 'Burgäschisee-Süd; Grabung 1957'	67	COR	21	800	1	4	12	29	Arsenic-copper (low)
2996	Seeberg 'Burgäschisee-Süd; Grabung 1957'	67	COR	21	800	1	4	12	29	Arsenic-copper (low)
2997	Seeberg 'Burgäschisee-Süd; Grabung 1957'	67	COR	21	800	1	4	12	29	Arsenic-copper (low)
2998	Seeberg 'Burgäschisee-Süd; Grabung 1957'	67	COR	21	800	1	4	12	29	Arsenic-copper (low)
2999	Seeberg 'Burgäschisee-Süd; Grabung 1957'	67	COR	21	800	1	4	12	29	Arsenic-copper (low)
3000	Seeberg 'Burgäschisee-Süd; Grabung 1957'	67	COR	21	800	1	4	12	29	Arsenic-copper (low)
3001	Seeberg 'Burgäschisee-Süd; Grabung 1957'	67	COR	21	800	1	4	12	29	Arsenic-copper (low)
3002	Seeberg 'Burgäschisee-Süd; Grabung 1957'	67	COR	21	800	1	4	12	29	Arsenic-copper (low)
3003	Seeberg 'Burgäschisee-Süd; Grabung 1957'	67	COR	21	800	1	5	14	31	Arsenic-copper (low)
3004	Seeberg 'Burgäschisee-Süd; Grabung 1957'	67	COR	21	800	1	4	12	29	Arsenic-copper (low)
3005	Seeberg 'Burgäschisee-Süd; Grabung 1957'	67	COR	21	800	1	4	12	29	Arsenic-copper (low)
3006	Seeberg 'Burgäschisee-Süd; Grabung 1957'	67	COR	21	800	1	4	12	29	Arsenic-copper (low)
3007	Seeberg 'Burgäschisee-Süd; Grabung 1957'	67	COR	21	800	1	4	12	29	Arsenic-copper (low)
3008	Seeberg 'Burgäschisee-Süd; Grabung 1957'	67	COR	21	800	1	4	12	29	Arsenic-copper (low)
3009	Seeberg 'Burgäschisee-Süd; Grabung 1957'	67	COR	21	800	1	4	12	29	Arsenic-copper (low)
3010	Seeberg 'Burgäschisee-Süd; Grabung 1957'	67	COR	21	800	1	4	12	29	Arsenic-copper (low)
3011	Seeberg 'Burgäschisee-Süd; Grabung 1957'	67	COR	21	800	1	4	12	29	Arsenic-copper (low)
3012	Seeberg 'Burgäschisee-Süd; Grabung 1957'	67	COR	21	800	1	4	12	29	Arsenic-copper (low)
3013	Seeberg 'Burgäschisee-Süd; Grabung 1957'	67	COR	21	800	1	4	12	29	Arsenic-copper (low)
3014	Seeberg 'Burgäschisee-Süd; Grabung 1957'	67	COR	21	800	1	4	12	29	Arsenic-copper (low)
3015	Seeberg 'Burgäschisee-Süd; Grabung 1957'	67	COR	21	800	1	4	12	29	Arsenic-copper (low)
3016	Seeberg 'Burgäschisee-Süd; Grabung 1957'	67	COR	21	800	1	4	12	29	Arsenic-copper (low)
3017	Seeberg 'Burgäschisee-Süd; Grabung 1957'	67	COR	21	800	1	4	12	30	Arsenic-copper (low)
3018	Seeberg 'Burgäschisee-Süd; Grabung 1957'	67	COR	21	800	1	4	12	30	Arsenic-copper (low)
3019	Seeberg 'Burgäschisee-Süd; Grabung 1957'	67	COR	21	800	1	4	12	29	Arsenic-copper (low)
3020	Seeberg 'Burgäschisee-Süd; Grabung 1957'	67	COR	21	800	1	4	12	29	Arsenic-copper (low)
3021	Seeberg 'Burgäschisee-Süd; Grabung 1957'	67	COR	21	800	1	4	12	29	Arsenic-copper (low)
3022	Seeberg 'Burgäschisee-Süd; Grabung 1957'	67	COR	21	800	1	4	12	29	Arsenic-copper (low)
3023	Seeberg 'Burgäschisee-Süd; Grabung 1957'	67	COR	21	800	1	4	12	29	Arsenic-copper (low)
3024	Seeberg 'Burgäschisee-Süd; Grabung 1957'	67	COR	21	800	1	4	12	29	Arsenic-copper (low)
3025	Seeberg 'Burgäschisee-Süd; Grabung 1957'	67	COR	21	800	1	4	12	29	Arsenic-copper (low)
3026	Seeberg 'Burgäschisee-Süd; Grabung 1957'	67	COR	21	800	1	4	12	29	Arsenic-copper (low)
3033	Mörigen	67		31	810	1	4	12	29	Arsenic-copper (low)

ANo	FIND_PLACE	AREA	CULT	PER	TYPE_2	CL_No_I	CL_No_II	CL_No_III	CL_No_IV	Metal_groups
3044	Mörigen	67	CWC	31	134	1	4	12	29	Arsenic-copper (low)
3090	Seeberg 'Burgäschisee-Süd; Grabung 1957'	67	COR	21	134	1	5	14	31	Arsenic-copper (low)
3240	Praha-Bubenec, 'Sibirska-Gasse'	112	BBC	35	215	1	4	11	28	Arsenic-copper (low)
3348	Brno-Lisen, 'Stare Zamky'	114	BAD	25	100	1	4	12	29	Arsenic-copper (low)
3352	Barca	122	BAD	25	100	1	4	11	28	Arsenic-copper (low)
3354	Tibava, 'Hrun za cintirom'	122	TC	10	130	1	4	11	28	Arsenic-copper (low)
3367	Tvorihráz	114		25	100	1	5	14	31	Arsenic-copper (low)
3375	Prace	114		25	100	1	4	12	29	Arsenic-copper (low)
3380	Vevcice	114		21	100	1	4	12	30	Arsenic-copper (low)
3382	Kojátky	114		21	100	1	4	12	30	Arsenic-copper (low)
3386	Senorady	114		25	100	1	4	12	30	Arsenic-copper (low)
3388	Koberice	114		10	100	1	4	12	30	Arsenic-copper (low)
3389	Kromeriz-Kotojedy	114		25	100	1	4	12	30	Arsenic-copper (low)
3401	Vitonice	114		25	134	1	4	11	28	Arsenic-copper (low)
3408	Plumlov	114		25	100	1	4	12	29	Arsenic-copper (low)
3612	Seewalchen, p. B. Vöcklabruck	73	MON	21	100	1	4	12	29	Arsenic-copper (low)
3613	Seewalchen, p. B. Völklabruck	73	MON	21	100	1	4	12	30	Arsenic-copper (low)
3615	Weyeregg, p. B. Völklabruck	73	MON	21	212	1	4	12	30	Arsenic-copper (low)
3617	Weyeregg, p. B. Völklabruck	73	MON	21	212	1	4	12	29	Arsenic-copper (low)
3730	Leobersdorf, p. B. Wiener-Neustadt	72	BAD	25	400	1	5	13	13	Arsenic-copper (low)
3731	Leobersdorf, p. B. Wiener-Neustadt	72	BAD	25	400	1	4	12	30	Arsenic-copper (low)
3960	Rouffach	62		31	100	1	4	12	29	Arsenic-copper (low)
3969	Niederschaeffolsheim	62		31	100	1	4	12	29	Arsenic-copper (low)
3970	Soufflenheim, 'Königsbruck'	62		31	100	1	4	12	29	Arsenic-copper (low)
4241	Ayent, 'Les Place'	67	BK	40	500	1	5	14	31	Arsenic-copper (low)
4318	Zwerndorf a. d. March, p. B. Gänserndorl	72	BAD	25	130	1	4	11	28	Arsenic-copper (low)
4341	Unterach-Mondsee, 'Station See', p. B. Völklabruck	73	MON	21	100	1	4	11	28	Arsenic-copper (low)
4342	Unterach-Mondsee, 'Station See', p. B. Völklabruck	73	MON	21	100	1	5	14	31	Arsenic-copper (low)
4343	Unterach-Mondsee, 'Station See', p. B. Völklabruck	73	MON	21	100	1	5	14	31	Arsenic-copper (low)
4344	Unterach-Mondsee, 'Station See', p. B. Völklabruck	73	MON	21	100	1	5	14	31	Arsenic-copper (low)
4345	Unterach-Mondsee, 'Station See', p. B. Völklabruck	73	MON	21	100	1	4	11	28	Arsenic-copper (low)
4346	Unterach-Mondsee, 'Station See', p. B. Völklabruck	73	MON	21	100	1	4	12	29	Arsenic-copper (low)
4347	Unterach-Mondsee, 'Station See', p. B. Völklabruck	73	MON	21	100	1	4	12	29	Arsenic-copper (low)
4348	Unterach-Mondsee, 'Station See', p. B. Völklabruck	73	MON	21	100	1	4	11	28	Arsenic-copper (low)
4349	Unterach-Mondsee, 'Station See', p. B. Völklabruck	73	MON	21	100	1	5	14	31	Arsenic-copper (low)
4350	Unterach-Mondsee, 'Station See', p. B. Völklabruck	73	MON	21	100	1	5	14	31	Arsenic-copper (low)
4470	Schlögelsbach b. Kilb, p. B. Melk	72		31	100	1	4	11	28	Arsenic-copper (low)
4638	Lichtenwörth, p. B. Wiener-Neustadt	72	BAD	25	400	1	4	12	29	Arsenic-copper (low)
4895	Seewalchen, p. B. Vöcklabruck	73	MON	21	300	1	4	11	28	Arsenic-copper (low)
4896	Seewalchen, p. B. Vöcklabruck	73	MON	21	810	1	4	12	29	Arsenic-copper (low)
4897	Seewalchen, p. B. Vöcklabruck	73	MON	21	810	1	4	12	29	Arsenic-copper (low)
4922	Ludkovice	114		21	100	1	4	12	30	Arsenic-copper (low)
5184	Púchov	114		21	130	1	4	12	30	Arsenic-copper (low)
5684	Horodnica	85		31	215	1	4	11	28	Arsenic-copper (low)
5687	Horodnica	85		31	800	1	4	11	28	Arsenic-copper (low)
5688	Horodnica	85		31	800	1	4	11	28	Arsenic-copper (low)
5689	Horodnica	85		31	800	1	4	11	28	Arsenic-copper (low)
5719	Hrkovce	121		25	100	1	4	12	29	Arsenic-copper (low)
6473	Fajsz	125		40	134	1	4	12	29	Arsenic-copper (low)
6485	Aszód	125		10	130	1	4	12	29	Arsenic-copper (low)
6530	Tiszapolgár-Basatanya	125	BOD	21	810	1	4	12	29	Arsenic-copper (low)
6542	Kerepes	125		21	130	1	4	12	29	Arsenic-copper (low)
6552	Veliki Gaj	125	BOD	21	130	1	4	12	30	Arsenic-copper (low)

ANo	FIND_PLACE	AREA	CULT	PER	TYPE_2	CL_No_I	CL_No_II	CL_No_III	CL_No_IV	Metal_groups
6555	Veliki Gaj	125	BOD	21	130	1	4	12	29	Arsenic-copper (low)
6556	Szilágy-Somlup	125		21	130	1	4	12	29	Arsenic-copper (low)
6563	Balassagyarmat	121		21	130	1	4	12	29	Arsenic-copper (low)
6569	Tiszapolgár-Basatanya	125	TC	10	410	1	4	12	29	Arsenic-copper (low)
6571	Tiszapolgár-Basatanya	125	TC	10	410	1	4	12	30	Arsenic-copper (low)
7051	Genève, 'Crêts St. Laurent'	67		21	100	1	4	12	29	Arsenic-copper (low)
7108	St. Blaise	67	CWC	31	200	1	4	12	30	Arsenic-copper (low)
7279	Mont	67		21	100	1	4	11	28	Arsenic-copper (low)
7293	Hitzkirch, 'Seematte'	67	COR	21	110	1	4	12	30	Arsenic-copper (low)
7496	Praha-Sárka	112		21	100	1	4	11	28	Arsenic-copper (low)
7543	Mlázovice	112		10	130	1	4	12	30	Arsenic-copper (low)
10366	Výcapy-Opatovce	121	NIT	40	300	1	5	14	31	Arsenic-copper (low)
10370	Výcapy-Opatovce	121	NIT	40	420	1	4	12	29	Arsenic-copper (low)
10378	Výcapy-Opatovce	121	NIT	40	420	1	4	12	29	Arsenic-copper (low)
10381	Výcapy-Opatovce	121	NIT	40	420	1	4	12	29	Arsenic-copper (low)
10387	Bojnice	121		25	100	1	4	12	29	Arsenic-copper (low)
10391	Hronské Kosihy	121		21	100	1	4	11	28	Arsenic-copper (low)
10411	Výcapy-Opatovce	121	NIT	40	420	1	4	11	28	Arsenic-copper (low)
10413	Výcapy-Opatovce	121	NIT	40	420	1	5	14	31	Arsenic-copper (low)
10415	Výcapy-Opatovce	121	NIT	40	420	1	4	11	28	Arsenic-copper (low)
10432	Výcapy-Opatovce	121	NIT	40	420	1	4	11	28	Arsenic-copper (low)
10447	Výcapy-Opatovce	121	NIT	40	420	1	4	11	28	Arsenic-copper (low)
10448	Výcapy-Opatovce	121	NIT	40	420	1	4	11	28	Arsenic-copper (low)
10450	Výcapy-Opatovce	121	NIT	40	300	1	4	12	29	Arsenic-copper (low)
10456	Výcapy-Opatovce	121	NIT	40	510	1	4	12	29	Arsenic-copper (low)
10464	Výcapy-Opatovce	121	NIT	40	420	1	4	11	28	Arsenic-copper (low)
10468	Výcapy-Opatovce	121	NIT	40	420	1	4	11	28	Arsenic-copper (low)
10490	Výcapy-Opatovce	121	NIT	40	300	1	4	11	28	Arsenic-copper (low)
10496	Výcapy-Opatovce	121	NIT	40	420	1	4	11	28	Arsenic-copper (low)
10500	Výcapy-Opatovce	121	NIT	40	420	1	4	11	28	Arsenic-copper (low)
10510	Výcapy-Opatovce	121	NIT	40	400	1	4	11	28	Arsenic-copper (low)
10511	Výcapy-Opatovce	121	NIT	40	400	1	4	11	28	Arsenic-copper (low)
10526	Výcapy-Opatovce	121	NIT	40	420	1	4	11	28	Arsenic-copper (low)
10532	Výcapy-Opatovce	121	NIT	40	420	1	4	11	28	Arsenic-copper (low)
10552	Výcapy-Opatovce	121	NIT	40	420	1	4	11	28	Arsenic-copper (low)
10567	Nitra, 'Partisanenstraße'	121	NIT	40	300	1	4	12	29	Arsenic-copper (low)
10572	Komjatice, 'Homoky'	121	NIT	40	420	1	4	11	28	Arsenic-copper (low)
10573	Komjatice, 'Homoky'	121	NIT	40	420	1	5	14	31	Arsenic-copper (low)
10651	Tvrdošovce, 'Körektó'	121	NIT	40	410	1	4	12	29	Arsenic-copper (low)
10657	Tvrdošovce, 'Körektó'	121	NIT	40	300	1	4	12	29	Arsenic-copper (low)
10732	Tibava, 'Hrun za cintirom'	122	TC	10	130	1	4	11	28	Arsenic-copper (low)
10733	Tibava, 'Hrun za cintirom'	122	TC	10	130	1	4	11	28	Arsenic-copper (low)
10827	Výcapy-Opatovce	121	NIT	40	420	1	4	11	28	Arsenic-copper (low)
10828	Výcapy-Opatovce	121	NIT	40	420	1	4	11	28	Arsenic-copper (low)
10829	Výcapy-Opatovce	121	NIT	40	420	1	4	11	28	Arsenic-copper (low)
10830	Výcapy-Opatovce	121	NIT	40	550	1	5	14	31	Arsenic-copper (low)
10850	Vysná Pokoradz	122		25	100	1	4	12	30	Arsenic-copper (low)
10939	Kisbér	123		31	122	1	4	11	28	Arsenic-copper (low)
11526	Oggau, 'Seestrasse', p. B. Eisenstadt-Umgebung	123	BBC	35	212	1	4	11	28	Arsenic-copper (low)
11592	Miedlingsdorf, p. B. Oberwarl	123		25	100	1	4	12	29	Arsenic-copper (low)
12438	Dunaföldvár	124	BBC	35	215	1	5	14	31	Arsenic-copper (low)
12475	Veliki Gaj	125	BOD	21	130	1	4	12	30	Arsenic-copper (low)
12999	Aranyosapáti	122		21	130	1	4	12	29	Arsenic-copper (low)

ANo	FIND_PLACE	AREA	CULT	PER	TYPE_2	CL_No_I	CL_No_II	CL_No_III	CL_No_IV	Metal_groups
13090	Beba Veche	125	PIT	40	600	1	4	12	29	Arsenic-copper (low)
13091	Beba Veche	125	PIT	40	600	1	4	12	29	Arsenic-copper (low)
13115	Bácsalmás	125		31	122	1	4	11	28	Arsenic-copper (low)
13122	Pitvaros	125	PIT	40	500	1	5	14	31	Arsenic-copper (low)
13123	Pitvaros	125	PIT	40	500	1	4	11	28	Arsenic-copper (low)
13124	Pitvaros	125	PIT	40	410	1	4	12	29	Arsenic-copper (low)
13126	Pitvaros	125		40	410	1	4	11	28	Arsenic-copper (low)
13128	Pitvaros	125	PIT	40	405	1	4	12	29	Arsenic-copper (low)
13208	Szeged	125		21	100	1	4	12	30	Arsenic-copper (low)
13209	Szeged	125		21	100	1	4	12	30	Arsenic-copper (low)
13402	Dunakömlöd	124		31	122	1	5	14	31	Arsenic-copper (low)
13404	Dunakömlöd	124		31	122	1	5	14	31	Arsenic-copper (low)
13409	Dunakömlöd	124		31	110	1	5	14	31	Arsenic-copper (low)
14391	Szirák	121		21	100	1	4	12	30	Arsenic-copper (low)
14419	Fajsz	125		31	122	1	4	12	30	Arsenic-copper (low)
14463	Seewalchen, p. B. Vöcklabruck	73	MON	21	100	1	4	11	28	Arsenic-copper (low)
14464	Weyeregg, 'Puschacher', p. B. Völklabruck	73	MON	21	100	1	4	12	29	Arsenic-copper (low)
14465	Misling am Attersee, 'Misling I' p. B. Vöcklabruck	73	MON	21	100	1	4	12	29	Arsenic-copper (low)
14466	Attersee, p. B. Vöcklabruck	73	MON	21	100	1	4	11	28	Arsenic-copper (low)
14469	Aufham, p. B. Vöcklabruck	73	MON	21	100	1	4	12	29	Arsenic-copper (low)
14470	Attersee, p. B. Vöcklabruck	73	MON	21	100	1	4	12	29	Arsenic-copper (low)
14473	Seewalchen, p. B. Vöcklabruck	73	MON	21	300	1	5	14	31	Arsenic-copper (low)
15908	Langquaid, Lkr. Kelheim	66		25	100	1	4	12	29	Arsenic-copper (low)
16523	Vully-le-Bas, 'Sur le Petit Bas'	67	BK	40	100	1	4	12	29	Arsenic-copper (low)
16524	Vully-le-Bas, 'Sur le Petit Bas'	67	BK	40	100	1	4	12	29	Arsenic-copper (low)
16600	Eschenz, 'Nili'	67		31	100	1	5	14	31	Arsenic-copper (low)
17689	Veselé	121	CWC	31	400	1	4	12	29	Arsenic-copper (low)
18029	Täuffelen-Gerolfingen	67		31	134	1	4	12	30	Arsenic-copper (low)
19922	Predmostí	114	BBC	35	215	1	4	12	30	Arsenic-copper (low)
19923	Predmostí	114	BBC	35	215	1	4	12	30	Arsenic-copper (low)
19924	Predmostí	114	BBC	35	215	1	4	12	29	Arsenic-copper (low)
21396	Jicín	112		21	100	1	4	12	29	Arsenic-copper (low)
21597	Hülen, 'Kapfenberg', Ostalbkreis	62	BBC	35	215	1	5	14	31	Arsenic-copper (low)
21797	Goldburghausen, 'Goldberg', Ostalbkreis	62	GIII	25	100	1	5	14	31	Arsenic-copper (low)
21798	Goldburghausen, 'Goldberg', Ostalbkreis	62	GIII	25	100	1	5	14	31	Arsenic-copper (low)
21801	Goldburghausen, 'Goldberg', Ostalbkreis	62	GIII	25	810	1	4	12	29	Arsenic-copper (low)
21802	Goldburghausen, 'Goldberg', Ostalbkreis	62	GIII	25	810	1	5	14	31	Arsenic-copper (low)
21804	Goldburghausen, 'Goldberg', Ostalbkreis	62	GIII	25	410	1	4	12	30	Arsenic-copper (low)
21805	Goldburghausen, 'Goldberg', Ostalbkreis	62	GIII	25	300	1	4	12	29	Arsenic-copper (low)
21808	Goldburghausen, 'Goldberg', Ostalbkreis	62	GIII	25	700	1	4	12	29	Arsenic-copper (low)
21851	Goldburghausen, 'Goldberg', Ostalbkreis	62	GIII	25	810	1	4	12	30	Arsenic-copper (low)
21853	Goldburghausen, 'Goldberg', Ostalbkreis	62	GIII	25	810	1	4	11	28	Arsenic-copper (low)
22140	Yvonand, 'Geilinger'	67	CWC	31	600	1	4	12	29	Arsenic-copper (low)
22141	Yvonand, 'Geilinger'	67	CWC	31	600	1	4	12	29	Arsenic-copper (low)
22143	Twann, 'Bahnhof; Grabung 1974'	67	COR	21	200	1	4	11	28	Arsenic-copper (low)
22151	Kottenheim, 'Oben auf'm Winnfeld', Kreis Mayen-Koblenz	51		31	130	1	5	14	31	Arsenic-copper (low)
22232	Bevaix	67	CWC	31	200	1	4	11	28	Arsenic-copper (low)
22238	St. Blaise	67	CWC	31	215	1	4	11	28	Arsenic-copper (low)
24213	Singen a. Hohentwiel, Kr. Konstanz	62	BK	40	400	1	5	14	31	Arsenic-copper (low)
29800	Sutz-Lattrigen	67	COR	21	212	1	4	12	29	Arsenic-copper (low)
34453	Drosa 'Küsterberg', Lkr. Anhalt-Bitterfeld	103	CWC	31	550	2	6	15	52	Arsenic-copper (low)
34544	Drosa 'Küsterberg', Lkr. Anhalt-Bitterfeld	103	CWC	31	550	2	6	15	53	Arsenic-copper (low)
34546	Drosa 'Küsterberg', Lkr. Anhalt-Bitterfeld	103	CWC	31	550	2	6	15	52	Arsenic-copper (low)

ANo	FIND_PLACE	AREA	CULT	PER	TYPE_2	CL_No_I	CL_No_II	CL_No_III	CL_No_IV	Metal_groups
44013	Strzyzów, 'Horodlo'	84	Mier	40	420	1	4	12	29	Arsenic-copper (low)
44043	Mierzanowice II	85	Mier	40	400	1	5	13	13	Arsenic-copper (low)
44047	Mierzanowice II	85	Mier	40	510	2	6	16	38	Arsenic-copper (low)
58001	Bojnice	121		25	100	1	5	13	13	Arsenic-copper (low)
58004	Hronské Kosihy	121		21	100	1	5	13	13	Arsenic-copper (low)
58012	Hronské Kosihy	121		21	100	1	5	13	13	Arsenic-copper (low)
58024	Makotrasy	111	BAA	21	134	1	4	11	28	Arsenic-copper (low)
58025	Makotrasy	111	BAA	21	700	2	6	16	38	Arsenic-copper (low)
58026	Makotrasy	111	BAA	21	700	2	6	16	38	Arsenic-copper (low)
58028	Makotrasy	111	BAA	21	700	1	5	13	13	Arsenic-copper (low)
58029	Makotrasy	111	BAA	21	700	1	5	13	13	Arsenic-copper (low)
58030	Makotrasy	111	BAA	21	700	1	5	13	13	Arsenic-copper (low)
58031	Makotrasy	111	BAA	21	134	1	5	13	13	Arsenic-copper (low)
62102	Chevroux	67		31	134	1	4	12	29	Arsenic-copper (low)
62110	Concise	67		21	510	2	6	16	38	Arsenic-copper (low)
62125	Morges	67		31	212	1	5	14	31	Arsenic-copper (low)
62164	Concise	67		31	212	1	5	13	13	Arsenic-copper (low)
62196	Linz-St. Peter, p. B. Linz	66	BAD	25	130	1	5	14	31	Arsenic-copper (low)
100013	Vikletice	111	CWC	31	405	2	6	16	38	Arsenic-copper (low)
100036	Unterach-Mondsee, 'Station See', p. B. Völklabruck	73	MON	21	500	2	6	16	38	Arsenic-copper (low)
100037	Unterach-Mondsee, 'Station See', p. B. Völklabruck	73	MON	21	500	2	6	16	38	Arsenic-copper (low)
100038	Unterach-Mondsee, 'Station See', p. B. Völklabruck	73	MON	21	500	2	6	16	38	Arsenic-copper (low)
100039	Unterach-Mondsee, 'Station See', p. B. Völklabruck	73	MON	21	810	2	6	16	38	Arsenic-copper (low)
100040	Unterach-Mondsee, 'Station See', p. B. Völklabruck	73	MON	21	810	2	6	16	38	Arsenic-copper (low)
100041	Unterach-Mondsee, 'Station See', p. B. Völklabruck	73	MON	21	810	2	6	16	38	Arsenic-copper (low)
100042	Unterach-Mondsee, 'Station See', p. B. Völklabruck	73	MON	21	212	2	6	16	38	Arsenic-copper (low)
100043	Unterach-Mondsee, 'Station See', p. B. Völklabruck	73	MON	21	212	2	6	16	38	Arsenic-copper (low)
100044	Unterach-Mondsee, 'Station See', p. B. Völklabruck	73	MON	21	212	2	6	16	38	Arsenic-copper (low)
100045	Unterach-Mondsee, 'Station See', p. B. Völklabruck	73	MON	21	212	2	6	16	38	Arsenic-copper (low)
100046	Unterach-Mondsee, 'Station See', p. B. Völklabruck	73	MON	21	212	2	6	16	38	Arsenic-copper (low)
100047	Unterach-Mondsee, 'Station See', p. B. Völklabruck	73	MON	21	212	2	6	16	38	Arsenic-copper (low)
100048	Unterach-Mondsee, 'Station See', p. B. Völklabruck	73	MON	21	100	2	6	16	38	Arsenic-copper (low)
100049	Unterach-Mondsee, 'Station See', p. B. Völklabruck	73	MON	21	100	1	5	13	13	Arsenic-copper (low)
100050	Unterach-Mondsee, 'Station See', p. B. Völklabruck	73	MON	21	100	2	6	16	38	Arsenic-copper (low)
100051	Unterach-Mondsee, 'Station See', p. B. Völklabruck	73	MON	21	100	2	6	16	38	Arsenic-copper (low)
100052	Unterach-Mondsee, 'Station See', p. B. Völklabruck	73	MON	21	100	2	6	16	38	Arsenic-copper (low)
100053	Misling am Attersee, 'Misling II', p. B. Völklabruck	73	MON	21	810	1	5	13	13	Arsenic-copper (low)
100054	Misling am Attersee, 'Misling II', p. B. Völklabruck	73	MON	21	200	2	6	16	38	Arsenic-copper (low)
100055	Misling am Attersee, 'Misling II', p. B. Völklabruck	73	MON	21	100	2	6	16	38	Arsenic-copper (low)
100056	Misling am Attersee, 'Misling II', p. B. Völklabruck	73	MON	21	100	2	6	16	38	Arsenic-copper (low)
100057	Misling am Attersee, 'Misling II', p. B. Völklabruck	73	MON	21	100	2	6	16	38	Arsenic-copper (low)
100058	Misling am Attersee, 'Misling II', p. B. Völklabruck	73	MON	21	100	1	5	13	13	Arsenic-copper (low)
100059	Misling am Attersee, 'Misling II', p. B. Völklabruck	73	MON	21	100	1	5	13	13	Arsenic-copper (low)
100060	Misling am Attersee, 'Misling II', p. B. Völklabruck	73	MON	21	200	1	5	13	13	Arsenic-copper (low)
100061	Misling am Attersee, 'Misling II', p. B. Völklabruck	73	MON	21	200	2	6	16	38	Arsenic-copper (low)
100064	Seewalchen, p. B. Vöcklabruck	73	MON	21	810	2	6	16	38	Arsenic-copper (low)
100065	Seewalchen, p. B. Vöcklabruck	73	MON	21	810	2	6	16	38	Arsenic-copper (low)
100067	Scharfling am Mondsee, 'St. Lorenz', p. B. Völklabruck	73	MON	21	100	1	5	13	13	Arsenic-copper (low)
100069	Weyeregg, p. B. Völklabruck	73	MON	21	100	1	5	13	13	Arsenic-copper (low)
100072	Bodman-Weiler, Lkr. Konstanz	62	PFY	21	100	1	5	13	13	Arsenic-copper (low)
100073	Gaienhofen, 'Hornstaad-Hörnle I, Schicht AH 4', Kr. Konstanz	62	PFY	21	100	2	6	16	38	Arsenic-copper (low)
100076	Egolzwil	67	COR	21	100	2	6	16	38	Arsenic-copper (low)
100079	Hüttenwilen, 'Nussbaumersee-Inseli'	67	PFY	21	100	2	6	16	38	Arsenic-copper (low)

ANo	FIND_PLACE	AREA	CULT	PER	TYPE_2	CL_No_I	CL_No_II	CL_No_III	CL_No_IV	Metal_groups
100080	Scharfling am Mondsee, p. B. Völklabruck	73	MON	21	700	2	6	16	38	Arsenic-copper (low)
100081	Unterach-Mondsee, 'Station See', p. B. Völklabruck	73	MON	21	700	2	6	16	38	Arsenic-copper (low)
100082	Unterach-Mondsee, 'Station See', p. B. Völklabruck	73	MON	21	500	2	6	16	38	Arsenic-copper (low)
2	Adlerberg b. Worms, Kr. Alzey-Worms	62	ABC	40	600	2	6	18	43	Arsenic-copper with Impurities
9	Altheim, Kr. Landshut	66	ALT	21	500	2	6	18	43	Arsenic-copper with Impurities
10	Altheim, Kr. Landshut	66	ALT	21	700	2	6	18	43	Arsenic-copper with Impurities
105	Goldburghausen, 'Goldberg', Ostalbkreis	62	GIII	25	100	2	6	18	43	Arsenic-copper with Impurities
174	Luckenpaint-Thalmassing,'Geiergucken', Lkr. Regensburg	66	BBC	35	215	2	6	15	33	Arsenic-copper with Impurities
184	Mettenheim, Lkr. Mühlheim a. Inn	66	BBC	35	810	2	6	18	43	Arsenic-copper with Impurities
192	Moosinning, 'Neue Kempfinger Brücke', Lkr. Erding	66	BBC	35	215	2	6	15	33	Arsenic-copper with Impurities
210	München-Sendling, 'Wolfratshauser Str.', Stkr. München	66	BBC	35	215	2	6	18	43	Arsenic-copper with Impurities
276	Safferstetten, Lkr. Passau	66	BBC	35	212	2	6	18	43	Arsenic-copper with Impurities
362	Steinfurth, Stadt Bad Nauheim, Wetteraukreis	62		40	100	2	6	18	43	Arsenic-copper with Impurities
479	Zimmern unter der Burg, Zollernalbkreis	62		31	130	2	6	16	36	Arsenic-copper with Impurities
1296	Debrecen	125		31	122	2	6	18	43	Arsenic-copper with Impurities
2834	Vinelz, 'Amt Erlau'	67	CWC	31	212	2	6	18	43	Arsenic-copper with Impurities
2926	Lüscherz	67	CWC	31	134	1	4	10	25	Arsenic-copper with Impurities
2927	Lüscherz	67	CWC	31	810	2	6	16	36	Arsenic-copper with Impurities
3193	Brandysek	111	BBC	35	215	2	6	18	43	Arsenic-copper with Impurities
3246	Praha-Kobylisy, 'Boranovicka-Gasse'	112	BBC	35	810	2	6	18	43	Arsenic-copper with Impurities
3343	Praha-Liben	112	BBC	35	600	2	6	18	43	Arsenic-copper with Impurities
3351	Vel'ké Kostol'any	121		21	130	1	4	10	25	Arsenic-copper with Impurities
3374	Prace	114		25	100	2	6	18	43	Arsenic-copper with Impurities
3619	Laa an der Thaya, p. B. Misterbach	72	BBC	35	810	2	6	18	43	Arsenic-copper with Impurities
3785	Handlová	121		21	130	1	4	10	25	Arsenic-copper with Impurities
4340	Unterach-Mondsee, 'Station See', p. B. Völklabruck	73	MON	21	100	2	6	18	43	Arsenic-copper with Impurities
4351	Unterach-Mondsee, 'Station See', p. B. Völklabruck	73	MON	21	100	2	6	18	43	Arsenic-copper with Impurities
5001	Hippersdorf, 'Plexenthal', p. B. Tulln	72	LLG	40	410	2	6	18	43	Arsenic-copper with Impurities
5003	Hippersdorf, 'Plexenthal', p. B. Tulln	72	LLG	40	410	2	7	20	68	Arsenic-copper with Impurities
5683	Horodnica	85		31	130	2	6	18	43	Arsenic-copper with Impurities
5840	Gemeinlebarn, p. B. St. Pölten	72	BK	40	405	2	7	20	68	Arsenic-copper with Impurities
5924	Gemeinlebarn, p. B. St. Pölten	72	BK	40	410	2	6	18	43	Arsenic-copper with Impurities
6470	Budakalász	124		25	212	2	6	18	43	Arsenic-copper with Impurities
6475	Jibert	125		21	130	2	6	18	43	Arsenic-copper with Impurities
7488	Praha-Sárka	112		21	100	1	4	10	25	Arsenic-copper with Impurities
7527	Kunetice	112		21	100	2	6	18	43	Arsenic-copper with Impurities
9256	Bagamér	125		25	100	2	6	18	43	Arsenic-copper with Impurities
10222	Praha-Podbaba	112		40	110	2	7	20	68	Arsenic-copper with Impurities
10367	Výcapy-Opatovce	121	NIT	40	420	2	6	18	43	Arsenic-copper with Impurities
10431	Výcapy-Opatovce	121	NIT	40	500	2	6	18	43	Arsenic-copper with Impurities
10451	Výcapy-Opatovce	121	NIT	40	300	2	6	18	43	Arsenic-copper with Impurities
10462	Výcapy-Opatovce	121	NIT	40	420	2	6	18	43	Arsenic-copper with Impurities
10463	Výcapy-Opatovce	121	NIT	40	420	2	6	18	43	Arsenic-copper with Impurities
10469	Výcapy-Opatovce	121	NIT	40	420	2	6	18	43	Arsenic-copper with Impurities
10470	Výcapy-Opatovce	121	NIT	40	420	2	6	18	43	Arsenic-copper with Impurities
10474	Výcapy-Opatovce	121	NIT	40	420	2	6	18	43	Arsenic-copper with Impurities
10488	Výcapy-Opatovce	121	NIT	40	420	2	6	18	43	Arsenic-copper with Impurities
10495	Výcapy-Opatovce	121	NIT	40	420	2	6	18	43	Arsenic-copper with Impurities
10502	Výcapy-Opatovce	121	NIT	40	420	2	6	18	43	Arsenic-copper with Impurities
10524	Výcapy-Opatovce	121	NIT	40	420	2	6	18	43	Arsenic-copper with Impurities
10531	Výcapy-Opatovce	121	NIT	40	420	2	6	18	43	Arsenic-copper with Impurities
10536	Výcapy-Opatovce	121	NIT	40	420	2	6	18	43	Arsenic-copper with Impurities
10537	Výcapy-Opatovce	121	NIT	40	400	2	6	18	43	Arsenic-copper with Impurities

ANo	FIND_PLACE	AREA	CULT	PER	TYPE_2	CL_No_I	CL_No_II	CL_No_III	CL_No_IV	Metal_groups
10565	Nitra, 'Partisanenstraße'	121	NIT	40	300	2	6	18	43	Arsenic-copper with Impurities
10646	Tvrdošovce, 'Körektó'	121	NIT	40	405	2	6	18	43	Arsenic-copper with Impurities
10649	Tvrdošovce, 'Körektó'	121	NIT	40	410	2	6	18	43	Arsenic-copper with Impurities
10656	Tvrdošovce, 'Körektó'	121	NIT	40	400	2	6	18	43	Arsenic-copper with Impurities
10660	Tvrdošovce, 'Körektó'	121	NIT	40	420	2	6	18	43	Arsenic-copper with Impurities
10664	Tvrdošovce, 'Körektó'	121	NIT	40	420	2	7	20	68	Arsenic-copper with Impurities
10666	Tvrdošovce, 'Körektó'	121	NIT	40	300	2	6	18	43	Arsenic-copper with Impurities
10670	Tvrdošovce, 'Körektó'	121	NIT	40	420	2	6	18	43	Arsenic-copper with Impurities
10848	Strecno	121		10	130	1	4	10	25	Arsenic-copper with Impurities
10849	Durciná	121		25	110	2	6	18	43	Arsenic-copper with Impurities
11197	Linz-Scharlinz, p. B. Linz	66	BBC	35	215	2	6	18	43	Arsenic-copper with Impurities
12505	Budapest-III-Obuda	124		31	122	2	6	18	43	Arsenic-copper with Impurities
12507	Senec	121		31	122	2	6	18	43	Arsenic-copper with Impurities
12526	Hajduböszörmény	125		21	212	2	6	18	43	Arsenic-copper with Impurities
12771	Alsónémedi	125	NGY	40	410	2	6	18	43	Arsenic-copper with Impurities
13092	Beba Veche	125	PIT	40	600	2	6	18	43	Arsenic-copper with Impurities
13098	Tápé, 'Lebö'	125		25	122	2	6	18	43	Arsenic-copper with Impurities
13119	Pitvaros	125	PIT	40	810	2	6	18	43	Arsenic-copper with Impurities
13120	Pitvaros	125	PIT	40	405	2	6	18	43	Arsenic-copper with Impurities
13127	Pitvaros	125	PIT	40	600	2	6	18	43	Arsenic-copper with Impurities
13405	Dunakömlöd	124		31	122	2	6	18	43	Arsenic-copper with Impurities
13406	Dunakömlöd	124		31	122	2	6	18	43	Arsenic-copper with Impurities
13408	Dunakömlöd	124		31	100	2	6	18	43	Arsenic-copper with Impurities
15843	München-Sendling, 'Lindwurmstr. 76', Stkr. München	66	BK	40	500	2	6	18	43	Arsenic-copper with Impurities
16526	St. Blaise	67	CWC	31	500	2	6	18	43	Arsenic-copper with Impurities
19925	Turovice I	114	BBC	35	215	2	6	18	43	Arsenic-copper with Impurities
19932	Stehelceves	111	BBC	35	215	2	7	20	69	Arsenic-copper with Impurities
20081	Hochstetten, 'Hartwald', Lkr. Karlsruhe	62	BK	40	110	2	6	18	43	Arsenic-copper with Impurities
21993	Wechingen, 'Lumpenhet', Lkr. Donau-Ries	65	BK	40	110	2	6	18	43	Arsenic-copper with Impurities
22230	Bevaix	67	CWC	31	212	2	6	18	43	Arsenic-copper with Impurities
22231	Bevaix	67	CWC	31	212	2	6	16	36	Arsenic-copper with Impurities
22276	Chevroux	67		31	212	2	6	18	43	Arsenic-copper with Impurities
22278	Concise	67		31	212	2	6	18	43	Arsenic-copper with Impurities
22288	Riekofen, Lkr. Regensburg	66	CHA	25	810	2	6	18	43	Arsenic-copper with Impurities
24208	Singen a. Hohentwiel, Kr. Konstanz	62	BK	40	212	2	6	16	36	Arsenic-copper with Impurities
32219	Gotha-Ostheim, Kr. Gotha	103	BER	25	550	2	6	16	36	Arsenic-copper with Impurities
32228	Mühlhausen, 'Aktivistenring', Lkr.Mühlhausen	103	BBC	35	215	2	6	15	33	Arsenic-copper with Impurities
32311	Saaldorf, 'Abtsdorfer See', Lkr. Berchtesgadener Land	66		31	100	2	6	16	36	Arsenic-copper with Impurities
32312	Wallerfing, 'Bachling', Lkr. Dingolfing-Landau	66	ALT	21	810	2	6	16	36	Arsenic-copper with Impurities
32315	Trieching, Gde. Pilsting, Lkr. Dinglofing-Landau	66	BBC	35	215	2	6	16	36	Arsenic-copper with Impurities
32322	Rastenberg, Lkr. Sömmerda	103		25	100	2	6	16	36	Arsenic-copper with Impurities
32326	Gotha-Ostheim, Kr. Gotha	103	BER	25	550	2	6	16	36	Arsenic-copper with Impurities
32328	Mühlhausen, 'Aktivistenring', Lkr.Mühlhausen	103	BBC	35	215	2	6	15	33	Arsenic-copper with Impurities
32331	Erfurt-Gispers, 'Nordhäuser Str', Stkr. Erfurt	103	CWC	31	550	2	6	15	33	Arsenic-copper with Impurities
33030	Bautzen-Niederkaina, 'Schafberg, Qu III', Lkr. Bautzen	105	CWC	31	510	2	6	15	33	Arsenic-copper with Impurities
33663	Poserna, Burgenlandkreis	104	CWC	31	410	2	6	15	33	Arsenic-copper with Impurities
34235	Nienburg, Salzlandkreis	103		31	130	2	6	16	36	Arsenic-copper with Impurities
34239	Schortewitz, 'Windmühlenberg',Lkr. Anhalt-Bitterfeld	103	CWC	31	410	2	6	16	36	Arsenic-copper with Impurities
34240	Schortewitz, 'Windmühlenberg',Lkr. Anhalt-Bitterfeld	103	CWC	31	410	2	6	16	36	Arsenic-copper with Impurities
34241	Drosa, 'Buchberg', Lkr. Anhalt-Bitterfeld	103	CWC	31	550	2	6	16	36	Arsenic-copper with Impurities
34431	Börnecke-Renneberg b. Blanckenburg, Lkr. Harz	103	CWC	31	410	2	6	16	36	Arsenic-copper with Impurities
34436	Preußlitz,'Ilgensteiner Mühlberg', Salzlandkreis	103	BAA	21	500	2	6	16	36	Arsenic-copper with Impurities
34437	Preußlitz,'Ilgensteiner Mühlberg', Salzlandkreis	103	BAA	21	500	2	6	16	36	Arsenic-copper with Impurities

ANo	FIND_PLACE	AREA	CULT	PER	TYPE_2	CL_No_I	CL_No_II	CL_No_III	CL_No_IV	Metal_groups
34438	Preußlitz,'Ilgensteiner Mühlberg', Salzlandkreis	103	BAA	21	410	2	6	16	36	Arsenic-copper with Impurities
34439	Preußlitz,'Ilgensteiner Mühlberg', Salzlandkreis	103	BAA	21	550	2	6	16	36	Arsenic-copper with Impurities
34440	Preußlitz,'Ilgensteiner Mühlberg', Salzlandkreis	103	BAA	21	500	2	6	16	36	Arsenic-copper with Impurities
34441	Preußlitz,'Ilgensteiner Mühlberg', Salzlandkreis	103	BAA	21	510	2	6	16	36	Arsenic-copper with Impurities
34442	Preußlitz,'Ilgensteiner Mühlberg', Salzlandkreis	103	BAA	21	510	2	6	16	36	Arsenic-copper with Impurities
34443	Preußlitz,'Ilgensteiner Mühlberg', Salzlandkreis	103	BAA	21	550	2	6	16	36	Arsenic-copper with Impurities
34444	Preußlitz,'Ilgensteiner Mühlberg', Salzlandkreis	103	BAA	21	550	2	6	16	36	Arsenic-copper with Impurities
34445	Preußlitz,'Ilgensteiner Mühlberg', Salzlandkreis	103	BAA	21	410	2	6	16	36	Arsenic-copper with Impurities
34446	Preußlitz,'Ilgensteiner Mühlberg', Salzlandkreis	103	BAA	21	510	2	6	16	36	Arsenic-copper with Impurities
34447	Preußlitz,'Ilgensteiner Mühlberg', Salzlandkreis	103	BAA	21	410	2	6	16	36	Arsenic-copper with Impurities
34501	Drosa 'Küsterberg', Lkr. Anhalt-Bitterfeld	103	CWC	31	550	2	6	15	33	Arsenic-copper with Impurities
34505	Drosa 'Küsterberg', Lkr. Anhalt-Bitterfeld	103	CWC	31	550	2	6	15	33	Arsenic-copper with Impurities
34508	Drosa 'Küsterberg', Lkr. Anhalt-Bitterfeld	103	CWC	31	550	2	6	15	33	Arsenic-copper with Impurities
34524	Drosa 'Küsterberg', Lkr. Anhalt-Bitterfeld	103	CWC	31	410	2	6	15	33	Arsenic-copper with Impurities
34525	Drosa 'Küsterberg', Lkr. Anhalt-Bitterfeld	103	CWC	31	410	2	6	15	33	Arsenic-copper with Impurities
34527	Drosa 'Küsterberg', Lkr. Anhalt-Bitterfeld	103	CWC	31	550	2	6	15	33	Arsenic-copper with Impurities
34543	Drosa 'Küsterberg', Lkr. Anhalt-Bitterfeld	103	CWC	31	550	2	6	15	33	Arsenic-copper with Impurities
34548	Drosa 'Küsterberg', Lkr. Anhalt-Bitterfeld	103	CWC	31	550	2	6	15	33	Arsenic-copper with Impurities
34552	Drosa 'Küsterberg', Lkr. Anhalt-Bitterfeld	103	CWC	31	550	2	6	15	33	Arsenic-copper with Impurities
34556	Drosa 'Küsterberg', Lkr. Anhalt-Bitterfeld	103	CWC	31	550	2	6	15	33	Arsenic-copper with Impurities
34566	Drosa 'Küsterberg', Lkr. Anhalt-Bitterfeld	103	CWC	31	550	2	6	15	33	Arsenic-copper with Impurities
34582	Drosa 'Küsterberg', Lkr. Anhalt-Bitterfeld	103	CWC	31	550	2	6	15	33	Arsenic-copper with Impurities
34584	Drosa 'Küsterberg', Lkr. Anhalt-Bitterfeld	103	CWC	31	550	2	6	15	33	Arsenic-copper with Impurities
34587	Drosa 'Küsterberg', Lkr. Anhalt-Bitterfeld	103	CWC	31	550	2	6	16	36	Arsenic-copper with Impurities
34592	Drosa 'Küsterberg', Lkr. Anhalt-Bitterfeld	103	CWC	31	550	2	6	15	33	Arsenic-copper with Impurities
34607	Drosa 'Küsterberg', Lkr. Anhalt-Bitterfeld	103	CWC	31	550	2	6	15	33	Arsenic-copper with Impurities
34608	Drosa 'Küsterberg', Lkr. Anhalt-Bitterfeld	103	CWC	31	550	2	6	15	33	Arsenic-copper with Impurities
44011	Strzyżów, 'Horodlo'	84	Mier	40	420	2	6	16	36	Arsenic-copper with Impurities
44015	Strzyżów, 'Horodlo'	84	Mier	40	550	2	6	16	36	Arsenic-copper with Impurities
44019	Strzyżów, 'Horodlo'	84	Mier	40	500	2	6	16	36	Arsenic-copper with Impurities
44020	Strzyżów, 'Horodlo'	84	Mier	40	500	2	6	16	36	Arsenic-copper with Impurities
44021	Strzyżów, 'Horodlo'	84	Mier	40	500	2	6	16	36	Arsenic-copper with Impurities
44022	Strzyżów, 'Horodlo'	84	Mier	40	400	2	6	16	36	Arsenic-copper with Impurities
44023	Strzyżów, 'Horodlo'	84	Mier	40	500	2	6	16	36	Arsenic-copper with Impurities
44025	Strzyżów, 'Horodlo'	84	Mier	40	400	2	6	16	36	Arsenic-copper with Impurities
49901	Schernau, Kr. Bad Kitzingen	65	MUE	10	810	2	6	16	36	Arsenic-copper with Impurities
58011	Besenová	121		10	100	1	4	10	25	Arsenic-copper with Impurities
58013	Besenová	121		10	130	1	4	10	25	Arsenic-copper with Impurities
58015	Handlová, 'Na Pstruhároch'	121		21	130	1	4	10	25	Arsenic-copper with Impurities
58027	Makotrasy	111	BAA	21	700	2	6	16	36	Arsenic-copper with Impurities
62108	Chevroux	67		31	212	2	6	15	33	Arsenic-copper with Impurities
62109	Chevroux	67		31	213	2	6	16	36	Arsenic-copper with Impurities
62113	Concise	67		31	212	2	6	15	33	Arsenic-copper with Impurities
62127	Onnens	67	CWC	31	212	2	6	15	33	Arsenic-copper with Impurities
62128	Genève, 'Crêts St. Laurent'	67		21	100	2	6	16	36	Arsenic-copper with Impurities
62137	St. Blaise	67	CWC	31	200	2	6	15	33	Arsenic-copper with Impurities
62146	Hauterive, 'Champréveyres'	67	COR	21	212	2	6	16	36	Arsenic-copper with Impurities
62148	Genève, 'Rue des Eaux Vives'	67		31	212	2	6	18	43	Arsenic-copper with Impurities
62160	St. Blaise	67	CWC	31	215	2	6	16	36	Arsenic-copper with Impurities
62161	St. Blaise	67	CWC	31	100	2	6	16	36	Arsenic-copper with Impurities
62168	Bevaix	67	CWC	31	200	2	6	16	36	Arsenic-copper with Impurities
62169	Bevaix	67	CWC	31	212	2	6	15	33	Arsenic-copper with Impurities
62181	Salez, 'Sennwald'	67	BK	40	110	2	6	16	36	Arsenic-copper with Impurities

ANo	FIND_PLACE	AREA	CULT	PER	TYPE_2	CL_No_I	CL_No_II	CL_No_III	CL_No_IV	Metal_groups
62189	Auvernier, 'La Saunerie; Grabung Strahm 1965'	67	CWC	31	810	2	6	15	33	Arsenic-copper with Impurities
62190	Yvonand, 'Geilinger'	67	CWC	31	600	2	6	16	36	Arsenic-copper with Impurities
62204	Mödling, p. B. Mödling	72		21	130	2	6	16	36	Arsenic-copper with Impurities
78142	Kosice-Sebastovce	122	KST	40	405	2	6	15	33	Arsenic-copper with Impurities
78143	Kosice-Sebastovce	122	KST	40	405	2	6	15	33	Arsenic-copper with Impurities
78145	Kosice-Sebastovce	122	KST	40	200	2	6	16	36	Arsenic-copper with Impurities
78146	Kosice-Sebastovce	122	KST	40	134	2	6	15	33	Arsenic-copper with Impurities
78159	Branc	121	NIT	40	420	2	6	15	33	Arsenic-copper with Impurities
78163	Branc	121	NIT	40	400	2	6	15	33	Arsenic-copper with Impurities
78168	Horné Lefantovce	121		25	100	2	6	16	36	Arsenic-copper with Impurities
78172	Tibava, 'Hrun za cintirom'	122	TC	10	130	2	6	16	36	Arsenic-copper with Impurities
78174	Tibava, 'Hrun za cintirom'	122	TC	10	130	2	6	16	36	Arsenic-copper with Impurities
98902	Makotrasy	111	BAA	21	134	2	6	16	36	Arsenic-copper with Impurities
98903	Makotrasy	111	BAA	21	700	2	6	16	36	Arsenic-copper with Impurities
100010	Giengen a. d. Brenz, 'Sachsenhausen', Klr. Heidenheim	62	BBC	35	810	2	7	20	70	Arsenic-copper with Impurities
100021	Bad Waldsee, 'Reute', Lkr. Ravensburg	62	PFY	21	212	2	6	16	36	Arsenic-copper with Impurities
100022	Bad Waldsee, 'Reute', Lkr. Ravensburg	62	PFY	21	213	2	6	16	36	Arsenic-copper with Impurities
100024	Praha-Nove Butovice	112	CWC	31	410	2	6	15	33	Arsenic-copper with Impurities
100025	Tousen	112	RIV	25	700	2	6	16	36	Arsenic-copper with Impurities
100026	Tousen	112	RIV	25	700	2	6	16	36	Arsenic-copper with Impurities
100029	Brixlegg, 'Mariahilfbergl', p.B. Brixlegg	71	MUE	10	800	2	6	16	36	Arsenic-copper with Impurities
100030	Brixlegg, 'Mariahilfbergl', p.B. Brixlegg	71	MUE	10	500	2	6	16	36	Arsenic-copper with Impurities
100033	Überlingen, Bodenseekreis	62	MUE	10	130	2	6	16	36	Arsenic-copper with Impurities
100035	Scharfling am Mondsee, p. B. Völklabruck	73	MON	21	510	2	6	16	36	Arsenic-copper with Impurities
100063	Bad Waldsee, 'Reute', Lkr. Ravensburg	62	PFY	21	510	2	6	16	36	Arsenic-copper with Impurities
100070	Hattingen, Kr. Tuttlingen	62		31	100	2	6	16	36	Arsenic-copper with Impurities
100074	Überlingen, 'Osthafen', Bodenseekreis	62		31	100	2	6	16	36	Arsenic-copper with Impurities
100075	Überlingen, 'Osthafen', Bodenseekreis	62		31	100	2	6	16	36	Arsenic-copper with Impurities
100077	Bottinghofen, 'Schlössli'	67	PFY	21	100	2	6	16	36	Arsenic-copper with Impurities
100089	Amesdorf-Warmsdorf, 'Der Hügel', Salzlandkreis	103	CWC	31	410	2	6	18	43	Arsenic-copper with Impurities
100090	Ampfurth-Schermcke, Lkr. Börde	103		25	100	2	6	16	36	Arsenic-copper with Impurities
100091	Ampfurth-Schermcke, Lkr. Börde	103		25	100	2	6	16	36	Arsenic-copper with Impurities
100093	Auleben, Lkr. Nordhausen	103	BAA	21	130	2	6	18	43	Arsenic-copper with Impurities
100094	Brachwitz, Saalekreis	103		25	122	2	6	18	43	Arsenic-copper with Impurities
100096	Egeln, 'Bleckendorf', Lkr. Salzlandkreis	103	CWC	31	215	2	6	18	43	Arsenic-copper with Impurities
100097	Egeln, 'Bleckendorf', Lkr. Salzlandkreis	103	CWC	31	810	2	6	18	43	Arsenic-copper with Impurities
100098	Esperstedt, Saalekreis	103	CWC	31	700	2	6	18	43	Arsenic-copper with Impurities
100099	Esperstedt, Saalekreis	103	CWC	31	700	2	6	18	43	Arsenic-copper with Impurities
100101	Halle, 'Advokatenweg'; Stkr. Halle	103		25	100	2	6	18	43	Arsenic-copper with Impurities
100104	Halle-Dölauer Heide, Stkr. Halle	103	SAL	25	410	2	6	18	43	Arsenic-copper with Impurities
100107	Halle-Trotha, Stkr. Halle	103	BBC	35	215	2	6	18	43	Arsenic-copper with Impurities
100110	Hecklingen, Lkr. Salzlandkreis	103	CWC	31	500	2	6	18	43	Arsenic-copper with Impurities
100111	Hecklingen, Lkr. Salzlandkreis	103	CWC	31	500	2	6	18	43	Arsenic-copper with Impurities
100113	Kleinheringen-Rödigen, Burgenlandkreis	103		25	100	2	6	18	43	Arsenic-copper with Impurities
100114	Merseburg, Saalekreis	103	SAL	25	410	2	6	18	43	Arsenic-copper with Impurities
100115	Merseburg, Saalekreis	103	SAL	25	410	2	6	18	43	Arsenic-copper with Impurities
100118	Petersberg, Saalekreis	103	CWC	31	410	2	6	18	43	Arsenic-copper with Impurities
100119	Petersberg, Saalekreis	103	CWC	31	410	2	6	16	36	Arsenic-copper with Impurities
100120	Petersberg, Saalekreis	103	CWC	31	410	2	6	18	43	Arsenic-copper with Impurities
100121	Petersberg, Saalekreis	103,00	CWC	31	410	2	6	18	43	Arsenic-copper with Impurities
100122	Quenstedt, Lkr. Mansfeld-Südharz	103	CWC	31	500	2	6	18	43	Arsenic-copper with Impurities
100123	Rastenberg, Lkr. Sömmerda	103		25	100	2	6	16	36	Arsenic-copper with Impurities
100125	Schafstädt, Saalekreis	103	CWC	31	410	2	6	18	43	Arsenic-copper with Impurities

ANo	FIND_PLACE	AREA	CULT	PER	TYPE_2	CL_No_I	CL_No_II	CL_No_III	CL_No_IV	Metal_groups
100126	Bad Bibra, 'Steinbach', Burgenlandkreis	103		25	100	2	6	18	43	Arsenic-copper with Impurities
100127	Westerhausen, Lkr. Harz	103	BBC	35	500	2	6	18	43	Arsenic-copper with Impurities
100128	Westerhausen, Lkr. Harz	103	BBC	35	500	2	6	18	43	Arsenic-copper with Impurities
100130	Zabitz, Lkr. Anhalt-Bitterfeld	103		31	130	2	6	18	43	Arsenic-copper with Impurities
100131	Zscheiplitz, Burgenlandkreis	103		21	122	2	6	18	43	Arsenic-copper with Impurities
27	Barbing, Lkr. Regensburg	66	BBC	35	215	2	7	19	58	Arsenic-lead-copper
32310	Saaldorf, 'Abtsdorfer See', Lkr. Berchtesgadener Land	66		31	100	2	6	15	51	Arsenic-lead-copper
32334	Freienbessingen, 'Neuland III', Kyffhäuserkreis	103	CWC	31	410	2	6	15	51	Arsenic-lead-copper
32335	Sömmerda-Orlishausen, Lkr. Sömmerda	103	CWC	31	400	2	6	15	50	Arsenic-lead-copper
33576	Landau a. d. Isar, 'Landauer Feld', Lkr. Dingolfing-Landau	66	BBC	35	215	2	6	15	51	Arsenic-lead-copper
33662	Poserna, Burgenlandkreis	104	CWC	31	410	2	6	15	51	Arsenic-lead-copper
33664	Poserna, Burgenlandkreis	104	CWC	31	410	2	6	15	51	Arsenic-lead-copper
34238	Altenburg, Lkr. Altenburger Land	103		31	130	2	6	15	51	Arsenic-lead-copper
34433	Latdorf, 'Pohlsberg', Salzlandkreis	103	CWC	31	800	2	6	15	51	Arsenic-lead-copper
34434	Latdorf, 'Pohlsberg', Salzlandkreis	103	CWC	31	800	2	6	15	51	Arsenic-lead-copper
34435	Latdorf, 'Pohlsberg', Salzlandkreis	103	CWC	31	800	2	6	15	51	Arsenic-lead-copper
34448	Drosa 'Küsterberg', Lkr. Anhalt-Bitterfeld	103	CWC	31	550	2	6	15	50	Arsenic-lead-copper
34449	Drosa 'Küsterberg', Lkr. Anhalt-Bitterfeld	103	CWC	31	550	2	6	15	51	Arsenic-lead-copper
34451	Drosa 'Küsterberg', Lkr. Anhalt-Bitterfeld	103	CWC	31	550	2	6	15	50	Arsenic-lead-copper
34452	Drosa 'Küsterberg', Lkr. Anhalt-Bitterfeld	103	CWC	31	550	2	6	15	50	Arsenic-lead-copper
34455	Drosa 'Küsterberg', Lkr. Anhalt-Bitterfeld	103	CWC	31	550	2	6	15	51	Arsenic-lead-copper
34456	Drosa 'Küsterberg', Lkr. Anhalt-Bitterfeld	103	CWC	31	550	2	6	15	51	Arsenic-lead-copper
34457	Drosa 'Küsterberg', Lkr. Anhalt-Bitterfeld	103	CWC	31	550	2	6	15	50	Arsenic-lead-copper
34458	Drosa 'Küsterberg', Lkr. Anhalt-Bitterfeld	103	CWC	31	550	2	6	15	51	Arsenic-lead-copper
34459	Drosa 'Küsterberg', Lkr. Anhalt-Bitterfeld	103	CWC	31	550	2	6	15	50	Arsenic-lead-copper
34460	Drosa 'Küsterberg', Lkr. Anhalt-Bitterfeld	103	CWC	31	550	2	6	15	51	Arsenic-lead-copper
34461	Drosa 'Küsterberg', Lkr. Anhalt-Bitterfeld	103	CWC	31	550	2	6	15	50	Arsenic-lead-copper
34462	Drosa 'Küsterberg', Lkr. Anhalt-Bitterfeld	103	CWC	31	550	2	6	15	51	Arsenic-lead-copper
34463	Drosa 'Küsterberg', Lkr. Anhalt-Bitterfeld	103	CWC	31	550	2	6	15	51	Arsenic-lead-copper
34464	Drosa 'Küsterberg', Lkr. Anhalt-Bitterfeld	103	CWC	31	550	2	6	15	50	Arsenic-lead-copper
34465	Drosa 'Küsterberg', Lkr. Anhalt-Bitterfeld	103	CWC	31	550	2	6	15	51	Arsenic-lead-copper
34466	Drosa 'Küsterberg', Lkr. Anhalt-Bitterfeld	103	CWC	31	550	2	6	15	51	Arsenic-lead-copper
34467	Drosa 'Küsterberg', Lkr. Anhalt-Bitterfeld	103	CWC	31	550	2	6	15	51	Arsenic-lead-copper
34468	Drosa 'Küsterberg', Lkr. Anhalt-Bitterfeld	103	CWC	31	550	2	6	15	51	Arsenic-lead-copper
34469	Drosa 'Küsterberg', Lkr. Anhalt-Bitterfeld	103	CWC	31	550	2	6	15	51	Arsenic-lead-copper
34470	Drosa 'Küsterberg', Lkr. Anhalt-Bitterfeld	103	CWC	31	550	2	6	15	51	Arsenic-lead-copper
34471	Drosa 'Küsterberg', Lkr. Anhalt-Bitterfeld	103	CWC	31	550	2	6	15	51	Arsenic-lead-copper
34472	Drosa 'Küsterberg', Lkr. Anhalt-Bitterfeld	103	CWC	31	550	2	6	15	51	Arsenic-lead-copper
34473	Drosa 'Küsterberg', Lkr. Anhalt-Bitterfeld	103	CWC	31	550	2	6	15	51	Arsenic-lead-copper
34474	Drosa 'Küsterberg', Lkr. Anhalt-Bitterfeld	103	CWC	31	550	2	6	15	51	Arsenic-lead-copper
34475	Drosa 'Küsterberg', Lkr. Anhalt-Bitterfeld	103	CWC	31	550	2	6	15	51	Arsenic-lead-copper
34476	Drosa 'Küsterberg', Lkr. Anhalt-Bitterfeld	103	CWC	31	550	2	6	15	51	Arsenic-lead-copper
34477	Drosa 'Küsterberg', Lkr. Anhalt-Bitterfeld	103	CWC	31	550	2	6	15	51	Arsenic-lead-copper
34478	Drosa 'Küsterberg', Lkr. Anhalt-Bitterfeld	103	CWC	31	550	2	6	15	51	Arsenic-lead-copper
34479	Drosa 'Küsterberg', Lkr. Anhalt-Bitterfeld	103	CWC	31	550	2	6	15	51	Arsenic-lead-copper
34480	Drosa 'Küsterberg', Lkr. Anhalt-Bitterfeld	103	CWC	31	550	2	6	15	51	Arsenic-lead-copper
34481	Drosa 'Küsterberg', Lkr. Anhalt-Bitterfeld	103	CWC	31	550	2	6	15	51	Arsenic-lead-copper
34482	Drosa 'Küsterberg', Lkr. Anhalt-Bitterfeld	103	CWC	31	550	2	6	15	51	Arsenic-lead-copper
34483	Drosa 'Küsterberg', Lkr. Anhalt-Bitterfeld	103	CWC	31	550	2	6	15	51	Arsenic-lead-copper
34484	Drosa 'Küsterberg', Lkr. Anhalt-Bitterfeld	103	CWC	31	550	2	6	15	51	Arsenic-lead-copper
34485	Drosa 'Küsterberg', Lkr. Anhalt-Bitterfeld	103	CWC	31	550	2	6	15	50	Arsenic-lead-copper
34486	Drosa 'Küsterberg', Lkr. Anhalt-Bitterfeld	103	CWC	31	550	2	6	15	50	Arsenic-lead-copper
34487	Drosa 'Küsterberg', Lkr. Anhalt-Bitterfeld	103	CWC	31	550	2	6	15	51	Arsenic-lead-copper

ANo	FIND_PLACE	AREA	CULT	PER	TYPE_2	CL_No_I	CL_No_II	CL_No_III	CL_No_IV	Metal_groups
34488	Drosa 'Küsterberg', Lkr. Anhalt-Bitterfeld	103	CWC	31	550	2	6	15	50	Arsenic-lead-copper
34489	Drosa 'Küsterberg', Lkr. Anhalt-Bitterfeld	103	CWC	31	550	2	6	15	51	Arsenic-lead-copper
34490	Drosa 'Küsterberg', Lkr. Anhalt-Bitterfeld	103	CWC	31	550	2	6	15	50	Arsenic-lead-copper
34491	Drosa 'Küsterberg', Lkr. Anhalt-Bitterfeld	103	CWC	31	550	2	6	15	51	Arsenic-lead-copper
34492	Drosa 'Küsterberg', Lkr. Anhalt-Bitterfeld	103	CWC	31	550	2	6	15	51	Arsenic-lead-copper
34495	Drosa 'Küsterberg', Lkr. Anhalt-Bitterfeld	103	CWC	31	550	2	6	15	51	Arsenic-lead-copper
34496	Drosa 'Küsterberg', Lkr. Anhalt-Bitterfeld	103	CWC	31	550	2	6	15	51	Arsenic-lead-copper
34498	Drosa 'Küsterberg', Lkr. Anhalt-Bitterfeld	103	CWC	31	550	2	6	15	50	Arsenic-lead-copper
34503	Drosa 'Küsterberg', Lkr. Anhalt-Bitterfeld	103	CWC	31	550	2	6	15	50	Arsenic-lead-copper
34511	Drosa 'Küsterberg', Lkr. Anhalt-Bitterfeld	103	CWC	31	550	2	6	15	50	Arsenic-lead-copper
34529	Drosa 'Küsterberg', Lkr. Anhalt-Bitterfeld	103	CWC	31	550	2	6	15	50	Arsenic-lead-copper
34532	Drosa 'Küsterberg', Lkr. Anhalt-Bitterfeld	103	CWC	31	550	2	6	15	51	Arsenic-lead-copper
34533	Drosa 'Küsterberg', Lkr. Anhalt-Bitterfeld	103	CWC	31	550	2	6	15	51	Arsenic-lead-copper
34534	Drosa 'Küsterberg', Lkr. Anhalt-Bitterfeld	103	CWC	31	550	2	6	15	51	Arsenic-lead-copper
34535	Drosa 'Küsterberg', Lkr. Anhalt-Bitterfeld	103	CWC	31	550	2	6	15	51	Arsenic-lead-copper
34536	Drosa 'Küsterberg', Lkr. Anhalt-Bitterfeld	103	CWC	31	550	2	6	15	51	Arsenic-lead-copper
34537	Drosa 'Küsterberg', Lkr. Anhalt-Bitterfeld	103	CWC	31	550	2	6	15	51	Arsenic-lead-copper
34538	Drosa 'Küsterberg', Lkr. Anhalt-Bitterfeld	103	CWC	31	550	2	6	15	51	Arsenic-lead-copper
34540	Drosa 'Küsterberg', Lkr. Anhalt-Bitterfeld	103	CWC	31	550	2	6	15	51	Arsenic-lead-copper
34545	Drosa 'Küsterberg', Lkr. Anhalt-Bitterfeld	103	CWC	31	550	2	6	15	50	Arsenic-lead-copper
34560	Drosa 'Küsterberg', Lkr. Anhalt-Bitterfeld	103	CWC	31	550	2	6	15	50	Arsenic-lead-copper
34568	Drosa 'Küsterberg', Lkr. Anhalt-Bitterfeld	103	CWC	31	550	2	6	15	50	Arsenic-lead-copper
34570	Drosa 'Küsterberg', Lkr. Anhalt-Bitterfeld	103	CWC	31	550	2	6	15	51	Arsenic-lead-copper
34572	Drosa 'Küsterberg', Lkr. Anhalt-Bitterfeld	103	CWC	31	550	2	6	15	50	Arsenic-lead-copper
34578	Drosa 'Küsterberg', Lkr. Anhalt-Bitterfeld	103	CWC	31	550	2	6	15	50	Arsenic-lead-copper
34580	Drosa 'Küsterberg', Lkr. Anhalt-Bitterfeld	103	CWC	31	550	2	6	15	50	Arsenic-lead-copper
34581	Drosa 'Küsterberg', Lkr. Anhalt-Bitterfeld	103	CWC	31	550	2	6	15	50	Arsenic-lead-copper
34585	Drosa 'Küsterberg', Lkr. Anhalt-Bitterfeld	103	CWC	31	550	2	6	15	50	Arsenic-lead-copper
34597	Drosa 'Küsterberg', Lkr. Anhalt-Bitterfeld	103	CWC	31	550	2	6	15	50	Arsenic-lead-copper
34603	Drosa 'Küsterberg', Lkr. Anhalt-Bitterfeld	103	CWC	31	550	2	6	15	50	Arsenic-lead-copper
34605	Drosa 'Küsterberg', Lkr. Anhalt-Bitterfeld	103	CWC	31	550	2	6	15	50	Arsenic-lead-copper
34609	Drosa 'Küsterberg', Lkr. Anhalt-Bitterfeld	103	CWC	31	550	2	6	15	50	Arsenic-lead-copper
2838	Vinelz, 'Amt Erlau'	67	CWC	31	212	2	6	16	37	Arsenic-nickel-copper
19920	Predmostí	114	BBC	35	215	2	6	16	37	Arsenic-nickel-copper
22239	St. Blaise	67	CWC	31	215	2	6	16	37	Arsenic-nickel-copper
22281	Tauberbischofsheim-Impfingen, Main-Tauber-Kreis	62	CWC	31	510	2	6	16	37	Arsenic-nickel-copper
62163	St-Blaise	67	CWC	31	212	2	6	16	37	Arsenic-nickel-copper
78144	Kosice-Sebastovce	122	KST	40	200	2	6	16	37	Arsenic-nickel-copper
100027	Tousen	112	RIV	25	700	2	6	16	37	Arsenic-nickel-copper
100066	Bischofshofen, 'Götschenberg', p. B. Völklabruck	73	MON	21	810	2	6	16	37	Arsenic-nickel-copper
100084	Bischofshofen, 'Götschenberg', p. B. Völklabruck	73	MON	21	700	2	6	16	37	Arsenic-nickel-copper
100085	Bischofshofen, 'Götschenberg', p. B. Völklabruck	73	MON	21	700	2	6	16	37	Arsenic-nickel-copper
100086	Bischofshofen, 'Götschenberg', p. B. Völklabruck	73	MON	21	700	2	6	16	37	Arsenic-nickel-copper
2026	Veliki Srediste	125	BOD	21	130	1	3	8	48	Pure
2051	Veliki Srediste	125	BOD	21	130	1	3	8	48	Pure
2597	Gilching, 'Steinbergäcker', Lkr. Starnberg	66	BK	40	600	1	3	8	48	Pure
2600	Manching, 'Hundsrückenacker i. d. Au', Lkr. Ingolstadt	66	BK	40	600	1	3	8	48	Pure
2876	Vinelz, 'Amt Erlau'	67	CWC	31	800	1	3	8	48	Pure
2886	Vinelz, 'Amt Erlau'	67	CWC	31	800	1	3	8	48	Pure
2892	Vinelz, 'Amt Erlau'	67	CWC	31	800	1	3	8	48	Pure
2898	Vinelz, 'Amt Erlau'	67	CWC	31	800	1	3	8	48	Pure
2899	Vinelz, 'Amt Erlau'	67	CWC	31	800	1	3	8	48	Pure
2911	Vinelz, 'Amt Erlau'	67	CWC	31	800	1	3	8	48	Pure

ANo	FIND_PLACE	AREA	CULT	PER	TYPE_2	CL_No_I	CL_No_II	CL_No_III	CL_No_IV	Metal_groups
2917	Vinelz, 'Amt Erlau'	67	CWC	31	800	1	3	8	48	Pure
2920	Vinelz, 'Amt Erlau'	67	CWC	31	800	1	3	8	48	Pure
3230	Zlonín	111		21	130	1	3	8	48	Pure
3346	Brno-Slatina	114	BBC	35	215	1	3	8	48	Pure
3368	Plavec	114		21	130	1	3	8	48	Pure
3369	Lovcice	114		21	130	1	3	8	48	Pure
3383	Uherské Hradište	114		10	130	1	3	8	48	Pure
3384	Hodonín	114		10	130	1	3	8	48	Pure
3387	Lesná	114		21	100	1	3	8	48	Pure
3635	Kremnica	121		25	100	1	3	8	48	Pure
3649	Lipova	125	MKC	31	130	1	3	8	48	Pure
3653	Banská Bystrica	121		10	130	1	3	8	48	Pure
3695	Lucenec	121		21	130	1	3	8	48	Pure
3741	Vrádište	114		21	130	1	3	8	48	Pure
3743	Sopron, 'Banfalva'	123		21	130	1	3	8	48	Pure
3750	Kosice	121		25	100	1	3	8	48	Pure
3780	Spitz, p. H. Krems	72		21	100	1	3	8	48	Pure
4240	Ayent, 'Les Place'	67	BK	40	500	1	3	8	48	Pure
4925	Stollhof, 'Lange Wand', p. B. Wiener Neustadt	72	BAD	25	100	1	3	8	49	Pure
4931	Stollhof, 'Lange Wand', p. B. Wiener Neustadt	72	BAD	25	510	1	3	8	49	Pure
4934	Stollhof, 'Lange Wand', p. B. Wiener Neustadt	72	BAD	25	410	1	3	8	49	Pure
4936	Stollhof, 'Lange Wand', p. B. Wiener Neustadt	72	BAD	25	410	1	3	8	49	Pure
4938	Stollhof, 'Lange Wand', p. B. Wiener Neustadt	72	BAD	25	410	1	3	8	49	Pure
6374	Mitterretzbach, p. B. Hollerbrunn	72		21	130	1	3	8	48	Pure
6392	Missingdorf, p. B. Horn	72		21	130	1	3	8	48	Pure
6396	Tiszanagyfalu	122		21	130	1	3	8	48	Pure
6397	Tiszanagyfalu	122		21	130	1	3	8	48	Pure
6398	Nagyrábé	125		21	130	1	3	8	48	Pure
6461	Gyor	123		21	100	1	3	8	48	Pure
6463	Budapest-III-Békásmegyer	124		21	100	1	3	8	48	Pure
6482	Budapest-Csepel	123		10	130	1	3	8	48	Pure
6491	Budakalász	123		10	130	1	3	8	48	Pure
6493	Tiszanagyfalu	122		21	130	1	3	8	48	Pure
6496	Vészto	125		21	130	1	3	8	48	Pure
6502	Levice	121		21	130	1	3	8	48	Pure
6503	Högyész	124		21	130	1	3	8	48	Pure
6504	Budapest-III-Békásmegyer	124		21	130	1	3	8	48	Pure
6505	Budapest-III-Békásmegyer	124		21	130	1	3	8	48	Pure
6509	Tiszanagyfalu	122		21	130	1	3	8	48	Pure
6510	Mátészalka	122		21	130	1	3	8	48	Pure
6511	Budapest-III-Békásmegyer	124		21	130	1	3	8	48	Pure
6515	Pomáz	124		21	130	1	3	8	48	Pure
6517	Högyész	124		21	130	1	3	8	48	Pure
6519	Tiszapolgár-Basatanya	125	TC	10	405	1	3	8	48	Pure
6520	Tiszapolgár-Basatanya	125	TC	10	405	1	3	8	48	Pure
6522	Tiszapolgár-Basatanya	125	TC	10	405	1	3	8	48	Pure
6523	Tiszapolgár-Basatanya	125		10	405	1	3	8	48	Pure
6531	Tiszapolgár-Basatanya	125	TC	10	400	1	3	8	48	Pure
6538	Veliki Gaj	125	BOD	21	100	1	3	8	48	Pure
6539	Budapest-III-Obuda	124		21	100	1	3	8	48	Pure
6540	Érd	124	BOD	21	100	1	3	8	48	Pure
6541	Hirics	124		21	130	1	3	8	48	Pure
6543	Érd	124	BOD	21	130	1	3	8	48	Pure

ANo	FIND_PLACE	AREA	CULT	PER	TYPE_2	CL_No_I	CL_No_II	CL_No_III	CL_No_IV	Metal_groups
6547	Rosiori	125		21	130	1	3	8	48	Pure
6551	Veliki Gaj	125	BOD	21	130	1	3	8	48	Pure
6553	Kerepes	125		21	130	1	3	8	48	Pure
6557	Ajak	122		21	130	1	3	8	48	Pure
6560	Érd	124	BOD	21	130	1	3	8	48	Pure
6568	Tiszapolgár-Basatanya	125	TC	10	410	1	3	8	48	Pure
6573	Tiszapolgár-Basatanya	125	TC	10	405	1	3	8	48	Pure
6574	Tiszapolgár-Basatanya	125	BOD	21	215	1	3	8	48	Pure
6643	Stollhof, 'Lange Wand', p. B. Wiener Neustadt	72	BAD	25	410	1	3	8	48	Pure
7544	Mazice	113		31	100	1	3	8	48	Pure
9153	Moldova Veche	125		21	130	1	3	8	48	Pure
9159	Deta	125		31	100	1	3	8	48	Pure
9161	Satchinez, Timisoara	125		21	130	1	3	8	48	Pure
9164	Orsova	125		21	130	1	3	8	48	Pure
9168	Bencecu de Jos	125		21	130	1	3	8	48	Pure
9170	Satchinez, Timisoara	125		21	130	1	3	8	48	Pure
9179	Zabrani	125		21	130	1	3	8	48	Pure
9190	Cermei	125		21	130	1	3	8	48	Pure
9193	Pecica	125		21	130	1	3	8	48	Pure
9215	Tarcea	125	BOD	21	130	1	3	8	48	Pure
9216	Tarcea	125	BOD	21	130	1	3	8	48	Pure
9217	Tarcea	125	BOD	21	130	1	3	8	48	Pure
9218	Tarcea	125	BOD	21	130	1	3	8	48	Pure
9219	Tarcea	125	BOD	21	130	1	3	8	48	Pure
9220	Tarcea	125	BOD	21	130	1	3	8	48	Pure
9221	Tarcea	125	BOD	21	130	1	3	8	48	Pure
9222	Tarcea	125	BOD	21	130	1	3	8	48	Pure
9223	Tarcea	125	BOD	21	130	1	3	8	48	Pure
9224	Tarcea	125	BOD	21	130	1	3	8	48	Pure
9225	Sisterea	125		21	130	1	3	8	48	Pure
9226	Adoni	124		21	130	1	3	8	48	Pure
9227	Josani	125		21	130	1	3	8	48	Pure
9228	Hencida	125		21	130	1	3	8	48	Pure
9229	Josani	125		21	130	1	3	8	48	Pure
9230	Hencida	125		21	130	1	3	8	48	Pure
9231	Josani	125		21	130	1	3	8	48	Pure
9232	Josani	125		21	130	1	3	8	48	Pure
9234	Hencida	125		21	130	1	3	8	48	Pure
9235	Josani	125		21	130	1	3	8	48	Pure
9236	Josani	125		21	130	1	3	8	48	Pure
9237	Josani	125		21	130	1	3	8	48	Pure
9239	Sacuieni	125		10	130	1	3	8	48	Pure
9241	Tauteu	125		10	130	1	3	8	48	Pure
9242	Tauteu	125		10	130	1	3	8	48	Pure
9247	Mezosas	125	BOD	21	134	1	3	8	48	Pure
9248	Valea lui Mihai	125	MKC	31	100	1	3	8	48	Pure
9249	Mezosas	125	BOD	21	100	1	3	8	48	Pure
9252	Josani	125		21	130	1	3	8	48	Pure
9274	Demecser, 'Borzsova patak'	122		21	130	1	3	8	48	Pure
9729	Bodman-Weiler, Lkr. Konstanz	62	PFY	21	550	1	3	8	48	Pure
10338	Pohranice	121		21	134	1	3	8	48	Pure
10395	Cachtice	121		21	130	1	3	8	48	Pure
10722	Tibava, 'Hrun za cintirom'	122	TC	10	130	1	3	8	48	Pure

ANo	FIND_PLACE	AREA	CULT	PER	TYPE_2	CL_No_I	CL_No_II	CL_No_III	CL_No_IV	Metal_groups
10723	Tibava, 'Hrun za cintirom'	122	TC	10	130	1	3	8	48	Pure
10725	Tibava, 'Hrun za cintirom'	122	TC	10	130	1	3	8	48	Pure
10726	Tibava, 'Hrun za cintirom'	122	TC	10	130	1	3	8	48	Pure
10727	Tibava, 'Hrun za cintirom'	122	TC	10	130	1	3	8	48	Pure
10728	Tibava, 'Hrun za cintirom'	122	TC	10	130	1	3	8	48	Pure
10729	Tibava, 'Hrun za cintirom'	122	TC	10	130	1	3	8	48	Pure
10816	Spišškà Novà Ves	122		21	130	1	3	8	48	Pure
10839	Hrádok, 'Opel'	121		10	130	1	3	8	48	Pure
10843	Marcelová	121		25	100	1	3	8	48	Pure
10844	Dolný Peter	121		25	100	1	3	8	48	Pure
10990	Luhacovice	114	LEN	10	130	1	3	8	48	Pure
10991	Uhersky Brod	114	LEN	10	130	1	3	8	48	Pure
10992	Uhersky Brod	114	LEN	10	130	1	3	8	48	Pure
11192	Linz-St. Peter, p. B. Linz	66	BAD	25	130	1	3	8	48	Pure
11193	Linz-St. Peter, p. B. Linz	66	BAD	25	130	1	3	8	48	Pure
11194	Linz-St. Peter, p. B. Linz	66	BAD	25	100	1	3	8	48	Pure
11195	Linz-St. Peter, p. B. Linz	66	BAD	25	100	1	3	8	48	Pure
12074	Sušice-Traplice	114		21	130	1	3	8	48	Pure
12076	Rasovice	112		21	510	1	3	8	48	Pure
12077	Rasovice	112		21	510	1	3	8	48	Pure
12078	Rasovice	112		21	510	1	3	8	48	Pure
12430	Székesfehérvár	124		21	130	1	3	8	48	Pure
12448	Budapest-III-Békásmegyer	124		21	130	1	3	8	48	Pure
12454	Komárom	123		21	130	1	3	8	48	Pure
12456	Tatabánya, 'Felsögalla'	123		21	100	1	3	8	48	Pure
12460	Budapest-III-Békásmegyer	124		21	130	1	3	8	48	Pure
12467	Siria, 'Világos'	125		21	130	1	3	8	48	Pure
12476	Veliki Gaj	125	BOD	21	100	1	3	8	48	Pure
12490	Budapest-III-Békásmegyer	124		21	130	1	3	8	48	Pure
12798	Kápolna	125		21	100	1	3	8	48	Pure
12821	Sárazsadány	122	BOD	21	100	1	3	8	48	Pure
12915	Budapest-III-Békásmegyer	124		21	100	1	3	8	48	Pure
12917	Tiszaug	125		21	100	1	3	8	48	Pure
12918	Budapest-III-Obuda	124		21	100	1	3	8	48	Pure
12920	Esztergom	123		21	100	1	3	8	48	Pure
12923	Somlóvásárhely	123		21	100	1	3	8	48	Pure
12964	Szendro	122	BOD	21	130	1	3	8	48	Pure
12971	Szendro	122	BOD	21	130	1	3	8	48	Pure
12975	Szendro	122		21	100	1	3	8	48	Pure
12980	Tiszamogyorós	122		21	130	1	3	8	48	Pure
12981	Fényeslitke	122	BOD	21	130	1	3	8	48	Pure
12982	Fényeslitke	122	BOD	21	130	1	3	8	48	Pure
12988	Székely, 'Szolohegy'	122	BOD	21	130	1	3	8	48	Pure
12990	Vencsellö	122		21	130	1	3	8	48	Pure
12996	Hajdusámson	125		21	130	1	3	8	48	Pure
12997	Fényeslitke	122	BOD	21	100	1	3	8	48	Pure
13000	Demecser, 'Borzsova patak'	122		21	130	1	3	8	48	Pure
13001	Nagyhalász	122		10	130	1	3	8	48	Pure
13002	Sárospatak	122		21	130	1	3	8	48	Pure
13003	Nagyhalász	122		21	134	1	3	8	48	Pure
13004	Fényeslitke	122	BOD	21	130	1	3	8	48	Pure
13005	Fényeslitke	122	BOD	21	130	1	3	8	48	Pure
13006	Fényeslitke	122	BOD	21	130	1	3	8	48	Pure

ANo	FIND_PLACE	AREA	CULT	PER	TYPE_2	CL_No_I	CL_No_II	CL_No_III	CL_No_IV	Metal_groups
13007	Nyírtura, 'Várhegy vidéke'	122		21	130	1	3	8	48	Pure
13008	Nagyhalász, 'Szölöhegy	122		21	130	1	3	8	48	Pure
13009	Apagy, 'Nagysziget dülo'	122	TC	10	130	1	3	8	48	Pure
13011	Tiszadob, 'Urkom dülo'	122		21	130	1	3	8	48	Pure
13012	Tiszadob, 'Urkom dülo'	122		21	410	1	3	8	48	Pure
13013	Vasmegyer	122		21	100	1	3	8	48	Pure
13014	Vasmegyer	122		21	130	1	3	8	48	Pure
13015	Vasmegyer, 'Kisvármi'	122		10	130	1	3	8	48	Pure
13020	Fényeslitke	122	BOD	21	130	1	3	8	48	Pure
13021	Fényeslitke	122	BOD	21	215	1	3	8	48	Pure
13022	Fényeslitke	122	BOD	21	200	1	3	8	48	Pure
13023	Fényeslitke	122	BOD	21	100	1	3	8	48	Pure
13024	Fényeslitke	122	BOD	21	100	1	3	8	48	Pure
13025	Fényeslitke	122	BOD	21	510	1	3	8	48	Pure
13027	Fényeslitke	122	BOD	21	134	1	3	8	48	Pure
13035	Ajak	122		21	130	1	3	8	48	Pure
13037	Nyírtura, 'Lengyelhegy	122		21	130	1	3	8	48	Pure
13038	Nyírgelse, 'Wald der Gräfin'	122		21	130	1	3	8	48	Pure
13039	Nyírlugos, 'Szennyespuszta	122		21	130	1	3	8	48	Pure
13040	Kisvárda	122		21	130	1	3	8	48	Pure
13042	Kisvárda, 'Umgebung von Kisvárda'	122		21	130	1	3	8	48	Pure
13043	Debrecen	125		10	130	1	3	8	48	Pure
13044	Nádudvar	125		21	130	1	3	8	48	Pure
13045	Balmaz jváros, 'Hortobágy-Máta'	125		21	130	1	3	8	48	Pure
13046	Berettyóújfalu	125		21	130	1	3	8	48	Pure
13047	Vámospércs	125		21	130	1	3	8	48	Pure
13048	Hajdúhadház	125	BOD	21	130	1	3	8	48	Pure
13050	Debrecen	125		21	130	1	3	8	48	Pure
13054	Monostorpályi, 'Szolohegy	125		21	130	1	3	8	48	Pure
13095	Beba Veche	125	PIT	40	405	1	3	8	48	Pure
13172	Coka	125	TZC	10	550	1	3	8	48	Pure
13196	Deszk, 'A'	125	TC	10	400	1	3	8	48	Pure
13197	Deszk, 'A'	125	TC	10	400	1	3	8	48	Pure
13198	Deszk, 'A'	125	TC	10	400	1	3	8	48	Pure
13201	Szeged	125		21	130	1	3	8	48	Pure
13202	Szeged	125		21	134	1	3	8	48	Pure
13203	Szeged	125		21	134	1	3	8	48	Pure
13204	Szeged	125		21	810	1	3	8	48	Pure
13205	Szeged	125		21	810	1	3	8	48	Pure
13206	Szeged	125		21	130	1	3	8	48	Pure
13207	Szeged	125		21	130	1	3	8	48	Pure
13210	Szeged	125		21	130	1	3	8	48	Pure
13211	Szeged	125		21	130	1	3	8	48	Pure
13212	Szeged	125		21	700	1	3	8	48	Pure
13213	Szeged	125		21	700	1	3	8	48	Pure
13215	Coka	125	TZC	10	130	1	3	8	48	Pure
13259	Hódmezövásárhely, 'Szakálhát-Nagy tanya	125	MKC	31	810	1	3	8	48	Pure
13260	Hódmezövásárhely, 'Szakálhát-Nagy tanya	125	MKC	31	810	1	3	8	48	Pure
13261	Hódmezövásárhely, 'Szakálhát-Nagy tanya	125	MKC	31	810	1	3	8	48	Pure
13262	Hódmezövásárhely, 'Szakálhát-Nagy tanya	125	MKC	31	810	1	3	8	48	Pure
13263	Hódmezövásárhely, 'Szakálhát-Nagy tanya	125	MKC	31	810	1	3	8	48	Pure
13266	Hódmezövásárhely, 'Szakálhát'	125	MKC	31	100	1	3	8	48	Pure
13267	Hódmezövásárhely, 'Cukor tanya'	125	MKC	31	405	1	3	8	48	Pure

ANo	FIND_PLACE	AREA	CULT	PER	TYPE_2	CL_No_I	CL_No_II	CL_No_III	CL_No_IV	Metal_groups
13268	Hódmezövásárhely, 'Cukor tanya'	125	MKC	31	410	1	3	8	48	Pure
13275	Hódmezövásárhely, 'Cukor tanya'	125	MKC	31	405	1	3	8	48	Pure
13285	Mezoberény	125		21	122	1	3	8	48	Pure
13339	Zengövárkony	124	LEN	10	410	1	3	8	48	Pure
13340	Zengövárkony	124	LEN	10	410	1	3	8	48	Pure
13341	Zengövárkony	124	LEN	10	410	1	3	8	48	Pure
13342	Zengövárkony	124	LEN	10	410	1	3	8	48	Pure
13343	Zengövárkony	124	LEN	10	410	1	3	8	48	Pure
13515	Adony	124		21	130	1	3	8	48	Pure
13815	Kóny	123		21	130	1	3	8	48	Pure
13816	Ravazd, 'Feketemáj'	123		21	100	1	3	8	48	Pure
14381	Balassagyarmat	121		10	130	1	3	8	48	Pure
14384	Polgár	125		21	100	1	3	8	48	Pure
17672	Kravany nad Dunajom	121		21	130	1	3	8	48	Pure
17677	Radosiná	121	LEN	10	410	1	3	8	48	Pure
17678	Radosiná	121	LEN	10	410	1	3	8	48	Pure
17679	Radosiná	121	LEN	10	810	1	3	8	48	Pure
21679	Font	67		21	510	1	3	8	48	Pure
49900	Schernau, Kr. Bad Kitzingen	65	MUE	10	510	1	3	8	48	Pure
58014	Cachtice	121		21	130	1	3	8	48	Pure
58019	Kláštor pod Znievom	121		21	130	1	3	8	48	Pure
1248	Geislingen, 'Binsdorf-Hasloch', Zollernalbkreis	62		31	100	1	3	8	22	Pure with Nickel
2833	Vinelz, 'Amt Erlau'	67	CWC	31	810	1	3	8	22	Pure with Nickel
2835	Vinelz, 'Amt Erlau'	67	CWC	31	810	1	3	8	22	Pure with Nickel
2836	Vinelz, 'Amt Erlau'	67	CWC	31	212	1	3	8	22	Pure with Nickel
2837	Vinelz, 'Amt Erlau'	67	CWC	31	212	1	3	8	22	Pure with Nickel
2841	Vinelz, 'Amt Erlau'	67	CWC	31	134	1	3	8	22	Pure with Nickel
2877	Vinelz, 'Amt Erlau'	67	CWC	31	800	1	3	8	22	Pure with Nickel
2878	Vinelz, 'Amt Erlau'	67	CWC	31	800	1	3	8	22	Pure with Nickel
2879	Vinelz, 'Amt Erlau'	67	CWC	31	800	1	3	8	22	Pure with Nickel
2880	Vinelz, 'Amt Erlau'	67	CWC	31	800	1	3	8	22	Pure with Nickel
2882	Vinelz, 'Amt Erlau'	67	CWC	31	800	1	3	8	22	Pure with Nickel
2883	Vinelz, 'Amt Erlau'	67	CWC	31	800	1	3	8	22	Pure with Nickel
2884	Vinelz, 'Amt Erlau'	67	CWC	31	800	1	3	8	22	Pure with Nickel
2885	Vinelz, 'Amt Erlau'	67	CWC	31	800	1	3	8	22	Pure with Nickel
2887	Vinelz, 'Amt Erlau'	67	CWC	31	800	1	3	8	22	Pure with Nickel
2888	Vinelz, 'Amt Erlau'	67	CWC	31	800	1	3	8	22	Pure with Nickel
2889	Vinelz, 'Amt Erlau'	67	CWC	31	800	1	3	8	22	Pure with Nickel
2890	Vinelz, 'Amt Erlau'	67	CWC	31	800	1	3	8	22	Pure with Nickel
2891	Vinelz, 'Amt Erlau'	67	CWC	31	800	1	3	8	22	Pure with Nickel
2893	Vinelz, 'Amt Erlau'	67	CWC	31	800	1	3	8	22	Pure with Nickel
2894	Vinelz, 'Amt Erlau'	67	CWC	31	800	1	3	8	22	Pure with Nickel
2895	Vinelz, 'Amt Erlau'	67	CWC	31	800	1	3	8	22	Pure with Nickel
2896	Vinelz, 'Amt Erlau'	67	CWC	31	800	1	3	8	22	Pure with Nickel
2897	Vinelz, 'Amt Erlau'	67	CWC	31	800	1	3	8	22	Pure with Nickel
2900	Vinelz, 'Amt Erlau'	67	CWC	31	800	1	3	8	22	Pure with Nickel
2901	Vinelz, 'Amt Erlau'	67	CWC	31	800	1	3	8	22	Pure with Nickel
2902	Vinelz, 'Amt Erlau'	67	CWC	31	800	1	3	8	22	Pure with Nickel
2903	Vinelz, 'Amt Erlau'	67	CWC	31	800	1	3	8	22	Pure with Nickel
2904	Vinelz, 'Amt Erlau'	67	CWC	31	800	1	3	8	22	Pure with Nickel
2905	Vinelz, 'Amt Erlau'	67	CWC	31	800	1	3	8	22	Pure with Nickel
2906	Vinelz, 'Amt Erlau'	67	CWC	31	800	1	3	8	22	Pure with Nickel
2907	Vinelz, 'Amt Erlau'	67	CWC	31	800	1	3	8	22	Pure with Nickel

ANo	FIND_PLACE	AREA	CULT	PER	TYPE_2	CL_No_I	CL_No_II	CL_No_III	CL_No_IV	Metal_groups
2908	Vinelz, 'Amt Erlau'	67	CWC	31	800	1	3	8	22	Pure with Nickel
2909	Vinelz, 'Amt Erlau'	67	CWC	31	800	1	3	8	22	Pure with Nickel
2910	Vinelz, 'Amt Erlau'	67	CWC	31	800	1	3	8	22	Pure with Nickel
2912	Vinelz, 'Amt Erlau'	67	CWC	31	800	1	3	8	22	Pure with Nickel
2913	Vinelz, 'Amt Erlau'	67	CWC	31	800	1	3	8	22	Pure with Nickel
2914	Vinelz, 'Amt Erlau'	67	CWC	31	800	1	3	8	22	Pure with Nickel
2915	Vinelz, 'Amt Erlau'	67	CWC	31	800	1	3	8	22	Pure with Nickel
2916	Vinelz, 'Amt Erlau'	67	CWC	31	800	1	3	8	22	Pure with Nickel
2918	Vinelz, 'Amt Erlau'	67	CWC	31	800	1	3	8	22	Pure with Nickel
2919	Vinelz, 'Amt Erlau'	67	CWC	31	800	1	3	8	22	Pure with Nickel
2921	Vinelz, 'Amt Erlau'	67	CWC	31	800	1	3	8	22	Pure with Nickel
2924	Lüscherz	67	CWC	31	212	1	3	8	22	Pure with Nickel
3086	Vinelz, 'Amt Erlau'	67	CWC	31	500	1	3	8	22	Pure with Nickel
3087	Vinelz, 'Amt Erlau'	67	CWC	31	810	1	3	8	22	Pure with Nickel
3089	Vinelz, 'Amt Erlau'	67	CWC	31	810	1	3	8	22	Pure with Nickel
3365	Holubice	114	BBC	35	215	1	3	8	22	Pure with Nickel
4639	Lichtenwörth, p. B. Wiener-Neustadt	72	BAD	25	400	1	3	8	22	Pure with Nickel
5682	Horodnica	85		31	100	1	3	8	22	Pure with Nickel
7790	Concise	67		21	510	1	3	8	22	Pure with Nickel
9240	Cotiglet	125		10	130	1	3	8	22	Pure with Nickel
9243	Tauteu	125		10	130	1	3	8	22	Pure with Nickel
9244	Tauteu	125		10	130	1	3	8	22	Pure with Nickel
9245	Tauteu	125		10	130	1	3	8	22	Pure with Nickel
9246	Tauteu	125		10	130	1	3	8	22	Pure with Nickel
10724	Tibava, 'Hrun za cintirom'	122	TC	10	130	1	3	8	22	Pure with Nickel
10730	Tibava, 'Hrun za cintirom'	122	TC	10	130	1	3	8	22	Pure with Nickel
10731	Tibava, 'Hrun za cintirom'	122	TC	10	130	1	3	8	22	Pure with Nickel
12966	Emöd	122	BOD	21	510	1	3	8	22	Pure with Nickel
12983	Berkesz, 'Csonkásdülö'	122	BOD	21	130	1	3	8	22	Pure with Nickel
13279	Mezoberény	125		21	130	1	3	8	22	Pure with Nickel
22050	Yvonand, 'La Peupleraie; Grabung 1973'	67	CWC	31	212	1	3	8	22	Pure with Nickel
22142	Auvernier, 'Brise-Lame; Grabung 1974'	67	CWC	31	600	1	3	8	22	Pure with Nickel
22249	Delley, 'Portalban'	67		31	100	1	3	8	22	Pure with Nickel
22272	Chevroux	67		31	212	1	3	8	22	Pure with Nickel
22275	Chevroux	67		31	212	1	3	8	22	Pure with Nickel
22280	Concise	67		21	510	1	3	8	22	Pure with Nickel
22282	Chevroux	67		21	510	1	3	8	22	Pure with Nickel
3235	Bosice	112		21	134	1	3	8	23	Pure with Silver
3647	Banská Bystrica	121		21	130	1	3	8	23	Pure with Silver
3651	Lucenec	121		21	130	1	3	8	23	Pure with Silver
4929	Stollhof, 'Lange Wand', p. B. Wiener Neustadt	72	BAD	25	510	1	3	8	23	Pure with Silver
4930	Stollhof, 'Lange Wand', p. B. Wiener Neustadt	72	BAD	25	510	1	3	8	23	Pure with Silver
4932	Stollhof, 'Lange Wand', p. B. Wiener Neustadt	72	BAD	25	510	1	3	8	23	Pure with Silver
4933	Stollhof, 'Lange Wand', p. B. Wiener Neustadt	72	BAD	25	410	1	3	8	23	Pure with Silver
4937	Stollhof, 'Lange Wand', p. B. Wiener Neustadt	72	BAD	25	410	1	3	8	23	Pure with Silver
4939	Stollhof, 'Lange Wand', p. B. Wiener Neustadt	72	BAD	25	410	1	3	8	23	Pure with Silver
4940	Stollhof, 'Lange Wand', p. B. Wiener Neustadt	72	BAD	25	410	1	3	8	23	Pure with Silver
4941	Stollhof, 'Lange Wand', p. B. Wiener Neustadt	72	BAD	25	410	1	3	8	23	Pure with Silver
4942	Stollhof, 'Lange Wand', p. B. Wiener Neustadt	72	BAD	25	410	1	3	8	23	Pure with Silver
4943	Stollhof, 'Lange Wand', p. B. Wiener Neustadt	72	BAD	25	500	1	3	8	23	Pure with Silver
6393	Jászladány	125		21	130	1	3	8	23	Pure with Silver
6394	Budapest-XVII-Rákoscsaba	125	BBC	35	215	1	3	8	23	Pure with Silver
6395	Pusztaistvánháza	125	BOD	21	215	1	3	8	23	Pure with Silver

ANo	FIND_PLACE	AREA	CULT	PER	TYPE_2	CL_No_I	CL_No_II	CL_No_III	CL_No_IV	Metal_groups
6468	Polgár	125		21	100	1	3	8	23	Pure with Silver
6471	Pusztaistvánháza	125	BOD	21	810	1	3	8	23	Pure with Silver
6472	Neszmély, 'Tekerespatak'	123	BOD	21	810	1	3	8	23	Pure with Silver
6480	Gyor	123		21	130	1	3	8	23	Pure with Silver
6481	Bratislava	121		21	130	1	3	8	23	Pure with Silver
6486	Dormánd	125		10	130	1	3	8	23	Pure with Silver
6512	Taksony	125		21	130	1	3	8	23	Pure with Silver
6513	Újszentmargita	125		21	130	1	3	8	23	Pure with Silver
6524	Tiszapolgár-Basatanya	125	TC	10	400	1	3	8	23	Pure with Silver
6537	Városlod	123		21	100	1	3	8	23	Pure with Silver
6549	Kerepes	125		21	130	1	3	8	23	Pure with Silver
6550	Székesfehérvár	124		21	130	1	3	8	23	Pure with Silver
6554	Rosiori	125		21	130	1	3	8	23	Pure with Silver
6559	Târgusor	125		21	130	1	3	8	23	Pure with Silver
6562	Siniob	125		21	130	1	3	8	23	Pure with Silver
6565	Jászladány	125		21	130	1	3	8	23	Pure with Silver
6570	Tiszapolgár-Basatanya	125	TC	10	410	1	3	8	23	Pure with Silver
6575	Tiszapolgár-Basatanya	125	BOD	21	810	1	3	8	23	Pure with Silver
6644	Stollhof, 'Lange Wand', p. B. Wiener Neustadt	72	BAD	25	410	1	3	8	23	Pure with Silver
10433	Výcapy-Opatovce	121	NIT	40	420	1	3	8	23	Pure with Silver
10444	Výcapy-Opatovce	121	NIT	40	300	1	3	8	23	Pure with Silver
10521	Výcapy-Opatovce	121	NIT	40	810	1	3	8	23	Pure with Silver
12435	Dorog, 'Kohlenbergwerk Drasche'	124		21	134	1	3	8	23	Pure with Silver
12436	Dorog, 'Kohlenbergwerk Drasche'	124		21	130	1	3	8	23	Pure with Silver
12809	Komárom	123		21	100	1	3	8	23	Pure with Silver
12904	Budakalász	124	BAD	25	500	1	3	8	23	Pure with Silver
12909	Budakalász	124	BAD	25	810	1	3	8	23	Pure with Silver
12910	Budakalász	124	BAD	25	810	1	3	8	23	Pure with Silver
12912	Budakalász	124	BAD	25	810	1	3	8	23	Pure with Silver
12969	Bükkaranyos	122		21	130	1	3	8	23	Pure with Silver
13171	Coka	125	TZC	10	300	1	3	8	23	Pure with Silver
13738	Vörs	124	BAD	25	500	1	3	8	23	Pure with Silver
22271	Chevroux	67		31	212	1	3	8	23	Pure with Silver
29848	Luderov	114	BBC	35	810	1	3	8	23	Pure with Silver
58017	Tekovská Nová Ves	121		10	130	1	3	8	23	Pure with Silver
58020	Malé Leváre, 'Topoly'	121		25	130	1	3	8	23	Pure with Silver
58023	Kláštor pod Znievom	121		21	130	1	3	8	23	Pure with Silver
72	Dillingen a. d. Donau,'Reitweg', Lkr. Dillingen	62	BBC	35	215	2	6	17	40	Silver-antimony-copper (high)
110	Groß-Gerau, 'Dominalwald', Lkr. Groß-Gerau	62		31	100	2	6	17	40	Silver-antimony-copper (high)
147	Kelsterbach, 'Siedl. Kesselschmiede', Lkr. Groß-Gerau	62	CWC	31	800	2	6	17	40	Silver-antimony-copper (high)
148	Kelsterbach, 'Siedl. Kesselschmiede', Lkr. Groß-Gerau	62	CWC	31	410	2	6	17	40	Silver-antimony-copper (high)
149	Kelsterbach, 'Siedl. Kesselschmiede', Lkr. Groß-Gerau	62	CWC	31	410	2	6	17	40	Silver-antimony-copper (high)
211	München-Sendling, 'Wolfratshauser Str.', Stkr. München	66	BBC	35	215	2	6	17	40	Silver-antimony-copper (high)
344	Singen a. Hohentwiel, Kr. Konstanz	62	BK	40	500	2	6	17	39	Silver-antimony-copper (high)
2642	Osterbuch, 'Altes Mahd', Lkr. Augsburg	62	BK	40	500	2	6	17	40	Silver-antimony-copper (high)
3247	Lhánice	114	BBC	35	215	2	6	17	40	Silver-antimony-copper (high)
3345	Bylany	112	BBC	35	810	2	6	17	40	Silver-antimony-copper (high)
3363	Ledce I	114	BBC	35	810	2	6	17	40	Silver-antimony-copper (high)
4984	Hradec Králové	112	BBC	35	215	2	6	17	40	Silver-antimony-copper (high)
6533	Gyor	123	BBC	35	215	1	4	10	26	Silver-antimony-copper (high)
11196	Linz-Scharlinz, p. B. Linz	66	BBC	35	215	2	6	17	40	Silver-antimony-copper (high)
11517	Loretto, p. B. Eisenstadt-Umgebung	123	LLG	40	500	2	6	17	40	Silver-antimony-copper (high)
13403	Dunakömlöd	124		31	122	2	6	17	40	Silver-antimony-copper (high)

ANo	FIND_PLACE	AREA	CULT	PER	TYPE_2	CL_No_I	CL_No_II	CL_No_III	CL_No_IV	Metal_groups
17683	Kúty	121		21	215	2	6	17	40	Silver-antimony-copper (high)
21329	Hrivice	111	CWC	31	500	2	6	17	40	Silver-antimony-copper (high)
22233	Colombier	67	CWC	31	200	2	6	17	40	Silver-antimony-copper (high)
22234	Colombier	67	CWC	31	200	2	6	17	40	Silver-antimony-copper (high)
22248	Bevaix	67		21	100	2	6	17	40	Silver-antimony-copper (high)
22264	Auvernier	67		21	100	2	6	17	40	Silver-antimony-copper (high)
24210	Singen a. Hohentwiel, Kr. Konstanz	62	BK	40	600	2	6	17	40	Silver-antimony-copper (high)
32313	Kirchheim, Lkr. München	66	BBC	35	215	2	6	17	40	Silver-antimony-copper (high)
32314	Großheubach, Lkr. Miltenberg	65		25	100	2	6	17	40	Silver-antimony-copper (high)
32316	Landau a. d. Isar, 'Arberstraße', Lkr. Dingolfing-Landau	66	BBC	35	215	2	6	17	40	Silver-antimony-copper (high)
32318	Straubing, 'Gabelsbergerstrasse', Lkr. Straubing-Bogen	66	BBC	35	212	2	6	17	40	Silver-antimony-copper (high)
32321	Nerkewitz, Saale-Holzland-Kreis	103		25	100	2	6	17	40	Silver-antimony-copper (high)
32323	Auleben, Lkr. Nordhausen	103	CWC	31	500	2	6	17	40	Silver-antimony-copper (high)
32324	Auleben, Lkr. Nordhausen	103	CWC	31	500	2	6	17	40	Silver-antimony-copper (high)
32325	Auleben, Lkr. Nordhausen	103	CWC	31	500	2	6	17	40	Silver-antimony-copper (high)
32330	Straubing, 'Wasserwerk', Lkr. Straubing-Bogen	66	MUE	10	400	2	6	17	40	Silver-antimony-copper (high)
32332	Erfurt-Gispers, 'Nordhäuser Str', Stkr. Erfurt	103	CWC	31	550	2	6	17	40	Silver-antimony-copper (high)
32333	Erfurt-Gispers, 'Nordhäuser Str', Stkr. Erfurt	103	CWC	31	550	2	6	17	40	Silver-antimony-copper (high)
33024	Treuen, Vogtlandkreis	104		25	100	2	6	17	40	Silver-antimony-copper (high)
33028	Jeßnitz, Lkr. Anhalt-Bitterfeld	105	CWC	31	500	2	6	17	39	Silver-antimony-copper (high)
33029	Bautzen-Niederkaina, 'Schafberg, Qu III', Lkr. Bautzen	105	CWC	31	410	2	6	17	40	Silver-antimony-copper (high)
34135	Seehausen, Kyffhäuserkreis	103		25	100	2	6	17	40	Silver-antimony-copper (high)
34454	Drosa 'Küsterberg', Lkr. Anhalt-Bitterfeld	103	CWC	31	550	2	6	17	40	Silver-antimony-copper (high)
34493	Drosa 'Küsterberg', Lkr. Anhalt-Bitterfeld	103	CWC	31	410	2	6	17	40	Silver-antimony-copper (high)
34494	Drosa 'Küsterberg', Lkr. Anhalt-Bitterfeld	103	CWC	31	550	2	6	17	40	Silver-antimony-copper (high)
34497	Drosa 'Küsterberg', Lkr. Anhalt-Bitterfeld	103	CWC	31	410	2	6	17	40	Silver-antimony-copper (high)
34554	Drosa 'Küsterberg', Lkr. Anhalt-Bitterfeld	103	CWC	31	550	2	6	17	40	Silver-antimony-copper (high)
34576	Drosa 'Küsterberg', Lkr. Anhalt-Bitterfeld	103	CWC	31	550	2	6	17	40	Silver-antimony-copper (high)
34619	Prießnitz, Burgenlandkreis	104		25	100	2	6	17	40	Silver-antimony-copper (high)
44000	Iwanowice, 'Babia Góra II	85	Mier	40	510	2	6	17	40	Silver-antimony-copper (high)
44016	Strzyzów, 'Horodlo	84	Mier	40	550	2	6	17	39	Silver-antimony-copper (high)
44027	Mierzanowice II	85	Mier	40	510	2	6	17	39	Silver-antimony-copper (high)
44029	Mierzanowice II	85	Mier	40	420	2	6	17	40	Silver-antimony-copper (high)
44030	Mierzanowice II	85	Mier	40	420	2	6	17	39	Silver-antimony-copper (high)
44031	Mierzanowice II	85	Mier	40	420	2	6	17	39	Silver-antimony-copper (high)
44032	Mierzanowice II	85	Mier	40	420	2	6	17	39	Silver-antimony-copper (high)
44033	Mierzanowice II	85	Mier	40	420	2	6	17	39	Silver-antimony-copper (high)
44035	Mierzanowice II	85	Mier	40	405	2	6	17	39	Silver-antimony-copper (high)
44036	Mierzanowice II	85	Mier	40	510	2	6	17	40	Silver-antimony-copper (high)
44037	Mierzanowice II	85	Mier	40	405	2	6	17	39	Silver-antimony-copper (high)
44039	Mierzanowice II	85	Mier	40	510	2	6	17	39	Silver-antimony-copper (high)
44040	Mierzanowice II	85	Mier	40	420	2	6	17	39	Silver-antimony-copper (high)
44041	Mierzanowice II	85	Mier	40	510	2	6	17	40	Silver-antimony-copper (high)
44042	Mierzanowice II	85	Mier	40	400	2	6	17	39	Silver-antimony-copper (high)
44044	Mierzanowice II	85	Mier	40	510	2	6	17	40	Silver-antimony-copper (high)
44045	Mierzanowice II	85	Mier	40	500	2	6	17	39	Silver-antimony-copper (high)
44046	Mierzanowice II	85	Mier	40	510	2	6	17	40	Silver-antimony-copper (high)
44048	Mierzanowice II	85	Mier	40	510	2	6	17	39	Silver-antimony-copper (high)
44049	Mierzanowice II	85	Mier	40	400	2	6	17	39	Silver-antimony-copper (high)
44050	Mierzanowice II	85	Mier	40	510	2	6	17	39	Silver-antimony-copper (high)
44051	Mierzanowice II	85	Mier	40	500	2	6	17	40	Silver-antimony-copper (high)
44052	Mierzanowice II	85	Mier	40	510	2	6	17	39	Silver-antimony-copper (high)
44053	Mierzanowice II	85	Mier	40	400	2	6	17	40	Silver-antimony-copper (high)

ANo	FIND_PLACE	AREA	CULT	PER	TYPE_2	CL_No_I	CL_No_II	CL_No_III	CL_No_IV	Metal_groups
44054	Mierzanowice II	85	Mier	40	400	2	6	17	39	Silver-antimony-copper (high)
44055	Mierzanowice II	85	Mier	40	500	2	6	17	40	Silver-antimony-copper (high)
44056	Mierzanowice II	85	Mier	40	510	2	6	17	40	Silver-antimony-copper (high)
58008	Oravsky Podzámok	121		25	100	2	6	17	39	Silver-antimony-copper (high)
62131	Corsier	67		21	100	2	6	17	40	Silver-antimony-copper (high)
62166	Bevaix	67		21	100	2	6	17	40	Silver-antimony-copper (high)
62177	Colombier	67	CWC	31	200	2	6	17	40	Silver-antimony-copper (high)
62203	Wien XXII, Aspern	72	BAD	25	215	2	6	17	40	Silver-antimony-copper (high)
100008	Unterheimbach, Hohenlohekreis	62	BK	40	100	2	6	17	39	Silver-antimony-copper (high)
100009	Sipplingen, Bodenseekreis	62	BK	40	100	2	6	17	39	Silver-antimony-copper (high)
100012	Kösching, Lkr. Eichstätt	66	CWC	31	300	2	6	17	39	Silver-antimony-copper (high)
100014	Sobesuky	111	CWC	31	405	2	6	17	40	Silver-antimony-copper (high)
100015	Brestany	111	CWC	31	400	2	6	17	40	Silver-antimony-copper (high)
100016	Cachovice	111	CWC	31	400	2	6	17	40	Silver-antimony-copper (high)
100017	Vikletice	111	CWC	31	410	2	6	17	39	Silver-antimony-copper (high)
100018	Sobesuky	111	CWC	31	550	2	6	17	39	Silver-antimony-copper (high)
100019	Brestany	111	CWC	31	410	2	6	17	40	Silver-antimony-copper (high)
100020	Vikletice	111	CWC	31	410	2	6	17	40	Silver-antimony-copper (high)
100023	Nehvizdky	112	CWC	31	500	2	6	17	39	Silver-antimony-copper (high)
100031	Gaienhofen, 'Hornstaad-Hörnle I, Schicht AH 1-3', Kr. Konstanz	62	MUE	10	500	2	6	17	39	Silver-antimony-copper (high)
100032	Gaienhofen, 'Hornstaad-Hörnle I, Schicht AH 1-3', Kr. Konstanz	62	MUE	10	800	2	6	17	39	Silver-antimony-copper (high)
100034	Bodman-Weiler, Lkr. Konstanz	62	PFY	21	550	2	6	17	39	Silver-antimony-copper (high)
100062	Unteruhldingen, Bodenseekreis	62		21	510	2	6	17	39	Silver-antimony-copper (high)
100068	Meersburg, 'Haltau', Bodenseekreis	62	PFY	21	134	2	6	17	40	Silver-antimony-copper (high)
100071	Bodman-Weiler, Lkr. Konstanz	62	PFY	21	100	2	6	17	40	Silver-antimony-copper (high)
100078	Bürglen, 'Schalmenacker'	67		31	100	2	6	17	40	Silver-antimony-copper (high)
100100	Goldschau, Burgenlandkreis	103		25	100	2	6	17	40	Silver-antimony-copper (high)
100105	Halle-Dölauer Heide, Stkr. Halle	103	CWC	31	700	2	6	17	40	Silver-antimony-copper (high)
100116	Mutschau-Köttichau, Burgenlandkreis	103	SAL	25	410	2	6	17	40	Silver-antimony-copper (high)
100129	Wischroda, Burgenlandkreis	103		21	410	2	6	17	40	Silver-antimony-copper (high)
102	Goldburghausen, 'Goldberg', Ostalbkreis	62	GIII	25	134	1	4	11	27	Silver-antimony-copper (low)
104	Goldburghausen, 'Goldberg', Ostalbkreis	62	GIII	25	100	1	4	11	27	Silver-antimony-copper (low)
225	Nieder-Ramstadt, 'Darmstädter Str. 22', Lkr. Darmstadt-Dieburg	62		31	100	1	5	14	32	Silver-antimony-copper (low)
480	Zipfen, Lkr. Darmstadt-Dieburg	62		31	100	1	4	10	24	Silver-antimony-copper (low)
2785	Vallamend	67		21	100	1	4	10	24	Silver-antimony-copper (low)
2787	Vinelz, 'Strandboden, Amt Erlau'	67	CWC	31	100	1	4	10	24	Silver-antimony-copper (low)
2881	Vinelz, 'Amt Erlau'	67	CWC	31	800	1	4	10	24	Silver-antimony-copper (low)
3194	Brandysek	111	BBC	35	215	1	4	11	27	Silver-antimony-copper (low)
3231	Litomerice	112		21	130	1	4	10	24	Silver-antimony-copper (low)
3236	Praha-Ruzyn	112	BBC	35	215	1	4	10	24	Silver-antimony-copper (low)
3243	Chrást	112	BBC	35	215	1	4	10	24	Silver-antimony-copper (low)
3244	Lysolaje	112	BBC	35	215	1	4	10	24	Silver-antimony-copper (low)
3245	Lysolaje	112	BBC	35	810	1	4	10	24	Silver-antimony-copper (low)
3347	Brno-Lisen, 'Stare Zamky'	114	BAD	25	810	1	4	10	24	Silver-antimony-copper (low)
3349	Brno-Lisen, 'Stare Zamky'	114	BAD	25	134	1	4	10	24	Silver-antimony-copper (low)
3350	Brno-Lisen, 'Stare Zamky'	114	BAD	25	122	1	4	10	24	Silver-antimony-copper (low)
3355	Kobylnice	114	BBC	35	215	1	4	10	24	Silver-antimony-copper (low)
3356	Ledce I	114	BBC	35	215	1	4	10	24	Silver-antimony-copper (low)
3361	Strelice, 'Strelice VI'	114	BBC	35	215	1	4	10	24	Silver-antimony-copper (low)
3362	Strelice, 'Strelice VI'	114	BBC	35	810	1	4	10	24	Silver-antimony-copper (low)
3364	Slapanice II	114	BBC	35	212	1	4	10	24	Silver-antimony-copper (low)
3366	Letonice III	114	BBC	35	212	1	4	10	24	Silver-antimony-copper (low)
3371	Kobylnice	114	BBC	35	215	1	4	10	24	Silver-antimony-copper (low)

ANo	FIND_PLACE	AREA	CULT	PER	TYPE_2	CL_No_I	CL_No_II	CL_No_III	CL_No_IV	Metal_groups
3472	Lechovice	114	BBC	35	215	1	4	10	24	Silver-antimony-copper (low)
3736	Bratislava	121		10	130	1	4	10	24	Silver-antimony-copper (low)
3742	Komárom	123		21	130	1	4	10	24	Silver-antimony-copper (low)
3954	Uttershausen, 'Auf der Espe', Schwalm-Eder-Kreis	53		31	100	1	4	10	24	Silver-antimony-copper (low)
3971	Schirrhein	62		31	110	1	4	10	24	Silver-antimony-copper (low)
4636	Lichtenwörth, p. B. Wiener-Neustadt	72	BAD	25	400	1	4	10	24	Silver-antimony-copper (low)
4637	Lichtenwörth, p. B. Wiener-Neustadt	72	BAD	25	400	1	4	10	24	Silver-antimony-copper (low)
4644	Lichtenwörth, p. B. Wiener-Neustadt	72	BAD	25	400	1	4	10	24	Silver-antimony-copper (low)
4923	Nedakonice	114		10	134	1	4	10	24	Silver-antimony-copper (low)
6526	Tiszapolgár-Basatanya	125	BOD	21	700	1	4	10	24	Silver-antimony-copper (low)
6572	Tiszapolgár-Basatanya	125	TC	10	410	1	4	10	24	Silver-antimony-copper (low)
7280	Bex	67		21	100	1	4	10	24	Silver-antimony-copper (low)
7497	Uherské Hradište	114		25	100	1	4	10	24	Silver-antimony-copper (low)
7509	Rosnice	112	BBC	35	215	1	4	10	24	Silver-antimony-copper (low)
7523	Dolný Chvatliny	112		21	100	1	4	10	24	Silver-antimony-copper (low)
8897	Oradea	125		21	130	1	4	10	24	Silver-antimony-copper (low)
9250	Salacea	125		25	100	1	4	11	27	Silver-antimony-copper (low)
10380	Výcapy-Opatovce	121	NIT	40	700	1	4	10	24	Silver-antimony-copper (low)
10417	Výcapy-Opatovce	121	NIT	40	405	1	4	11	27	Silver-antimony-copper (low)
10418	Výcapy-Opatovce	121	NIT	40	405	1	4	11	27	Silver-antimony-copper (low)
10445	Výcapy-Opatovce	121	NIT	40	420	1	4	11	27	Silver-antimony-copper (low)
10491	Výcapy-Opatovce	121	NIT	40	420	1	4	11	27	Silver-antimony-copper (low)
10655	Tvrdošovce, 'Körektó'	121	NIT	40	410	1	4	11	27	Silver-antimony-copper (low)
10658	Tvrdošovce, 'Körektó'	121	NIT	40	405	1	4	11	27	Silver-antimony-copper (low)
10661	Tvrdošovce, 'Körektó'	121	NIT	40	420	1	4	11	27	Silver-antimony-copper (low)
10821	Brusno	121		25	100	1	4	10	24	Silver-antimony-copper (low)
10995	Vlcnov	114	SOP	21	100	1	4	10	24	Silver-antimony-copper (low)
10996	Suchá Loz	114		40	110	1	4	11	27	Silver-antimony-copper (low)
11110	Mlázovice	112		10	130	1	4	10	24	Silver-antimony-copper (low)
11111	Mlázovice	112		10	130	1	4	10	24	Silver-antimony-copper (low)
11527	Oggau, 'Seestrasse', p. B. Eisenstadt-Umgebung	123	BBC	35	212	1	4	11	27	Silver-antimony-copper (low)
12907	Budakalász	124	BAD	25	500	1	4	10	24	Silver-antimony-copper (low)
12911	Budakalász	124	BAD	25	500	1	4	10	24	Silver-antimony-copper (low)
13010	Apagy, 'Nagysziget dülo'	122	TC	10	130	1	4	10	24	Silver-antimony-copper (low)
13030	Fényeslitke	122	BOD	21	100	1	4	11	27	Silver-antimony-copper (low)
13053	Debrecen	125		10	130	1	4	10	24	Silver-antimony-copper (low)
13099	Tápé, 'Lebö'	125		25	122	1	4	11	27	Silver-antimony-copper (low)
13100	Tápé, 'Lebö'	125		25	100	1	4	11	27	Silver-antimony-copper (low)
13125	Pitvaros	125		40	405	1	4	11	27	Silver-antimony-copper (low)
13407	Dunakömlöd	124		31	100	1	4	11	27	Silver-antimony-copper (low)
19921	Predmostí	114	BBC	35	215	1	4	10	24	Silver-antimony-copper (low)
19930	Vresovice III	114	BBC	35	215	1	4	10	24	Silver-antimony-copper (low)
19934	Prosimerice	114	BBC	35	215	1	4	10	26	Silver-antimony-copper (low)
21576	Griesheim, Ortenaukreis	62	BK	40	110	1	4	10	24	Silver-antimony-copper (low)
44014	Strzyzów, 'Horodlo'	84	Mier	40	420	1	4	11	27	Silver-antimony-copper (low)
58007	Malé Leváre, 'Topoly'	121		25	100	1	4	10	24	Silver-antimony-copper (low)
58021	Mikula	121		21	130	1	4	10	24	Silver-antimony-copper (low)
311	Singen a. Hohentwiel, Kr. Konstanz	62	BK	40	212	2	6	17	41	Tin-rich-copper (incl. Ag, Sb etc)
319	Singen a. Hohentwiel, Kr. Konstanz	62	BK	40	212	2	6	17	41	Tin-rich-copper (incl. Ag, Sb etc)
473	Westhofen, Kr. Alzey-Worms	62	ABC	40	810	2	6	17	41	Tin-rich-copper (incl. Ag, Sb etc)
796	Dürnten	67		31	100	2	6	17	41	Tin-rich-copper (incl. Ag, Sb etc)
2862	Trachselwald	67		21	100	2	6	17	41	Tin-rich-copper (incl. Ag, Sb etc)
3037	Mörigen	67		31	212	2	6	17	41	Tin-rich-copper (incl. Ag, Sb etc)

ANo	FIND_PLACE	AREA	CULT	PER	TYPE_2	CL_No_I	CL_No_II	CL_No_III	CL_No_IV	Metal_groups
3237	Roztoky, 'Levy Hradec'	112	BBC	35	215	2	6	17	41	Tin-rich-copper (incl. Ag, Sb etc)
3241	Neratovice	112	BBC	35	215	2	6	17	41	Tin-rich-copper (incl. Ag, Sb etc)
3242	Lysolaje	112	BBC	35	215	2	6	17	41	Tin-rich-copper (incl. Ag, Sb etc)
3248	Bohdalice, 'Pavlovice'	114	BBC	35	810	2	7	20	69	Tin-rich-copper (incl. Ag, Sb etc)
3249	Lhánice	114	BBC	35	810	2	6	17	41	Tin-rich-copper (incl. Ag, Sb etc)
3357	Mistrín I	114	BBC	35	215	2	6	17	41	Tin-rich-copper (incl. Ag, Sb etc)
3359	Brno-Holásky II	114	BBC	35	215	2	6	17	41	Tin-rich-copper (incl. Ag, Sb etc)
3370	Hodonín	114		21	130	2	6	17	41	Tin-rich-copper (incl. Ag, Sb etc)
4194	Sion, 'Maladaires'	67	BK	40	600	2	6	17	41	Tin-rich-copper (incl. Ag, Sb etc)
4196	Sion, 'Maladaires'	67	BK	40	600	2	6	17	41	Tin-rich-copper (incl. Ag, Sb etc)
4238	Ayent, 'Les Place'	67	BK	40	500	2	6	17	41	Tin-rich-copper (incl. Ag, Sb etc)
4239	Ayent, 'Les Place'	67	BK	40	500	2	6	17	41	Tin-rich-copper (incl. Ag, Sb etc)
5964	Unterwölbling, p. B. St. Pölten	72	BK	40	410	2	6	17	41	Tin-rich-copper (incl. Ag, Sb etc)
10375	Výcapy-Opatovce	121	NIT	40	420	2	6	17	41	Tin-rich-copper (incl. Ag, Sb etc)
10446	Výcapy-Opatovce	121	NIT	40	420	2	6	17	41	Tin-rich-copper (incl. Ag, Sb etc)
13093	Beba Veche	125	PIT	40	600	2	6	17	41	Tin-rich-copper (incl. Ag, Sb etc)
15793	Nattenhausen, Lkr. Günzburg	62		40	410	2	6	17	41	Tin-rich-copper (incl. Ag, Sb etc)
19933	Stehelceves	111	BBC	35	215	2	6	17	41	Tin-rich-copper (incl. Ag, Sb etc)
19935	Ledce III	114	BBC	35	215	2	6	17	41	Tin-rich-copper (incl. Ag, Sb etc)
21404	Slany, 'Slánská Hora'	111	BBC	35	215	2	7	20	69	Tin-rich-copper (incl. Ag, Sb etc)
22149	Bacherach, 'aus dem Rhein', Kr. Mainz-Bingen	51		40	110	2	6	17	41	Tin-rich-copper (incl. Ag, Sb etc)
22277	Onnens	67	CWC	31	212	2	6	17	41	Tin-rich-copper (incl. Ag, Sb etc)
32317	Altenmarkt, 'Sportplatz', Lkr. Deggendorf	66	BBC	35	215	2	7	20	71	Tin-rich-copper (incl. Ag, Sb etc)
33025	Plauen, Vogtlandkreis	104		21	130	2	6	17	41	Tin-rich-copper (incl. Ag, Sb etc)
33026	Bautzen-Niederkaina, 'Schafberg, Qu III', Lkr. Bautzer	105		31	405	2	6	17	41	Tin-rich-copper (incl. Ag, Sb etc)
33577	Landau a. d. Isar, 'Landauer Feld', Lkr. Dingolfing-Landau	66	BBC	35	810	2	6	17	40	Tin-rich-copper (incl. Ag, Sb etc)
44010	Iwanowice, 'Babia Góra II	85	Mier	40	400	2	6	17	41	Tin-rich-copper (incl. Ag, Sb etc)
44024	Strzyzów, 'Horodlo'	84	Mier	40	500	2	6	17	41	Tin-rich-copper (incl. Ag, Sb etc)
78155	Branc	121	NIT	40	300	2	6	17	41	Tin-rich-copper (incl. Ag, Sb etc)
100028	Praha-Kopylisy	112	BBC	35	215	2	6	17	41	Tin-rich-copper (incl. Ag, Sb etc)
100083	Zürich, 'Mozartstrasse'	67	PFY	21	700	2	6	17	41	Tin-rich-copper (incl. Ag, Sb etc)
100106	Halle-Dölauer Heide, Stkr. Halle	103	CWC	31	700	2	6	17	41	Tin-rich-copper (incl. Ag, Sb etc)
100109	Hecklingen, Lkr. Salzlandkreis	103	CWC	31	410	2	6	17	41	Tin-rich-copper (incl. Ag, Sb etc)
100112	Kayna-Grosskayna, 'Janushügel', Burgenlandkreis	103	SAL	25	410	2	6	17	41	Tin-rich-copper (incl. Ag, Sb etc)
100117	Mutschau-Köttichau, Burgenlandkreis	103	CWC	31	410	2	6	17	41	Tin-rich-copper (incl. Ag, Sb etc)

Correlation tables of the Trace Element Analyses, Find places, Clusters and Metal Groups of the Bell Beaker metal finds

ANo	FIND_PLACE	AREA	CULT	PER	TYPE_2	OBJECT	CL_No_I	CL_No_II	CL_No_III	CL_No_IV	Metal_groups
3239	Praha-Prosek	112	BBC	35	215	Dagger, tanged	1	3	9	9	Almost pure
3358	Mistrín I	114	BBC	35	810	Awl, quadrangular	1	3	9	9	Almost pure
3360	Brno-Holásky II	114	BBC	35	200	Dagger, fragm.	1	3	9	9	Almost pure
3471	Vyskov	114	BBC	35	600	Pin, oar-headed	1	3	9	9	Almost pure
19931	Luderov	114	BBC	35	810	Awl, quadrangular	1	3	9	9	Almost pure
7546	Kostov	111	BBC	35	100	Axe, flat axe	1	4	10	26	Antimony-silver-copper
32319	Buttelstedt, 'Ziegeleigrube Florstedt',Kr. Weimar	103	BBC	35	215	Dagger, tanged	2	6	15	52	Arsenic-copper (high)
3240	Praha-Bubenec, 'Sibirska-Gasse'	112	BBC	35	215	Dagger, tanged	1	4	11	28	Arsenic-copper (low)
11526	Oggau, 'Seestrasse', p. B. Eisenstadt-Umgebung	123	BBC	35	212	Dagger, riveted	1	4	11	28	Arsenic-copper (low)
12438	Dunaföldvár	124	BBC	35	215	Dagger, tanged	1	5	14	31	Arsenic-copper (low)
19922	Predmostí	114	BBC	35	215	Dagger, tanged	1	4	12	30	Arsenic-copper (low)
19923	Predmostí	114	BBC	35	215	Dagger, tanged	1	4	12	30	Arsenic-copper (low)
19924	Predmostí	114	BBC	35	215	Dagger, tanged	1	4	12	29	Arsenic-copper (low)
21597	Hülen, 'Kapfenberg', Ostalbkreis	62	BBC	35	215	Dagger, tanged, 1 rivet, small	1	5	14	31	Arsenic-copper (low)
174	Luckenpaint-Thalmassing,'Geiergucken', Lkr. Regensburg	66	BBC	35	215	Dagger, tanged, small	2	6	15	33	Arsenic-copper with Impurities
184	Mettenheim, Lkr. Mühlheim a. Inn	66	BBC	35	810	Awl	2	6	18	43	Arsenic-copper with Impurities
192	Moosinning, 'Neue Kempfinger Brücke', Lkr. Erding	66	BBC	35	215	Dagger, tanged	2	6	15	33	Arsenic-copper with Impurities
210	München-Sendling, 'Wolfratshauser Str.', Stkr. München	66	BBC	35	215	Dagger, tanged	2	6	18	43	Arsenic-copper with Impurities
276	Safferstetten, Lkr. Passau	66	BBC	35	212	Dagger, 5 rivets	2	6	18	43	Arsenic-copper with Impurities
3193	Brandysek	111	BBC	35	215	Dagger, tanged	2	6	18	43	Arsenic-copper with Impurities
3246	Praha-Kobylisy, 'Boranovicka-Gasse'	112	BBC	35	810	Awl, quadrangular	2	6	18	43	Arsenic-copper with Impurities
3343	Praha-Liben	112	BBC	35	600	Pin shaft	2	6	18	43	Arsenic-copper with Impurities
3619	Laa an der Thaya, p. B. Misterbach	72	BBC	35	810	Awl	2	6	18	43	Arsenic-copper with Impurities
11197	Linz-Scharlinz, p. B. Linz	66	BBC	35	215	Dagger, tanged	2	6	18	43	Arsenic-copper with Impurities
19925	Turovice I	114	BBC	35	215	Dagger, tanged	2	6	18	43	Arsenic-copper with Impurities
19932	Stehelceves	111	BBC	35	215	Dagger, tanged	2	7	20	69	Arsenic-copper with Impurities
32228	Mühlhausen, 'Aktivistenring', Lkr.Mühlhausen	103	BBC	35	215	Dagger, tanged	2	6	15	33	Arsenic-copper with Impurities
32315	Triechīng, Gde. Pilsting, Lkr. Dinglofing-Landau	66	BBC	35	215	Dagger, tanged	2	6	16	36	Arsenic-copper with Impurities
32328	Mühlhausen, 'Aktivistenring', Lkr.Mühlhausen	103	BBC	35	215	Dagger, tanged	2	6	15	33	Arsenic-copper with Impurities
100010	Giengen a. d. Brenz, 'Sachsenhausen', Klr. Heidenheim	62	BBC	35	810	Awl	2	7	20	70	Arsenic-copper with Impurities
100107	Halle-Trotha, Stkr. Halle	103	BBC	35	215	Dagger, tanged	2	6	18	43	Arsenic-copper with Impurities
100127	Westerhausen, Lkr. Harz	103	BBC	35	500	Sheet metal	2	6	18	43	Arsenic-copper with Impurities
100128	Westerhausen, Lkr. Harz	103	BBC	35	500	Sheet metal	2	6	18	43	Arsenic-copper with Impurities
27	Barbing, Lkr. Regensburg	66	BBC	35	215	Dagger, tanged	2	7	19	58	Arsenic-lead-copper
33576	Landau a. d. Isar, 'Landauer Feld', Lkr. Dingolfing-Landau	66	BBC	35	215	Dagger, tanged	2	6	15	51	Arsenic-lead-copper
19920	Predmostí	114	BBC	35	215	Dagger, tanged	2	6	16	37	Arsenic-nickel-copper
3346	Brno-Slatina	114	BBC	35	215	Dagger, tanged	1	3	8	48	Pure
3365	Holubice	114	BBC	35	215	Dagger, tanged	1	3	8	22	Pure with Nickel
6394	Budapest-XVII-Rákoscsaba	125	BBC	35	215	Dagger, tanged	1	3	8	23	Pure with Silver
29848	Luderov	114	BBC	35	810	Awl, quadrangular	1	3	8	23	Pure with Silver
72	Dillingen a. d. Donau,'Reitweg', Lkr. Dillingen	62	BBC	35	215	Dagger, tanged	2	6	17	40	Silver-antimony-copper (high)
211	München-Sendling, 'Wolfratshauser Str.', Stkr. München	66	BBC	35	215	Dagger, tanged	2	6	17	40	Silver-antimony-copper (high)
3247	Lhánice	114	BBC	35	215	Dagger, tanged	2	6	17	40	Silver-antimony-copper (high)
3345	Bylany	112	BBC	35	810	Awl	2	6	17	40	Silver-antimony-copper (high)
3363	Ledce I	114	BBC	35	810	Awl, quadrangular	2	6	17	40	Silver-antimony-copper (high)
4984	Hradec Králové	112	BBC	35	215	Dagger, tanged	2	6	17	40	Silver-antimony-copper (high)
6533	Gyor	123	BBC	35	215	Dagger, tanged	1	4	10	26	Silver-antimony-copper (high)
11196	Linz-Scharlinz, p. B. Linz	66	BBC	35	215	Dagger, tanged	2	6	17	40	Silver-antimony-copper (high)
32313	Kirchheim, Lkr. München	66	BBC	35	215	Dagger, tanged	2	6	17	40	Silver-antimony-copper (high)
32316	Landau a. d. Isar, 'Arberstraße', Lkr. Dingolfing-Landau	66	BBC	35	215	Dagger, tanged	2	6	17	40	Silver-antimony-copper (high)
32318	Straubing, 'Gabelsbergerstrasse', Lkr. Straubing-Bogen	66	BBC	35	212	Dagger, riveted	2	6	17	40	Silver-antimony-copper (high)
3194	Brandysek	111	BBC	35	215	Dagger, tanged	1	4	11	27	Silver-antimony-copper (low)
3236	Praha-Ruzyn	112	BBC	35	215	Dagger, tanged	1	4	10	24	Silver-antimony-copper (low)
3243	Chrást	112	BBC	35	215	Dagger, tanged	1	4	10	24	Silver-antimony-copper (low)

3244	Lysolaje	112	BBC	35	215	Dagger, tanged	1	4	10	24	Silver-antimony-copper (low)
3245	Lysolaje	112	BBC	35	810	Awl, quadrangular	1	4	10	24	Silver-antimony-copper (low)
3355	Kobylnice	114	BBC	35	215	Dagger, tanged	1	4	10	24	Silver-antimony-copper (low)
3356	Ledce I	114	BBC	35	215	Dagger, tanged	1	4	10	24	Silver-antimony-copper (low)
3361	Strelice, 'Strelice VI'	114	BBC	35	215	Dagger, tanged	1	4	10	24	Silver-antimony-copper (low)
3362	Strelice, 'Strelice VI'	114	BBC	35	810	Awl	1	4	10	24	Silver-antimony-copper (low)
3364	Slapanice II	114	BBC	35	212	Dagger, 4 rivets	1	4	10	24	Silver-antimony-copper (low)
3366	Letonice III	114	BBC	35	212	Dagger, 4 rivets	1	4	10	24	Silver-antimony-copper (low)
3371	Kobylnice	114	BBC	35	215	Dagger, tanged	1	4	10	24	Silver-antimony-copper (low)
3472	Lechovice	114	BBC	35	215	Dagger, tanged	1	4	10	24	Silver-antimony-copper (low)
7509	Rosnice	112	BBC	35	215	Dagger, tanged	1	4	10	24	Silver-antimony-copper (low)
11527	Oggau, 'Seestrasse', p. B. Eisenstadt-Umgebung	123	BBC	35	212	Dagger, riveted	1	4	11	27	Silver-antimony-copper (low)
19921	Predmostí	114	BBC	35	215	Dagger, tanged	1	4	10	24	Silver-antimony-copper (low)
19930	Vresovice III	114	BBC	35	215	Dagger, riveted	1	4	10	24	Silver-antimony-copper (low)
19934	Prosimerice	114	BBC	35	215	Dagger, tanged	1	4	10	26	Silver-antimony-copper (low)
3237	Roztoky, 'Levy Hradec'	112	BBC	35	215	Dagger, tanged	2	6	17	41	Tin-rich-copper (incl. Ag, Sb etc)
3241	Neratovice	112	BBC	35	215	Dagger, tanged	2	6	17	41	Tin-rich-copper (incl. Ag, Sb etc)
3242	Lysolaje	112	BBC	35	215	Dagger, tanged	2	6	17	41	Tin-rich-copper (incl. Ag, Sb etc)
3248	Bohdalice, 'Pavlovice'	114	BBC	35	810	Awl, quadrangular	2	7	20	69	Tin-rich-copper (incl. Ag, Sb etc)
3249	Lhánice	114	BBC	35	810	Awl, quadrangular	2	6	17	41	Tin-rich-copper (incl. Ag, Sb etc)
3357	Mistrín I	114	BBC	35	215	Dagger, tanged	2	6	17	41	Tin-rich-copper (incl. Ag, Sb etc)
3359	Brno-Holásky II	114	BBC	35	215	Dagger, tanged, fragm.	2	6	17	41	Tin-rich-copper (incl. Ag, Sb etc)
19933	Stehelceves	111	BBC	35	215	Dagger, tanged	2	6	17	41	Tin-rich-copper (incl. Ag, Sb etc)
19935	Ledce III	114	BBC	35	215	Dagger, tanged	2	6	17	41	Tin-rich-copper (incl. Ag, Sb etc)
21404	Slany, 'Slánská Hora'	111	BBC	35	215	Dagger, tanged	2	7	20	69	Tin-rich-copper (incl. Ag, Sb etc)
32317	Altenmarkt, 'Sportplatz', Lkr. Deggendorf	66	BBC	35	215	Dagger, tanged	2	7	20	71	Tin-rich-copper (incl. Ag, Sb etc)
33577	Landau a. d. Isar, 'Landauer Feld', Lkr. Dingolfing-Landau	66	BBC	35	810	Awl	2	6	17	40	Tin-rich-copper (incl. Ag, Sb etc)
100028	Praha-Kopylisy	112	BBC	35	215	Dagger, tanged	2	6	17	41	Tin-rich-copper (incl. Ag, Sb etc)

www.ingramcontent.com/pod-product-compliance
Lightning Source LLC
LaVergne TN
LVHW072252300926
849160LV00018B/1650